Springer INdAM Series

Volume 56

This series will publish textbooks, multi-authors books, thesis and monographs in English language resulting from workshops, conferences, courses, schools, seminars, doctoral thesis, and research activities carried out at INDAM - Istituto Nazionale di Alta Matematica, http://www.altamatematica.it/en. The books in the series will discuss recent results and analyze new trends in mathematics and its applications.

THE SERIES IS INDEXED IN SCOPUS

Stéphane Menozzi • Andrea Pascucci •
Sergio Polidoro
Editors

Kolmogorov Operators and Their Applications

Springer

Editors
Stéphane Menozzi
Laboratoire de Mathématiques et
Modélisation d'Évry, I.B.G.B.I.
Université d'Évry Val d'Essonne
Évry Cedex, France

Andrea Pascucci
Dipartimento di Matematica
Alma Mater Studiorum - Università di
Bologna
Bologna, Italy

Sergio Polidoro
Dipartimento di Scienze Fisiche,
Informatiche e Matematiche
Università degli Studi di Modena e Reggio
Emilia
Modena, Italy

ISSN 2281-518X ISSN 2281-5198 (electronic)
Springer INdAM Series
ISBN 978-981-97-0224-4 ISBN 978-981-97-0225-1 (eBook)
https://doi.org/10.1007/978-981-97-0225-1

This Springer imprint is published by the registered company Springer Nature Singapore Pte Ltd.
The registered company address is: 152 Beach Road, #21-01/04 Gateway East, Singapore 189721, Singapore

Foreword

Kolmogorov Equations play a pivotal role in bridging the theories of Partial Differential Equations and Stochastic Differential Equations that emerge across various research domains. The objective of the INdAM Meeting on "Kolmogorov Operators and their Applications", held in Cortona from June 13 to 17, 2022, was to convene established researchers whose work revolves around Kolmogorov operators and their applications. This gathering aimed to stimulate the exploration of novel research directions and foster new collaborations.

Degenerate Kolmogorov operators are highly degenerate evolution operators that exhibit invariance with respect to non-Euclidean geometric structures. The simplest instance of this family of differential operators appears when considering the Langevin process $(V_t, X_t)_{t\geq 0}$ in the phase space $\mathbb{R}^d \times \mathbb{R}^d$

$$\begin{cases} V_t = v_0 + W_t, \\ X_t = x_0 + \int_0^t V_s \, ds. \end{cases} \tag{1}$$

Here $(W_t)_{t\geq 0}$ denotes a d-dimensional Wiener process. The density $p = p(t, v, x, v_0, x_0)$ of $(V_t, X_t)_{t\geq 0}$ is the fundamental solution to the strongly degenerate Kolmogorov equation $\mathcal{L}p = 0$, being

$$\mathcal{L}p := \tfrac{1}{2}\Delta_v p + v \cdot \nabla_x p + \partial_t p = 0, \qquad t \geq 0, \quad (v, x) \in \mathbb{R}^{2d}. \tag{2}$$

In 1934 Kolmogorov provided us with the explicit expression of p

$$p(t, v, x, v_0, x_0) = \frac{3^{d/2}}{(2\pi t^2)^d} \exp\left(-\frac{|v - v_0|^2}{t} - 3\frac{(v - v_0)\cdot(x - x_0 + t v_0)}{t^2} - 3\frac{|x - x_0 + t v_0|^2}{t^3} \right),$$

and pointed out that it is a smooth function, despite the strong degeneracy of the opertor $\mathcal{L}$. As it is suggested by the smoothness of the density p, the operator $\mathcal{L}$ is hypoelliptic, that is every distributional solution $f \in L^1_{\mathrm{loc}}(\Omega)$ to the equation

$\mathcal{L}f = g$, in some open set $\Omega \subset \mathbb{R}^{2d+1}$, we have that

$$g \in C^\infty(\Omega) \quad \Rightarrow \quad f \in C^\infty(\Omega).$$

The study of this category of differential operators has seen substantial growth in recent years, driven by both theoretical and practical considerations. They hold significant relevance in the realms of Partial Differential Equations and the theory of Stochastic Processes. In terms of their numerous real-world applications, degenerate Kolmogorov operators find utility in areas such as Kinetic Theory and the Theory of Financial Markets. These research domains are of great interest, featuring numerous open problems and unexplored issues. Currently, researchers are investigating existence, uniqueness, and regularity problems associated with this family of equations, considering both classical and weak theories, as well as the presence of fractional derivatives.

This volume comprises contributions from several speakers at the conference, encompassing a wide array of topics addressed during the event. These topics have transdisciplinary implications, spanning various fields within mathematics. The editors of this volume extend their heartfelt gratitude to the authors for their outstanding contributions.

Université d'Évry Val d'Essonne — Stéphane Menozzi
Alma Mater Studiorum - Università di Bologna — Andrea Pascucci
Università degli Studi di Modena e Reggio Emilia — Sergio Polidoro

Contents

Local Regularity for the Landau Equation (with Coulomb Interaction Potential)

François Golse and Cyril Imbert

Abstract This is a survey of our recent work with A. Vasseur [`arXiv:2206.05155`] on the local regularity of some class of weak solutions of the space homogeneous Landau equation with Coulomb singularity. Our main result is that any axisymmetric solution in this class is smooth outside the axis of symmetry.

1 The Landau Equation and the Regularity Problem

The Landau equation discussed here is a variant of the Boltzmann equation of the kinetic theory of gases proposed by Landau [11] in the context of plasma physics. It has been known for quite a long time (see for instance [16]) that the Boltzmann collision integral cannot be defined for particle interacting via a repulsive Coulomb potential, because of a logarithmic divergence, whose coefficient is precisely the Landau collision integral: see for instance §41 in [12], or [3].

As in the case of all kinetic models, the unknown in the Landau equation is the velocity distribution function $f \equiv f(t, x, v)$, that is the number density of particles located at the position x with velocity v at time t. Henceforth, we shall restrict our attention to velocity distribution functions that are independent of the x-variable, a situation referred to as the space-homogeneous case.

The (space-homogeneous) Landau equation with unknown $f \equiv f(t, v) \geq 0$ reads

$$\partial_t f(t, v) = \operatorname{div}_v \int_{\mathbf{R}^3} a(v - w)(\nabla_v - \nabla_w)(f(t, v) f(t, w)) dw , \quad v \in \mathbf{R}^3 ,$$

F. Golse (✉)
École polytechnique, IP Paris, CMLS, Palaiseau, France
e-mail: francois.golse@polytechnique.edu

C. Imbert
CNRS and École normale supérieure, Université PSL, DMA, Paris, France
e-mail: cyril.imbert@ens.psl.eu

S. Menozzi et al. (eds.), *Kolmogorov Operators and Their Applications*,
Springer INdAM Series 56, https://doi.org/10.1007/978-981-97-0225-1_1

In the right-hand side above, Landau collision kernel a is the matrix-valued function defined by the explicit formula

$$a(z) := \tfrac{1}{8\pi}\nabla^2|z| = \tfrac{1}{8\pi|z|}\Pi(z)\,,\qquad \text{where } \Pi(z) := I - \left(\tfrac{z}{|z|}\right)^{\otimes 2}\,.$$

It is instructive to write a nonconservative form of the Landau equation. It reads

$$\partial_t f(t,v) = \operatorname{trace}\Big((a \star_v f(t,v))\nabla_v^2 f(t,v)\Big) + f(t,v)^2\,,$$

where $\star_v$ designates the convolution product in the v-variable. What is remarkable in this form of the Landau equation is the local term $f(t,v)^2$, which comes from the identity

$$\operatorname{div}(\operatorname{div} a(z)) = -\delta_0(z)\,,\qquad \text{in } \mathcal{D}'(\mathbf{R}^3)\,.$$

The presence of the term $f(t,v)^2$ on the right-hand side obviously raises the following question: does the Cauchy problem with $f\big|_{t=0} = f_{in}$ admit global classical solutions (defined for all $t \geq 0$), or is there a finite-time blow-up for classical solutions of the Landau equation?

If one considers solutions $f(t) > 0$ independent of v, the Landau equation reduces in this case to the Riccati equation $f'(t) = f(t)^2$, whose solution $f(t) = \frac{f_{in}}{1-tf_{in}}$ blows up at time $1/f_{in}$. This is somehow uninteresting because, on physical grounds, the velocity distribution function $f(t,\cdot)$ is expected to be a probability density, or at least an element of $L^1(\mathbf{R}^3)$, which excludes positive constants.

It may seem more relevant to think of the Landau equation as a variant of the semilinear heat equation

$$\partial_t u(t,x) = \Delta_x u(t,x) + \alpha u(t,x)^2\,,\qquad x \in \mathbf{R}^d\,.$$

In this case again, there is a finite-time blow-up phenomenon, which easily follows from Kaplan's clever argument [9]. Pick ϕ to be a ground state for the Dirichlet Laplacian in the unit ball B of $\mathbf{R}^d$:

$$\begin{cases} -\Delta\phi = \lambda_0\phi\,, & \phi > 0 \text{ on } B\,,\\ \phi\big|_{\partial B} = 0\,. \end{cases}$$

Consider the quantity

$$L(t) := \frac{\int_B u(t,x)\phi(x)dx}{\int_B \phi(x)dx}\,.$$

Using the Green formula and the Jensen inequality shows that

$$u(t,\cdot) \geq 0 \text{ on } B \implies \dot{L}(t) \geq -\lambda_0 L(t) + \alpha L^2(t),$$

and a comparison argument based on the Riccati equation again proves that positive classical solutions of the semilinear heat equation for which $L(0) > \lambda_0/\alpha$ blow up in finite time.

However, comparing the Landau equation with the semilinear equation is misleading for the following reason. If $f(t,v)$ grows very fast as $t \to t^* < +\infty$ for v near some point v_0, one can hope that the diffusion matrix $a \star_v f(t,v)$ will grow accordingly, and that the diffusion term $\operatorname{trace}\big((a \star_v f(t,v))\nabla_v^2 f(t,v)\big)$ will ultimately offset the effect of the local quadratic nonlinearity $f(t,v)^2$.

For this reason, several authors have considered an "Isotropic" Landau Equation

$$\partial_t u = ((-\Delta)^{-1}u)\Delta u + \alpha u^2, \qquad 0 \leq \alpha \leq 1.$$

(The term "isotropic" comes from the fact that the matrix field a in Landau's collision integral is replaced with $\frac{1}{\pi|z|}\mathrm{I}_{\mathbf{R}^3}$.) See for instance [10] for a proof of global regularity for all $\alpha \in [0, \frac{74}{75})$. Although this result is an interesting contribution to the understanding of the competition between the smoothing effect of the diffusion term and the promotion of blow-up by the quadratic nonlinearity, the total mass of the solution is dissipated by the dynamics of the isotropic model for $0 \leq \alpha < 1$, which is not very satisfying on physical grounds. More recently, the case $\alpha = 1$ has been considered in [8], where the global regularity of radially symmetric, nonincreasing solutions is proved. This is more satisfying, since the total mass of the solution is conserved by the dynamics of the isotropic Landau equation with $\alpha = 1$, as in the case of the true Landau equation—and yet not fully satisfying because this model fails to conserve the energy of the solution, at variance with the Landau equation. Energy conservation is somehow related to the existence of nontrivial equilibrium solutions of the Landau equation, i.e. solutions of

$$\operatorname{trace}\Big((a \star_v f(v))\nabla_v^2 f(t,v)\Big) + f(v)^2 = 0$$

with finite mass and energy. We shall return to this later, but the existence of such equilibria shows that one cannot argue that either the diffusion term or the quadratic nonlinearity dominates the dynamics.

Other approaches to the regularity issue for the Landau equation itself, instead of a model equation, have been attempted. Silvestre [15] proved the regularity of H-solutions (a class of weak solutions of the Landau equation obtained by Villani in [16]) belonging to $L^\infty((0,+\infty); L^p(\mathbf{R}^3; (1+|v|)^k dv))$ with $p > 3/2$ and $k > 8$. Desvillettes, He and Jiang [5] have obtained a new Lyapunov functional for the Landau equation, and proved the global existence of regular solutions for near equilibrium initial data; they also prove that H-solutions of the Landau equation cannot become singular after some finite time that can be computed explicitly in terms of the initial data. Still another approach to the regularity issue consists in showing that the (potential) singular set of a H-solution must be "small" in some sense. For instance, with Gualdani and Vasseur [6], we proved that the set of singular times of Villani solutions of the Landau equation has Hausdorff dimension $\leq 1/2$. This result is of course reminiscent of the upper bound on the Hausdorff dimension of the set of singular times [13] of Leray solutions of the Navier-Stokes equations in three space dimensions, and of the more recent partial regularity results for the Navier-Stokes equations obtained by Scheffer [14] and Caffarelli-Kohn-Nirenberg [2].

This is survey article with a (hopefully) streamlined presentation of the local regularity results under appropriate symmetry assumptions obtained in collaboration with Vasseur in [7]. There are also partial regularity results in [6, 7] without any additional symmetry assumptions. These results are based in part on the same arguments as the local regularity results discussed here, and on another important ingredient briefly sketched in the last section of the present paper. We have chosen to refrain from giving a thorough presentation of these partial regularity results in the present survey for the sake of brevity. The interested reader is referred to Sect. 7 for a quick description of the partial regularity obtained for the Landau equation with Coulomb interaction, and to [6, 7] for complete proofs of the main results in that section.

2 Fundamental Properties of the Landau Collision Integral

In the first part of section, we recall some very classical properties of the collision integral, such as the conservation laws of mass, momentum and energy, together with the Boltzmann H Theorem, with the expression of the entropy production rate. All these properties have been well known since Landau published his equation in [11].

We shall conclude this section with a much more recent, yet equally fundamental inequality comparing the Landau entropy production rate with the Fisher information.

2.1 Conservation Laws and H-Theorem

Write the Landau collision integral as

$$C(F)(v) := \operatorname{div}_v \int_{\mathbf{R}^3} a(v-w)(\nabla_v - \nabla_w)(F(v)F(w))dw \,.$$

Lemma 1 *For all $F \in C^1(\mathbf{R}^3)$ such that $F(v) \geq 0$ and $\nabla F(v)$ are rapidly decaying as $|v| \to \infty$, the Landau collision integral $C(F)$ is the distribution* *defined by*

$$\langle C(F), \phi \rangle = -\tfrac{1}{2} \iint_{\mathbf{R}^3 \times \mathbf{R}^3} (\nabla\phi(v) - \nabla\phi(w)) \cdot a(v-w)(\nabla_v - \nabla_w)(F(v)F(w))dvdw$$

for all test functions $\phi \in C^\infty(\mathbf{R}^3)$ such that $\phi(v)$ and $\nabla\phi(v)$ have at most polynomial growth as $|v| \to \infty$.

(This formula is based on the definition of the divergence of a vector field in the sense of distributions, viz.

$$\langle C(F), \phi \rangle = -\iint_{\mathbf{R}^3 \times \mathbf{R}^3} \nabla\phi(v) \cdot a(v-w)(\nabla_v - \nabla_w)(F(v)F(w))dvdw\,;$$

the conclusion follows from symmetrizing the integrand with the substitution $(v, w) \mapsto (w, v)$.)

Corollary 1 *Under the assumptions of Lemma 1*

$$\langle C(F), 1 \rangle = \langle C(F), v_1 \rangle = \langle C(F), v_2 \rangle = \langle C(F), v_3 \rangle = \langle C(F), |v|^2 \rangle = 0\,.$$

(Indeed, if $\phi \equiv 1$, one has $\nabla\phi = 0$, so that the expression in the right-hand side of the formula in Lemma 1 is obviously 0. By the same token, if $\phi = v_j$ for $j = 1, 2, 3$, then $\nabla\phi$ is a constant vector field, hence $\nabla\phi(v) - \nabla\phi(w) = 0$. Finally, if $\phi(v) = |v|^2$, one has

$$\nabla\phi(v) - \nabla\phi(w) = 2(v - w)$$

and we conclude after observing that $a(v-w)^* \cdot (v-w) = a(v-w) \cdot (v-w) = 0$.)

Corollary 2 (H-Theorem) *Under the assumptions of Lemma 1, if moreover $F(v) > 0$ is such that $\ln F(v)$ has at most polynomial growth as $|v| \to \infty$, then*

$$\begin{aligned}\langle C(F), \ln F \rangle - - \tfrac{1}{16\pi} \iint_{\mathbf{R}^3 \times \mathbf{R}^3} & \frac{F(v)F(w)}{|v-w|} \\ & \times \left| \Pi(v-w) \left(\frac{\nabla F(v)}{F(v)} - \frac{\nabla F(w)}{F(w)} \right) \right|^2 dvdw \\ = - \iint_{\mathbf{R}^3 \times \mathbf{R}^3} & \left| \Pi(v-w)(\nabla_v - \nabla_w)\sqrt{F(v)F(w)} \right|^2 \frac{dvdw}{4\pi|v-w|} \\ = - \iint_{\mathbf{R}^3 \times \mathbf{R}^3} & \tfrac{1}{4\pi} \left| \Pi(v-w)(\nabla_v - \nabla_w) \right. \\ & \left. \times \sqrt{F(v)F(w)/|v-w|} \right|^2 dvdw\,.\end{aligned}$$

(The first equality follows from the identity

$$(\nabla_v - \nabla_w)(F(v)F(w)) = F(v)F(w)(\nabla \ln F(v) - \nabla \ln F(w))$$

in the right-hand side of the equality in Lemma 1, and the second from the identity

$$\sqrt{F(v)F(w)}(\nabla \ln F(v) - \nabla \ln F(w)) = 2(\nabla_v - \nabla_w)\sqrt{F(v)F(w)}\,.$$

The last equality is based on observing that $\Pi(v - w)(\nabla_v - \nabla_w)\Phi(|v - w|) = 0$ for all $\Phi \in C^1((0, +\infty))$.)

2.2 *The Entropy Production and the Fisher Information*

In 2015, Desvillettes came up with a remarkable inequality relating the entropy production rate for the Landau equation with the Fisher information.

Theorem 1 ([4]) *For each* $f \in L^1(\mathbf{R}^3; (1 + |v|^2)dv)$ *such that* $f \geq 0$ *a.e. on* $\mathbf{R}^3$ *and* $f \ln f \in L^1(\mathbf{R}^3)$, *one has*

$$\int_{\mathbf{R}^3} \frac{\left|\nabla\sqrt{f(v)}\right|^2 dv}{(1+|v|^2)^{3/2}} \leq C_D + C_D \int_{\mathbf{R}^6} \frac{\left|\Pi(v-w)(\nabla_v - \nabla_w)\sqrt{f(v)f(w)}\right|^2}{|v-w|} dv dw$$

for some constant C_D *depending only on the conserved moments of* f, *and on its* H *function:*

$$C_D \equiv C_D\left[\int_{\mathbf{R}^3} (1, v, |v|^2, |\ln f(v)|) f(v) dv\right] > 0\,.$$

Interestingly, the Desvillettes theorem puts the Landau equation in the same class as the Navier-Stokes in three space dimensions, in terms of Lebesgue exponents, as shown by Table 1.

There are several applications of the Desvillettes inequality, including for instance the propagation of moments for weak solutions of the Landau equations,

Table 1 Analogy between the Navier-Stokes equations, with unknown the fluid velocity field $u(t, x) \in \mathbf{R}^3$, and the Landau equation with unknown the distribution function $f(t, v)$

Equations	Unknowns	Dissipation rates
Navier-Stokes	$u \in L^\infty_t L^2_x$	$\nabla_x u \in L^2_t L^2_x$
Landau	$\sqrt{f} \in L^\infty_t L^2((1 + \|v\|)^2 dv)$	$\nabla_v \sqrt{f} \in L^2_t L^2(1 + \|v\|)^{-3} dv)$

i.e. the fact that quantities of the form

$$\int_{\mathbf{R}^3} |v|^m f(t,v)dv$$

remain bounded over finite time intervals if they are bounded for $t = 0$. (We shall not use this fact here, since we are primarily interested in local properties of weak solutions of the Landau equation.)

3 Weak Solutions

In [16], Villani proposed the following notion of weak solution of the Landau equation.

Definition 1 A H-solution of the Landau equation[1] on the time interval $[0, T)$ with initial data $f_{in} \geq 0$ is an element

$$f \in C([0,T), \mathcal{D}'(\mathbf{R}^3)) \cap L^1([0,T); L^1(\mathbf{R}^3; (1+|v|)^{-1}dv)$$

such that

$$\Pi(v-w)(\nabla_v - \nabla_w)\sqrt{f(t,v)f(t,w)/|v-w|} \in L^2([0,T) \times \mathbf{R}^3_v \times \mathbf{R}^3_w),$$

satisfying the Landau equation in the sense of distributions in the form

$$\begin{aligned}&\int_{\mathbf{R}^3} f_{in}(v)\phi(0,v)dv - \int_{\mathbf{R}^3} f(t,v)\phi(t,v)dv + \int_0^t\int_{\mathbf{R}^3} f(t,v)\partial_t\phi(s,v)dvds\\ &= \int_0^t\iint_{\mathbf{R}^6}\sqrt{\tfrac{f(s,v)f(s,w)}{|v-w|}}\Phi(s,v,w)\cdot\Pi(v-w)(\nabla_v - \nabla_w)\sqrt{\tfrac{f(s,v)f(s,w)}{|v-w|}}dvdwds\,,\end{aligned}$$

for all $\phi \in C_c^\infty([0,+\infty) \times \mathbf{R}^3)$, where $\Phi(t,v,w) := \nabla_v\phi(t,v) - \nabla_w\phi(t,w) \in \mathbf{R}^3$, together with the conservation of mass, momentum and energy, i.e.

$$\int_{\mathbf{R}^3} f(t,v)\begin{pmatrix}1\\ v\\ |v|^2\end{pmatrix}dv = \int_{\mathbf{R}^3} f_{in}(v)\begin{pmatrix}1\\ v\\ |v|^2\end{pmatrix}dv \quad \text{for all } t \in [0,T)\,,$$

[1] In the Coulomb interaction case.

and the Boltzmann H-Theorem, in the form

$$f(t,v) \geq 0 \text{ a.e. on } [0,T) \times \mathbf{R}^3, \text{ and } \int_{\mathbf{R}^3} f(t,v) \ln f(t,v) dv \leq \int_{\mathbf{R}^3} f_{in}(v) \ln f_{in}(v) dv$$

for all $t \in [0, T)$.

Observe that, by

$$\begin{aligned}
&\left(\iint_{\mathbf{R}^6} \sqrt{\tfrac{f(s,v)f(s,w)}{|v-w|}} \Phi(s,v,w) \cdot \Pi(v-w)(\nabla_v - \nabla_w) \sqrt{\tfrac{f(s,v)f(s,w)}{|v-w|}} dvdw \right)^2 \\
&\leq \iint_{\mathbf{R}^6} \left| \Pi(v-w)(\nabla_v - \nabla_w) \sqrt{\tfrac{f(s,v)f(s,w)}{|v-w|}} \right|^2 dvdw \\
&\quad \times \iint_{\mathbf{R}^6} \tfrac{|\Phi(s,v,w)|^2}{|v-w|} f(s,v) f(s,w) dvdw \\
&\leq \sup_{s,v,w} \tfrac{|\Phi(s,v,w)|^2}{|v-w|} \| f_{in} \|^2_{L^1(\mathbf{R}^3)} \iint_{\mathbf{R}^6} \left| \Pi(v-w)(\nabla_v - \nabla_w) \sqrt{\tfrac{f(s,v)f(s,w)}{|v-w|}} \right|^2 dvdw
\end{aligned}$$

by the Cauchy-Schwarz inequality. Since the right-hand side of this last inequality is integrable on $[0, T]$, we conclude from the Cauchy-Schwarz inequality (applied to integrals in the time variable) that, for each H-solution f of the Landau equation

$$t \mapsto \int_{\mathbf{R}^3} f(t,v) \phi(t,v) dv$$

belongs to $C^{0,1/2}([0, +\infty))$ for each $\phi \in C_c^\infty(\mathbf{R}^3)$.

Villani proved the following global existence theorem for the Landau equation (see Theorem 4 (i) of [16]).

Theorem 2 *Let* $f_{in} \in L^1(\mathbf{R}^3; (1 + |v|^2)dv)$ *satisfy*

$$f_{in} \geq 0 \text{ a.e. on } \mathbf{R}^3, \quad \text{and } f_{in} \ln f_{in} \in L^1(\mathbf{R}^3).$$

There exists a H*-solution* f *of the Landau equation on* $[0, +\infty)$ *such that* $f\big|_{t=0} = f_{in}$.

We need to explain the basics of Villani's construction. To do this, we introduce a truncating procedure for the Landau collision kernel. Pick $\chi \in C^\infty(\mathbf{R})$ such that

$$\mathbf{1}_{[-1,1]} \leq \chi \leq \mathbf{1}_{[-2,2]}, \qquad \text{with } |\chi'| \leq 2,$$

and define the truncated collision kernel by the formula

$$a_\delta(z) := (1 - \chi(|z|/\delta))a(z), \quad z \in \mathbf{R}^3$$

for each $\delta > 0$. Set

$$\xi_\delta(v) := \chi(\delta|v|), \quad \text{and} \quad \zeta_\delta(v) := \frac{1}{X}\frac{1}{\delta^3}\chi\left(\frac{|v|}{\delta}\right), \quad \text{where} \quad X := \int_{\mathbf{R}^3} \chi(|v|)dv.$$

For each $\delta > 0$, the regularized problem

$$\begin{cases} (\partial_t - \frac{\delta}{2}\Delta_v) f_\delta(t, v) = \operatorname{div}_v \int_{\mathbf{R}^3} a_\delta(v-w)(\nabla_v - \nabla_w)(f_\delta(t, v) f_\delta(t, w))dw, \\ f_\delta\big|_{t=0} = \zeta_\delta \star (\xi_\delta f_{in}) + \frac{\delta}{(2\pi)^{3/2}} e^{-|v|^2/2}, \end{cases}$$

has a unique smooth solution by standard fixed-point arguments. Besides, the maximum principle implies that

$$f_\delta(t, v) \geq C_\delta(t) e^{-|v|^2/2}$$

so that $\ln f_\delta(t, v) \geq -\frac{1}{2}|v|^2 + \ln C_\delta(t)$, and formal computations with the Boltzmann H-Theorem are legitimate.

Theorem 3 *Under the same assumptions as in Theorem 2, the family f_δ of solutions of the regularized Landau equation is relatively compact in $L^1_{loc}([0, +\infty) \times \mathbf{R}^3)$, and any of its limit points as $\delta \to 0^+$ is a H-solution of the Landau equation with initial data f_{in}.*

Henceforth, we refer to any H-solution of the Landau equation obtained as a limit point of the family of solutions of the regularized Landau equation as in Theorem 3 as a *Villani solution* of the Landau equation. In the sequel, we focus on this specific type of H-solutions to the Landau equation, since they enjoy useful additional properties, which general H-solutions are not known to satisfy (to the best of our knowledge).

4 Local Regularity

Our main results in this paper address the local regularity of Villani solutions of the Landau equation enjoying additional symmetry properties.

Let f be a Villani solution of the Landau equation with initial data f_{in} measurable on $\mathbf{R}^3$ satisfying

$$f_{in} \geq 0 \text{ a.e. and } \quad \int_{\mathbf{R}^3} (1 + |v|^2 + |\ln f_{in}(v)|) f_{in}(v) dv < \infty$$

(see Theorem 3 for the existence of a Villani solution).

Our first result deals with the regularity of axisymmetric Villani solutions of the Landau equation.

Theorem 4 *If f is axisymmetric, i.e. if there exists $\omega \in \mathbf{S}^2$ such that*

$$f(t, v) = f(t, \mathcal{R}(v - (v \cdot \omega)\omega) + (v \cdot \omega)\omega),$$

for all $\mathcal{R} \in SO((\mathbf{R}\omega)^\perp)$, then $f \in C^\infty((0, +\infty) \times (\mathbf{R}^3 \setminus \mathbf{R}\omega))$.

In plain words, axisymmetric Villani solutions of the Landau equation are smooth except maybe on the axis of symmetry.

Our next main result is a straightforward consequence of Theorem 4, and treats the case of radial solutions.

Theorem 5 *If f is radially symmetric, i.e. of the form*

$$f(t, v) = F(t, |v|)$$

then $f \in C^\infty((0, +\infty) \times (\mathbf{R}^3 \setminus \{0\}))$.

Thus radial Villani solutions of the Landau equation are smooth except maybe at the origin. This is an obvious consequence of Theorem 4, since a radial solution is axisymmetric around any straight line through the origin.

Indeed, according to Theorem 4, the set of singularities of such a solution is included in the intersection of all the straight lines through the origin, which is precisely the origin itself.

After [7] appeared on the `arXiv` preprint server, A. Bobylev wrote a very thorough and interesting discussion [1] of the regularity problem for radial solutions of the Landau equation. The interested reader is strongly advised to study [1] to understand the various scenarios, and competing effects promoting or preventing blow-up for radial solutions of the Landau equation.

5 Mathematical Tools

The proof of Theorem 4 is rather involved. In this section, we shall briefly outline the main steps, and describe the mathematical methods used in this proof.

5.1 Ellipticity Bound

In this brief section, we return to the truncated collision kernel a_δ introduced after the statement of Theorem 2. Obviously, a_δ is bounded and smooth for each $\delta > 0$. Is the ellipticity preserved by this truncation? This is precisely the question addressed in this section.

Lemma 2 *Assume that $f \equiv f(v) \geq 0$ a.e. is a measurable function satisfying the bounds*

$$0 < m_0 \leq \int_{\mathbf{R}^3} f(v)dv \leq M_0\,, \qquad \int_{\mathbf{R}^3} \begin{pmatrix} |v|^2 \\ \ln f(v) \end{pmatrix} f(v)dv \leq \begin{pmatrix} E_0 \\ H_0 \end{pmatrix}.$$

Then, there exists $c_0[m_0, M_0, E_0, H_0] > 0$ and $\delta_0[m_0, M_0, E_0, H_0] \in (0, 1)$ such that

$$0 < \delta < \delta_0 \implies f \star a_\delta(v) \geq \tfrac{c_0}{(1+|v|)^3} I \quad \text{for all } v \in \mathbf{R}^3\,.$$

5.2 Locally Bounded Solutions Are Smooth

Henceforth, we shall use parabolic cylinders, denoted as follows:

$$Q_r(t_0, v_0) := (t_0 - r^2, t_0] \times B_r(v_0)\,.$$

Because of the parabolic nature of the Landau equation, in order to prove the local smoothness of weak solutions, only some very low regularity is needed. Smoothness follows by some bootstrap procedure.

Lemma 3 *Let f be a Villani solution of the Landau equation. Assume that $f \in L^\infty(Q_R(t_0, v_0))$ for some $R, t_0 > 0$ and some $v_0 \in \mathbf{R}^3$. Then f is a.e. equal to function of class C^∞ on $Q_{R/2}(t_0, v_0)$.*

A consequence of Desvillettes' theorem (Theorem 1) is that Villani solutions of the Landau equation are solutions in the sense of distributions. Thus the Villani solution f satisfies

$$\partial_t f = \operatorname{div}_v(A\nabla_v f + fB)$$

where

$$A(t, v) := f(t, \cdot) \star a(v) \quad \text{and}$$
$$B(t, v) = -f(t, \cdot) \star \operatorname{div} a(v) = 2\left(f(t, \cdot) \star \tfrac{z}{|z|^3}\right)(v)\,.$$

Thus, if $0 \le f \in L^\infty(Q_R(t_0, v_0))$, then

$$f(t,\cdot) \star a = \underbrace{f(t,\cdot) \star a_\delta}_{L^\infty_t L^1_v \star_v L^\infty_v} + \underbrace{f(t,\cdot) \star (a - a_\delta)}_{=(f \mathbf{1}_{Q_R(t_0,v_0)}) \star_v (a-a_\delta)} \in L^\infty(Q_{R-2\delta}(t_0, v_0)),$$

and, by the same token, $fB \in L^\infty(Q_{R-2\delta}(t_0, v_0))$. On the other hand, reducing $\delta > 0$ so that $\delta < \delta_0$ as in Lemma 2, one has

$$f(t,\cdot) \star a(v) \ge f(t,\cdot) \star a_\delta(v) \ge \tfrac{c_0}{(1+|v|)^3} I, \qquad v \in \mathbf{R}^3,$$

so that f belongs to a parabolic De Giorgi class, and is therefore Hölder continuous, with a bound of the form

$$[f]_{C^\alpha(Q_{\beta_2 R}(t_0,v_0))} \le C \|f\|_{L^\infty(Q_{\beta_1 R}(t_0,v_0))} (1 + \|B\|_{L^\infty(Q_{\beta_1 R}(t_0,v_0))})$$

for all β_1, β_2 such that $0 < \beta_2 < \beta_1 < 1$. This implies that A and B belong to the Hölder class C^α on $Q_{\beta_3 R}(t_0, v_0)$ for all $\beta_3 \in (0, 1)$, again by decomposing a as $a = (a - a_\delta) + a_\delta$. Then, we can use the Schauder estimates to conclude that $\partial_t f$ and $\nabla_v f$ are of Hölder class C^α on $Q_{\beta_4 R}(t_0, v_0)$ for all $0 < \beta_4 < \beta_3 < 1$. Differentiating both sides of the Landau equation in all variables, the proof of Lemma 3 follows by a standard bootstrap argument.

5.3 *Truncated Entropy Inequality*

In the sequel, we shall use the De Giorgi method to prove a sufficient condition for local boundedness, and therefore local regularity according to Lemma 3. The De Giorgi method involves considering truncations of the distribution functions. The following notation will be convenient:

$$f_+^\kappa := (f - \kappa)_+ = \max(f - \kappa, 0), \qquad \text{for all } \kappa > 0.$$

We shall also consider the truncated logarithm

$$\ln_+ z := \max(\ln z, 0) = \ln \max(z, 1).$$

Definition 2 For each $g \ge 0$ and each $\kappa > 0$, the truncated entropy generating function is (Fig. 1)

$$h_+^\kappa(g) := \kappa h_+\left(\tfrac{g}{\kappa}\right), \qquad \text{where} \qquad h_+(z) := z \ln_+ z - (z - 1)_+ .$$

At this point, we shall use a property of Villani solutions of the Landau equation, pertaining to the evolution of the truncated entropy of the distribution

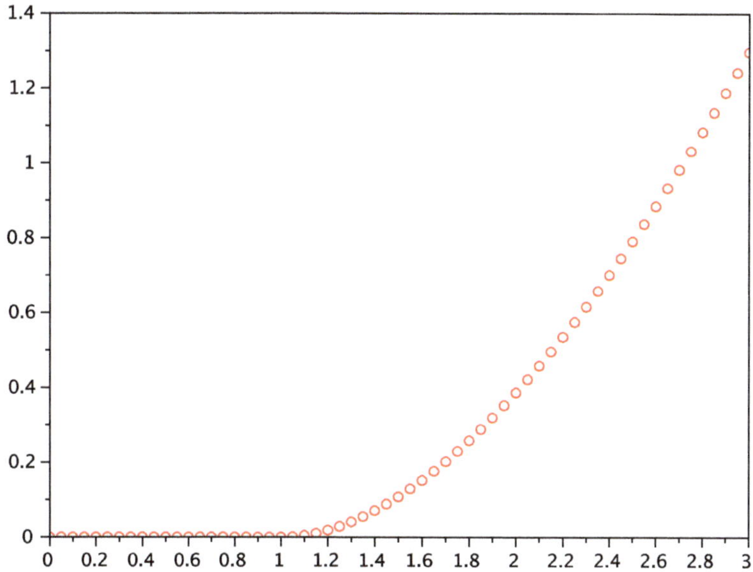

Fig. 1 Graph of the function h_+

function, which we do not know to be satisfied by all H-solutions. This property is fundamental for our analysis, and this is the reason for restricting our attention to Villani solutions.

Let f be a Villani solution of the Landau equation on $\mathbf{R}^3$.

For all $\Psi \in C_c^\infty((0,T)\times\mathbf{R}^3)$, there exists a Lebesgue-negligible set $\mathcal{N}\subset(0,T)$ such that

$$\begin{aligned}
&\int_{\mathbf{R}^3} h_+^\kappa(f(t_2,v))\phi(t,v)dv \\
&\le \int_{\mathbf{R}^3} h_+^\kappa(f(t_1,v))\phi(t,v)dv + \int_{t_1}^{t_2}\int_{\mathbf{R}^3} h_+^\kappa(f(t,v))\partial_t\phi(t,v)dvdt \\
&-\int_{t_1}^{t_2}\int_{\mathbf{R}^3}(A\nabla_v f - (\mathrm{div}_v A)f)\Big(\underbrace{\phi\frac{\nabla_v f_+^\kappa}{f}}_{\mathcal{T}_1} + \underbrace{\ln_+(\tfrac{f}{\kappa})\nabla_v\phi}_{\mathcal{T}_2}\Big)(t,v)dvdt
\end{aligned}$$

for all $\kappa > 0$ and all $t_1 < t_2 \notin \mathcal{N}$, with

$$\phi := \Psi^2, \qquad A(t,v) := \int_{\mathbf{R}^3} f(t,w)a(v-w)dw, \quad v \in \mathbf{R}^3.$$

This inequality is obtained by writing the evolution of the truncated entropy for solutions of the regularized Landau equation used in Villani's construction of H-solutions, passing to the limit as the regularization parameter $\delta \to 0$.

At this point, we forget about the regularized Landau equation, and think of Villani solutions of the Landau equation as H-solution satisfying the truncated entropy inequality above. In fact, what we are going to use is a consequence of the inequality above, where the test function Ψ is chosen so that

$$\mathbf{1}_{B_r(v_0)} < \Psi < \mathbf{1}_{B_{r+\delta}(v_0)}\,, \qquad \text{with} \quad \|\nabla\Psi\|_{L^\infty} \le \frac{c_*}{\delta} \quad \text{and} \quad \|\nabla^2\Psi\|_{L^\infty} \le \frac{c_*}{\delta^2}\,.$$

Choosing Ψ as above, and using the lower bound for the entropy production rate provided by the Desvillettes theorem (Theorem 1), one arrives at the following local estimate, which will be at the core of our analysis of the Landau equation.

Lemma 4 (Key Local Estimate) *Let $v_0 \in \mathbf{R}^3$, and assume that $f_{in} \ge 0$ a.e. on $\mathbf{R}^3$ is a measurable function satisfying*

$$0 < m_0 \le \int_{\mathbf{R}^3} f_{in}(v)dv \le M_0\,, \qquad \int_{\mathbf{R}^3} \begin{pmatrix} |v|^2 \\ \ln f(v) \end{pmatrix} f_{in}(v)dv \le \begin{pmatrix} E_0 \\ H_0 \end{pmatrix}.$$

There exists $C_0 \equiv C_0[|v_0|, m_0, M_0, E_0, H_0] \ge 1$ such that any Villani solution of the Landau equation with initial data f_{in} satisfies the inequality

$$\begin{aligned} &\operatorname*{ess\,sup}_{t_0-r^2<t\le t_0} \int_{B_r(v_0)} h_+^\kappa(f(t,v))dv + \int_{Q_r(t_0,v_0)} \frac{|\nabla_v f_+^\kappa(t,v)|^2}{f(t,v)} dtdv \\ &\le C_0 \int_{Q_{r+\delta}(t_0,v_0)} \left(\kappa + \tfrac{1}{\delta^2} + \tfrac{1}{\delta^2} f \star_v \tfrac{1}{|\cdot|}\right) f \left(\ln_+\left(\tfrac{f}{\kappa}\right) + \ln_+\left(\tfrac{f}{\kappa}\right)^2\right) dtdv \\ &\quad + C_0 \int_{Q_{r+\delta}(t_0,v_0)} \left(\tfrac{1}{\delta}(f \wedge \kappa) \star_v \tfrac{1}{|\cdot|^2}\right) f \ln_+\left(\tfrac{f}{\kappa}\right) dtdv \end{aligned}$$

for all $t_0 \in (0,T)$ and all $r, \delta \in (0,1)$, provided that

$$t_0 > (r+\delta)^2\,, \qquad \kappa \in \mathbf{Q} \cap [1, +\infty)\,.$$

5.4 Scalings

We have observed in Table 1 the analogy between H-solutions of the Landau equation and Leray solutions of the Navier-Stokes equations in three space dimensions. However, there is at least one major difference between these equations, namely the scaling invariance.

First, we mention the translation invariance of the Landau equation, which is obvious (and a consequence of the Galilean invariance in classical mechanics).

Thus, for the Landau equation

$$f\ H\text{-solution} \implies f(t_0+\cdot, v_0+\cdot)\ H\text{-solution}$$

for all $t_0 > 0$ and all $v_0 \in \mathbf{R}^3$.

At variance with the Navier-Stokes equation there is a two-parameter group of scaling transformations leaving the Landau equation with Coulomb interaction invariant: for all $\lambda > 0$ and all $\mu > 0$, set

$$f_{\lambda,\mu}(t,v) := \lambda f(\lambda t, \mu v), \qquad (t,v) \in (0,+\infty) \times \mathbf{R}^3 .$$

Then

$$f\ H\text{-solution of Landau's equation} \implies f_{\lambda,\mu}\ H\text{-solution of Landau's equation}$$

(We recall that if $u \equiv u(t,x) \in \mathbf{R}^3$ is a Leray solution of the Navier-Stokes equations, then, for each $\epsilon > 0$, the rescaled velocity field $u_\epsilon(t,x) := \epsilon u(\epsilon^2 t, \epsilon x)$ is also a Leray solution of the Navier-Stokes equations.)

Here is how the scaling transformation $f \mapsto f_{\lambda,\mu}$ acts on the quantities appearing on the left-hand side of the key local estimate, viz. the truncated entropy and the Fisher information of the truncated distribution function:

$$\int_{B_r(0)} h_+^{\lambda\kappa}(f_{\lambda,\mu}(t,v))dv = \tfrac{\lambda}{\mu^3} \int_{B_{\mu r}(0)} h_+^{\kappa}(f(\lambda t, \tilde v))d\tilde v\,,$$

$$\int_{-\tau}^0 \int_{B_r} \tfrac{|\nabla_v (f_{\lambda,\mu}(t,v)-\lambda\kappa)_+|^2}{f_{\lambda,\mu}(t,v)} dt dv = \tfrac{1}{\mu} \int_{-\lambda\tau}^0 \int_{B_{\mu r}} \tfrac{|\nabla_{\tilde v}(f(\tilde t,\tilde v)-\kappa)_+|^2}{f(\tilde t,\tilde v)} d\tilde t d\tilde v\,.$$

It is obviously desirable to keep both these terms of the same order of magnitude under the scaling transforms as one zooms in near a point (t_0, v_0) to study the local regularity of the distribution function. Among all the scaling transformations leaving the Landau equation with Coulomb interaction invariant, we are therefore led to consider the special case $\lambda/\mu^3 = 1/\mu$. Henceforth, we set

$$\mu = \epsilon \quad \text{and} \quad \lambda = \epsilon^2 .$$

Thus, let $f \equiv f(t,v)$ be a global Villani solution of the Landau equation. Pick $(t_0, v_0) \in (0,+\infty) \times \mathbf{R}^3$. Henceforth, we seek to study this solution near (t_0, v_0) by considering its translated and scaled variant

$$f_\epsilon(t,v) := \epsilon^2 f(t_0 + \epsilon^2 t, v_0 + \epsilon v)\,, \qquad v \in \mathbf{R}^3\,, \quad t > -t_0/\epsilon^2 .$$

If one returns to Table 1, one easily checks that this scaling transformation is exactly the same as in the case of the Navier-Stokes equations, since

$$\sqrt{f_\epsilon(t,v)} = \epsilon\sqrt{f(t_0+\epsilon^2 t, v_0+\epsilon v)}$$

is the quantity analogous to the rescaled Navier-Stokes velocity field

$$u_\epsilon(t,x) = \epsilon u(t_0+\epsilon^2 t, x_0+\epsilon x)\,.$$

Set

$$\kappa_\epsilon := \epsilon^2\kappa\,,\qquad \delta_\epsilon := \delta/\epsilon\,,\qquad r_\epsilon := r/\epsilon\,,\qquad f_{\epsilon,+}^{\kappa_\epsilon} := (f_\epsilon-\kappa_\epsilon)_+\,.$$

Assume that

$$\epsilon \in (0,\min(\tfrac{1}{2},\sqrt{t_0}))\,,\quad \kappa_\epsilon\in[1,2]\cap\mathbf{Q}\,,\quad r_\epsilon\in(0,2]\,,\quad \delta_\epsilon\in(0,1]\,.$$

Corollary 3 (Scaled Local Estimate) *Under the same assumptions and with the same notations as in Lemma 4, there exists* $C_0'\equiv C_0'[|v_0|,m_0,M_0,E_0,H_0]\geq 1$, ***independent*** *of* ϵ, *such that*

$$\underset{-r_\epsilon^2<t\le 0}{\text{ess sup}}\int_{B_{r_\epsilon}} h_+^{\kappa_\epsilon}(f_\epsilon(t,v))dv + \int_{Q_{r_\epsilon}} \frac{|\nabla_v f_{\epsilon+}^{\kappa_\epsilon}(t,v)|^2}{f_\epsilon(t,v)}dtdv$$

$$\le C_0'\int_{Q_{r_\epsilon+\delta_\epsilon}}\left(\kappa_\epsilon+\tfrac{1}{\delta_\epsilon^2}+\tfrac{1}{\delta_\epsilon^2}f_\epsilon\star\frac{\mathbf{1}_{B_1^c}}{|\cdot|}\right)f_\epsilon(\ln_+(\tfrac{f_\epsilon}{\kappa_\epsilon})+\ln_+(\tfrac{f_\epsilon}{\kappa_\epsilon})^2)dtdv$$

$$+\frac{C_0'}{\delta_\epsilon^2}\int_{Q_{r_\epsilon+\delta_\epsilon}} f_\epsilon\star\frac{\mathbf{1}_{B_1}}{|\cdot|}f_\epsilon(\ln_+(\tfrac{f_\epsilon}{\kappa_\epsilon})+\ln_+(\tfrac{f_\epsilon}{\kappa_\epsilon})^2)dtdv\,.$$

This is a rather straightforward consequence of Lemma 4, splitting the term coming from the Landau collision kernel as

$$\frac{1}{|z|}=\frac{\mathbf{1}_{|z|<1}}{|z|}+\frac{\mathbf{1}_{|z|\ge 1}}{|z|}\,,$$

and transforming the integrals by the substitution

$$(t,v)\mapsto(t_0+\epsilon^2 t, v_0+\epsilon v)\,.$$

5.5 *Local Regularity Criterion*

The last mathematical tool used in our study of the local regularity for the Landau equation with Coulomb singularity is the De Giorgi iteration method—specifically, the part of De Giorgi's argument leading to a local bound—and not the reduction of oscillations (in fact, this latter part of De Giorgi's analysis has already been used as a black box procedure in the proof of Lemma 3 before the bootstrap argument based on Schauder's estimates).

Lemma 5 (De Giorgi's First Lemma) *Let* $f_\epsilon(t, v) = \epsilon^2 f(t_0 + \epsilon^2 t, v_0 + \epsilon v)$ *be a scaled Villani solution of the Landau equation with initial data* $f_{in} := f(0, \cdot)$ *satisfying*

$$0 < m_0 \le \int_{\mathbf{R}^3} f_{in}(v)dv \le M_0\,, \qquad \int_{\mathbf{R}^3} \begin{pmatrix} |v|^2 \\ \ln f_{in}(v) \end{pmatrix} f_{in}(v)dv \le \begin{pmatrix} E_0 \\ H_0 \end{pmatrix}.$$

Assume

$$\operatorname*{ess\,sup}_{Q_1(0,0)} f_\epsilon(t, \cdot) \star \frac{\mathbf{1}_{B_1(0)^c}}{|\cdot|} \le Z_\epsilon\,, \qquad \text{where } Z_\epsilon \ge 1\,.$$

Then, there exists $\eta_{DG} \equiv \eta_{DG}[m_0, M_0, E_0, H_0, |v_0|] \in (0, 1)$ *such that*

$$\operatorname*{ess\,sup}_{-4<t\le 0} \int_{B_2} (f_\epsilon(t, v) - 1)_+ dv + \int_{Q_2} |\nabla_v \sqrt{f_\epsilon}|^2 \mathbf{1}_{f_\epsilon \ge 1} dt dv \le \frac{\eta_{DG}}{Z_\epsilon^{3/2}}$$

$$\implies f_\epsilon \le 2 \text{ a.e. on } Q_{1/2}\,.$$

Of course, the proof of Lemma 5 is based on simultaneous truncations of the values of the distribution function f, and parabolic zooming transformations near (t_0, v_0). Specifically choose a sequence r_j of radii shrinking from 1 to $\frac{1}{2}$, and a sequence of levels κ_j increasing from 1 to 2, as follows:

$$r_j := \tfrac{1}{2}(1 + 2^{-j})\,, \qquad \kappa_j := 2 - 2^{-j}\,, \quad j \ge 0\,.$$

Then we apply Corollary 3 with $r_\epsilon = r_{j+1}$ and $\delta_\epsilon = r_j - r_{j+1}$, while $\kappa_\epsilon = \frac{1}{2}(\kappa_j + \kappa_{j+1})$. Then we proceed with the De Giorgi iteration method, involving a nonlinearization procedure as usual, and arrive at the statement in Lemma 5.

6 A Sketch of the Proof of Theorem 4

The proof of Theorem 4 is based on two ingredients:

(a) a local $L^2_{t,v}$-bound away from the axis of symmetry, and
(b) a local $L^\infty_{t,v}$ bound for the diffusion matrix.

Henceforth we fix $(t_0, v_0) \in (0, +\infty) \times \mathbf{R}^3$ such that v_0 is at a distance $\rho_0 > 0$ from the axis of symmetry of the distribution function f.

6.1 Local $L^2_{t,v}$ Bound

Roughly speaking, away from the axis of symmetry, an axisymmetric function is a function of two variables (the distance to the axis and the height). Therefore, such functions benefit from better Sobolev embeddings than in the generic three dimensional case. This elementary idea is the basis for the following lemma.

Lemma 6 *Let $f \in L^\infty((0,T); L^1(B_R))$ with $\nabla_v f \in L^2((0,T); L^1(B_R))$ be of the form*

$$f(t,v) = g\Big(t, \underbrace{\sqrt{v_1^2+v_2^2}}_{V_1}, \underbrace{v_3}_{V_2}\Big).$$

Let $t_0 \in (0,T)$, let $V^0 = (V_1^0, V_2^0) \in \mathbf{R}^2$, and let $0 < r < V_1^0 - \rho_0$; then

$$\int_{t_0-r^2}^{t_0} \int_{|V-V_0|\le r} g(t,V)^2 dV dt$$

$$\le \frac{C_S(B_r)^2}{(2\pi\rho_0)^2} \int_{t_0-r^2}^{t_0} \left(\|f(t,\cdot)\|^2_{L^1(B_r(0,0,V_2^0))} + \|\nabla f(t,\cdot)\|^2_{L^1(B_r(0,0,V_2^0))} \right) dt\,,$$

where $C_S(B_r)$ is the Sobolev constant for $W^{1,1}(B_r) \subset L^2(B_r)$ in $\mathbf{R}^2$.

(Notice that the right-hand side of this inequality involves L^1 norms for the 3-dimensional Lebesgue measure over the 3-dimensional ball $B_r(0,0,V_2^0)$, whereas the integral on the left-hand side is over a 2-dimensional domain. This is the reason why the right-hand side is expressed in terms of f, and the left-hand side in terms of the function g.)

6.2 Local $L^\infty_{t,v}$ Bound of the Diffusion Matrix

The quantity $f(t,\cdot) \star \frac{1}{|\cdot|}$ appears twice on the right-hand side of the inequality in Corollary 3. Now this quantity is interesting, because of the following observation. For each $\epsilon > 0$ and each Villani solution of the Landau equation, it holds

$$f_\epsilon(t,\cdot) \star \frac{1}{|\cdot|}(v) = f(t_0+\epsilon^2 t, \cdot) \star \frac{1}{|\cdot|}(v_0+\epsilon v)\,,$$

for all $(t_0, v_0) \in (0, +\infty) \times \mathbf{R}^3$, all $v \in \mathbf{R}^3$ and all $t > -t_0/\epsilon^2$, so that

$$\left\| f_\epsilon \star_v \frac{1}{|\cdot|} \right\|_{L^\infty(Q_r(0,0))} = \left\| f \star_v \frac{1}{|\cdot|} \right\|_{L^\infty(Q_{\epsilon r}(t_0,v_0))} .$$

In other words, the quantity $f(t,\cdot) \star \frac{1}{|\cdot|}$ is L^∞-critical.

The next lemma provides an L^∞ control on this quantity for rescaled axisymmetric Villani solutions of the Landau equation.

Lemma 7 *Let $f \geq 0$ be measurable on $(0,T) \times \mathbf{R}^3$ and of the form*

$$f(t,v) := F\left(t, \sqrt{v_1^2 + v_2^2}, v_3\right) .$$

Let $(t_0, v_0) \in (0,T) \times \mathbf{R}^3$ be such that

$$\rho_0 := \sqrt{v_1^2 + v_2^2} > 0 ,$$

and let

$$f_\epsilon(t,v) := \epsilon^2 f(t_0 + \epsilon^2 t, v_0 + \epsilon v) .$$

Then, there exists an absolute constant $C_* > 0$ such that, for $0 < \epsilon < \frac{\rho_0}{6}$, it holds

$$\operatorname*{ess\,sup}_{Q_3} f_\epsilon(t,\cdot) \star \frac{1}{|\cdot|} \leq \frac{C_*}{\rho_0} \int_{\mathbf{R}^3} f(1 + \ln_+ f)(t,v)dv + C_* \rho_0^2$$
$$=: Z_*[M_0, E_0, H_0, \rho_0] .$$

6.3 *Deducing Local Regularity from Lemmas 6 and 7*

Write the scaled key local estimate of Corollary 3 with $r_\epsilon = 2$ and $\delta_\epsilon = \kappa_\epsilon = \frac{1}{2}$, using the elementary inequality

$$\ln_+(2y) + (\ln_+(2y))^2 \leq 6y , \qquad y > 0 .$$

Then, by Lemma 7

$$\operatorname*{ess\,sup}_{-4<t\leq 0}\int_{B_2(0)} h_+^{1/2}(f_\epsilon(t,v))dv + \int_{Q_2(0,0)} \frac{|\nabla_v f_{\epsilon+}^{1/2}(t,v)|^2}{f_\epsilon(t,v)} dtdv$$
$$\leq 6C_0'(\tfrac{3}{4}+\tfrac{1}{2}Z_*)\int_{Q_3(0,0)} f_\epsilon(t,v)^2 dtdv\,.$$

Then

$$\int_{Q_3(0,0)} f_\epsilon(t,v)^2 dtdv$$
$$\leq 2\pi(\rho_0+3)\int_{t_0-9\epsilon^2}^{t_0}\int_{\rho_0-3\epsilon}^{\rho_0+3\epsilon}\int_{v_3^0-3\epsilon}^{v_3^0+3\epsilon} g(s,\rho,w_3)^2 ds d\rho dw_3 \to 0$$

as $\epsilon \to 0^+$ by Lebesgue's Theorem, since

$$(t_0-9\epsilon^2,t_0)\times(\rho_0-3\epsilon,\rho_0+3\epsilon)\times(v_3^0-3\epsilon,v_3^0+3\epsilon)$$

is a set of set of vanishing measure as $\epsilon \to 0$, and since

$$g \in L^1((t_0-9\epsilon_0^2,t_0)\times(\rho_0-3\epsilon_0,\rho_0+3\epsilon_0)\times(v_3^0-3\epsilon_0,v_3^0+3\epsilon_0))$$

by Lemma 6 for some $\epsilon_0 > 0$ fixed.

Therefore, there exists $\epsilon > 0$ small enough such that

$$\operatorname*{ess\,sup}_{-4<t\leq 0}\int_{B_2(0)} h_+^{1/2}(f_\epsilon(t,v))dv + \int_{Q_2(0,0)} \frac{|\nabla_v f_{\epsilon+}^{1/2}(t,v)|^2}{f_\epsilon(t,v)} dtdv$$

falls below the threshold $\eta_{DG}/Z_*^{3/2}$ in the De Giorgi local regularity criterion (Lemma 5).

Lemma 5 implies that, for this value of ϵ it holds $f_\epsilon \leq 2$ a.e. on $Q_{1/2}(0,0)$, so that

$$f \leq 2/\epsilon \text{ a.e. on } Q_{\epsilon/2}(t_0,v_0)\,.$$

By Lemma 3, this implies that f is of class C^∞ in $Q_{\epsilon/4}(t_0,v_0)$.

7 Conclusion and Final Remarks

We have proved that axisymmetric, or radially symmetric Villani solutions of the Landau-Coulomb equation are regular away from the axis of symmetry, or from the origin in the case of radial solutions.

Our proof based on the De Giorgi method applied to a localized variant of the inequality derived from the H-Theorem for the Landau equation. One key ingredient in this proof is the **upper** bound for the diffusion matrix (Lemma 7 above), which is a critical quantity in $L^\infty_{t,x}$.

The same approach with an additional ingredient, viz. a control of the local mass $\|f_\epsilon\|_{L^\infty_t L^1_v(Q_R(0,0))}$ for $R < \frac{1}{2}$ in terms of $\|\nabla_v \sqrt{f_\epsilon}\|_{L^2(Q_2(0,0))}$ and of $\|\mathfrak{R}_\epsilon\|_{L^2(Q_2(0,0)\times\mathbf{R}^3)}$ with

$$\mathfrak{R}_\epsilon := \Pi(v-w)(\nabla_v - \nabla_w)\sqrt{f_\epsilon(t,v)f_\epsilon(t,w)/|v-w|}$$

shows that, for all Villani solutions f of the Landau equation with Coulomb interaction, the 7/2-dimensional parabolic Hausdorff measure[2] of its singular set satisfies

$$\mathcal{H}^{7/2}_{\text{parabolic}}(\mathbf{S}[f]) = 0\,,$$

where $\mathbf{S}[f]$ is the singular set of f, i.e.

$$\mathbf{S}[f] := \left\{(t,v) \in (0,+\infty) \times \mathbf{R}^3 \text{ s.t. } f \notin L^\infty(Q_r(t,v)) \text{ for all } r < \sqrt{t}\right\}.$$

This can be seen as a refinement on our earlier result with M.P. Gualdani and A. Vasseur recalled in the introduction (the fact that the Hausdorff dimension of the set of singular times for any Villani solution of the Landau equation with Coulomb interaction is at most 1/2), also based on the De Giorgi method.

---

[2] For all $s \in [0,5]$, the s-dimensional parabolic Hausdorff measure of a Borel set $X \subset [0,+\infty) \times \mathbf{R}^3$ is defined as

$$\mathcal{H}^{7/2}_{\text{parabolic}}(X) = \lim_{\delta\to 0^+} \inf\left\{\sum_{k\geq 1} r_k^s \text{ s.t. } X \subset \bigcup_{k\geq 1} Q_{r_k}(t_k,x_k) \text{ with } 0 < r_k < \delta\right\}.$$

References

1. Bobylev, A.V.: Radially symmetric models of the Landau kinetic equation and high energy tails. J. Stat. Phys. **190**, paper no. 48 (2023)
2. Caffarelli, L., Kohn, R.V., Nirenberg, L.: Partial regularity of suitable weak solutions of the Navier-Stokes equations Commun. Pure Appl. Math. **35**(6), 771–831 (1982)
3. Degond, P., Lucquin-Desreux, B.: The Fokker-Planck asymptotics of the Boltzmann collision operator in the Coulomb case. Math. Models Methods Appl. Sci. **2**(2), 167–182 (1992)
4. Desvillettes, L.: Entropy dissipation estimates for the Landau equation in the Coulomb case and applications. J. Funct. Anal. **269**(5), 1359–1403 (2015)
5. Desvillettes, L., He, L.-B., Jiang, J.-C.: A new monotonicity formula for the spatially homogeneous Landau equation with Coulomb potential and its applications. Preprint `arXiv:2011.00386`
6. Golse, F., Gualdani, M.P., Imbert, C., Vasseur, A.: Partial regularity in time for the space-homogenous Landau equation with Coulomb potential. Ann. Sci. Éc. Norm. Supér. (4) **55**(6), 1575–1611 (2022)
7. Golse, F., Imbert, C., Vasseur, A.: Local regularity for the space-homogeneous Landau equation with very soft potentials. Preprint `arXiv:2206.05155` [math.AP]
8. Gualdani, M.P., Guillen, N.: Estimates for radial solutions of the homogeneous Landau equation with Coulomb potential. Anal. PDE **9**(8), 1772–1809 (2016)
9. Kaplan, S.: On the growth of solutions of quasi-linear parabolic equations. Commun. Pure Appl. Math. **16**(3), 305–330 (1963)
10. Krieger, J., Strain, R.M.: Global solutions to a nonlocal diffusion equation with quadratic nonlinearity. Commun. Partial Differ. Equ. **37**(4), 647–689 (2012)
11. Landau, L.D.: Kinetic equation for the case of Coulomb interaction. Phys. Zs. Sov. Union **10**, 154–164 (1936)
12. Landau, L.D., Lifshitz, E.M., Pitayevskii, L.P.: Course of Theoretical Physics. Physical Kinetics, vol. 10. Pergamon Press Ltd., Oxford (1981)
13. Leray, J.: Sur le mouvement d'un liquide visqueux emplissant l'espace. Acta Math. **63**, 193–248 (1934)
14. Scheffer, V.: Partial regularity of solutions to the Navier-Stokes equations. Pac. J. Math. **66**(2), 535–552 (1976)
15. Silvestre, L.: Upper bounds for parabolic equations and the Landau equation. J. Differ. Equ. **262**(3), 3034–3055 (2017)
16. Villani, C.: On a new class of weak solutions to the spatially homogeneous Boltzmann and Landau equations. Arch. Ration. Mech. Anal. **143**(3), 273–307 (1998)

L^2 Hypocoercivity Methods for Kinetic Fokker-Planck Equations with Factorised Gibbs States

Emeric Bouin, Jean Dolbeault, and Luca Ziviani

Abstract This contribution deals with L^2 hypocoercivity methods for kinetic Fokker-Planck equations with integrable local equilibria and a *factorisation* property that relates the Fokker-Planck and the transport operators. Rates of convergence in presence of a global equilibrium, or decay rates otherwise, are estimated either by the corresponding rates in the diffusion limit, or by the rates of convergence to local equilibria, under moment conditions. On the basis of the underlying functional inequalities, we establish a classification of decay and convergence rates for large times, which includes for instance sub-exponential local equilibria and sub-exponential potentials.

1 Introduction

Hypocoercivity refers to the method developed by C. Villani in order to capture large time asymptotics in kinetic equations, see [32, 36], which borrows ideas from Hörmander's *hypoelliptic* theory and from the *carré du champ* method introduced by D. Bakry and M. Emery in [3]. For this reason, the Fisher information plays an important role and, to some extent, we can consider it as an H^1-theory. Here we shall focus more on the notion of L^2-hypocoercivity inspired by Hérau [26] and introduced in [20, 22] in a simple case of a kinetic Fokker-Planck equation, which puts the emphasis on the underlying diffusion limit. The heuristic idea is simple: while the Fokker-Planck diffusion operator controls the rate of convergence towards local equilibria in the velocity space, the equilibration of the spatial density (its convergence to the spatial density of a global equilibrium or its decay when no such equilibrium exists) can be interpreted as a diffusion in the position space, at least

E. Bouin · J. Dolbeault (✉) · L. Ziviani
CEREMADE (CNRS UMR no 7534), PSL University, Université Paris-Dauphine, Paris, France
e-mail: bouin@ceremade.dauphine.fr; dolbeaul@ceremade.dauphine.fr; luca.ziviani@dauphine.eu
https://www.ceremade.dauphine.fr/~bouin/; https://www.ceremade.dauphine.fr/~dolbeaul/; https://www.ceremade.dauphine.fr/fr/membres/detail-cv/profile/luca-ziviani.html

S. Menozzi et al. (eds.), *Kolmogorov Operators and Their Applications*,
Springer INdAM Series 56, https://doi.org/10.1007/978-981-97-0225-1_2

in a certain parabolic scaling, which results of the interplay of the diffusion in the velocity direction and the transport and the mixing in the phase space induced by the transport operator. The advantage of this approach is that rates are fully determined by the functional inequalities associated to the diffusion operator on the velocity space and to the diffusion limit in the position space.

This paper is organized as follows. We start by recalling the abstract L^2 hypocoercivity method in Theorem 1 before applying it to the framework of non-Maxwellian local equilibria and a compatible transport operator in Corollary 1. Although an adaptation of the standard theory in presence of microscopic and macroscopic coercivity associated respectively to the Fokker-Planck operator and the diffusion limit, this result is new and covers for instance the case of relativistic transport. This framework is also well adapted to situations with weaker notions of coercivity corresponding to either an external potential with slower growth at infinity or to local equilibria with fatter tails than the Maxwellian. After reviewing various families of interpolation inequalities which dictate the asymptotic behaviours of the solutions to the Fokker-Planck equations, we extend the L^2 hypocoercivity method to the kinetic Fokker-Planck equations and establish a classification in terms the (microscopic) local equilibria and the (macroscopic) equilibria associated with the Fokker-Planck diffusion limit.

Reviewing extensively the literature on the asymptotic behaviour of the solutions of Fokker-Planck, degenerate Fokker-Planck and kinetic Fokker-Planck equations goes beyond our scope. Let us simply quote some papers directly related to our methods, with more details to be given later.

Concerning *Fokker-Planck* equations, coercivity for the diffusion operator means spectral gap and Poincaré inequality, and thus exponential decay of the solutions. This is a basic example of application of the carré du champ method: see [5]. *Weak Poincaré* inequalities are natural in the absence of spectral gaps as explained in [6] and have been quite systematically explored: see [4, 29, 35] and earlier references therein. However, such methods require strong assumptions on the initial data. This is the reason why we adopt an alternative approach based on moments and weighted functional inequalities, where the extreme case corresponds to Nash's inequality in absence of an external potential. See Table 1.

As for *kinetic Fokker-Planck* equations, hypocoercivity primarily refers to the method exposed in [36]. We can also quote [34] for a detailed presentation of the commutator method and of Bakry-Emery type computations applied to estimates of the relaxation to equilibrium. More results and further references can also be found in [19]. In [28] S. Hu and X. Wang introduced a weak hypocoercivity approach *à la* Villani, using a weak Poincaré inequality, and proved subexponential convergence to equilibrium. This was later extended to a class of degenerate diffusion processes in [24] by M. Grothaus and F.-Y. Wang using weak Poincaré inequalities for the symmetric and antisymmetric part of the generator, with non-exponential rates of convergence. In the same vein, C. Cao proved quantitative convergence rates for the kinetic Fokker-Planck equation with more general confinement forces in [16, 17].

Alternatively the method of [22] was extended to cases without potentials in [9] while cases of weak or very weak potentials were considered respectively in [8] and [10]. Here the idea is to introduce moments and weighted interpolation inequalities to prove non-exponential decay or convergence rates. In these papers the effort has been mostly focused on the role of the external potential and fat tail local equilibria were not taken into consideration. However, it is known from [30] that fat-tail local equilibria can be responsible of a fractional diffusion limit, which may govern decay rates in the case without external potential as shown in [12], but this is not always the case. Non-Maxwellian local equilibria have been less explored than the Maxwellian case, but one has to mention [14, 15] for such an extension of the earlier works [1, 18], with slightly different methods based on weak norms, Lions' lemma and time-averages.

2 From Microscopic and Macroscopic Coercivity to Hypocoercivity

Let us start by an expository section which collects some known results and introduces our present purpose. Let us consider the general evolution equation

$$\frac{dF}{dt} + \mathsf{T}F = \mathsf{L}F \tag{1}$$

where F is the density of a probability distribution defined on a real or complex Hilbert space $\mathscr{H}$ with scalar product $\langle\cdot,\cdot\rangle$ and norm $\|\cdot\|$. We assume that T and L are two linear operators, respectively anti-Hermitian and Hermitian: $\mathsf{T}^* = -\mathsf{T}$ and $\mathsf{L}^* = \mathsf{L}$, where * denotes the adjoint with respect to $\langle\cdot,\cdot\rangle$. We are interested in the decay rate of F or in the convergence to a steady state $F_\star$. We assume that $F_\star$ is unique, up to normalization. Since (1) is linear, we can always replace F by $F - F_\star$ and study the convergence to 0 of an eventually sign changing function F. We have in mind that L is an elliptic degenerate operator. If Π is the orthogonal projection onto the kernel of L, we assume that L has the *microscopic coercivity* property in the sense that it is *coercive* on $(1-\Pi)\,\mathscr{H}$, where 1 is here a shorthand notation for the identity that will make sense in the functional setting of interest. In other words, we claim that

$$\frac{1}{2}\,\frac{d}{dt}\,\|F\|^2 = \langle \mathsf{L}F, F\rangle \le -\lambda_m\,\|(1-\Pi)F\|^2 \tag{H1}$$

for some $\lambda_m > 0$. This is not enough to conclude that $\|F(t,\cdot)\|$ decays exponentially as we have no decay rate on Ker(L), but if the operators L and T do not commute, we can hope that some of the decay properties on $(1-\Pi)\,\mathscr{H}$ are transferred on $\Pi\,\mathscr{H}$. This points towards the computation of various commutators and the whole machinery of Hörmander's hypoellipticity theory. A micro/macro approach offers a simpler framework, that has the advantage of clarifying the role played by various functional inequalities in estimating decay rates of F. The underlying ideas rely on the formal *macroscopic limit* of the scaled evolution equation

$$\varepsilon \frac{dF}{dt} + \mathsf{T}F = \frac{1}{\varepsilon} \mathsf{L}F$$

on the Hilbert space $\mathscr{H}$, which is a typical parabolic scaling when ε is a small parameter. Using a formal expansion of a solution $F_\varepsilon = F_0 + \varepsilon\, F_1 + \varepsilon^2\, F_2 + O(\varepsilon^3)$ as $\varepsilon \to 0_+$ and solving the equation order by order, we obtain

$$\begin{aligned} &\text{at order } O\big(\varepsilon^{-1}\big): && \mathsf{L}F_0 = 0\,, \\ &\text{at order } O\big(\varepsilon^{0}\big): && \mathsf{T}F_0 = \mathsf{L}F_1\,, \\ &\text{at order } O\big(\varepsilon^{1}\big): && \tfrac{dF_0}{dt} + \mathsf{T}F_1 = \mathsf{L}F_2\,. \end{aligned}$$

The first and second equation respectively read as $F_0 = \Pi F_0$ and $F_1 = -\,(\mathsf{T}\Pi)\,F_0$. After projection on $\mathrm{Ker}(\mathsf{L})$, the third equation becomes $\frac{d}{dt}\,(\Pi F_0) -\, \Pi\mathsf{T}\,(\mathsf{T}\Pi)\,F_0 = \Pi\mathsf{L}F_2 = 0$ that we can also write as

$$\frac{\partial F_0}{\partial t} + (\mathsf{T}\Pi)^*\,(\mathsf{T}\Pi)\,F_0 = 0\,. \tag{2}$$

Assuming *macroscopic coercivity*, i.e., the property that the operator $(\mathsf{T}\Pi)^*\,(\mathsf{T}\Pi)$ is coercive on $(1-\Pi)\,\mathscr{H}$, we obtain

$$\frac{1}{2}\,\frac{d}{dt}\,\|F_0\|^2 = -\,\|(\mathsf{T}\Pi)\,F_0\|^2 \le -\,\lambda_M\,\|F_0\|^2 \tag{H2}$$

for some $\lambda_M > 0$. In order to derive (2), we implicitly used the fact that all terms are of order ε, which relies on the *parabolic macroscopic dynamics* condition

$$\Pi\mathsf{T}\Pi\,F = 0\,. \tag{H3}$$

As in the *hypocoercivity* method of [22], let us consider the operator

$$\mathsf{A} := \big(1 + (\mathsf{T}\Pi)^*(\mathsf{T}\Pi)\big)^{-1}(\mathsf{T}\Pi)^*$$

where the $(\mathsf{T}\Pi)^*(\mathsf{T}\Pi)$ term is of course reminiscent of (2), and, the Lyapunov functional, or *entropy*,

$$\mathsf{H}[F] := \frac{1}{2}\,\|F\|^2 + \delta\,\mathrm{Re}\langle \mathsf{A}F,\,F\rangle\,. \tag{3}$$

The parameter $\delta > 0$, to be determined, as to be thought as a small parameter so that $\mathsf{H}[F]$ is a perturbation of $\frac{1}{2}\,\|F\|^2$. The following estimate is by now classical but deserves some emphasis. Let us consider $G = \mathsf{A}F$, i.e., the solution of $(\mathsf{T}\Pi)^* F = G + (\mathsf{T}\Pi)^* \mathsf{T}\Pi\, G$. As in [22, Lemma 1], by a Cauchy-Schwarz inequality, we learn that

$$\begin{aligned}\langle \mathsf{TA}F, F\rangle = \langle G, (\mathsf{T}\Pi)^* F\rangle = \|G\|^2 + \|\mathsf{T}\Pi G\|^2 = \|\mathsf{A}F\|^2 + \|\mathsf{TA}F\|^2 \\ \le \|\mathsf{TA}F\|\,\|(\mathrm{Id}-\Pi)F\| \le \frac{1}{2\,\mu}\,\|\mathsf{TA}F\|^2 + \frac{\mu}{2}\,\|(\mathrm{Id}-\Pi)F\|^2 .\end{aligned}$$

Applied with either $\mu = 1/2$ or $\mu = 1$, this estimate proves that $\|\mathsf{A}F\| \le \frac{1}{2}\,\|(\mathrm{Id} - \Pi)F\|$ and $\|\mathsf{TA}F\| \le \|(\mathrm{Id} - \Pi)F\|$. Incidentally, this proves that

$$|\langle \mathsf{TA}F, F\rangle| = |\langle \mathsf{TA}F, (\mathrm{Id}-\Pi)F\rangle| \le \|(\mathrm{Id}-\Pi)F\|^2 , \tag{4a}$$

$$|\langle \mathsf{A}F, F\rangle| \le \frac{1}{2}\,\|\Pi F\|\,\|(\mathrm{Id}-\Pi)F\| \le \frac{1}{4}\,\|F\|^2 . \tag{4b}$$

We read from (4b) that $\mathsf{H}[F]$ and $\|F\|^2$ are equivalent with

$$\frac{2-\delta}{4}\,\|F\|^2 \le \mathsf{H}[F] \le \frac{2+\delta}{4}\,\|F\|^2 .$$

However, the twist introduced in $\mathsf{H}[F]$ by $\langle \mathsf{A}F, F\rangle$ makes it exponentially decaying in t if F solves (1). We can indeed compute

$$-\frac{d}{dt}\mathsf{H}[F] = \mathsf{D}[F]$$

where

$$\begin{aligned}\mathsf{D}[F] := &- \langle \mathsf{L}F, F\rangle + \delta\,\langle \mathsf{AT}\Pi F, F\rangle \\ &- \delta\,\Big(\mathrm{Re}\langle \mathsf{TA}F, F\rangle - \mathrm{Re}\langle \mathsf{AT}(1-\Pi)F, F\rangle + \mathrm{Re}\langle \mathsf{AL}F, F\rangle\Big) .\end{aligned} \tag{5}$$

By (H1), we know that $-\langle \mathsf{L}F, F\rangle \ge \lambda_m\,\|(1-\Pi)F\|^2$. On the other hand, (H2) amounts to

$$\big\langle (\mathsf{T}\Pi)^*\,(\mathsf{T}\Pi)\,F, F\big\rangle \ge \lambda_M\,\|F\|^2 \quad \text{if} \quad F \in \mathrm{Ker}(\mathsf{L})$$

and, by construction, the operator A is therefore such that

$$\langle \mathsf{AT}\Pi F, F\rangle \ge \frac{\lambda_M}{1+\lambda_M}\,\|\Pi F\|^2 . \tag{6}$$

The first two terms in the definition of $\mathsf{D}[F]$ can be combined to prove that

$$\mathsf{D}[F] \geq \lambda_m \|(1-\Pi)F\|^2 + \frac{\delta\,\lambda_M}{1+\lambda_M}\|\Pi F\|^2$$

$$- \delta\left(\mathrm{Re}\langle \mathsf{TA}F, F\rangle - \mathrm{Re}\langle \mathsf{AT}(1-\Pi)F, F\rangle + \mathrm{Re}\langle \mathsf{AL}F, F\rangle\right). \tag{7}$$

Under the additional assumption that the last term in the above identity involves only *bounded auxiliary operators* in the sense that

$$\|\mathsf{AT}(1-\Pi)F\| + \|\mathsf{AL}F\| \leq C_M \|(1-\Pi)F\|, \tag{H4}$$

one obtains the *entropy—entropy production* inequality

$$\mathsf{D}[F] \geq \lambda\,\mathsf{H}[F]$$

for some explicit constant $\lambda > 0$. The precise statement goes as follows. It has been established in [22] in the case of a real Hilbert space $\mathscr{H}$ and extended to complex Hilbert spaces in [9].

Theorem 1 ([9, 22]) *Let* L *and* T *be closed linear operators in the complex Hilbert space* $(\mathcal{H}, \langle\cdot,\cdot\rangle)$. *We assume that* L *is Hermitian and* T *is anti-Hermitian, and that* (H1)–(H4) *hold for some positive constants* λ_m, λ_M, *and* C_M. *Then there is some* $\delta_\star \in (0,2)\cap(0,\lambda_m)$ *such that, for any* $\delta \in (0,\delta_\star)$, *there are explicit constants* $\lambda > 0$ *and* $\mathscr{C} > 1$ *for which, if* F *solves* (1) *with initial datum* $F_0 \in \mathcal{H}$, *then*

$$\mathsf{H}[F(t,\cdot)] \leq \mathsf{H}[F_0]\,e^{-\lambda t} \quad \textit{and} \quad \|F(t,\cdot)\|^2 \leq \mathscr{C}\,e^{-\lambda t}\,\|F_0\|^2 \quad \forall\, t \geq 0. \tag{8}$$

The estimates of $\lambda > 0$ and $\mathscr{C} > 1$ in [9, Proposition 4] have been improved in [2, Proposition 2] as follows. With $X := \|(\mathrm{Id}-\Pi)F\|$ and $Y := \|\Pi F\|$. Using (4b), we read from (3) that

$$\mathsf{H}[F] \leq \frac{1}{2}\left(X^2 + Y^2\right) + \frac{\delta}{2}\,X\,Y$$

while it follows from (7), (4a) and (H4) that

$$\mathsf{D}[F] - \lambda\,\mathsf{H}[F] \geq \left(\lambda_m - \delta - \frac{\lambda}{2}\right)X^2 - \delta\left(C_M + \frac{\lambda}{2}\right)X\,Y + \left(\frac{\delta\,\lambda_M}{1+\lambda_M} - \frac{\lambda}{2}\right)Y^2.$$

With $K_M := \frac{\lambda_M}{1+\lambda_M} < 1$ and $\delta_\star := \frac{4\,K_M\,\lambda_m}{4\,K_M + C_M^2} < \lambda_m$, a simple discriminant condition shows that for any $\delta \in (0,\delta_\star)$, the right-hand side is nonnegative for the largest (positive) solution of

$$\delta^2\left(C_M + \frac{\lambda}{2}\right)^2 - 4\left(\lambda_m - \delta - \frac{\lambda}{2}\right)\left(\frac{\delta\,\lambda_M}{1+\lambda_M} - \frac{\lambda}{2}\right) = 0.$$

We refer to [2] for further details and to [23] for more considerations on the functional framework.

In the framework of kinetic equations, T and L are respectively the *transport operator* and the *collision operator* acting on a distribution function $f(t,x,v)$ where $t \geq 0$ is the time, x is the position and v is the velocity. To fix ideas, we shall assume that $x, v \in \mathbb{R}^d$ and consider

- a transport operator defined by the Poisson bracket as

$$\mathsf{T}f := \nabla_v \mathscr{E} \cdot \nabla_x f - \nabla_x \mathscr{E} \cdot \nabla_v f \tag{9}$$

  corresponding to the Hamiltonian energy

$$(x,v) \mapsto \mathscr{E}(x,v) := \frac{1}{\beta} \langle v \rangle^\beta + \phi(x)\,,$$

  where ϕ denotes an external, given potential,
- a collision operator of Fokker-Planck type given by

$$\mathsf{L}f := \nabla_v \cdot \left(\nabla_v f + v \, \langle v \rangle^{\beta-2} \, f \right) . \tag{10}$$

Here we use the notation

$$\langle v \rangle := \sqrt{1 + |v|^2}\,.$$

Unless $\beta = 2$, our choice of the transport operator differs from the transport operator corresponding to Newton's equations, namely $v \cdot \nabla_x - \nabla_x \phi \cdot \nabla_v$, which has been widely studied in the literature. More general dependences of $\mathcal{E}$ and L on v than $\langle v \rangle^\beta$, with for instance a power law asymptotic growth as $|v| \to +\infty$, could be considered under minor changes. Our purpose is to study the asymptotic behaviour of the solution of

$$\frac{\partial f}{\partial t} + \mathsf{T}f = \mathsf{L}f\,, \quad f(t=0,\cdot,\cdot) = f_0 \tag{11}$$

with T and L given respectively by (9) and (10) as $t \to +\infty$. With these choices and under the condition that $e^{-\phi}$ is integrable, a remarkable property is that the *Gibbs state*

$$f_\star(x,v) := \frac{1}{Z} \, e^{-\mathscr{E}(x,v)} \quad \text{where} \quad Z = \int_{\mathbb{R}^d} e^{-\phi} \, dx \int_{\mathbb{R}^d} e^{-\frac{1}{\beta} \langle v \rangle^\beta} \, dv \tag{12}$$

is a stationary solution of mass $\| f_\star \|_{L^1(\mathbb{R}^d \times \mathbb{R}^d,\, dx\, dv)} = 1$. We consider L as an operator on $L^2\big(\mathbb{R}^d, e^{\langle v \rangle^\beta} dv\big)$ acting on functions depending on the velocity variable v and extend it to the Hilbert space $L^2(\mathbb{R}^d \times \mathbb{R}^d, d\mu)$ of functions depending on x

and v where

$$d\mu := \frac{dx\,dv}{f_\star(x,v)}\,.$$

Since $f_\star$ is integrable, notice that $\mathrm{L}^1(\mathbb{R}^d \times \mathbb{R}^d, dx\,dv) \subset \mathrm{L}^2(\mathbb{R}^d \times \mathbb{R}^d, d\mu)$ by a Cauchy-Schwarz inequality. After these preliminaries, we observe that $f_\star$ is local equilibrium, i.e., belongs to Ker(L) which is generated by functions of the type

$$f_\rho(x,v) := \frac{\rho(x)\,e^{-\frac{1}{\beta}\langle v\rangle^\beta}}{\int_{\mathbb{R}^d} e^{-\frac{1}{\beta}\langle v\rangle^\beta}\,dv} \tag{13}$$

where $\rho \in \mathrm{L}^2\big(\mathbb{R}^d, e^{-\phi}\,dx\big) \supset \mathrm{L}^1(\mathbb{R}^d, dx)$ is an arbitrary function. The property $\mathsf{T} f_\star = \mathsf{L} f_\star = 0$ sometimes appears in the physics literature as a *factorization* property. The orthogonal projector onto Ker(L) is defined as the projection on local equilibria by

$$\Pi f = f_\rho(x,v) \quad \text{where} \quad \rho = \int_{\mathbb{R}^d} f\,dv\,.$$

Notice that f and f_ρ have the same spatial density because $\int_{\mathbb{R}^d} f_\rho\,dv = \rho$. Under the assumption that the measure $e^{-\phi}\,dx$ admits a Poincaré inequality, that is, there is some positive constant λ_ϕ for which

$$\int_{\mathbb{R}^d} |\nabla u|^2\,e^{-\phi}\,dx \ge \lambda_\phi \int_{\mathbb{R}^d} |u|^2\,e^{-\phi}\,dx \;\forall\, u \in \mathscr{D}(\mathbb{R}^d) \text{ such that } \int_{\mathbb{R}^d} u\,e^{-\phi}\,dx = 0\,, \tag{14}$$

Theorem 1 applies as follows.

Corollary 1 *Assume that ϕ is such that* (14) *holds for some $\lambda_1 > 0$ and $\beta \ge 1$. If f solves* (11) *for some nonnegative function $f_0 \in \mathrm{L}^2(\mathbb{R}^d \times \mathbb{R}^d, d\mu)$ with $\|f_0\|_{\mathrm{L}^1(\mathbb{R}^d\times\mathbb{R}^d,\,dx\,dv)} = 1$, then for some $\delta > 0$, there exists $\lambda > 0$ and $\mathscr{C} > 1$ such that* (8) *holds with $\|\cdot\| := \|\cdot\|_{\mathrm{L}^2(\mathbb{R}^d\times\mathbb{R}^d,\,d\mu)}$.*

The case $\beta = 2$ is by now standard and covered in various papers: see [20, 22] for an L^2 hypocoercivity approach and [25–27] as well as references therein for earlier results based on hypoelliptic methods. To our knowledge the case $\beta \neq 2$ has not been studied yet by L^2-hypocoercivity methods, but convergence results are known from [15, 18] using other methods. Of particular interest in physics is the case $\beta = 1$ where $\mathscr{E}$ is the standard energy for relativistic particles, up to physical constants (mass and speed of light are taken equal to 1), while the corresponding L operator is not much more than a caricature of a relativistic collision operator. Concerning L and from the point of view of phenomenological models, it is however interesting to consider local equilibria given by (13) and it makes sense to assume that stationary solutions have the *factorization* property. Throughout this paper, we will make this simplifying assumption.

The strategy of the proof of Corollary 1 goes as follows. With $\beta \geq 1$, we have the Poincaré inequality: there is some $\lambda_m > 0$ such that, for all $g \in \mathrm{L}^2(\mathbb{R}^d, e^{-\frac{1}{\beta}\langle v\rangle^\beta}\,dv)$ such that $\int_{\mathbb{R}^d} g\, e^{-\frac{1}{\beta}\langle v\rangle^\beta}\,dv = 0$, we have

$$\int_{\mathbb{R}^d} |\nabla g|^2\, e^{-\frac{1}{\beta}\langle v\rangle^\beta}\,dv \geq \lambda_m \int_{\mathbb{R}^d} |g|^2\, e^{-\frac{1}{\beta}\langle v\rangle^\beta}\,dv\,. \tag{15}$$

This is for instance a consequence of Persson's lemma based on the observation that $\psi(v) := \frac{1}{\beta}\langle v\rangle^\beta$ is such that

$$\liminf_{|v|\to+\infty}\left(\frac{1}{4}\,|\nabla\psi(v)|^2 - \frac{1}{2}\,\Delta\psi(v)\right) > 0\,.$$

See for instance [11, Appendix A.1] for details. As a consequence, (H1) holds. The macroscopic coercivity condition (H2) follows from (14). The *parabolic macroscopic dynamics* condition (H3) is a simple consequence of the definitions of T and Π. Hence the only assumption that deserves some attention is (H4), which is obtained by elliptic estimates. A detailed proof is given in Sect. 4.6.

In the framework of (9) and (10), an elementary computation shows that (2) written for f_ρ defined by (13) reduces to the *Fokker-Planck* equation

$$\frac{\partial\rho}{\partial t} = \sigma\left(\Delta\rho + \nabla\cdot(\rho\,\nabla\phi)\right) \tag{16}$$

with diffusion coefficient σ given in terms of β by

$$\sigma = \frac{1}{d}\int_{\mathbb{R}^d} |v|^2\,\langle v\rangle^{2\beta-4}\,e^{-\frac{1}{\beta}\langle v\rangle^\beta}\,dv\,. \tag{17}$$

In order to simplify the discussion, we shall assume that

$$\phi(x) = \frac{1}{\alpha}\,\langle x\rangle^\alpha \quad \forall\,x \in \mathbb{R}^d\,.$$

Corollary 1 corresponds to $\alpha \geq 1$. Our purpose is to investigate the decay rates of (11) in terms of $\beta > 0$ and $\alpha > 0$. Let us start by studying the asymptotic behaviour of a solution ρ of (16) depending on the various cases for the potential ϕ. For completeness, we will also consider the limit case as $\alpha \to 0$ and distinguish several cases depending on whether we take $\phi = 0$ a.e., or (in the case of the Fokker-Planck equation), depending on $\gamma > 0$,

$$\phi(x) = \gamma\,\log\langle x\rangle \quad \forall\,x \in \mathbb{R}^d\,.$$

Up to minor technicalities, general potentials ϕ with asymptotic power law or logarithmic growths as $|x| \to +\infty$ could also be covered.

3 Fokker-Planck Equations with Various External Potentials, Moments and Functional Inequalities

We collect some results on the asymptotic behaviour of the solutions of (16) as $t \to +\infty$ based on various functional inequalities. In this section we omit the discussion of optimality cases and estimates on sharp constants in the functional inequalities. By default, constants in the inequalities are always taken to their optimal value. Table 1 collects the results in a synthetic picture, although without all details on the assumptions.

3.1 Strong Confinement Case: Poincaré Inequality

If $\phi(x) = \frac{1}{\alpha}\langle x\rangle^\alpha$ with $\alpha \geq 1$, then (14) holds with $\lambda_\phi = \lambda_M > 0$. We apply it to $u = \rho/e^{-\phi}$. A solution ρ of (16) with initial datum ρ_0 at $t = 0$ satisfies

$$\begin{aligned}\frac{d}{dt}\|\rho(t,\cdot)\|^2_{\mathrm{L}^2(\mathbb{R}^d,\,e^\phi\,dx)} &= -2\,\sigma\,\|\nabla\rho(t,\cdot)\|^2_{\mathrm{L}^2(\mathbb{R}^d,\,e^\phi\,dx)}\\ &\leq -2\,\lambda_M\,\sigma\,\|\rho(t,\cdot)\|^2_{\mathrm{L}^2(\mathbb{R}^d,\,e^\phi\,dx)}\,,\end{aligned}$$

which yields the estimate

$$\|\rho(t,\cdot)\|^2_{\mathrm{L}^2(\mathbb{R}^d,\,e^\phi\,dx)} \leq \|\rho_0\|^2_{\mathrm{L}^2(\mathbb{R}^d,\,e^\phi\,dx)}\,e^{-2\,\lambda_M\,\sigma\,t}\quad \forall\,t \geq 0\,.$$

3.2 Weak Confinement Case: Weighted Poincaré Inequality

The following results are taken from [8, Appendices A and B]. We assume here that $\alpha \in (0,1)$ and consider a solution of (16) with nonnegative initial datum $\rho_0 \in \mathrm{L}^1(\mathbb{R}^d, dx)$ such that $\|\rho_0\|_{\mathrm{L}^1(\mathbb{R}^d)} = 1$. The function $u = \rho\,e^\phi$ is a solution of the *Ornstein-Uhlenbeck* equation (also known as the *backward Kolmogorov* equation)

$$\frac{\partial u}{\partial t} = \sigma\left(\Delta u - \nabla\phi\cdot\nabla u\right). \tag{18}$$

With $k \geq 0$, let us compute

$$\frac{d}{dt}\int_{\mathbb{R}^d} |u(t,x)|^2 \langle x\rangle^k e^{-\phi}\,dx + 2\sigma\int_{\mathbb{R}^d} |\nabla_x u(t,x)|^2 \langle x\rangle^k e^{-\phi}\,dx \leq \int_{\mathbb{R}^d}\left(a_k - b_k \langle x\rangle^{\alpha-2}\right)|u(t,x)|^2 \langle x\rangle^k e^{-\phi}\,dx$$

for some $a_k \in \mathbb{R}$, $b_k \in (0,+\infty)$. As a consequence, there exists a constant $\mathcal{K}(k) > 0$ such that

$$\int_{\mathbb{R}^d} \langle x\rangle^k |\rho(t,x)|^2 e^{\phi}\,dx \leq \mathcal{K}(k)\int_{\mathbb{R}^d} \langle x\rangle^k |\rho_0|^2 e^{\phi}\,dx \quad \forall\, t \geq 0\,.$$

See [8, Proposition 4 and Appendix B.2] for details. With $k = 0$, we notice that $a_0 = b_0 = 0$ and use the *weighted Poincaré inequality*

$$\int_{\mathbb{R}^d} |\nabla_x u(t,x)|^2 e^{-\phi}\,dx \geq \mathscr{C}_\alpha^{\mathrm{wP}}\int_{\mathbb{R}^d} |u(t,x)-\bar u|^2 \frac{e^{-\phi}}{\langle x\rangle^{2(1-\alpha)}}\,dx \quad \text{where} \quad \bar u = \frac{\int_{\mathbb{R}^d} u\,e^{-\phi}\,dx}{\int_{\mathbb{R}^d} e^{-\phi}\,dx} \tag{19}$$

(notice that the average $\bar u$ is computed with respect to the measure of the l.h.s.) to prove that

$$\frac{d}{dt}\int_{\mathbb{R}^d} |u(t,x)-\bar u|^2 e^{-\phi}\,dx = -2\sigma\int_{\mathbb{R}^d} |\nabla_x u(t,x)|^2 e^{-\phi}\,dx \leq -2\sigma\,\mathscr{C}_\alpha^{\mathrm{wP}}\int_{\mathbb{R}^d} |u(t,x)-\bar u|^2 \frac{e^{-\phi}}{\langle x\rangle^{2(1-\alpha)}}\,dx\,.$$

With $k \geq 2(1-\alpha)$ and $\theta = k/\big(k+2(1-\alpha)\big)$, Hölder's inequality

$$\int_{\mathbb{R}^d} |u-\bar u|^2 e^{-\phi}\,dx \leq \left(\int_{\mathbb{R}^d} |u-\bar u|^2 \frac{e^{-\phi}}{\langle x\rangle^{2(1-\alpha)}}\,dx\right)^{\theta} \left(\int_{\mathbb{R}^d} |u-\bar u|^2 \langle x\rangle^k e^{-\phi}\,dx\right)^{1-\theta}$$

allows us to prove that

$$\int_{\mathbb{R}^d} |\rho(t,x) - \rho_\star(x)|^2 \, e^{\phi} \, dx$$

$$\le \left(\|\rho_0 - \rho_\star\|_{\mathrm{L}^2(\mathbb{R}^d,\, e^{\phi}\, dx)}^{-4\,(1-\alpha)/k} + \frac{4\,(1-\alpha)\,\sigma\,\mathscr{C}_\alpha^{\mathrm{wP}}}{k\,\mathcal{K}_*^{2\,(1-\alpha)/k}}\, t \right)^{-\frac{k}{2\,(1-\alpha)}} \quad \forall\, t \ge 0$$

with $\mathcal{K}_* := \mathcal{K}(k)^2 \int_{\mathbb{R}^d} \langle x\rangle^k \, |\rho_0|^2 \, e^{\phi} \, dx + \int_{\mathbb{R}^d} \langle x\rangle^k \, e^{-\phi} \, dx \, \|\rho_0\|^2_{\mathrm{L}^1(\mathbb{R}^d)}$.

3.3 Weak Confinement, a Limit Case: Hardy-Poincaré Inequality

The results in this case are new. In the limit as $\alpha \to 0_+$, we can assume that $\phi(x) = \gamma\,\log\langle x\rangle$ with $\gamma > d$ so that $f_\star$ defined by (12) is integrable. Let $u = \rho\, e^{\phi}$ be a solution of (18). With $k \ge 0$, let us compute

$$\begin{aligned}
&\frac{d}{dt} \int_{\mathbb{R}^d} |u(t,x)|^2 \, \langle x\rangle^k \, e^{-\phi} \, dx + 2\,\sigma \int_{\mathbb{R}^d} |\nabla_x u(t,x)|^2 \, \langle x\rangle^k \, e^{-\phi} \, dx \\
&\quad = k \int_{\mathbb{R}^d} |u(t,x)|^2 \, \langle x\rangle^{k-2} \left(d + (k-\gamma-2)\,\tfrac{|x|^2}{\langle x\rangle^2} \right) e^{-\phi} \, dx \\
&\quad \le k \int_{\mathbb{R}^d} |u(t,x)|^2 \, \langle x\rangle^{k-2} \left(d + k - \gamma - 2 - (k-\gamma-2)\,\langle x\rangle^{-2} \right) e^{-\phi} \, dx \,.
\end{aligned} \tag{20}$$

By arguing as in [8, Proposition 4 and Appendix B.2], this is enough to prove that there exists a constant $\mathcal{K}(k) > 0$ such that

$$\int_{\mathbb{R}^d} \langle x\rangle^k \, |u(t,x)|^2 \, e^{-\phi} \, dx \le \mathcal{K}(k) \int_{\mathbb{R}^d} \langle x\rangle^k \, |\rho_0|^2 \, e^{\phi} \, dx \quad \forall\, t \ge 0$$

if $k \in (\gamma - d, \gamma + 2 - d)$. Notice that a better range of k can be obtained as follows. Since $\langle x\rangle^k \, e^{-\phi} = \langle x\rangle^{k-\gamma}$, we learn from [7, 21] that for some positive constant $\mathscr{C}_{\gamma-k}^{\mathrm{HP}}$, we have the *Hardy-Poincaré* inequality

$$\int_{\mathbb{R}^d} |\nabla_x u|^2 \, \langle x\rangle^k \, e^{-\phi} \, dx \ge \mathscr{C}_{\gamma-k}^{\mathrm{HP}} \int_{\mathbb{R}^d} |u - \bar{u}|^2 \, \langle x\rangle^{k-2} \, e^{-\phi} \, dx \tag{21}$$

for an appropriate choice of $\bar{u}$ depending on $k - \gamma$. In any case, Inequality (20) written with $k = 0$, that is,

$$\frac{d}{dt} \int_{\mathbb{R}^d} |u(t,x) - \bar{u}|^2 \, e^{-\phi} \, dx \le -2\,\sigma \int_{\mathbb{R}^d} |\nabla_x u(t,x)|^2 \, e^{-\phi} \, dx$$

and then (21) combined with Hölder's inequality applied as in the case of the weighted Poincaré inequality (with $\alpha = 0$) show that

$$\int_{\mathbb{R}^d} |\rho(t,x) - \rho_\star(x)|^2 \, e^\phi \, dx \le \int_{\mathbb{R}^d} |\rho_0 - \rho_\star|^2 \, e^\phi \, dx \, (1 + c\,t)^{-\frac{k}{2}} \quad \forall\, t \ge 0$$

for some constant c which depends on d, γ, σ, k, $\int_{\mathbb{R}^d} |\rho_0|^2 \, \langle x \rangle^{k-\gamma} \, dx$ and $\|\rho_0\|^2_{\mathrm{L}^1(\mathbb{R}^d)}$.

3.4 Very Weak Confinement Case: Caffarelli-Kohn-Nirenberg Inequality

According to [10, Theorem 1], if $1 \le \gamma < d$ and $\phi(x) = \gamma \log\langle x \rangle$, a solution ρ of (16) with nonnegative initial datum $\rho_0 \in \mathrm{L}^1(\mathbb{R}^d, \langle x \rangle^k \, dx) \cap \mathrm{L}^2\big(\mathbb{R}^d, e^\phi \, dx\big)$ with $k = \max\{2, \gamma/2\}$ satisfies the estimate

$$\begin{aligned} M_k(t) &:= \int_{\mathbb{R}^d} \langle x \rangle^k \, \rho(t,x) \, dx \\ &\le 2^{\frac{k-2}{2}} \left(M_0 + \left(\big(M_k(0) - M_0\big)^{2/k} + 2\,\sigma \, \big(d + k - 2 - \gamma\big) \, M_0^{2/k} t \right)^{k/2} \right). \end{aligned}$$

With $e^{-\phi} = \langle x \rangle^{-\gamma}$ and $u = \rho \, \langle x \rangle^\gamma$, a solution u of (18) satisfies the estimate

$$\frac{d}{dt} \int_{\mathbb{R}^d} |u(t,x)|^2 \, \langle x \rangle^{-\gamma} \, dx = -\,2\,\sigma \int_{\mathbb{R}^d} |\nabla u(t,x)|^2 \, \langle x \rangle^{-\gamma} \, dx$$

Combined with the *inhomogeneous Caffarelli-Kohn-Nirenberg inequality*

$$\int_{\mathbb{R}^d} |u|^2 \, \langle x \rangle^{-\gamma} \, dx \le \mathscr{C}^{\mathrm{CKN}}_{k,\gamma} \left(\int_{\mathbb{R}^d} |\nabla u|^2 \, \langle x \rangle^{-\gamma} \, dx \right)^a \left(\int_{\mathbb{R}^d} u \, \langle x \rangle^{k-\gamma} \, dx \right)^{2(1-a)}$$

$$\text{with} \quad a = \frac{d + 2k - \gamma}{d + 2 + 2k - \gamma},$$

this proves the decay estimate

$$\|\rho(t,\cdot)\|^2_{\mathrm{L}^2(\mathbb{R}^d,\, e^\phi \, dx)} \le \|\rho_0\|^2_{\mathrm{L}^2(\mathbb{R}^d,\, e^\phi \, dx)} \, (1 + c\,t)^{-\frac{d-\gamma}{2}} \quad \forall\, t \ge 0$$

where the constant c depends on d, γ, σ, $\|\rho_0\|_{\mathrm{L}^2(\mathbb{R}^d,\, e^\phi \, dx)}$, $M_0 = \|u_0\|_1$, and $M_k(0) = \big\| |x|^k \, \rho_0 \big\|_1$. For more details, as well as a proof of the Caffarelli-Kohn-Nirenberg inequality, see [10, Appendix B].

3.5 No Potential Case: Nash's Inequality

We assume that $\phi = 0$ so that (16) is the standard heat equation. By *Nash's inequality*

$$\|u\|_{\mathrm{L}^2(\mathbb{R}^d)} \le \mathscr{C}_{\mathrm{Nash}}\, \|\nabla u\|_{\mathrm{L}^2(\mathbb{R}^d)}^{\frac{d}{d+2}}\, \|u\|_{\mathrm{L}^1(\mathbb{R}^d)}^{\frac{2}{d+2}} \quad \forall\, u \in \mathrm{H}^1(\mathbb{R}^d, dx)\,,$$

a solution ρ of (16) with initial datum ρ_0 at $t = 0$ satisfies

$$\frac{d}{dt}\, \|\rho(t,\cdot)\|^2_{\mathrm{L}^2(\mathbb{R}^d)} = -\,2\,\sigma\, \|\nabla\rho(t,\cdot)\|^2_{\mathrm{L}^2(\mathbb{R}^d)}\,.$$

Hence $y(t) := \|\rho(t,\cdot)\|^2_{\mathrm{L}^2(\mathbb{R}^d)}$ solves the differential inequality $y' \le -\,2\,\sigma\,\mathscr{C}_{\mathrm{Nash}}^{-1}\, \|\rho_0\|_{\mathrm{L}^1(\mathbb{R}^d)}^{-\frac{4}{d}}\, y^{1+\frac{2}{d}}$ which, after integration, yields the estimate

$$\|\rho(t,\cdot)\|^2_{\mathrm{L}^2(\mathbb{R}^d)} \le \left(\|\rho_0\|_{\mathrm{L}^2(\mathbb{R}^d)}^{-4/d} + \frac{4\,\sigma}{d\,\mathscr{C}_{\mathrm{Nash}}}\, \|\rho_0\|_{\mathrm{L}^1(\mathbb{R}^d)}^{-4/d}\, t \right)^{-d/2} \quad \forall\, t \ge 0\,.$$

3.6 A Short Summary

In case of the Fokker-Planck equation (16), Table 1 summarizes what is known on decay rates based on moment estimates and interpolation inequalities. Cases in gray will be further considered in the case of kinetic equations.

4 Kinetic Fokker-Planck Equations and Hypocoercivity Results

4.1 State of the Art

Some known results are collected in Table 2. They are exclusively concerned with the classical transport operator

$$\mathsf{T} f := v \cdot \nabla_x f - \nabla_x \phi \cdot \nabla_v f\,,$$

i.e., coincide with our framework if $\beta = 2$ (at the level of the transport operator).

In Table 2, if $\beta \ge 1$, *Micro* Poincaré refers to a Poincaré inequality written in the velocity variable v, which controls the convergence towards a local equilibrium while, *Macro* Poincaré refers to a Poincaré inequality written in the position variable

Table 1 Short summary of the behaviours as $t \to +\infty$ of the solution of (16) depending on the choice of ϕ, with some references. On the left side ($\phi = 0$ or $\gamma < d$), there is no global stationary solution and we study decay rates. On the right side, we investigate the convergence rates to a global stationary solution. Under additional or different constraints on the initial data, other behaviours can be obtained based for instance on *weak Poincaré inequalities*: see [35, Theorem 2.1], [4, Theorem 1.4] and [29]

Potential	$\phi = 0$	$\phi(x) = \gamma \log\langle x\rangle \gamma < d$	$\phi(x) = \gamma \log\langle x\rangle \gamma > d$	$\phi(x) = \frac{1}{\alpha}\langle x\rangle^\alpha \alpha \in (0,1)$	$\phi(x) = \frac{1}{\alpha}\langle x\rangle^\alpha \alpha \geq 1$
Inequality	Nash	Caffarelli-Kohn-Nirenberg	Hardy-Poincaré	Weighted Poincaré	Poincaré
Asymptotic behavior	$t^{-d/2}$ decay	$t^{-(d-\gamma)/2}$ decay	$t^{-k/2}$ convergence	$t^{-\frac{k}{2(1-\alpha)}}$ convergence	$e^{-\lambda t}$ convergence
References	[33]	[10]		[8]	(*)

(*) The use of the Poincaré inequality in relation with the Fokker-Planck equation has a long history, which we cannot cover entirely here: we can for instance refer to [33], and to [5, Chapter 4] for an overview in the context of Markov processes

Table 2 Rough classification of the asymptotic behaviour of the solutions of $\partial_t f + v \cdot \nabla_x f - \nabla_x \phi \cdot \nabla_v f = f_\star \nabla_v \big(f_\star^{-1} \nabla_v f \big)$ as $t \to +\infty$ where $f_\star(x,v) = Z^{-1} \exp(-\phi(x) - \psi(v))$. Additional assumptions on the initial datum $f_0 = f(t=0,\cdot,\cdot)$ are needed: for instance in the case $\alpha \geq 1$ and $\beta \in (0,1)$, the initial datum is such that $f_0 \in \mathrm{L}^2\big(\mathbb{R}^d \times \mathbb{R}^d, \langle v \rangle^{(1-\beta)\sigma} \, d\mu\big)$. In the case $\phi = 0$ and $\beta \in (0,1)$, we assume that $f_0 \in \mathrm{L}^2\big(\mathbb{R}^d \times \mathbb{R}^d, \langle v \rangle^{\ell/2} \, d\mu\big)$. If $\alpha \in (0,1)$ and $\beta \in (0,1)$, using the weak Poincaré inequality requires specific bounds. Further cases and more detailed assumptions can be found in the references collected above

Potential	$\phi = 0$	$\phi(x) = \frac{1}{\alpha} \langle x \rangle^\alpha$ $\alpha \in (0,1)$	$\phi(x) = \frac{1}{\alpha} \langle x \rangle^\alpha$ $\alpha \geq 1$, or $\mathbb{T}^d$ Macro Poincaré
$\psi(v) = \frac{1}{\beta} \langle v \rangle^\beta$ $\beta \geq 1$ Micro Poincaré	$t^{-d/2}$ decay [9]	e^{-t^b}, $b < 1 \beta = 2$ convergence [16]	$e^{-\lambda t}$ convergence [1, 14, 15, 18, 20, 22, 26, 31, 32]
$\psi(v) = \frac{1}{\beta} \langle v \rangle^\beta$ $\beta \in (0,1)$	$t^{-\zeta}$ $\zeta = \min\{\frac{d}{2}, \frac{\ell}{2(1-\beta)}\}$ decay, [8]	$t^{-\zeta}$ convergence [13]	$t^{-\zeta}$ convergence [13]
Limit as $\beta \to 0_+$ $\psi(v) = -(d+\varepsilon) \log\langle v \rangle$	$\varepsilon \in (0,2)$ fractional diffusion limit, [12]	[13]	$t^{-\zeta}$ if $\varepsilon > 2$ convergence [13]

x, which controls the convergence of the solution in the macroscopic or diffusion limit, towards a global equilibrium, or to 0 if there is no such equilibrium. Cases in gray will be further considered in the case of the transport operator given by (9).

4.2 *Notation and Basic Observations*

From here on, we assume that $\psi(v) = \frac{1}{\beta} \langle v \rangle^\beta$ and $\phi(x) = \frac{1}{\alpha} \langle x \rangle^\alpha$ for some $\beta > 0$ and $\alpha > 0$, use the notation

$$\rho_\star := \frac{e^{-\phi}}{\int_{\mathbb{R}^d} e^{-\phi} \, dx}, \quad \rho_f := \int_{\mathbb{R}^d} f \, dv \quad \text{and} \quad u_f := \frac{\rho_f}{\rho_\star},$$

and consider the transport operator given by (9). We recall that $f_\star$ is defined by (12). The following observations can be omitted at first reading and will be used only in Sects. 4.4–4.6 for proving Theorem 2. We can write

$$\Pi f = \rho_f \, \frac{e^{-\psi}}{\int_{\mathbb{R}^d} e^{-\psi} \, dv} = u_f \, f_\star, \quad \mathsf{T}\Pi f = \langle v \rangle^{\beta-2} \, (v \cdot \nabla_x u_f) \, f_\star$$

$$\text{and} \quad \Pi\mathsf{T} f = \left(\nabla_x \cdot \int_{\mathbb{R}^d} v \, \langle v \rangle^{\beta-2} \, f \, dv \right) \frac{f_\star}{\rho_\star}.$$

If $f = u\, f_\star \in \mathrm{Ker}(\mathsf{L})$ and σ is defined by (17), then

$$(\mathsf{T}\Pi)^* (\mathsf{T}\Pi)\, f = -\frac{\sigma}{\rho_\star}\, \nabla_x \cdot \big(\rho_\star\, \nabla_x u\big)\, f_\star = -\,\sigma\, \big(\Delta_x u - \nabla_x \phi \cdot \nabla_x u\big)\, f_\star\,,$$

Solving $g = \big(1 + (\mathsf{T}\Pi)^*(\mathsf{T}\Pi)\big)^{-1} f$ means that $g = u\, f_\star$ where $u = u_g$ solves

$$u - \sigma\, \big(\Delta_x u - \nabla_x \phi \cdot \nabla_x u\big) = u_f\,. \tag{22}$$

In order to justify integrations by parts (see Sect. 4.4 below), one can notice that $c\, f_\star$ with an arbitrary $c \in \mathbb{R}$ can be used as a barrier function, so that we can assume that u_f is bounded as $|x| \to +\infty$. Standard elliptic estimates apply to the solution u of (22) and one can conclude using density arguments.

4.3 Main Result

Our goal is to get a classification similar to the results summarized in Table 2 for $\beta \neq 2$ in the transport operator defined by (9), i.e., with $\mathsf{T} f := \nabla_v \mathscr{E} \cdot \nabla_x f - \nabla_x \mathscr{E} \cdot \nabla_v f$. As far as we know, this transport operator has not been studied yet in the framework of hypocoercivity methods, except for some recent results in the bounded domain case in [1, 14] or when $\alpha \geq 1$ in [15, 18] which are based on weak norms and Lions' lemma.

Theorem 2 *Let $f = f(t, x, v)$ be a solution of* (11) *with transport and collision operators given respectively by* (9) *and* (10) *for some $\beta > 0$ and $\alpha > 0$. With $f_\star$ defined by* (12), *we assume that the initial datum satisfies*

$$0 \leq f_0 \leq C\, f_\star \tag{23}$$

for a suitable constant $C > 0$. Depending on β and α, we have the following convergence and decay estimates.

1. *Assume $\beta \geq 1$ and $\alpha \geq 1$. Then there exist constants $\mathscr{C} > 0$ and $\lambda > 0$ such that any solution f of* (11) *with initial datum $f_0 \in \mathrm{L}^2(\mathbb{R}^d \times \mathbb{R}^d, d\mu)$ satisfies*

$$\|f - f_\star\|^2_{\mathrm{L}^2(\mathbb{R}^d \times \mathbb{R}^d,\, d\mu)} \leq \mathscr{C} e^{-\lambda t}\, \|f_0 - f_\star\|^2_{\mathrm{L}^2(\mathbb{R}^d \times \mathbb{R}^d,\, d\mu)} \quad \forall\, t \geq 0\,.$$

2. *Assume $\beta \in (0, 1)$ and $\alpha \geq 1$. Then there exists a constant $\mathscr{C}_\ell > 0$ such that any solution f of* (11), *with initial datum $f_0 \in \mathrm{L}^2\big(\mathbb{R}^d \times \mathbb{R}^d, \langle v\rangle^\ell\, d\mu\big)$ for some $\ell > 0$, satisfies*

$$\|f - f_\star\|^2_{\mathrm{L}^2(\mathbb{R}^d \times \mathbb{R}^d,\, d\mu)} \leq \mathscr{C}_\ell\, (1 + t)^{-\frac{\ell}{2(1-\beta)}}\, \|f_0 - f_\star\|^2_{\mathrm{L}^2(\mathbb{R}^d \times \mathbb{R}^d,\, d\mu)} \quad \forall\, t \geq 0\,.$$

3. *Assume $\beta \geq 1$ and $\alpha \in (0,1)$. Then there exists a constant $\mathscr{C}_k > 0$ such that any solution f of* (11)*, with initial datum $f_0 \in \mathrm{L}^2\big(\mathbb{R}^d \times \mathbb{R}^d, \langle x\rangle^k \, d\mu\big)$ for some $k > 0$, satisfies*

$$\|f - f_\star\|^2_{\mathrm{L}^2(\mathbb{R}^d \times \mathbb{R}^d,\, d\mu)} \leq \mathscr{C}_k \,(1+t)^{-\frac{k}{2(1-\alpha)}} \,\|f_0 - f_\star\|^2_{\mathrm{L}^2(\mathbb{R}^d \times \mathbb{R}^d,\, d\mu)} \quad \forall t \geq 0\,.$$

4. *Assume $\beta \in (0,1)$ and $\alpha \in (0,1)$. Then there exist a constant $\mathscr{C}_{k,\ell} > 0$ such that any solution f of* (11)*, with initial datum $f_0 \in \mathrm{L}^2\big(\mathbb{R}^d \times \mathbb{R}^d, \langle x\rangle^k \, d\mu\big) \cap \mathrm{L}^2\big(\mathbb{R}^d \times \mathbb{R}^d, \langle v\rangle^\ell \, d\mu\big)$ for some $k > 0$ and $\ell > 0$, satisfies*

$$\|f - f_\star\|^2_{\mathrm{L}^2(\mathbb{R}^d \times \mathbb{R}^d,\, d\mu)} \leq \mathscr{C}_{k,\ell} \,(1+t)^{-\zeta} \,\|f_0 - f_\star\|^2_{\mathrm{L}^2(\mathbb{R}^d \times \mathbb{R}^d,\, d\mu)} \quad \forall t \geq 0\,.$$

where $\zeta = \min\left\{\frac{k}{2(1-\alpha)}, \frac{\ell}{2(1-\beta)}\right\}$

5. *Assume $\beta \geq 1$ and $\phi = 0$. Then there exist a constant $\mathscr{K} > 0$ depending on $\|f_0\|_{\mathrm{L}^1(\mathbb{R}^d \times \mathbb{R}^d,\, dx\, dv)}$ such that any solution f of* (11)*, with initial datum $f_0 \in \mathrm{L}^2(\mathbb{R}^d \times \mathbb{R}^d, d\mu)$, satisfies*

$$\|f\|^2_{\mathrm{L}^2(\mathbb{R}^d \times \mathbb{R}^d,\, d\mu)} \leq \mathscr{K} \,(1+t)^{-\frac{d}{2}} \,\|f_0\|^2_{\mathrm{L}^2(\mathbb{R}^d \times \mathbb{R}^d,\, d\mu)} \quad \forall t \geq 0\,.$$

6. *Assume $\beta \in (0,1)$ and $\phi = 0$. Then there exist a constant $\mathscr{K}_\ell > 0$ such that any solution f of* (11)*, with initial datum $f_0 \in \mathrm{L}^2\big(\mathbb{R}^d \times \mathbb{R}^d, \langle v\rangle^\ell \, d\mu\big)$ for some $\ell > 0$, satisfies*

$$\|f\|^2_{\mathrm{L}^2(\mathbb{R}^d \times \mathbb{R}^d,\, d\mu)} \leq \mathscr{K}_\ell \,(1+t)^{-\zeta} \,\|f_0\|^2_{\mathrm{L}^2(\mathbb{R}^d \times \mathbb{R}^d,\, d\mu)} \quad \forall t \geq 0\,,$$

where $\zeta = \min\left\{\frac{d}{2}, \frac{\ell}{2(1-\beta)}\right\}$.

In the statement of Theorem 2, even if it is not specified, the constants may depend on norms of f_0. See Table 3 for a summary of the results. Assumption (23) is a simplifying assumption which can be removed in various cases: see for instance [8, 16, 22, 31]. It allows an immediate conservation of moments along the flow, see Lemma 3 below.

Table 3 Summary of the results of Theorem 2. See the statement for the precise meaning of the rates and the assumptions

Potential	$\phi = 0$	$\phi(x) = \frac{1}{\alpha}\langle x\rangle^\alpha$ $\alpha \in (0,1)$	$\phi(x) = \frac{1}{\alpha}\langle x\rangle^\alpha \alpha \geq 1$
$\psi(v) = \frac{1}{\beta}\langle v\rangle^\beta$ $\beta \geq 1$ Micro Poincaré	$t^{-d/2}$ decay	$t^{-\frac{k}{2(1-\alpha)}}$ convergence	$e^{-\lambda t}$ convergence
$\psi(v) = \frac{1}{\beta}\langle v\rangle^\beta$ $\beta \in (0,1)$	$t^{-\min\left\{\frac{d}{2}, \frac{\ell}{2(1-\beta)}\right\}}$ convergence	$t^{-\min\left\{\frac{k}{2(1-\alpha)}, \frac{\ell}{2(1-\beta)}\right\}}$ convergence	$t^{-\frac{\ell}{2(1-\beta)}}$ convergence

Remark The results of Theorem 2 can be extended to functions ψ and ϕ depending monotonously on $|v|$ and $|x|$ respectively, which behave like $\langle v\rangle^\beta$ and $\langle x\rangle^\alpha$ as $|v| \to +\infty$ and $|x| \to +\infty$. Typically, one has to assume that for any $v \in \mathbb{R}^d$,

$$C_1 \langle v\rangle^\beta \le \phi(v) \le C_2 \langle v\rangle^\beta\,, \quad C_3 |v| \langle v\rangle^{\beta-1} \le v\cdot\nabla_v\psi(v) \le C_4 |v| \langle v\rangle^{\beta-1}$$
$$\text{and} \quad |\mathrm{Hess}(\psi)|(v) \le C_5 \langle v\rangle^{\beta-2}$$

for some positive constants C_i, with $i = 1, \ldots, 5$, and similar estimates for ϕ.

4.4 An Estimate of the Entropy Production

Let us introduce the weighted norm defined by

$$\|f\|_\beta := \|f\|_{\mathrm{L}^2(\mathbb{R}^d\times\mathbb{R}^d,\,\langle v\rangle^{-2(1-\beta)_+}\,d\mu)}$$

where $(1-\beta)_+$ denotes the positive part of $1-\beta$. As a consequence $\|f\|_2$ denotes the standard norm with no weight and we keep using the notation $\langle\cdot,\cdot\rangle$ for the associated scalar product. We can rephrase the Poincaré inequality (15) corresponding to the case $\beta \ge 1$, and the weighted Poincaré inequality (19) rewritten in the variable v with $\beta \in (0, 1)$ instead of α and $\lambda_m = \mathscr{C}_\beta$, as

$$-\langle \mathsf{L}f, f\rangle \ge \lambda_m \|(1-\Pi)f\|_\beta^2\,.$$

In the language of Theorem 1, this inequality replaces (H1) while (H3) is still satisfied. Next we use the notation of Sect. 2 for A, H and D, with T and L given respectively by (9) and (10).

Lemma 1 *For any $\beta > 0$, there is a positive constant κ such that*

$$D[f] \ge \kappa\left(\|(1-\Pi)f\|_\beta^2 + \langle AT\Pi f, \Pi f\rangle\right). \tag{24}$$

Proof We recall that by (5), D is defined as

$$\mathsf{D}[f] := -\langle \mathsf{L}f, f\rangle + \delta\,\langle \mathsf{AT}\Pi f, f\rangle - \delta\Big(\mathrm{Re}\langle \mathsf{TA}f, f\rangle$$
$$-\,\mathrm{Re}\langle \mathsf{AT}(1-\Pi)f, f\rangle + \mathrm{Re}\langle \mathsf{AL}f, f\rangle\Big)\,.$$

In order to prove Lemma 1, we have to give estimates on the last three terms using $\|(1-\Pi)f\|_\beta$ and $\langle \mathsf{AT}\Pi f, \Pi f\rangle$. We obtain these estimates in four steps, as follows.

Step 1. *Expressions of* $\langle \mathsf{AT}\Pi f, \Pi f\rangle$. We consider the function $u = u(x)$ implicitly defined by $u\, f_\star = \big(1 + (\mathsf{T}\Pi)^*(\mathsf{T}\Pi)\big)^{-1}\, \Pi f$, that is, the solution of (22) that can be rewritten as

$$u - \frac{\sigma}{\rho_\star}\, \nabla_x \cdot \big(\rho_\star\, \nabla_x u\big) = u_f = \frac{\rho_f}{\rho_\star}\,. \tag{25}$$

We deduce from

$$\begin{aligned} \mathsf{AT}\Pi f &= \big(1 + (\mathsf{T}\Pi)^*(\mathsf{T}\Pi)\big)(\mathsf{T}\Pi)^*(\mathsf{T}\Pi) f = \Pi f - \big(1 + (\mathsf{T}\Pi)^*(\mathsf{T}\Pi)\big)^{-1} f \\ &= \Pi f - u\, f_\star = \left(\frac{\rho_f}{\rho_\star} - u\right) f_\star \end{aligned}$$

and (25) that

$$\begin{aligned} \langle \mathsf{AT}\Pi f, \Pi f\rangle &= -\,\sigma \int_{\mathbb{R}^d} \big(\nabla_x \cdot \big(\rho_\star\, \nabla_x u\big)\big)\, \rho_f\, dx \\ &= -\,\sigma \int_{\mathbb{R}^d} \nabla_x \cdot \big(\rho_\star\, \nabla_x u\big) \left(u - \frac{\sigma}{\rho_\star}\, \nabla_x \cdot \big(\rho_\star\, \nabla_x u\big)\right) dx\,, \end{aligned}$$

that is,

$$\langle \mathsf{AT}\Pi f, \Pi f\rangle = \sigma \int_{\mathbb{R}^d} |\nabla_x u|^2\, \rho_\star\, dx + \sigma^2 \int_{\mathbb{R}^d} \big|\nabla_x \cdot \big(\rho_\star\, \nabla_x u\big)\big|^2\, \rho_\star^{-1}\, dx\,. \tag{26}$$

Testing (25) with $u\, \rho_\star$, we learn after an integration by parts that

$$\int_{\mathbb{R}^d} |u|^2\, \rho_\star\, dx + \sigma \int_{\mathbb{R}^d} |\nabla u|^2\, \rho_\star\, dx = \int_{\mathbb{R}^d} u\, \rho_f\, dx \le \left(\int_{\mathbb{R}^d} |u|^2\, \rho_\star\, dx\right)^{1/2} \|\Pi f\|$$

where the inequality arises from a Cauchy-Schwarz estimate using $\|\Pi f\|^2 = \int_{\mathbb{R}^d} |\rho_f|^2\, \rho_\star^{-1}\, dx$. Hence

$$\int_{\mathbb{R}^d} |u|^2\, \rho_\star\, dx \le \|\Pi f\|^2 \quad \text{and} \quad \sigma \int_{\mathbb{R}^d} |\nabla u|^2\, \rho_\star\, dx \le \|\Pi f\|^2\,. \tag{27}$$

Step 2. *An estimate of* $|\langle \mathsf{TA} f, f\rangle|$. We know from (4a) that $|\langle \mathsf{TA} f, f\rangle| \le \|(1 - \Pi) f\|_2^2$ if $\beta \ge 1$. With σ defined by (17), we claim that

$$\big|\mathrm{Re}\langle \mathsf{TA} f, f\rangle\big| \le \frac{1}{\sigma}\, \|(1 - \Pi) f\|_\beta^2 \tag{28}$$

also holds if $\beta \in (0, 1)$. In this later case, let us consider the function $w = w(x)$ implicitly defined by $w\, f_\star = \mathsf{A} f$, that is, the solution of

$$w - \frac{\sigma}{\rho_\star}\nabla_x \cdot (\rho_\star \nabla_x w) = -\frac{1}{\rho_\star}\nabla_x \cdot \int_{\mathbb{R}^d} v\,\langle v\rangle^{\beta-2} f\,dv\,.$$

Testing with $w\,\rho_\star$ we obtain

$$\begin{aligned}\sigma \int_{\mathbb{R}^d} |\nabla_x w|^2\,\rho_\star\,dx &\le \int_{\mathbb{R}^d} |w|^2\,\rho_\star\,dx + \sigma \int_{\mathbb{R}^d} |\nabla_x w|^2\,\rho_\star\,dx \\ &= \int_{\mathbb{R}^d} \nabla_x w \cdot \left(\int_{\mathbb{R}^d} v\,\langle v\rangle^{\beta-2} f\,dv\right) dx\,.\end{aligned}$$

By applying the Cauchy-Schwarz inequality and after squaring, we obtain

$$\sigma^2 \int_{\mathbb{R}^d} |\nabla_x w|^2\,\rho_\star\,dx \le \int_{\mathbb{R}^d} \left(\int_{\mathbb{R}^d} v\,\langle v\rangle^{\beta-2}(1-\Pi) f\,dv\right)^2 \rho_\star^{-1}\,dx \le \|(1-\Pi) f\|_\beta^2$$

using $|v|/\langle v\rangle \le 1$ so that $\int_{\mathbb{R}^d} |v|^2\,\langle v\rangle^{-2}\,e^{-\psi}\,dv / \int_{\mathbb{R}^d} e^{-\psi}\,dv \le 1$. Altogether, we prove (28) with

$$\begin{aligned}\|\mathsf{TA} f\,\langle v\rangle^{(1-\beta)_+}\|_2^2 &= \|\mathsf{T}(w\,f_\star)\,\langle v\rangle^{(1-\beta)_+}\|_2^2 \\ &= \iint_{\mathbb{R}^d\times\mathbb{R}^d} \left|\tfrac{v}{\langle v\rangle}\cdot \nabla_x w\right|^2 \langle v\rangle^{2(\beta-1)_+}\,f_\star\,dx\,dv \\ &\le \frac{1}{\sigma^2}\,\|(1-\Pi) f\|_\beta^2\,.\end{aligned}$$

Step 3. We claim that, for some explicit constant $C_\beta > 0$,

$$\left|\mathrm{Re}\langle \mathsf{AL}(1-\Pi) f, f\rangle\right| \le \frac{C_\beta}{\sigma}\,\|(1-\Pi) f\|_\beta\,\langle \mathsf{AT}\Pi f, \Pi f\rangle^{1/2}. \tag{29}$$

This follows from a direct computation. Consider $u = u(x)$ defined by (25) and observe that $\mathsf{A}^*\Pi f = \mathsf{T}(u\,f_\star)$. Then $\langle \mathsf{AL}(1-\Pi) f, \Pi f\rangle = \big\langle (1-\Pi) f, \mathsf{LT}(u\,f_\star)\big\rangle$. Since

$$\mathsf{LT}(u\,f_\star) = \mathsf{L}\left(\langle v\rangle^{\beta-2}\,v\cdot\nabla_x u\,f_\star\right) = \big(\xi(v)\cdot\nabla_x u\big)\,f_\star$$

where $\xi(v) := \nabla_v \cdot \mathrm{Hess}_v(\psi) - \nabla_v\psi \cdot \mathrm{Hess}_v(\psi)$ is a vector valued function of v, it follows that

$$\|\mathsf{LT}(u\, f_\star)\langle v\rangle^{(1-\beta)_+}\|_2^2 \le C_\beta \int_{\mathbb{R}^d} |\nabla_x u|^2\, \rho_\star\, dx$$
$$\text{with } C_\beta = \int_{\mathbb{R}^d} \frac{|\xi(v)|^2\, \langle v\rangle^{2(1-\beta)_+}\, e^{-\psi}}{\int_{\mathbb{R}^d} e^{-\psi}\, dv}\, dv\,.$$

Then (29) follows from (26).

Step 4. We claim that, for some explicit constant $C > 0$,

$$\left|\langle \mathsf{AT}(1-\Pi)f, \Pi f\rangle\right| \le C\, \|(1-\Pi)f\|_\beta\, \langle \mathsf{AT}\Pi f, \Pi f\rangle^{1/2}. \tag{30}$$

With $u = u(x)$ defined by (25), we have $(\mathsf{AT})^*\Pi f = -\,\mathsf{T}^2(u\, f_\star)$ and

$$\langle \mathsf{AT}(1-\Pi)f, \Pi f\rangle = \langle (1-\Pi)f, -\mathsf{T}^2(u\, f_\star)\rangle\,.$$

Using the expression (9) for T, a computation yields

$$\mathsf{T}^2(u\, f_\star) = \Big(\nabla_v\psi \cdot \mathrm{Hess}_x(u)\cdot\nabla_v\psi - \nabla_x u\cdot \mathrm{Hess}_v(\psi)\cdot\nabla_x\phi\Big)\, f_\star\,.$$

Adding and subtracting $(\nabla_v\psi\cdot\nabla_x u)\,(\nabla_v\psi\cdot\nabla_x\phi)$, we can rewrite

$$\begin{aligned}\mathsf{T}^2(uf_\star) &= \nabla_v\psi\cdot\big(\mathrm{Hess}(u) - \nabla_x u\otimes\nabla_x\phi\big)\cdot\nabla_v\psi\, f_\star\\ &\quad - \nabla_x u\cdot\big(\mathrm{Hess}(\psi) - \nabla_v\psi\otimes\nabla_v\psi\big)\cdot\nabla_x\phi\, f_\star\end{aligned}$$

where $\nabla_x u\otimes\nabla_x\phi$ and $\nabla_v\psi\otimes\nabla_v\psi$ are respectively the matrices with entries $\big(\partial_{x_i}u\,\partial_{x_j}\phi\big)_{i,j}$ and $\big(\partial_{v_i}\psi\,\partial_{v_j}\psi\big)_{i,j}$. We estimate independently the two terms in the expression of $\mathsf{T}^2(uf_\star)$.

(1) The second term is estimated by

$$\begin{aligned}&\big\|\nabla_x u\cdot\big(\mathrm{Hess}(\psi) - \nabla_v\psi\otimes\nabla_v\psi\big)\cdot\nabla_x\phi\, f_\star\,\langle v\rangle^{(1-\beta)_+}\big\|_2^2\\ &\le \iint_{\mathbb{R}^d\times\mathbb{R}^d} |\nabla_x u|^2\,\big|\mathrm{Hess}(\psi) - \nabla_v\psi\otimes\nabla_v\psi\big|^2\,|\nabla_x\phi|^2\,\langle v\rangle^{2(1-\beta)_+}\, f_\star\, dx\, dv\\ &\le C_{\beta,2}\int_{\mathbb{R}^d}|\nabla_x u|^2\,|\nabla_x\phi|^2\,\rho_\star\, dx\end{aligned}$$

where $C_{\beta,2} = \int_{\mathbb{R}^d}\big|\mathrm{Hess}(\psi) - \nabla_v\psi\otimes\nabla_v\psi\big|^2\,\langle v\rangle^{2(1-\beta)_+}\, e^{-\psi}\, dv\big/\int_{\mathbb{R}^d} e^{-\psi}\, dv$. Using the fact that $|\nabla_x\phi|^2$ is bounded for $\alpha\in(0,1)$ and [22, Lemma 8] if $\alpha\ge 1$, there is some constant $c_\alpha > 0$ such that the solution of (25) satisfies

$$\int_{\mathbb{R}^d} |\nabla_x u|^2 \, |\nabla_x \phi|^2 \, \rho_\star \, dx \le c_\alpha \int_{\mathbb{R}^d} |\nabla_x u|^2 \, \rho_\star \, dx \,, \tag{31}$$

which, after using (26), is enough to obtain the bound

$$\begin{aligned}&\left\| \nabla_x u \cdot \big(\mathrm{Hess}(\psi) - \nabla_v \psi \otimes \nabla_v \psi\big) \cdot \nabla_x \phi \, f_\star \, \langle v \rangle^{(1-\beta)_+} \right\|_2^2 \\ &\le \frac{c_\alpha \, C_{\beta,2}}{\sigma} \, \|(1-\Pi) f\|_\beta \, \langle \mathsf{AT}\Pi f, \Pi f \rangle^{1/2} \,.\end{aligned}$$

(2) For the first term, we have

$$\begin{aligned}&\left\| \nabla_v \psi \cdot \big(\mathrm{Hess}(u) - \nabla_x u \otimes \nabla_x \phi\big) \cdot \nabla_v \psi \, f_\star \, \langle v \rangle^{(1-\beta)_+} \right\|_2^2 \\ &\le \iint_{\mathbb{R}^d \times \mathbb{R}^d} |\nabla_v \psi|^4 \, \big| \mathrm{Hess}(u) - \nabla_x u \otimes \nabla_x \phi \big|^2 \, \langle v \rangle^{2(1-\beta)_+} \, f_\star \, dx \, dv \\ &\le C_{\beta,3} \int_{\mathbb{R}^d} \big| \mathrm{Hess}(u) - \nabla_x u \otimes \nabla_x \phi \big|^2 \, \rho_\star \, dx\end{aligned}$$

where $C_{\beta,3} = \int_{\mathbb{R}^d} |\nabla_v \psi|^4 \, \langle v \rangle^{2(1-\beta)_+} \, e^{-\psi} \, dv \big/ \int_{\mathbb{R}^d} e^{-\psi} \, dv$. Notice that $\mathrm{Hess}(u) - \nabla_x u \otimes \nabla_x \phi$ is the matrix with entries $\partial_{x_i x_j} u - \partial_{x_i} u \, \partial_{x_j} \phi = \partial_{x_i}\big(\rho_\star \, \partial_{x_j} u\big) \, \rho_\star^{-1}$ for $i, j = 1, \ldots, d$. Hence

$$\begin{aligned}\int_{\mathbb{R}^d} \big| \mathrm{Hess}(u) - \nabla_x u \otimes \nabla_x \phi \big|^2 \, \rho_\star \, dx &= \sum_{i,j=1}^d \int_{\mathbb{R}^d} \big(\partial_{x_i}(\rho_\star \, \partial_{x_j} u)\big)^2 \, \rho_\star^{-1} \, dx \\ &= \int_{\mathbb{R}^d} \big| \nabla_x \cdot (\rho_\star \, \nabla_x u) \big|^2 \, \rho_\star^{-1} \, dx + \int_{\mathbb{R}^d} |\nabla_x u|^2 \, \Delta_x \phi \, \rho_\star \, dx \\ &\quad - \int_{\mathbb{R}^d} \mathrm{Hess}(\phi) : \nabla_x u \otimes \nabla_x u \, \rho_\star \, dx \,.\end{aligned} \tag{32}$$

To prove (32), it is indeed enough to notice that

$$\begin{aligned}\int_{\mathbb{R}^d} \big(\partial_{x_i}(\rho_\star \, \partial_{x_j} u)\big)^2 \, \rho_\star^{-1} \, dx &= -\int_{\mathbb{R}^d} \partial_{x_i} \phi \, \partial_{x_j} u \, \big(\partial_{x_i}(\rho_\star \, \partial_{x_j} u)\big) \, dx \\ &\quad + \int_{\mathbb{R}^d} \partial^2_{x_i x_j} u \, \big(\partial_{x_i}(\rho_\star \, \partial_{x_j} u)\big) \, dx \,.\end{aligned}$$

The observation on the solutions of (22) in Sect. 4.2 applies. By integrating by parts, the two integrals are

$$\int_{\mathbb{R}^d} \partial^2_{x_i x_j} u \left(\partial_{x_i}(\rho_\star\, \partial_{x_j} u)\right) dx = \int_{\mathbb{R}^d} \partial^2_{x_i x_i} u \left(\partial_{x_j}(\rho_\star\, \partial_{x_j} u)\right) dx\,,$$

$$\begin{aligned} -\int_{\mathbb{R}^d} \partial_{x_i}\phi\, \partial_{x_j} u \left(\partial_{x_i}(\rho_\star\, \partial_{x_j} u)\right) dx &= \int_{\mathbb{R}^d} \partial^2_{x_i x_i}\phi\, \partial_{x_j} u \left(\rho_\star\, \partial_{x_j} u\right) dx \\ &\quad + \int_{\mathbb{R}^d} \partial_{x_i}\phi\, \partial^2_{x_i x_j} u \left(\rho_\star\, \partial_{x_j} u\right) dx \\ &= \int_{\mathbb{R}^d} \partial^2_{x_i x_i}\phi\, \partial_{x_j} u \left(\rho_\star\, \partial_{x_j} u\right) dx \\ &\quad - \int_{\mathbb{R}^d} \partial_{x_i}\phi\, \partial_{x_i} u \partial_{x_j} \left(\rho_\star\, \partial_{x_j} u\right) dx \\ &\quad - \int_{\mathbb{R}^d} \partial^2_{x_j x_i}\phi\, \partial_{x_i} u \left(\rho_\star\, \partial_{x_j} u\right) dx\,. \end{aligned}$$

Putting everything together, we get

$$\begin{aligned} \int_{\mathbb{R}^d} \left(\partial_{x_i}(\rho_\star\, \partial_{x_j} u)\right)^2 \rho_\star^{-1}\, dx &= \int_{\mathbb{R}^d} \partial^2_{x_i x_i}\phi\, (\partial_{x_j} u)^2\, \rho_\star\, dx \\ &\quad + \int_{\mathbb{R}^d} \partial_{x_i}(\rho_\star\, \partial_{x_i} u)\, \partial_{x_j}\left(\rho_\star\, \partial_{x_j} u\right) dx \\ &\quad - \int_{\mathbb{R}^d} \partial^2_{x_j x_i}\phi\, \partial_{x_i} u\, \partial_{x_j} u\, \rho_\star\, dx \end{aligned}$$

and (32) is obtained by summing over i and j. Finally all integrals are estimated by $\langle \mathsf{AT}\Pi f, \Pi f\rangle$ using (26) and the improved Poincaré inequality (31), which completes the proof of (30).

In all cases, we conclude that $\|\mathsf{T}^2(u\, f_\star)\langle v\rangle^{(1-\beta)_+}\|_2^2 \le C\,\langle \mathsf{AT}\Pi f, \Pi f\rangle$ using (26), for some explicit constant $C > 0$. This completes the proof of (30).

Conclusion By (28), (29) and (30), we control $|\langle \mathsf{TA} f, f\rangle|$, $\left|\mathrm{Re}\langle \mathsf{AL}(1-\Pi) f, f\rangle\right|$ and $\left|\langle \mathsf{AT}(1-\Pi) f, \Pi f\rangle\right|$. A discriminant condition on δ completes the proof of Lemma 1 as in the proof of Theorem 1.

□

4.5 Moment Estimates

Lemma 2 *Let $u = u(x)$ be defined in terms of f as in (25) and assume For any $k \geq 0$, there exists a constant $C_k > 0$ such that*

$$M_k := \int_{\mathbb{R}^d} |u|^2 \langle x\rangle^k \rho_\star \, dx \leq C_k \iint_{\mathbb{R}^d\times\mathbb{R}^d} |\Pi f|^2 \langle x\rangle^k f_\star^{-1} \, dx \, dv \,. \tag{33}$$

Proof The case $k = 0$ is true from (27). By squaring equation (25) and testing with $\langle x\rangle^k \rho_\star$, we obtain

$$\int_{\mathbb{R}^d} u^2 \langle x\rangle^k \rho_\star \, dx + 2\sigma \int_{\mathbb{R}^d} \nabla_x\big(u \langle x\rangle^k\big) \cdot \nabla_x u \, \rho_\star \, dx + \sigma^2 \int_{\mathbb{R}^d} |\nabla_x(\rho_\star \nabla_x u)|^2 \frac{\langle x\rangle^k}{\rho_\star} \, dx$$
$$= \int_{\mathbb{R}^d} \rho_f^2 \frac{\langle x\rangle^k}{\rho_\star} \, dx \,.$$

Moreover, after an integration by parts, we have

$$\begin{aligned}
2\sigma \int_{\mathbb{R}^d} \nabla_x(u \langle x\rangle^k) \cdot \nabla_x u \, \rho_\star \, dx &= 2\sigma \int_{\mathbb{R}^d} |\nabla_x u|^2 \langle x\rangle^k \rho_\star \, dx \\
&\quad - \sigma \int_{\mathbb{R}^d} |u|^2 (\Delta_x \langle x\rangle^k - \nabla_x \langle x\rangle^k \cdot \nabla_x \phi) \, \rho_\star \, dx \\
&= 2\sigma \int_{\mathbb{R}^d} |\nabla_x u|^2 \langle x\rangle^k \rho_\star \, dx \\
&\quad - \sigma \, k \, (k + d - 2) \, M_{k-2} + \sigma \, k \, (k-2) \, M_{k-4} \\
&\quad + \sigma \, k \, M_{k+\alpha-2} - \sigma \, k \, M_{k+\alpha-4} \,.
\end{aligned}$$

After dropping the positive terms we get

$$M_k \leq \int_{\mathbb{R}^d} \frac{\rho_f^2}{\rho_\star} \langle x\rangle^k \, dx + \sigma \, k \, (k + d - 2) \, M_{k-2} + \sigma \, k \, M_{k+\alpha-4} \,.$$

Inequality (33) follows by induction and interpolation. □

Under the simplifying assumption (23), we also obtain moment estimates directly for the distribution function f. Here there is space for improvements.

Lemma 3 *Let $f = f(t, x, v)$ be a solution of (11) with transport and collision operators given respectively by (9) and (10) for some $\beta > 0$ and $\alpha > 0$. Assume that the initial datum f_0 satisfies the bound (23). Then for any $k > 0$ and for any*

$\ell > 0$ *there exist positive constants* C_k *and* C_ℓ *such that, for any* $t \geq 0$

$$J_k(t) := \iint_{\mathbb{R}^d\times\mathbb{R}^d} |f(t,x,v)|^2 \langle x\rangle^k f_\star^{-1}\,dx\,dv \leq C_k\,, \tag{34}$$

$$K_\ell(t) := \iint_{\mathbb{R}^d\times\mathbb{R}^d} |f(t,x,v)|^2 \langle v\rangle^\ell f_\star^{-1}\,dx\,dv \leq C_\ell\,. \tag{35}$$

Proof Since $f_\star$ is a stationary solution, the maximum principle yields

$$f(t,\cdot,\cdot) \leq C\, f_\star \quad \forall t \geq 0.$$

Therefore (34) and (35) follow by taking

$$C_k = C^2 \iint_{\mathbb{R}^d\times\mathbb{R}^d} f_\star \langle x\rangle^k\,dx\,dv \quad \text{and} \quad C_\ell = C^2 \iint_{\mathbb{R}^d\times\mathbb{R}^d} f_\star \langle v\rangle^\ell\,dx\,dv\,.$$

This completes the proof of Lemma 3. □

Notice that, with the elementary estimates

$$2^{\frac{k}{2}-1}\left(1+r^k\right) \leq \langle r\rangle^k \leq 1+r^k \qquad \text{if} \quad k \in (0,2)\,,$$

$$1+r^k \leq \langle r\rangle^k \leq 2^{\frac{k}{2}-1}\left(1+r^k\right) \qquad \text{if} \quad k \geq 2\,,$$

we have the simple moment estimate

$$\begin{aligned} M_{k,\eta} := \int_{\mathbb{R}^d} \langle x\rangle^k e^{-\frac{1}{\eta}\langle x\rangle^\eta}\,dx &\leq \max\left\{1,\, 2^{\frac{k}{2}-1}\right\} |\mathbb{S}^{d-1}| \int_0^\infty r^{d-1}\left(1+r^k\right) e^{-\frac{r^\eta}{\eta}}\,dr \\ &= \max\left\{1,\, 2^{\frac{k}{2}-1}\right\} \frac{2\,\pi^{\frac{d}{2}}}{\Gamma(\frac{d}{2})}\,\eta^{\frac{d-\eta}{\eta}} \left(\Gamma\big(\tfrac{d}{\eta}\big) + \eta^{\frac{k}{\eta}}\,\Gamma\big(\tfrac{d+k}{\eta}\big)\right) \end{aligned}$$

for any $k > 0$ and Γ is the Euler Gamma function. As a consequence, $f_\star$ defined by (12) with $Z = M_{0,\beta}\, M_{0,\alpha}$ is such that

$$\iint_{\mathbb{R}^d\times\mathbb{R}^d} \langle x\rangle^k f_\star\,dx\,dv = \frac{M_{k,\alpha}}{M_{0,\alpha}} \quad \text{and} \quad \iint_{\mathbb{R}^d\times\mathbb{R}^d} \langle v\rangle^\ell f_\star\,dx\,dv = \frac{M_{k,\beta}}{M_{0,\beta}}\,.$$

4.6 *Proof of Theorem 2*

In this section we will work in the framework of Theorem 2, i.e. we will consider a solution $f = f(t,x,v)$ to the kinetic Fokker-Planck equation (11) with initial datum $0 \leq f_0 \leq C\, f_\star$, for a certain $C > 0$. As sign plays no role, up to replacing f

with $f - f_\star$ when $f_\star$ is integrable, we may assume that

$$\iint_{\mathbb{R}^d\times\mathbb{R}^d} f(t,x,v)\,dx\,dv = 0 \quad \forall\, t \geq 0\,. \tag{36}$$

We distinguish various cases depending on the values of β and α.

• Case $\beta \geq 1$ and $\alpha \geq 1$

Thanks to Lemma 1, we have

$$\mathsf{D}[f] \geq \kappa\left(\|(1-\Pi)f\|^2_{\mathrm{L}^2(\mathbb{R}^d\times\mathbb{R}^d,\,d\mu)} + \langle \mathsf{AT}\Pi f, \Pi f\rangle\right).$$

for some $\kappa > 0$. Because of the assumption $\alpha \geq 1$ the operator $(\mathsf{T}\Pi)^*(\mathsf{T}\Pi)$ is coercive, that is (H2) hold. As a consequence we also have (6), i.e.,

$$\langle \mathsf{AT}\Pi f, \Pi f\rangle \geq \frac{\lambda_M}{1+\lambda_M}\,\|\Pi f\|^2_{\mathrm{L}^2(\mathbb{R}^d\times\mathbb{R}^d,\,d\mu)}\,.$$

Therefore

$$\mathsf{D}[f] \geq \frac{\kappa\,\lambda_M}{1+\lambda_M}\,\|f\|^2_{\mathrm{L}^2(\mathbb{R}^d\times\mathbb{R}^d,\,d\mu)}$$

holds for some $\kappa > 0$, which gives exponential convergence:

$$\begin{aligned}\|f(t,\cdot,\cdot)\|^2_{\mathrm{L}^2(\mathbb{R}^d\times\mathbb{R}^d,\,d\mu)} &\leq \frac{4}{2-\delta}\,\mathsf{H}[f(t,\cdot,\cdot)]\\ &\leq \frac{4}{2-\delta}\,\mathsf{H}[f_0]\,e^{-\lambda t} \text{ where } \lambda = \frac{\kappa\,\lambda_M}{1+\lambda_M}\,.\end{aligned}$$

Exactly the same proof applies in the case of Corollary 1, with λ_M now given by the Poincaré inequality associated with the measure $e^{-\phi}\,dx$, of which the case $\phi(x) = \frac{1}{\alpha}\,\langle x\rangle^\alpha$ with $\alpha \geq 1$ is a special case.

• Case $\beta \in (0,1)$ and $\alpha \geq 1$

In this case, on the one hand we still have macroscopic coercivity (H2) due to the fact that $\alpha \geq 1$, but on the other hand, a loss of weight now appears for the microscopic component because of $\beta \in (0,1)$. Inequality (24) now reads as

$$\mathsf{D}[f] \geq \kappa\left(\|(1-\Pi)f\|^2_{\mathrm{L}^2(\mathbb{R}^d\times\mathbb{R}^d,\,\langle v\rangle^{-2(1-\beta)}\,d\mu)} + \frac{\lambda_M}{1+\lambda_M}\,\|f\|^2_{\mathrm{L}^2(\mathbb{R}^d\times\mathbb{R}^d,\,d\mu)}\right).$$

In order to recover the $L^2(\mathbb{R}^d \times \mathbb{R}^d, d\mu)$ norm we need to interpolate with the conservation of moments. Let $\ell > 0$ and notice that $f_0 \in L^2(\mathbb{R}^d \times \mathbb{R}^d, \langle v\rangle^\ell d\mu)$ by our assumption on the initial datum (23). Setting $a = \ell/\big(\ell + 2(1-\beta)\big)$, by Hölder's inequality and Lemma 3, we have

$$\begin{aligned}
\|(1-\Pi)f\|^2_{L^2(\mathbb{R}^d\times\mathbb{R}^d,\, d\mu)} &\le \|(1-\Pi)f\|^{2a}_{L^2(\mathbb{R}^d\times\mathbb{R}^d,\, \langle v\rangle^{-2(1-\beta)}\, d\mu)} \\
&\qquad \|(1-\Pi)f\|^{2(1-a)}_{L^2(\mathbb{R}^d\times\mathbb{R}^d,\, \langle v\rangle^{\ell}\, d\mu)} \\
&\le \|(1-\Pi)f\|^{2a}_{L^2(\mathbb{R}^d\times\mathbb{R}^d,\, \langle v\rangle^{-2(1-\beta)}\, d\mu)}\, K_\ell(t)^{1-a} \\
&\le C_\ell^{1-a}\, \|(1-\Pi)f\|^{2a}_{L^2(\mathbb{R}^d\times\mathbb{R}^d,\, \langle v\rangle^{-2(1-\beta)}\, d\mu)}\,.
\end{aligned}$$

As a consequence, for a certain constant $C > 0$ we have

$$\begin{aligned}
\mathsf{D}[f] &\ge C\left(\|(1-\Pi)f\|^{2\left(1+\frac{2(1-\beta)}{\ell}\right)}_{L^2(\mathbb{R}^d\times\mathbb{R}^d,\, d\mu)} + \|\Pi f\|^2_{L^2(\mathbb{R}^d\times\mathbb{R}^d,\, d\mu)}\right) \\
&\ge C\, \|f\|^{2\left(1+\frac{2(1-\beta)}{\ell}\right)}_{L^2(\mathbb{R}^d\times\mathbb{R}^d,\, d\mu)}\,.
\end{aligned}$$

By the Bihari-LaSalle estimate, we finally have

$$\mathsf{H}[f] \le \mathsf{H}[f_0]\left(1 + C\,\mathsf{H}[f_0]^{\frac{2(1-\beta)}{\ell}}\, t\right)^{-\frac{\ell}{2(1-\beta)}}\,.$$

• **Case $\beta \ge 1$ and $\alpha \in (0,1)$**

In this case we have the symmetrical situation compared to the previous one. The dissipation of entropy is

$$\mathsf{D}[f] \ge \kappa\left(\|(1-\Pi)f\|^2_{L^2(\mathbb{R}^d\times\mathbb{R}^d,\, d\mu)} + \langle \mathsf{AT}\Pi f, \Pi f\rangle\right)$$

where now $\langle \mathsf{AT}\Pi f, \Pi f\rangle$ does not produce macroscopic coercivity, but is given in terms of u by (26). Fix $k > 0$ and assume that $f_0 \in L^2(\mathbb{R}^d \times \mathbb{R}^d, \langle x\rangle^k\, d\mu)$, then from Lemmas 2 and 3, we have that the moments $M_k(t) = \int_{\mathbb{R}^d} |u|^2 \langle x\rangle^k \rho_\star\, dx$ are uniformly bounded in time. With $b = k/\big(k + 2(1-\alpha)\big) \in (0,1)$, using Hölder's inequality and the weighted Poincaré inequality with non-classical average of [8, Cor. 10], we obtain

$$
\begin{aligned}
\|\Pi f\|^2_{\mathrm{L}^2(\mathbb{R}^d\times\mathbb{R}^d,d\mu)} &= \int_{\mathbb{R}^d} |u|^2\,\rho_\star\,dx + 2\,\sigma\int_{\mathbb{R}^d} |\nabla_x u|^2\,\rho_\star\,dx \\
&\quad + \sigma^2\int_{\mathbb{R}^d} |\nabla_x\cdot(\rho_\star\,\nabla_x u)|^2\,\rho_\star\,dx \\
&\le \left(\int_{\mathbb{R}^d} |u|^2\,\langle x\rangle^{-2(1-\alpha)}\,\rho_\star\,dx\right)^b \left(\int_{\mathbb{R}^d} |u|^2\,\langle x\rangle^k\,\rho_\star\,dx\right)^{1-b} \\
&\quad + 2\,\langle \mathsf{AT}\Pi f,\,\Pi f\rangle \\
&\le \left(\mathscr{C}^{\mathrm{wP}}_\alpha\right)^b \left(\int_{\mathbb{R}^d} |\nabla_x u|^2\,\rho_\star\,dx\right)^b \left(\int_{\mathbb{R}^d} |u|^2\,\langle x\rangle^k\,\rho_\star\,dx\right)^{1-b} \\
&\quad + 2\,\langle \mathsf{AT}\Pi f,\,\Pi f\rangle \\
&\le C\,\langle \mathsf{AT}\Pi f,\,\Pi f\rangle^b + 2\,\langle \mathsf{AT}\Pi f,\,\Pi f\rangle =: \Phi\big(\langle \mathsf{AT}\Pi f,\,\Pi f\rangle\big)
\end{aligned}
$$

where C depends on $\mathscr{C}^{\mathrm{wP}}_\alpha$ and the bound on $\|f_0\|_{\mathrm{L}^2(\mathbb{R}^d\times\mathbb{R}^d,\langle x\rangle^k\,d\mu)}$. In the weighted Poincaré inequality we used (36), hence $\int_{\mathbb{R}^d} u\,\rho_\star\,dx = \iint_{\mathbb{R}^d\times\mathbb{R}^d} f\,dx\,dv = 0$. Now we have

$$
\begin{aligned}
\mathsf{H}[f] &\le \frac{2+\delta}{4}\,\|f\|^2_{\mathrm{L}^2(\mathbb{R}^d\times\mathbb{R}^d,d\mu)} \\
&\le \frac{2+\delta}{4}\left(\|(1-\Pi)f\|^2_{\mathrm{L}^2(\mathbb{R}^d\times\mathbb{R}^d,d\mu)} + \Phi\big(\langle \mathsf{AT}\Pi f,\,\Pi f\rangle\big)\right) \\
&\le \frac{2+\delta}{4}\,\Phi\left(\|(1-\Pi)f\|^2_{\mathrm{L}^2(\mathbb{R}^d\times\mathbb{R}^d,d\mu)} + \langle \mathsf{AT}\Pi f,\,\Pi f\rangle\right) \\
&\le \frac{2+\delta}{4}\,\Phi\left(\kappa^{-1}\,\mathsf{D}[f]\right).
\end{aligned}
$$

Therefore the decay of $\mathsf{H}[f]$ is estimated by the decay of the solution $z(t)$ of

$$
z' = \frac{dz}{dt} = -\,\kappa\,\Phi^{-1}\left(\frac{4\,z}{2+\delta}\right), \quad z(0) = \mathsf{H}[f_0]\,.
$$

In view of the expression of Φ, we conclude that z monotonically converges to 0 as $t\to+\infty$ and, as a consequence z' also converges to 0. This implies that after some time $t_0\ge 0$, we have

$$
\Phi\left(-\,\kappa^{-1}\,z'\right) \le C\left(-\,\kappa^{-1}\,z'\right)^b,
$$

Where C denotes a positive constant that may change from line to line. Altogether, we end up with the differential inequality

$$z' \leq -C\,z^{1/b}\,.$$

Integrating and using $\frac{b}{1-b} = \frac{k}{2(1-\alpha)}$, we obtain that $z(t) \leq C\,(1+t)^{-\frac{k}{2(1-\alpha)}}$.

• Case $\beta \in (0,1)$ and $\alpha \in (0,1)$

If $\beta \in (0,1)$ and $\alpha \in (0,1)$ we have neither microscopic coercivity nor macroscopic coercivity. The dissipation of entropy is

$$\mathsf{D}[f] \geq \kappa\left(\|(1-\Pi)f\|^2_{\mathrm{L}^2(\mathbb{R}^d\times\mathbb{R}^d,\langle v\rangle^{-2(1-\beta)}\,d\mu)} + \langle \mathsf{AT}\Pi f, \Pi f\rangle\right)$$

and we have to interpolate with moments in both variables x and v. As in the previous cases, we have

$$\|\Pi f\|^2_{\mathrm{L}^2(\mathbb{R}^d\times\mathbb{R}^d,\,d\mu)} \leq C\,\langle \mathsf{AT}\Pi f, \Pi f\rangle^b + 2\,\langle \mathsf{AT}\Pi f, \Pi f\rangle = \Phi\big(\langle \mathsf{AT}\Pi f, \Pi f\rangle\big)$$

and

$$\begin{aligned}\|(1-\Pi)f\|^2_{\mathrm{L}^2(\mathbb{R}^d\times\mathbb{R}^d,\,d\mu)} &\leq C\,\|(1-\Pi)f\|^{2a}_{\mathrm{L}^2(\mathbb{R}^d\times\mathbb{R}^d,\langle v\rangle^{-2(1-\beta)}\,d\mu)}\\ &=: \Psi\left(\|(1-\Pi)f\|^2_{\mathrm{L}^2(\mathbb{R}^d\times\mathbb{R}^d,\langle v\rangle^{-2(1-\beta)}\,d\mu)}\right)\end{aligned}$$

with $a = \frac{\ell}{\ell+2(1-\beta)}$ and $b = \frac{k}{k+2(1-\alpha)}$. As above we have

$$\begin{aligned}\mathsf{H}[f] &\leq \frac{2+\delta}{4}\left(\Psi\left(\|(1-\Pi)f\|^2_{\mathrm{L}^2(\mathbb{R}^d\times\mathbb{R}^d,\langle v\rangle^{-2(1-\beta)}\,d\mu)}\right) + \Phi\big(\langle \mathsf{AT}\Pi f, \Pi f\rangle\big)\right)\\ &\leq \frac{2+\delta}{4}\left(\Psi\big(\kappa^{-1}\,\mathsf{D}[f]\big) + \Phi\big(\kappa^{-1}\,\mathsf{D}[f]\big)\right)\,.\end{aligned}$$

Notice that the function $t \mapsto \Psi(t)+\Phi(t)$ is increasing, concave and $\Psi(0)+\Phi(0) = 0$. Moreover we have

$$\frac{d}{dt}\mathsf{H}[f(t,\cdot,\cdot)] \leq -\kappa\,(\Psi+\Phi)^{-1}\left(\frac{4}{2+\delta}\,\mathsf{H}[f(t,\cdot,\cdot)]\right)\,.$$

As a consequence, $\mathsf{H}[f(t,\cdot,\cdot)]$ can be estimated by the solution z of

$$z' = -\kappa\,(\Psi+\Phi)^{-1}\left(\frac{4\,z}{2+\delta}\right)\,.$$

For the same reasons as before, z' converges to 0 as $t \to +\infty$. Using the explicit expressions of Φ and Ψ, we see that there exists some $t_0 \geq 0$ such that, for any $t \geq t_0$,

$$(\Psi + \Phi)\left(-\kappa^{-1} z'\right) \leq C\left(-\kappa^{-1} z'\right)^{\zeta}$$

for some $C > 0$, where $\zeta = \min\{a, b\}$. This inequality leads to $z' \leq -C\, z^{1/\zeta}$ and therefore to

$$z(t) \leq C\,(1+t)^{-\min\left\{\frac{k}{2(1-\alpha)}, \frac{\ell}{2(1-\beta)}\right\}} \quad \forall\, t \geq 0$$

by the Bihari-LaSalle estimate.

• Case $\beta \geq 1$ and $\phi = 0$

In absence of a global equilibrium, we can still consider (24) written with $\rho_\star = 1$. Identity (26) now reads as

$$\langle \mathsf{AT}\Pi f, \Pi f\rangle = \sigma \int_{\mathbb{R}^d} |\nabla_x u|^2\, dx + \sigma^2 \int_{\mathbb{R}^d} |\Delta_x u|^2\, dx\,.$$

Because of Nash's inequality and the conservation of mass, we have

$$\begin{aligned}
\|\Pi f\|^2_{\mathrm{L}^2(\mathbb{R}^d\times\mathbb{R}^d, d\mu)} &= \|u\|^2_{\mathrm{L}^2(dx)} + 2\,\sigma\,\|\nabla_x u\|^2_{\mathrm{L}^2(dx)} + \sigma^2\,\|\Delta_x u\|^2_{\mathrm{L}^2(dx)} \\
&\leq C_{\mathrm{Nash}}\,\|u\|^{\frac{4}{d+2}}_{\mathrm{L}^1(dx)}\,\|\nabla_x u\|^{\frac{2d}{d+2}}_{\mathrm{L}^2(dx)} + 2\,\langle \mathsf{AT}\Pi f, \Pi f\rangle \\
&\leq C\,\langle \mathsf{AT}\Pi f, \Pi f\rangle^{\frac{d}{d+2}} + 2\,\langle \mathsf{AT}\Pi f, \Pi f\rangle\,.
\end{aligned}$$

In the asymptotic regime of small $\|\Pi f\|^2_{\mathrm{L}^2(\mathbb{R}^d\times\mathbb{R}^d,\, d\mu)}$, we have

$$\langle \mathsf{AT}\Pi f, \Pi f\rangle \geq C\,\|\Pi f\|^{2\left(1+\frac{2}{d}\right)}$$

for some suitable constant $C > 0$, and

$$\mathsf{D}[f] \geq C\left(\|(1-\Pi) f\|^2_{\mathrm{L}^2(\mathbb{R}^d\times\mathbb{R}^d, d\mu)} + \|\Pi f\|^{2\left(1+\frac{2}{d}\right)}_{\mathrm{L}^2(\mathbb{R}^d\times\mathbb{R}^d, d\mu)}\right) \geq C\,\|f\|^{2\left(1+\frac{2}{d}\right)}.$$

By the Bihari-LaSalle estimate, we can finally conclude

$$\mathsf{H}[f(t,\cdot,\cdot)] \leq \mathsf{H}[f_0]\left(1 + C\,\mathsf{H}[f_0]^{\frac{2}{d}}\, t\right)^{-\frac{d}{2}} \;\forall\, t \geq 0\,.$$

• Case $\beta \in (0, 1)$ and $\phi = 0$

We proceed as in the previous case. The macroscopic part obeys the same estimate

$$\langle \mathsf{AT}\Pi f, \Pi f\rangle \gtrsim C\, \|\Pi f\|^{2\left(1+\frac{2}{d}\right)}.$$

The microscopic component has to be interpolated with moments:

$$\|(1-\Pi) f\|^2_{\mathrm{L}^2(\mathbb{R}^d\times\mathbb{R}^d,\,d\mu)} \le C\, \|(1-\Pi) f\|^{2b}_{\mathrm{L}^2(\mathbb{R}^d\times\mathbb{R}^d,\,\langle v\rangle^{-2(1-\beta)}\,d\mu)}\,.$$

We conclude that, as $\|f\|_{\mathrm{L}^2(\mathbb{R}^d\times\mathbb{R}^d,\,d\mu)} \to 0$,

$$\begin{aligned}\mathsf{D}[f] &\ge C\left(\|(1-\Pi) f\|^{2\left(1+\frac{2(1-\beta)}{k}\right)}_{\mathrm{L}^2(\mathbb{R}^d\times\mathbb{R}^d,\,d\mu)} + \|\Pi f\|^{2\left(1+\frac{2}{d}\right)}_{\mathrm{L}^2(\mathbb{R}^d\times\mathbb{R}^d,\,d\mu)}\right)\\ &\ge C\,\|f\|^{2\left(1+\frac{1}{\zeta}\right)}_{\mathrm{L}^2(\mathbb{R}^d\times\mathbb{R}^d,\,d\mu)}\end{aligned}$$

where $\zeta = \min\{\frac{k}{2(1-\beta)}, \frac{d}{2}\}$. By the Bihari-LaSalle estimate we conclude

$$\mathsf{H}[f] \le \mathsf{H}[f_0]\left(1 + C\,\mathsf{H}[f_0]^{\frac{1}{\zeta}}\,t\right)^{-\zeta}.$$

Acknowledgments L.Z. received funding from the European Union's Horizon 2020 research and innovation program under the Marie Sklodowska-Curie grant agreement No. 945332. J.D. thanks S. Menozzi, A. Pascucci and S. Polidoro for the organization of the meeting on *Kolmogorov operators and their applications* which has turned to be a great opportunity to write down and collect the results of the present contribution. All authors thank an anonymous referee for his/her valuable suggestions and remarks.

References

1. Albritton, D., Armstrong, S., Mourrat, J.C., Novack, M.: Variational methods for the kinetic Fokker–Planck equation. Anal. PDE. arXiv: 1902.04037. https://msp.org/soon/coming.php?jpath=apde
2. Arnold, A., Dolbeault, J., Schmeiser, C., Wöhrer, T.: Sharpening of decay rates in Fourier based hypocoercivity methods. In: Salvarani, F. (ed.) Recent Advances in Kinetic Equations and Applications. INdAM Series, vol. 48, pp. 1–50 (2021). https://doi.org/10.1007/978-3-030-82946-9_1
3. Bakry, D., Émery, M.: Diffusions hypercontractives. In: Séminaire de probabilités, XIX, 1983/84. Lecture Notes in Mathematics, vol. 1123, pp. 177–206. Springer, Berlin (1985). https://doi.org/10.1007/BFb0075847

4. Bakry, D., Cattiaux, P., Guillin, A.: Rate of convergence for ergodic continuous Markov processes: Lyapunov versus Poincaré. J. Funct. Anal. **254**(3), 727–759 (2008). https://doi.org/10.1016/j.jfa.2007.11.002
5. Bakry, D., Gentil, I., Ledoux, M.: Analysis and geometry of Markov diffusion operators. Grundlehren der mathematischen Wissenschaften [Fundamental Principles of Mathematical Sciences], vol. 348. Springer, Cham (2014). https://doi.org/10.1007/978-3-319-00227-9
6. Ben-Artzi, J., Einav, A.: Weak Poincaré inequalities in the absence of spectral gaps. Ann. Henri Poincaré **21**(2), 359–375 (2020). https://doi.org/10.1007/s00023-019-00858-4
7. Bonforte, M., Grillo, G., Vázquez, J.L.: Special fast diffusion with slow asymptotics: entropy method and flow on a Riemann manifold. Arch. Ration. Mech. Anal. **196**(2), 631–680 (2010). https://doi.org/10.1007/s00205-009-0252-7
8. Bouin, E., Dolbeault, J., Lafleche, L., Schmeiser, C.: Hypocoercivity and sub-exponential local equilibria. Monatsh. Math. **194**, 41–65 (2020). https://doi.org/10.1007/s00605-020-01483-8
9. Bouin, E., Dolbeault, J., Mischler, S., Mouhot, C., Schmeiser, C.: Hypocoercivity without confinement. Pure Appl. Anal. **2**(2), 203–232 (2020). https://doi.org/10.2140/paa.2020.2.203
10. Bouin, E., Dolbeault, J., Schmeiser, C.: Diffusion and kinetic transport with very weak confinement. Kinet. Relat. Models **13**(2), 345–371 (2020). https://doi.org/10.3934/krm.2020012
11. Bouin, E., Dolbeault, J., Schmeiser, C.: A variational proof of Nash's inequality. Rend. Lincei Mat. Appl. **31**(1), 211–223 (2020). https://doi.org/10.4171/rlm/886
12. Bouin, E., Dolbeault, J., Lafleche, L.: Fractional hypocoercivity. Commun. Math. Phys. **390**(3), 1369–1411 (2022). https://doi.org/10.1007/s00220-021-04296-4
13. Bouin, E., Dolbeault, J., Ziviani, L.: Decay estimates for kinetic Fokker–Planck equations with heavy tailed stationary solutions. (Work in progress)
14. Brigati, G.: Time averages for kinetic Fokker–Planck equations. Kinet. Relat. Models **16**(4), 524–539 (2023). https://doi.org/10.3934/krm.2022037
15. Brigati, G., Stoltz, G.: How to construct decay rates for kinetic Fokker–Planck equations? (2023, preprint) arXiv: 2302.14506
16. Cao, C.: The kinetic Fokker-Planck equation with weak confinement force. Commun. Math. Sci. **17**(8), 2281–2308 (2019). https://doi.org/10.4310/CMS.2019.v17.n8.a9
17. Cao, C.: The kinetic Fokker-Planck equation with general force. J. Evol. Equ. **21**(2), 2293–2337 (2021). https://doi.org/10.1007/s00028-021-00684-4
18. Cao, Y., Lu, J., Wang, L.: On explicit L^2-convergence rate estimate for underdamped Langevin dynamics. Archive for Rational Mechanics and Analysis. **247**(5), (2023). https://doi.org/10.1007/s00205-023-01922-4
19. Dolbeault, J., Li, X.: Φ-Entropies: convexity, coercivity and hypocoercivity for Fokker-Planck and kinetic Fokker-Planck equations. Math. Models Methods Appl. Sci. **28**(13), 2637–2666 (2018). https://doi.org/10.1142/S0218202518500574
20. Dolbeault, J., Mouhot, C., Schmeiser, C.: Hypocoercivity for kinetic equations with linear relaxation terms. Comptes Rendus Math. **347**(9–10), 511 – 516 (2009). https://doi.org/10.1016/j.crma.2009.02.025
21. Dolbeault, J., Toscani, G.: Fast diffusion equations: matching large time asymptotics by relative entropy methods. Kinet. Relat. Models **4**(3), 701–716 (2011). https://doi.org/10.3934/krm.2011.4.701
22. Dolbeault, J., Mouhot, C., Schmeiser, C.: Hypocoercivity for linear kinetic equations conserving mass. Trans. Am. Math. Soc. **367**(6), 3807–3828 (2015). https://doi.org/10.1090/s0002-9947-2015-06012-7
23. Grothaus, M., Stilgenbauer, P.: Hilbert space hypocoercivity for the Langevin dynamics revisited. Methods Funct. Anal. Topol. **22**(2), 152–168 (2016). http://mfat.imath.kiev.ua/article/?id=847
24. Grothaus, M., Wang, F.Y.: Weak Poincaré inequalities for convergence rate of degenerate diffusion processes. Ann. Probab. **47**(5), 2930–2952 (2019). https://doi.org/10.1214/18-AOP1328
25. Helffer, B., Nier, F.: Hypoelliptic Estimates and Spectral Theory for Fokker-Planck Operators and Witten Laplacians. Lecture Notes in Mathematics, vol. 1862. Springer, Berlin (2005). https://doi.org/10.1007/b104762

26. Hérau, F.: Hypocoercivity and exponential time decay for the linear inhomogeneous relaxation Boltzmann equation. Asymptot. Anal. **46**(3–4), 349–359 (2006). https://content.iospress.com/articles/asymptotic-analysis/asy741
27. Hérau, F., Nier, F.: Isotropic hypoellipticity and trend to equilibrium for the Fokker-Planck equation with a high-degree potential. Arch. Ration. Mech. Anal. **171**(2), 151–218 (2004). https://doi.org/10.1007/s00205-003-0276-3
28. Hu, S., Wang, X.: Subexponential decay in kinetic Fokker-Planck equation: weak hypocoercivity. Bernoulli **25**(1), 174–188 (2019). https://doi.org/10.3150/17-bej982
29. Kavian, O., Mischler, S., Ndao, M.: The Fokker-Planck equation with subcritical confinement force. J. Math. Pures Appl. **151**, 171–211 (2021). https://doi.org/10.1016/j.matpur.2021.04.007
30. Mellet, A., Mischler, S., Mouhot, C.: Fractional diffusion limit for collisional kinetic equations. Arch. Rational Mech. Anal. **199**(2), 493–525 (2011). https://doi.org/10.1007/s00205-010-0354-2
31. Mischler, S., Mouhot, C.: Exponential stability of slowly decaying solutions to the kinetic-Fokker-Planck equation. Arch. Ration. Mech. Anal. **221**(2), 677–723 (2016). https://doi.org/10.1007/s00205-016-0972-4
32. Mouhot, C., Neumann, L.: Quantitative perturbative study of convergence to equilibrium for collisional kinetic models in the torus. Nonlinearity **19**(4), 969–998 (2006). https://doi.org/10.1088/0951-7715/19/4/011
33. Nash, J.: Continuity of solutions of parabolic and elliptic equations. Am. J. Math. **80**, 931–954 (1958). https://doi.org/10.2307/2372841
34. Piazzoli, D.: Relaxation to equilibrium for kinetic Fokker–Planck equation. Master's Thesis, University of Cambridge (2019). https://doi.org/10.17863/CAM.32020
35. Röckner, M., Wang, F.Y.: Weak Poincaré inequalities and -convergence rates of Markov semigroups. J. Funct. Anal. **185**(2), 564–603 (2001). https://doi.org/10.1006/jfan.2001.3776
36. Villani, C.: Hypocoercivity. Mem. Am. Math. Soc. **202**(950), iv+141 (2009). https://doi.org/10.1090/S0065-9266-09-00567-5

New Perspectives on Recent Trends for Kolmogorov Operators

Francesca Anceschi, Mirco Piccinini, and Annalaura Rebucci

Abstract After carrying out an overview on the non Euclidean geometrical setting suitable for the study of Kolmogorov operators with rough coefficients, we list some properties of the functional space $\mathcal{W}$, mirroring the classical H^1 theory for uniformly elliptic operators. Then we provide the reader with the proof of a new Sobolev embedding for functions in $\mathcal{W}$. Additionally, after reviewing recent results regarding weak regularity theory, we discuss some of their latest applications to real life problems arising both in Physics and in Economics. Finally, we conclude our analysis stating some recent results regarding the study of fractional kinetic operators.

F. Anceschi
Dipartimento di Ingegneria Industriale e Scienze Matematiche, Università Politecnica delle Marche, Ancona, Italy
e-mail: f.anceschi@staff.univpm.it

M. Piccinini (✉)
Dipartimento di Scienze Matematiche, Fisiche e Informatiche, Università degli Studi di Parma, Parma, Italy
e-mail: mirco.piccinini@unipr.it

A. Rebucci
Max Planck Institute for Mathematics in the Sciences, Leipzig, Germany
e-mail: annalaura.rebucci@mis.mpg.de

S. Menozzi et al. (eds.), *Kolmogorov Operators and Their Applications*,
Springer INdAM Series 56, https://doi.org/10.1007/978-981-97-0225-1_3

1 Introduction

This work is devoted to the study of ultraparabolic second order partial differential equations of Kolmogorov type of the form

$$\mathscr{L}u(x,t) := \sum_{i,j=1}^{m_0} \partial_{x_i}\big(a_{ij}(x,t)\partial_{x_j}u(x,t)\big) + \sum_{i,j=1}^{N} b_{ij}x_j\partial_{x_i}u(x,t) - \partial_t u(x,t)$$
$$+ \sum_{i=1}^{m_0} b_i(x,t)\partial_i u(x,t) + c(x,t)u(x,t) = f(x,t), \qquad (1.1)$$

where $z = (x,t) = (x_1,\ldots,x_N,t) \in \mathbb{R}^{N+1}$ and $1 \le m_0 \le N$. Furthermore, matrices $A_0 = (a_{ij}(x,t))_{i,j=1,\ldots,m_0}$ and $B = (b_{ij})_{i,j=1,\ldots,N}$ satisfy the following structural assumptions.

(H1) The matrix A_0 is symmetric with real measurable entries. Moreover, there exist two positive constants λ and Λ such that

$$\lambda|\xi|^2 \le \sum_{i,j=1}^{m_0} a_{ij}(x,t)\xi_i\xi_j \le \Lambda|\xi|^2$$

for every $(x,t) \in \mathbb{R}^{N+1}$ and $\xi \in \mathbb{R}^{m_0}$. The matrix B has constant entries.

Despite the degeneracy of $\mathscr{L}$ whenever $m_0 < N$, its first order part is a strongly regularizing operator. Indeed, it is known that, under suitable structural assumptions on the matrix B, the *principal part operator* $\mathscr{L}_0$ of $\mathscr{L}$

$$\mathscr{L}_0u(x,t) := \sum_{i=1}^{m_0} \partial^2_{x_i}u(x,t) + \sum_{i,j=1}^{N} b_{ij}x_j\partial_{x_i}u(x,t) - \partial_t u(x,t), \qquad (x,t) \in \mathbb{R}^{N+1},$$

is hypoelliptic (i. e. every distributional solution u to $\mathscr{L}_0u = f$ defined in some open set $\Omega \subset \mathbb{R}^{N+1}$ belongs to $C^\infty(\Omega)$ and it is a classical solution whenever $f \in C^\infty(\Omega)$). Hence, in the sequel, we rely on the following assumption.

(H2) The *principal part operator* $\mathscr{L}_0$ *of* $\mathscr{L}$ is hypoelliptic and homogeneous of degree 2 with respect to the family of dilations $(\delta_r)_{r>0}$ introduced in (1.12).

This latter assumption is clearly satisfied whenever $\mathscr{L}_0$ is uniformly parabolic, which corresponds to the choice $m_0 = N$ and $B \equiv \mathbb{O}$. Actually, in this case the principal part operator $\mathscr{L}_0$ coincides with the heat operator, which is known to be hypoelliptic. Moreover, [53, Propositions 2.1 and 2.2] imply that assumption **(H2)** is equivalent to assume there exists a basis of $\mathbb{R}^N$ with respect to which B takes the form

$$B = \begin{pmatrix} \mathbb{O} & \mathbb{O} & \dots & \mathbb{O} & \mathbb{O} \\ B_1 & \mathbb{O} & \dots & \mathbb{O} & \mathbb{O} \\ \mathbb{O} & B_2 & \dots & \mathbb{O} & \mathbb{O} \\ \vdots & \vdots & \ddots & \vdots & \vdots \\ \mathbb{O} & \mathbb{O} & \dots & B_\kappa & \mathbb{O} \end{pmatrix}, \tag{1.2}$$

where every B_j is a $m_j \times m_{j-1}$ matrix of rank m_j, with $j = 1, 2, \dots, \kappa$,

$$m_0 \geq m_1 \geq \dots \geq m_\kappa \geq 1 \quad \text{and} \quad \sum_{j=0}^{\kappa} m_j = N. \tag{1.3}$$

Thus, from now on we assume B has the canonical form (1.2). Moreover, by introducing the *spatial homogeneous dimension of* $\mathbb{R}^{N+1}$, a quantity defined as

$$Q = m_0 + 3m_1 + \dots + (2\kappa + 1)m_\kappa, \tag{1.4}$$

and the *homogeneous dimension of* $\mathbb{R}^{N+1}$ defined as $Q+2$, we are now in a position to state our last assumption regarding the integrability of b, c and of f.

(H3) $c, f \in L^q_{loc}(\Omega)$, with $q > \frac{Q+2}{2}$, and $b \in \left(L^\infty_{loc}(\Omega)\right)^{m_0}$.

From now on, we denote by $D = (\partial_{x_1}, \dots, \partial_{x_N})$, $\langle \cdot, \cdot \rangle$, and div the gradient, the inner product and the divergence in $\mathbb{R}^N$, respectively. Moreover, $D_{m_0} = (\partial_{x_1}, \dots, \partial_{x_{m_0}})$ and div_{m_0} stand for the partial gradient and the partial divergence in the first m_0 components, respectively.

If a_{ij} are the coefficients appearing in (1.1) for every $i, j = 1, \dots, m_0$ and $a_{ij} \equiv 0$ whenever $i > m_0$, or $j > m_0$, then we introduce

$$A(x,t) = \left(a_{ij}(x,t)\right)_{1 \leq i,j \leq N}, \quad b(x,t) := \left(b_1(x,t), \dots, b_{m_0}(x,t), 0, \dots, 0\right),$$

$$Y := \sum_{i,j=1}^{N} b_{ij} x_j \partial_{x_i} u(x,t) - \partial_t u(x,t).$$

Then we are in a position to rewrite operators $\mathscr{L}$ and $\mathscr{L}_0$ in a compact form

$$\mathscr{L}u = \text{div}(ADu) + Yu + \langle b, Du \rangle + cu \quad \text{and} \quad \mathscr{L}_0 u = \Delta_{m_0} u + Yu,$$

and to introduce a useful example for the class of ultraparabolic operators of type (1.1).

Examplee 1.1 A notable prototype belonging to the class (1.1) is the kinetic Kolmogorov-Fokker-Planck equation

$$\begin{aligned}
&\nabla_v \cdot (u(v,x,t)\, v + \nabla_v u(v,x,t)) + v \cdot \nabla_x u(v,x,t) \\
&- \partial_t u(v,x,t) = f(v,x,t),
\end{aligned} \tag{1.5}$$

where $(v, x, t) \in \mathbb{R}^{2n+1}$. Equation (1.5) can be obtained from (1.1) by choosing $N = 2n$, $\kappa = 1$, $m_0 = m_1 = n$, with a suitable choice of c and b. From the physical point of view, Fokker-Planck equations like the one in (1.5) provide a continuous description of the dynamics of the distribution of Brownian test particles immersed in a fluid in thermodynamical equilibrium. More precisely, the distribution function u of a test particle evolves according to the linear equation in (1.5), provided that the test particle is much heavier than the molecules of the fluid. In particular, Eq. (1.5) is the backward Kolmogorov equation of the stochastic process

$$\begin{cases} dV_t = \sqrt{2} dW_t - V_t dt, \\ dX_t = V_t dt, \end{cases} \tag{1.6}$$

where $(W_t)_{t \geq 0}$ denotes a $n-$dimensional Wiener process. We refer the reader to [6], and the reference therein, for an exhaustive treatment of Fokker-Planck equations, and their applications.

1.1 Geometrical Setting

First of all, we describe the most suitable geometrical setting for the study of $\mathscr{L}$, which is not an Euclidean one. Indeed, as it was firstly observed by Lanconelli and Polidoro in [53], operator $\mathscr{L}_0$ is invariant with respect to left translations in the Lie group $\mathbb{K} = (\mathbb{R}^{N+1}, \circ)$, whose group law is defined as

$$(x,t) \circ (\xi, \tau) = (\xi + E(\tau)x, t + \tau), \qquad (x,t), (\xi,\tau) \in \mathbb{R}^{N+1}, \tag{1.7}$$

where the exponential of the group is

$$E(s) = \exp(-sB), \qquad s \in \mathbb{R}. \tag{1.8}$$

We observe that $\mathbb{K}$ is a non-commutative group with zero element $(0, \ldots, 0, 0)$ and inverse

$$(x,t)^{-1} = (-E(-t)x, -t).$$

Additionally, if for a given $\zeta \in \mathbb{R}^{N+1}$ we denote by ℓ_ζ the left traslation on $\mathbb{K} = (\mathbb{R}^{N+1}, \circ)$ defined as follows

$$\ell_\zeta : \mathbb{R}^{N+1} \to \mathbb{R}^{N+1}, \quad \ell_\zeta(z) = \zeta \circ z,$$

then it is possible to show $\mathscr{L}_0$ is left invariant with respect to the Lie product $\circ$, i.e.

$$\mathscr{L}_0 \circ \ell_\zeta = \ell_\zeta \circ \mathscr{L}_0 \qquad \text{or, equivalently,} \qquad \mathscr{L}_0 \left(u(\zeta \circ z)\right) = \left(\mathscr{L}_0 u\right)(\zeta \circ z),$$

for every u sufficiently smooth.

Now, let us focus on assumption **(H2)**. Indeed, the hypoellipticity of $\mathscr{L}_0$ is implied by Hörmander's rank condition, which was introduced for the first time in [42] and reads as:

$$\text{rank Lie}\left(\partial_{x_1}, \ldots, \partial_{x_{m_0}}, Y\right)(x,t) = N+1, \qquad \forall\,(x,t) \in \mathbb{R}^{N+1}, \tag{1.9}$$

where $\text{Lie}\left(\partial_{x_1}, \ldots, \partial_{x_{m_0}}, Y\right)$ denotes the Lie algebra generated by the first order differential operators $\left(\partial_{x_1}, \ldots, \partial_{x_{m_0}}, Y\right)$ computed at (x,t). On a side note, it is worth noting that requiring

$$C(t) > 0, \quad \forall\, t > 0, \tag{1.10}$$

it is equivalent to assume the hypoellipticity for $\mathscr{L}_0$ as in **(H2)**, see [53, Proposition A.1], where $E(\cdot)$ is defined in (1.8) and the covariance matrix is defined as

$$C(t) = \int_0^t E(s)\, A_0\, E^T(s)\, ds.$$

As far as we are concerned with the second half of assumption **(H2)**, $\mathscr{L}_0$ is invariant with respect to a family of dilations $(\delta_r)_{r>0}$ if

$$\mathscr{L}_0\left(u \circ \delta_r\right) = r^2 \delta_r \left(\mathscr{L}_0 u\right), \quad \forall\, r > 0, \tag{1.11}$$

and for every function u sufficiently smooth. As pointed out in [53, Proposition 2.2], it is possible to read this dilation invariance property in the expression of the matrix B in (1.2). Specifically, $\mathscr{L}_0$ satisfies (1.11) if and only if B takes the form (1.2). In this case, it holds

$$\delta_r = \text{diag}\left(r\mathbb{I}_{m_0}, r^3\mathbb{I}_{m_1}, \ldots, r^{2\kappa+1}\mathbb{I}_{m_\kappa}, r^2\right), \qquad r > 0. \tag{1.12}$$

Taking this definition into account, we introduce a homogeneous norm of degree 1 with respect to $(\delta_r)_{r>0}$ and a corresponding invariant quasi-distance with respect to the group operation (1.7).

Definition 1.1 Let $\alpha_1, \dots, \alpha_N$ be positive integers such that

$$\mathrm{diag}\left(r^{\alpha_1}, \dots, r^{\alpha_N}, r^2\right) = \delta_r. \tag{1.13}$$

If $\|z\| = 0$, then we set $z = 0$; if $z \in \mathbb{R}^{N+1} \setminus \{0\}$, then we define $\|z\| = r$, where r is the unique positive solution to the equation

$$\frac{x_1^2}{r^{2\alpha_1}} + \frac{x_2^2}{r^{2\alpha_2}} + \dots + \frac{x_N^2}{r^{2\alpha_N}} + \frac{t^2}{r^4} = 1.$$

Accordingly, we define the quasi-distance d by

$$d(z, w) = \|z^{-1} \circ w\|, \qquad z, w \in \mathbb{R}^{N+1}. \tag{1.14}$$

We remark that the Lebesgue measure is invariant with respect to the translation group associated to $\mathscr{L}_0$, since $\det E(t) = e^{t \text{ trace } B} = 1$. Moreover, by definition, the semi-norm $\|\cdot\|$ is homogeneous of degree 1 with respect to $(\delta_r)_{r>0}$. Indeed,

$$\|\delta_r(x, t)\| = r\|(x, t)\| \qquad \forall r > 0, \ \forall\, (x, t) \in \mathbb{R}^{N+1}.$$

Since in $\mathbb{R}^{N+1}$ all the norms which are 1-homogeneous with respect to $(\delta_r)_{r>0}$ are equivalent, the one introduced in Definition 1.1 is equivalent to other norms, such as the following one

$$\|(x, t)\|_1 = |t|^{\frac{1}{2}} + |x|, \quad |x| = \sum_{j=1}^{N} |x_j|^{\frac{1}{\alpha_j}}$$

where the exponents α_j, for $j = 1, \dots, N$ were introduced in (1.13). Nevertheless, Definition 1.1 is usually preferred since its level sets are smooth surfaces. For further information on this matter, we refer to [59].

Remark 1.1 In the case of the kinetic Kolmogorov-Fokker-Planck equation (1.5) in the absence of friction, i.e.

$$\begin{aligned} \mathscr{K}_0 u(v, x, t) &:= \Delta_v u(v, x, t) + v \cdot \nabla_x u(v, x, t) - \partial_t u(v, x, t) \\ &= f(v, x, t), \end{aligned} \tag{1.15}$$

where $(v, x, t) \in \mathbb{R}^{2n+1}$, the Lie group has a quite natural interpretation. Indeed the composition law (1.7) agrees with the *Galilean* change of variables

$$(v, x, t) \circ (v_0, x_0, t_0) = (v_0 + v, x_0 + x + t v_0, t_0 + t). \tag{1.16}$$

It is easy to see that $\mathscr{K}_0$ is invariant with respect to the above change of variables. Specifically, if $w(v, x, t) = u(v_0 + v, x_0 + x + tv_0, t_0 + t)$ and $g(v, x, t) = f(v_0 + v, x_0 + x + tv_0, t_0 + t)$, then

$$\mathscr{K}_0 u = f \quad \Longleftrightarrow \quad \mathscr{K}_0 w = g \quad \text{for every} \quad (v_0, x_0, t_0) \in \mathbb{R}^{2n+1}. \tag{1.17}$$

Moreover, $\mathscr{K}_0$ is invariant with respect to the dilation (1.12), which in this case takes the simpler form $\delta_r(v, x, t) := (rv, r^3 x, r^2 t)$. Let us remark that the dilation acts as the usual parabolic scaling with respect to variables v and t. Moreover, the term r^3 in front of x is due to the fact that the velocity v is the derivative of the position x with respect to time t.

Hence, considering the group law "$\circ$" and the family of dilations $(\delta_r)_{r>0}$, we are now in a position to introduce a suitable family of cylinders, starting from the unit past cylinders

$$\mathcal{Q}_1 := B_1 \times B_1 \times \ldots \times B_1 \times (-1, 0), \qquad \widetilde{\mathcal{Q}}_1 := B_1 \times B_1 \times \ldots \times B_1 \times (-1, 0]$$

defined through open balls

$$B_1 = \{x^{(j)} \in \mathbb{R}^{m_j} : |x| \leq 1\},$$

where $j = 0, \ldots, \kappa$ and $|\cdot|$ denotes the Euclidean norm in $\mathbb{R}^{m_j}$. Then, for every $z_0 \in \mathbb{R}^{N+1}$ and $r > 0$, we set

$$\mathcal{Q}_r(z_0) := z_0 \circ (\delta_r(\mathcal{Q}_1)) = \{z \in \mathbb{R}^{N+1} : z = z_0 \circ \delta_r(\zeta), \zeta \in \mathcal{Q}_1\}.$$

We point out that this definition of slanted cylinders admits an equivalent ball representation in the origin, see [84, equation (21)], that it is sometimes preferred. More specifically, there exists a positive constant $\overline{c}$ such that

$$\begin{aligned} &B_{r_1}(0) \times B_{r_1^3}(0) \times \ldots \times B_{r_1^{2\kappa+1}}(0) \times (-r_1^2, 0] \\ &\quad \subset \mathcal{Q}_r \subset B_{r_2}(0) \times B_{r_2^3}(0) \times \ldots \times B_{r_2^{2\kappa+1}}(0) \times (-r_2^2, 0], \end{aligned}$$

where $r_1 = r/\overline{c}$ and $r_2 = \overline{c}r$. Moreover, since $\det \delta_r = r^{Q+2}$, it is true that

$$\text{meas}\,(\mathcal{Q}_r(z_0)) = r^{Q+2}\text{meas}\,(\mathcal{Q}_1(z_0)), \qquad \forall\, r > 0, z_0 \in \mathbb{R}^{N+1},$$

where Q is the homogeneous dimension defined in (1.4). Finally, by Cinti et al. [27, Lemma 6] there exists a positive constant $\widetilde{c} \in (0, 1)$ such that

$$z \circ \mathcal{Q}_{\widetilde{c}(r-\rho)} \subseteq \mathcal{Q}_r, \qquad \text{for every } 0 < \rho < r \leq 1 \quad \text{and} \quad z \in \mathcal{Q}_\rho. \tag{1.18}$$

We refer to [6, 18] and the references therein for further information on this subject.

Finally, we conclude this subsection by recalling the useful notion of Hölder continuous function in this non Euclidean setting.

Definition 1.2 Let α be a positive constant, $\alpha \le 1$, and let Ω be an open subset of $\mathbb{R}^{N+1}$. We say that a function $f : \Omega \longrightarrow \mathbb{R}$ is Hölder continuous with exponent α in Ω with respect to the group $\mathbb{K} = (\mathbb{R}^{N+1}, \circ)$, defined in (1.7), (in short: Hölder continuous with exponent α, $f \in C^{\alpha}_{K}(\Omega)$) if there exists a positive constant $C > 0$ such that

$$|f(z) - f(\zeta)| \le C\, d(z,\zeta)^{\alpha} \qquad \text{for every} \quad z, \zeta \in \Omega,$$

where d is the distance defined in (1.14).

To every bounded function $f \in C^{\alpha}_{K}(\Omega)$ we associate the semi-norm

$$[f]_{C^{\alpha}(\Omega)} = \sup_{\substack{z,\zeta \in \Omega \\ z \neq \zeta}} \frac{|f(z) - f(\zeta)|}{d(z,\zeta)^{\alpha}}.$$

Moreover, we say a function f is locally Hölder continuous, and we write $f \in C^{\alpha}_{K,\mathrm{loc}}(\Omega)$, if $f \in C^{\alpha}_{K}(\Omega')$ for every compact subset Ω' of Ω.

Furthermore, we also refer the interested reader to the interesting paper [66], where the most general and definitive notion of Hölder spaces was introduced.

1.2 Fundamental Solution

Another useful tool in the study of (1.1) with rough coefficients is the fundamental solution of its principal part operator $\mathscr{L}_0$. Indeed, under assumption **(H2)**, Hörmander explicitly constructed in [42] the fundamental solution of $\mathscr{L}_0$, that is

$$\Gamma(z,\zeta) = \Gamma(\zeta^{-1} \circ z, 0), \quad \forall z, \zeta \in \mathbb{R}^{N+1}, \ z \neq \zeta,$$

where

$$\Gamma((x,t),(0,0)) = \begin{cases} \frac{(4\pi)^{-\frac{N}{2}}}{\sqrt{\det C(t)}} \exp\left(-\frac{1}{4}\langle C^{-1}(t)x, x\rangle - t\,\mathrm{tr}(B)\right), & \text{if } t > 0, \\ 0, & \text{if } t < 0. \end{cases}$$

Since assumption **(H2)** implies that condition (1.10) holds true, the function Γ is well-defined. Moreover, since $\mathscr{L}_0$ is dilation invariant with respect to $(\delta_r)_{r>0}$, also Γ is a homogeneous function of degree $-Q$, namely

$$\Gamma\left(\delta_r(z), 0\right) = r^{-Q}\, \Gamma\left(z, 0\right), \qquad \forall z \in \mathbb{R}^{N+1} \setminus \{0\}, \ r > 0.$$

This property implies a L^p estimate for Newtonian potentials (see for instance [27]). Hence, by defining the $\Gamma-$*potential* of a function $f \in L^1(\mathbb{R}^{N+1})$ as

$$\Gamma(f)(z) = \int_{\mathbb{R}^{N+1}} \Gamma(z,\zeta) f(\zeta) \mathrm{d}\zeta, \qquad z \in \mathbb{R}^{N+1}, \tag{1.19}$$

we are able to introduce a function $\Gamma(D_{m_0} f) : \mathbb{R}^{N+1} \longrightarrow \mathbb{R}^{m_0}$, that is well-defined for any $f \in L^p(\mathbb{R}^{N+1})$, at least in the distributional sense, see [27]. Indeed

$$\Gamma(D_{m_0} f)(z) := - \int_{\mathbb{R}^{N+1}} D_{m_0}^{(\zeta)} \Gamma(z,\zeta)\, f(\zeta)\, \mathrm{d}\zeta, \tag{1.20}$$

where $D_{m_0}^{(\zeta)} \Gamma(z,\zeta)$ is the gradient with respect to $\xi_1, \ldots, \xi_{m_0}$. Thus, by directly applying [27, Proposition 3], with $\alpha = 1$ and $\alpha = 2$ when considering the Γ-potential for f and $D_0 f$, respectively, it is possible to derive explicit potential estimates for (1.19) and (1.20).

Corollary 1.1 *Let $f \in L^p(\mathcal{Q}_r)$. There exists a positive constant $c = c(T, B)$ such that*

$$\|\Gamma(f)\|_{L^{p**}(\mathcal{Q}_r)} \le c \|f\|_{L^p(\mathcal{Q}_r)}, \qquad \|\Gamma(D_{m_0} f)\|_{L^{p*}(\mathcal{Q}_r)} \le c \|f\|_{L^p(\mathcal{Q}_r)},$$

where $\frac{1}{p*} = \frac{1}{p} - \frac{1}{Q+2}$ and $\frac{1}{p**} = \frac{1}{p} - \frac{2}{Q+2}$.

1.3 Plan of the Paper

This work is organized as follows. Section 2 is devoted to the study of the functional space $\mathcal{W}$. In particular, after recalling all known results regarding the space $\mathcal{W}$, we prove a Sobolev embedding for functions belonging to it. In Sect. 3 we provide an overview on the De Giorgi-Nash-Moser weak regularity theory in this framework. Section 4 is devoted to the application of these results to the study of real life problems arising both in Physics and Economics. Finally, we conclude with Sect. 5, where we discuss recent trends of fractional kinetic operators.

2 Functional Setting

From now on, we consider a set $\Omega = \Omega_{m_0} \times \Omega_{N-m_0+1}$ of $\mathbb{R}^{N+1}$, where Ω_{m_0} is a bounded C^1 domain of $\mathbb{R}^{m_0}$ and Ω_{N-m_0+1} is a bounded domain of $\mathbb{R}^{N-m_0+1}$. Then, according to the scaling introduced in (1.12), we split the coordinate $x \in \mathbb{R}^N$ as

$$x = \left(x^{(0)}, x^{(1)}, \ldots, x^{(\kappa)}\right), \qquad x^{(0)} \in \mathbb{R}^{m_0}, \quad x^{(j)} \in \mathbb{R}^{m_j}, \quad j \in \{1, \ldots, \kappa\}, \tag{2.1}$$

where every m_j is a positive integer satisfying conditions exposed in (1.3).

Furthermore, we denote by $\mathcal{D}(\Omega)$ the set of C^∞ functions compactly supported in Ω and by $\mathcal{D}'(\Omega)$ the set of distributions in Ω. From now on, $H^1_{x^{(0)}}$ is the Sobolev space of functions $u \in L^2(\Omega_{m_0})$ with distribution gradient $D_{m_0}u$ lying in $(L^2(\Omega_{m_0}))^{m_0}$, i.e.

$$H^1_{x^{(0)}}(\Omega_{m_0}) := \left\{u \in L^2(\Omega_{m_0}) : D_{m_0}u \in (L^2(\Omega_{m_0}))^{m_0}\right\}, \tag{2.2}$$

paired with the norm

$$\|u\|^2_{H^1_{x^{(0)}}(\Omega_{m_0})} := \|u\|^2_{L^2(\Omega_{m_0})} + \|D_{m_0}u\|^2_{L^2(\Omega_{m_0})}. \tag{2.3}$$

Now, let $H^1_{c,x^{(0)}}$ denote the closure of $C_c^\infty(\Omega_{m_0})$ in the norm of $H^1_{x^{(0)}}$, and recall that $C_c^\infty(\overline{\Omega}_{m_0})$ is dense in $H^1_{x^{(0)}}$ since Ω_{m_0} is a bounded C^1 domain by assumption. Moreover, $H^1_{c,x^{(0)}}$ is a reflexive Hilbert space and thus we may consider its dual space

$$\left(H^1_{c,x^{(0)}}\right)^* = H^{-1}_{x^{(0)}} \quad \text{and} \quad \left(H^{-1}_{x^{(0)}}\right)^* = H^1_{c,x^{(0)}},$$

where the adopted notation is the classical one. Hence, we denote by $H^{-1}_{x^{(0)}}$ the dual of $H^1_{c,x^{(0)}}$ acting on functions in $H^1_{c,x^{(0)}}$ through the duality pairing $\langle\cdot|\cdot\rangle := \langle\cdot|\cdot\rangle_{H^1_{x^{(0)}}, H^1_{c,x^{(0)}}}$. From now on, we consider the shorthand notation L^2H^{-1} to denote $L^2\left(\Omega_{N-m_0+1}; H^{-1}_{c,x^{(0)}}\right)$.

Then in a standard manner, see [2, 7, 57], we introduce the space of functions $\mathcal{W}$ as the closure of $C^\infty(\overline{\Omega})$ in the norm

$$\|u\|^2_{\mathcal{W}} = \|u\|^2_{L^2\left(\Omega_{N-m_0+1}; H^1_{x^{(0)}}\right)} + \|Yu\|^2_{L^2\left(\Omega_{N-m_0+1}; H^{-1}_{x^{(0)}}\right)}, \tag{2.4}$$

which can explicitly be computed as

$$\|u\|^2_{\mathcal{W}} = \int_{\Omega_{N-m_0+1}} \|u(\cdot, y, t)\|^2_{H^1_{x^{(0)}}} \,\mathrm{d}y\,\mathrm{d}t + \int_{\Omega_{N-m_0+1}} \|Yu(\cdot, y, t)\|^2_{H^{-1}_{x^{(0)}}} \,\mathrm{d}y\,\mathrm{d}t,$$

where $y = (x^{(1)}, \ldots, x^{(\kappa)})$. In particular, it is possible to infer that $\mathcal{W}$ is a Banach space and we recall that the dual of $L^2(\Omega_{N-m_0+1}; H^1_{c,x^{(0)}})$ satisfies

$$\left(L^2(\Omega_{N-m_0+1}; H^1_{c,x^{(0)}})\right)^* = L^2(\Omega_{N-m_0+1}; H^{-1}_{c,x^{(0)}}) \quad \text{and}$$
$$\left(L^2(\Omega_{N-m_0+1}; H^{-1}_{c,x^{(0)}})\right)^* = L^2(\Omega_{N-m_0+1}; H^1_{c,x^{(0)}}).$$

This functional setting was firstly proposed in [2] for the study of weak regularity theory for the Kolmogorov-Fokker-Planck equation, see Example 1.1. Later on, it was considered in [57] for the study of well-posedness results for a Dirichlet problem in the kinetic setting. Finally, two of the authors extended this framework in [7] to the ultraparabolic setting considered in this work.

Notice that the major issue one has to tackle with when dealing with $\mathcal{W}$ is the duality pairing between L^2H^1 and L^2H^{-1}. Usually, when considering bounded domains, this problem is overcome observing that for every open subset $A \subset \mathbb{R}^n$ and for every function $g \in H^{-1}(A)$, there exist two functions $H_0, H_1 \in L^2(A)$ such that

$$g = \mathrm{div}_{m_0} H_1 + H_0 \qquad \text{and} \qquad \|H_0\|_{L^2(A)} + \|H_1\|_{L^2(A)} \leq 2\|g\|_{H^{-1}(A)},$$

see, for example, [51, Chapter 4].

Moreover, in recent years the community focused on providing suitable functional inequalities for functions belonging to $\mathcal{W}$. In particular, there was the necessity to prove a suitable Poincaré and Sobolev embedding in this framework. As far as we are concerned with the first one, the proof of a weak Poincaré inequality was recently achieved in [7, 39]. Therefore, we will only recall its statement and a scheme of its proof. On the other hand, the Sobolev embedding is still not available in literature and for this reason we provide the reader with its full proof.

2.1 Poincaré Inequality

An useful tool we have at our disposal when considering functions belonging to $\mathcal{W}$ is a weak Poincaré inequality proved for the first time in [39] for the kinetic case and later on extended in [7] to the setting considered in this work. The idea is to firstly derive a local Poincaré inequality in terms of an error function h defined as the solution of a suitable Cauchy problem

$$\begin{cases} \widetilde{\mathscr{K}} h = u \widetilde{\mathscr{K}} \psi, & \text{in } \mathbb{R}^N \times (-\rho^2, 0) \\ h = 0, & \text{in } \mathbb{R}^N \times \{-\rho^2\} \end{cases}$$

where $\rho > 0$, ψ is a given cut-off function and $\widetilde{\mathscr{K}}$ is an auxiliary operator defined as

$$\widetilde{\mathscr{K}} u(x,t) := -\sum_{i=1}^{m_0} \partial_{x_i}^2 u(x,t) - \sum_{i,j=1}^{N} b_{ij} x_j \partial_{x_i} u(x,t) + \partial_t u(x,t), \qquad (x,t) \in \mathbb{R}^{N+1}.$$

We observe that the auxiliary operator $\widetilde{\mathscr{K}}$ is chosen in accordance with the definition of $\mathcal{W}$, where only the partial gradient D_{m_0} and the Lie derivative Y appear. On one hand, completing the proof by explicitly controlling the error h through the L^∞ norm of the function u allows us to avoid studying functional properties of weak solutions to (1.1) and to obtain a purely functional result. On the other hand, it is immediately clear that our inequality only holds for *bounded* functions belonging to $\mathcal{W}$.

In order to state this result, we first need to introduce the following sets

$$\begin{aligned} \mathcal{Q}_{zero} &= \{(x,t) : |x_j| \le \eta^{\alpha_j}, j = 1, \ldots, N, -1 - \eta^2 < t \le -1\}, \\ \mathcal{Q}_{ext} &= \{(x,t) : |x_j| \le 2^{\alpha_j} R, j = 1, \ldots, N, -1 - \eta^2 < t \le 0\}, \end{aligned} \tag{2.5}$$

where $R > 1$, $\eta \in (0,1)$, exponents α_j, for $j = 1, \ldots, N$, are defined in (1.13); $\mathcal{Q}_{zero}$ and $\mathcal{Q}_{ext}$ are introduced via the ball representation.

Theorem 2.1 (Weak Poincaré Inequality) *Let $\eta \in (0,1)$; let $\mathcal{Q}_{zero}$ and $\mathcal{Q}_{ext}$ be defined as in* (2.5)*. Then there exist $R > 1$ and $\vartheta_0 \in (0,1)$ such that for any non-negative function $u \in \mathcal{W}$ such that $u \le M$ in $\mathcal{Q}_1 = B_1 \times B_1 \times \ldots \times B_1 \times (-1,0)$ for a positive constant M and*

$$|\{u = 0\} \cap \mathcal{Q}_{zero}| \ge \frac{1}{4} |\mathcal{Q}_{zero}|,$$

we have

$$\|(u - \vartheta_0 M)_+\|_{L^2(\mathcal{Q}_1)} \le C_P \left(\|D_{m_0} u\|_{L^2(\mathcal{Q}_{ext})} + \|Yu\|_{L^2 H^{-1}(\mathcal{Q}_{ext})} \right),$$

where $C > 0$ is a constant only depending on Q.

The notation we consider here needs to be intended in the sense of (2.4). In particular, we have that $L^2 H^{-1}(\mathcal{Q}_{ext})$ is short for

$$L^2(B_{2^3 R} \times \ldots \times B_{2^{2\kappa+1} R} \times (-1 - \eta^2, 0], H^{-1}_{x^{(0)}}(B_{2R})),$$

where we split $x = \left(x^{(0)}, x^{(1)}, \ldots, x^{(\kappa)}\right)$ according to (2.1).

2.2 Sobolev-Type Inequality

Here, we give proof to a Sobolev-type inequality for functions belonging to $\mathcal{W}$. We refer to [71] and [73] for interpolations results in the non Euclidean geometrical setting of Kolmogorov equations.

Now, in order to prove our desired result we firstly recall its classical formulation, see [20], for functions belonging to the space $H^1_{x^{(0)}}(\Omega_{m_0})$, that we earlier introduced in (2.2) alongside with its norm (2.3). In order to do this, we recall that when $2 < m_0$ we may introduce the Sobolev exponent

$$2^* = \frac{2m_0}{m_0 - 2}, \qquad \text{such that } \frac{1}{2^*} = \frac{1}{2} - \frac{1}{m_0}, \;\; 2^* > 2.$$

Theorem 2.2 *Let Ω_{m_0} be a bounded open subset of $\mathbb{R}^{m_0}$ with C^1 boundary and $m_0 > 2$. If $u \in H^1_{x^{(0)}}(\Omega_{m_0})$, then $u \in L^q(\Omega_{m_0})$ with $q \in [2, 2^*]$, and the following estimate holds*

$$\|u\|_{L^q(\Omega_{m_0})} \leq c_1 \left(\frac{1}{d_{\Omega_{m_0}}} \|u\|_{L^2(\Omega_{m_0})} + \|D_{m_0} u\|_{L^2(\Omega_{m_0})} \right),$$

where $d_{\Omega_{m_0}} := \sup\{\|\omega - w\|, \;\; \omega, w \in \Omega_{m_0}\}$ and c_1 is a positive constant.

Remark 2.1 In this framework it is mandatory to consider a Sobolev embedding for $H^1(\Omega_{m_0})$ involving the L^2 norm of the function, alongside with the L^2 norm of the gradient of the function. The motivation for this is two fold. Firstly, the norm we consider in (2.3) is given by the sum of these two contributions, i.e. $\|u\|_2 + \|D_{m_0} u\|_2$. Secondly, it is not true that such norm is equivalent to the norm of the gradient in the first m_0 variables in $L^2(\Omega_{N-m_0+1}; H^1(\Omega_{m_0}))$. For instance, the interested reader may consider a function $u \in \mathcal{W}$ that does not depend on $x^{(0)}$ to find an explicit counterexample.

Then, by following the approach proposed in [5], we prove a new Sobolev-type embedding for functions belonging to $\mathcal{W}$.

Theorem 2.3 *Let $\Omega = \Omega_{N-m_0+1} \times \Omega_{m_0}$ be a bounded open subset of $\mathbb{R}^{N+1}$, where Ω_{m_0} is a bounded C^1 domain of $\mathbb{R}^{m_0}$ and Ω_{N-m_0+1} is a bounded Lipschitz domain of $\mathbb{R}^{N-m_0+1}$. Let $u \in \mathcal{W}$ and $m_0 > 2$. Then for every $q \in [2, 2^*]$ the following* *inequality holds*

$$\int_{\Omega_{m_0}} \left| \fint_{\Omega_{N-m_0+1}} u(x^{(0)}, y, t)\, \mathrm{d}y\, \mathrm{d}t \right|^q \mathrm{d}x^{(0)}$$

$$\leq c_1 \left(\frac{1}{d^q_{\Omega_{m_0}}} \left(\int_{\Omega_{m_0}} \fint_{\Omega_{N-m_0+1}} |u(x^{(0)}, y, t)|^2\, \mathrm{d}y\, \mathrm{d}t\, \mathrm{d}x^{(0)} \right)^{\frac{q}{2}} + \right.$$

$$\left. + \left(\int_{\Omega_{m_0}} \fint_{\Omega_{N-m_0+1}} |D_{m_0} u(x^{(0)}, y, t)|^2\, \mathrm{d}y\, \mathrm{d}t\, \mathrm{d}x^{(0)} \right)^{\frac{q}{2}} \right).$$

Proof Let $u \in L^2(\Omega_{N-m_0+1}; H^1_{x^{(0)}}(\Omega_{m_0}))$. We define the mean of u with respect to the variables y and t as

$$(u)_{y,t}(x^{(0)}) := \fint_{\Omega_{N-m_0+1}} u(x^{(0)}, y, t)\, \mathrm{d}y\, \mathrm{d}t.$$

Then, $(u)_{y,t} \in H^1_{x^{(0)}}(\Omega_{m_0})$. Indeed, by integral's monotonicity property and Jensen's Inequality, we have

$$\int_{\Omega_{m_0}} |(u)_{y,t}(x^{(0)})|^2 \mathrm{d}x^{(0)} \leq \int_{\Omega_{m_0}} \fint_{\Omega_{N-m_0+1}} |u(x^{(0)}, y, t)|^2\, \mathrm{d}y\, \mathrm{d}t\, \mathrm{d}x^{(0)}, \quad \text{and}$$

$$\int_{\Omega_{m_0}} |D_{m_0}(u)_{y,t}(x^{(0)})|^2 \mathrm{d}x^{(0)} \leq \int_{\Omega_{m_0}} \fint_{\Omega_{N-m_0+1}} |D_{m_0} u(x^{(0)}, y, t)|^2\, \mathrm{d}y\, \mathrm{d}t\, \mathrm{d}x^{(0)}.$$

Now, let $q \in [2, 2^*]$. Then, by applying Theorem 2.2 to $(u)_{y,t}$ together with Jensen's inequality we get

$$\int_{\Omega_{m_0}} |(u)_{y,t}(x^{(0)})|^q \mathrm{d}x^{(0)}$$

$$\leq c_1 \left(\frac{1}{d_{\Omega_{m_0}}} \left(\int_{\Omega_{m_0}} |(u)_{y,t}(x^{0)})|^2 \mathrm{d}x^{(0)} \right)^{\frac{1}{2}} + \left(\int_{\Omega_{m_0}} |D_{m_0}(u)_{y,t}(x^{0)})|^2 \mathrm{d}x^{(0)} \right)^{\frac{1}{2}} \right)^q$$

$$\leq c_1 \left(\frac{1}{d^q_{\Omega_{m_0}}} \left(\int_{\Omega_{m_0}} \fint_{\Omega_{N-m_0+1}} |u(x^{(0)}, y, t)|^2\, \mathrm{d}y\, \mathrm{d}t\, \mathrm{d}x^{(0)} \right)^{\frac{q}{2}} + \right.$$

$$\left. + \left(\int_{\Omega_{m_0}} \fint_{\Omega_{N-m_0+1}} |D_{m_0} u(x^{(0)}, y, t)|^2\, \mathrm{d}y\, \mathrm{d}t\, \mathrm{d}x^{(0)} \right)^{\frac{q}{2}} \right).$$

□

Remark 2.2 We observe that in the case $m_0 \leq 2$ an analog result of Theorem 2.3 can be proved. Indeed, when $m_0 = 1$ the function $(u)_{y,t} \in H^1_{x^{(0)}}(\Omega_{m_0})$ is absolutely continuous, whereas the case when $m_0 = 2$ can be treated via the Rellich-Kondrachov embedding since $(u)_{y,t} \in H^1_{x^{(0)}}(\Omega_{m_0}) \subset L^q(\Omega_{m_0})$ for any $q \in [2, \infty)$.

3 De Giorgi-Nash-Moser Weak Regularity Theory

The extension of the De Giorgi-Nash-Moser weak regularity theory to the class of ultraparabolic equations in divergence form of type (1.1) had been an open problem for decades. A first breakthrough in this direction was obtained by Pascucci and Polidoro in [72], where the authors proved the Moser's iterative scheme for *strong weak solutions*, i.e. weak solutions $u \in L^2$ such that $D_{m_0}u, Yu \in L^2$. Then, later on, this result was extended to the nondilation invariant case in [11, 27].

Based on these local boundedness results, the local Hölder continuity for strong weak solutions was later on addressed by Wang and Zhang in [83, 84] for the specific case of Kolmogorov-Fokker-Planck equation (1.5). Subsequently, the procedure was extended to the ultraparabolic setting in the paper [85]. The method considered in these works is based on the combination of Sobolev and Poincaré inequalities constructed for strong weak solutions, alongside with qualitative properties of a suitably chosen G function. Then the local Hölder continuity is recovered by providing an estimate of the oscillations following Kruzkhov's level set method [52].

In recent years, the interest of the community began to focus on the extension of these regularity results to weak solutions belonging to the space $\mathcal{W}$, that are defined as follows.

Definition 3.1 A function $u \in \mathcal{W}$ is a weak solution to (1.1) with source term $f \in L^2(\Omega)$ if for every non-negative test function $\varphi \in \mathcal{D}(\Omega)$, we have

$$\int_{\Omega} -\langle ADu, D\varphi\rangle - uY\varphi + \langle b, Du\rangle\varphi + cu\varphi = \int_{\Omega} f\varphi. \tag{3.1}$$

In the sequel, we will also consider weak sub-solutions to (1.1), namely functions $u \in \mathcal{W}$ that satisfy the following inequality

$$\int_{\Omega} -\langle ADu, D\varphi\rangle - uY\varphi + \langle b, Du\rangle\varphi + cu\varphi \overset{(\leq)}{\geq} \int_{\Omega} f\varphi, \tag{3.2}$$

for every non-negative test function $\varphi \in \mathcal{D}(\Omega)$. A function u is a super-solution to (1.1) if it satisfies (3.2) with $(\leq)$.

A first attempt in this direction is represented by the seminal paper [38], where the authors propose a non-constructive proof of a Harnack inequality for weak solutions to the kinetic Kolmogorov-Fokker-Planck equation (1.5). This approach is based on classical energy estimates and apriori fractional estimates proved in [19]. Driven by the aim of simplifying and extending the proof proposed in [85], various authors recently suggested alternative proofs both for the Harnack inequality and the Hölder continuity in the kinetic setting, see [39, 40, 45, 77, 78, 86].

It was only recently that the weak regularity theory was extended to the ultraparabolic case in [7] by two of the authors, and their main results can be stated after the introduction of these two sets:

$$\begin{aligned} \mathcal{Q}_+ &= \delta_\omega\left(\widetilde{\mathcal{Q}}_1\right) = B_\omega \times B_{\omega^3} \times \ldots \times B_{\omega^{2\kappa+1}} \times (-\omega^2, 0] \quad \text{and} \\ \widetilde{\mathcal{Q}}_- &= (0, \ldots, 0, -1 + 2\rho^2) \circ \delta_\rho\left(\mathcal{Q}_1\right) \\ &= B_\rho \times B_{\rho^3} \times \ldots \times B_{\rho^{2\kappa+1}} \times (-1 + \rho^2, -1 + 2\rho^2). \end{aligned}$$

Theorem 3.1 (Harnack inequality) *Let u be a non-negative weak solution to $\mathscr{L}u = f$ in $\Omega \supset \widetilde{\mathcal{Q}}_1$ under the assumptions **(H1)–(H2)–(H3)**. Then we have*

$$\sup_{\widetilde{\mathcal{Q}}_-} u \le C \left(\inf_{\mathcal{Q}_+} u + \| f \|_{L^q(\mathcal{Q}^0)} \right),$$

where $0 < \omega < 1$ is given by Theorem 3.4 and $0 < \rho < \frac{\omega}{\sqrt{2}}$. Finally, the constants C, ω, ρ only depend on the homogeneous dimension Q defined in (1.4), *q and on the ellipticity constants λ and Λ*

Theorem 3.2 (Hölder Regularity) *There exists $\alpha \in (0, 1)$ only depending on dimension Q, λ, Λ such that all weak solutions u to* (1.1) *under assumption **(H1)-(H2)-(H3)** in $\Omega \supset \mathcal{Q}_1$ satisfy*

$$[u]_{C^\alpha(\mathcal{Q}_{\frac{1}{2}})} \le C \left(\|u\|_{L^2(\mathcal{Q}_1)} + \|f\|_{L^q(\mathcal{Q}_1)} \right),$$

where the constant C only depends on the homogeneous dimension Q defined in (1.4), *q and the ellipticity constants λ and Λ.*

Note that the estimates above can be stated and scaled in any arbitrary cylinder $\mathcal{Q}_r(z_0)$. Furthermore, these results are comparable with the ones obtained in [85], but framework and methodology are different. Indeed, in [7] the authors apply the technique proposed by Guerand and Imbert in [39], based on the combination of a weak Poincaré inequality (Theorem 2.1) for functions belonging to the space $\mathcal{W}$, with a $L^2 - L^\infty$ estimate for weak sub-solutions (Theorem 3.3) and a weak Harnack inequality for weak super-solutions (Theorem 3.4). For the sake of completeness, these two tools will be briefly discussed in the following.

3.1 Local Boundedness Estimates

The proof of this result is obtained via the extension of the Moser iterative scheme introduced in [63] for the parabolic setting and it is based on the iterative combination of a Caccioppoli and a Sobolev inequality. When dealing with the ultraprabolic setting of our interest, the degeneracy of the diffusion part only allows us to estimate the partial gradient $D_{m_0}u$ of the solution through a Caccioppoli type inequality, also known as energy estimate. Moreover, in accordance with our definition of weak solution, u does not lie in a classical Sobolev space. Nevertheless, as firstly observed in [72], it is true that

$$\mathscr{L}_0 u = (\mathscr{L}_0 - \mathscr{L})\, u + f = \mathrm{div}_{m_0}\left(\left(\mathbb{I}_{m_0} - A\right) D_{m_0} u\right) + f.$$

Hence, as pointed out at [72, p. 396], it seems quite natural to consider a representation formula for solutions to (1.1) in terms of the fundamental solution of the principal part operator $\mathscr{L}_0$ to prove a Sobolev embedding for solutions to (1.1). This is very convenient because we have at our disposal an explicit expression of the fundamental solution of $\mathscr{L}_0$, alongside with potential estimates for it, see Sect. 1.2.

In literature, we find various extensions of the Moser's iterative scheme to Kolmogorov operators of the type $\mathscr{L}$, see for instance [27, 72, 85]. The most recent one is proved in [7] for the functional setting $\mathcal{W}$ and it reads as follows.

Theorem 3.3 *Let $z_0 \in \Omega$ and $0 < \frac{r}{2} \le \rho < r \le 1$, be such that $\overline{\mathcal{Q}_r(z_0)} \subseteq \Omega$. Let u be a non-negative weak solution to $\mathscr{L}u = f$ in Ω under assumptions* ***(H1)-(H2)-(H3)****. Then for every $p \ge 1$ there exists two positive constants $C = C\left(p, \lambda, \Lambda, Q, \| b \|_{L^q(\mathcal{Q}_r(z_0))}, \| c \|_{L^q(\mathcal{Q}_r(z_0))}\right)$, such that*

$$\sup_{\mathcal{Q}_\rho(z_0)} u_l^p \le \frac{C}{(r-\rho)^{\frac{Q+2}{\beta}}} \| u_l^p \|_{L^\beta(\mathcal{Q}_r(z_0))},$$

where $\beta = \frac{q}{q-1}$, q introduced in ***(H3)*** *and $u_l := u + \| f \|_{L^q(\mathcal{Q}_r)}$. The same statement holds true if u is a non-negative weak sub-solution to* (1.1) *for $p \ge 1$; if u is a non-negative weak super-solution to* (1.1) *for $0 < p < \frac{1}{2}$. In particular, by choosing $p = 1$, for every sub-solution to* (1.1) *it holds*

$$\sup_{\mathcal{Q}_\rho(z_0)} u \le \frac{C}{(r-\rho)^{\frac{Q+2}{\beta}}} \left(\| u \|_{L^\beta(\mathcal{Q}_r(z_0))} + \| f \|_{L^q(\mathcal{Q}_r)} \right),$$

Firstly, we observe the above statement holds true also for weak sub and super solutions, but for a different range of p. This depends on the chosen method for the proof of the Caccioppoli type inequality, see also [72, Remark 1.3], and it is a classical feature of local boundedness results of this type, see for instance [85].

Moreover, this result holds true also under a less restrictive assumption on the lower order coefficients, i.e. $c, f \in L^q_{loc}(\Omega)$ and $b \in \left(L^q_{loc}(\Omega)\right)^{m_0}$ for some

$q > \frac{3}{4}(Q+2)$ with $\operatorname{div} b \geq 0$ in Ω. The additional requirement on the sign of the divergence is the price to pay in order to lower the integrability requirement on b. Indeed, the non-standard structure of the space $\mathcal{W}$ is responsible for several underlying difficulties while proving a local boundedness result. Among these, we find the impossibility to lower the integrability requirements on the term b up to $\frac{Q+2}{2}$, the hypoelliptic counterpart of the parabolic homogeneous dimension $N/2$. It is our belief that further improvements in the integrability requirements for lower order coefficients may be obtained by taking advantage of Theorem 2.3 in the proof of Theorem 3.3. Indeed, it would allow us to get rid of the employment of Stein's Potential Estimates ([27, Theorem 2]), which are responsible for the inability to lower the integrability exponent to $\frac{Q+2}{2}$ in the existing literature, see for instance [7] for further informations on this matter.

3.2 *Weak Harnack Inequality*

The method of the proof of this result is an extension of the classical one proposed in [52] for the elliptic and parabolic setting, and later on followed by Guerand and Imbert in [39] for the Kolmogorov-Fokker-Planck equation. It is based on the combination of the fact that super-solutions to (1.1) expand positivity along times with a suitable covering argument. Note that this method is very convenient for the study of the weak regularity theory, because it only relies on the functional structure of the space $\mathcal{W}$ and on the non-Euclidean geometrical setting. To the best of our knowledge, the following weak Harnack inequality is the only available result of this type for solutions to (1.1) in the framework $\mathcal{W}$. Moreover, we underline that our statement holds true for solution or super-solutions depending on the sign of c. This is mainly due to the method of proof followed in [7], and the extension to this result to super-solutions without any sign assumption on c is still an open problem.

Theorem 3.4 (Weak Harnack Inequality) *Let $R_0 > 0$. Let $\mathcal{Q}^0 = B_{R_0} \times B_{R_0} \times \ldots \times B_{R_0} \times (-1, 0]$ and let u be a non-negative weak solution to $\mathscr{L} u = f$ in $\Omega \supset \mathcal{Q}^0$ under assumptions **(H1)-(H3)**. Then we have*

$$\left(\int_{\mathcal{Q}_-} u^p \right)^{\frac{1}{p}} \leq C \left(\inf_{\mathcal{Q}_+} u + \| f \|_{L^q(\mathcal{Q}^0)} \right),$$

where $\mathcal{Q}_+ = B_\omega \times B_{\omega^3} \times \ldots \times B_{\omega^{2\kappa+1}} \times (-\omega^2, 0]$ and $\mathcal{Q}_- = B_\omega \times B_{\omega^3} \times \ldots \times B_{\omega^{2\kappa+1}} \times (-1, -1+\omega^2]$. Moreover, the constants C, p, ω and R_0 only depend on the homogeneous dimension Q defined in (1.4), *q and on the ellipticity constants λ and Λ. Additionally, if the term c is of positive sign, the statement holds true also for non-negative super-solutions to* (1.1).

3.3 Applications of the Harnack Inequality

It is widely known that invariant Harnack inequalities are one of the most powerful tools in regularity theory. In this subsection, we briefly discuss some of the most important applications of Theorem 3.1, when considering Kolmogorov operators with rough coefficients.

First of all, invariant Harnack inequalities can be used to construct Harnack chains. A Harnack chain connecting any two given starting point z_0 and ending point z_k of our domain is a set $\{z_0, z_1, \ldots, z_k\}$ of interior points such that there exist k positive constants $C_1, \ldots, C_k$ for which

$$u(z_j) \leq C_j u(z_{j-1}), \quad j = 1, \ldots, k$$

for every non-negative solution to $\mathscr{L}u = f$ in Ω. Hence, a Harnack chain is a set of points through which the quantitative information provided by a Harnack inequality is able to travel. This tool had widely been employed over the years, especially in combination with techniques from control theory and optimization, for the study of qualitative properties of classical solutions to $\mathscr{L}u = f$, where the regularity of the coefficients was assumed to be equal, or better than Hölder continuous, see for instance [53].

Ever since Theorem 3.1 had been proved, various qualitative results regarding bounds for the weak fundamental solution [8, 54] where established adapting those techniques to the rough coefficients case. Moreover, we recall the work [10], where the authors prove a geometric characterization for the set where the Harnack inequality holds true and strong maximum principle in the Kolmogorov-Fokker-Planck case. Note that, since the proof of this two latter results is based on Harnack chains, control theory and an invariant Harnack inequality, their proof may be extended to the more general case (1.1) of our interest.

4 Applications to Physics and Economics

In this section, we provide the reader with a motivation for the study of the class of operators (1.1), by illustrating some applications of Kolmogorov equations (and associated regularity results) to real life problems arising in various research fields, such as Economics and Physics. In particular, we first address applications to option pricing, with a specific focus on American and Asian options. We then present new results regarding a relativistic Kolmogorov-Fokker-Planck equation. Finally, we analyze the link between the linear Kolmogorov-Fokker-Planck equation (1.5) and the Boltzmann equation.

4.1 American and Asian Options

In mathematical finance, equations of the form (1.1) appear in various models for the pricing of financial instruments, such as Asian and American options (see for instance, [14, 70]), as well as in the theory of stochastic utility [4] and stochastic volatility models [41]. Here, we focus on applications of (1.1) to Asian and American options, and we present an overview of the most recent results in this direction. For a more comprehensive analysis of applications of Kolmogorov operators $\mathscr{L}$ to finance and stochastic theory we refer to the monograph [70] by Pascucci.

Asian options are a family of *path-dependent derivatives*, whose payoff depends on the average of the underlying stock price over a certain time interval. In the Black & Scholes framework, the price of the underlying Stock S_t and of the bond B_t are described by the processes

$$S_t = S_0 e^{\mu t + \sigma W_t}, \qquad B_t = B_0 e^{rt}, \qquad 0 \le t \le T,$$

where μ, r, T, and σ are given constants. In this setting, the price $(Z_t)_{0 \le t \le T}$ of a continuous path-dependent option is a function $Z_t = Z(S_t, A_t, t)$ that depends on the *price of the stock* S_t, on the *time to maturity* t and on an *average* A_t of the stock price

$$A_t = \int_0^t f(S_\tau)\, \mathrm{d}\tau, \qquad t \in [0, T],$$

and it is computed by solving the following Cauchy problem

$$\begin{cases} \frac{1}{2}\sigma^2 S^2 \frac{\partial^2 Z}{\partial S^2} + f(S)\frac{\partial Z}{\partial A} + r\left(S\frac{\partial Z}{\partial S} - Z\right) & \\ \quad + \frac{\partial Z}{\partial t} = 0 & (S, A, t) \in \mathbb{R}^+ \times \mathbb{R}^+ \times (0, T), \\ Z(S, A, T) = \varphi(S, A) & (S, A) \in \mathbb{R}^+ \times \mathbb{R}^+. \end{cases} \tag{4.1}$$

We remark that the initial data φ in (4.1) corresponds to the pay-off of the option. Moreover, depending on the choice of the function $f(S)$, we find a different Kolmogorov type equation with locally Hölder continuous coefficients. In [12] the first author, Muzzioli and Polidoro, prove the existence and uniqueness of the fundamental solution associated to Geometric Average Asian Options, i.e. when $f(S) = \log(S)$, and Arithmetic Average Asian Options, i.e. when $f(S) = S$ in this framework.

On the other hand, an American option with pay-off ψ is a contract which grants the holder the right to receive the payment of the sum $\psi(X_t)$ at a chosen time $t \in [0, T]$, where $X = \left(X_t^x\right)$ is a N-dimensional diffusion process which solves the stochastic differential equation

$$dX_t^x = BX_t^x dt + \sigma(t, X_t^x) dW_t,$$

with $X_{t_0}^{t_0,x} = x$ for $(x, t_0) \in \mathbb{R}^N \times [0, T]$ and where, as usual, $(W_t)_{t\geq 0}$ denotes a n-dimensional Wiener process, $1 \leq n \leq N$.

In particular, Eq. (1.1) is relevant in connection to the problem of determining the arbitrage-free price of American options. Indeed, there are significant classes of American options, whose corresponding diffusion process X is associated to Kolmogorov-type operators which are not uniformly parabolic and are of the kind (1.1). Two such examples are given by American style options (see [14]) and by the American options priced in the stochastic volatility introduced in the article [41]. Moreover, in virtue of the classical arbitrage theory (see, for instance [70]), the arbitrage free price at time t of the American option when assuming the risk-free interest rate is zero is given by the following optimal stopping problem

$$u(x,t) = \sup_{\tau \in [t,T]} E\left[\psi\left(X_\tau^{t,x}\right)\right],$$

where the supremum is taken over all stopping times $\tau \in [t, T]$ of X. In [69], it is proved that the function u in (4.1) is a solution to the obstacle problem

$$\begin{cases} \max\{\mathscr{L}u - f, \psi - u\} & (x,t) \in \mathbb{R}^N \times [0,T] \\ u(x,t) = g & (x,t) \in \mathbb{R}^N \times \{0\}, \end{cases} \tag{4.2}$$

where $\mathscr{L}$ is the operator (1.1) in trace form, the obstacle ψ corresponds to the pay-off of the option and it is a Lipschitz continuous function in $\overline{\Omega}$ satisfying a weak convexity condition with respect to the variables $x_1, \ldots, x_n$ (see [30, Assumption H4.]).

In virtue of its importance in finance, the mathematical study of the obstacle problem (4.2) was already initiated in the papers [30, 34, 64]. More precisely, the main result of [30] is the existence of a strong solution to problem (4.2) in certain bounded cylindrical domains and in the strips $\mathbb{R}^N \times [0, T]$ through the adaptation of a classical penalization technique. On the other hand, the main purpose of papers [34, 64] is to prove some new regularity results for solutions to (4.2). In particular, [34] concerns the optimal interior regularity for solutions to the problem (4.2), while [64] contains new results regarding the regularity near the initial state for solutions to the Cauchy-Dirichlet problem and to (4.2).

However, all the results contained in [30, 34, 64] only hold true for strong solutions and continuous obstacles satisfying the aforementioned convexity condition. Only very recently, the first and the third author initiated in [9] the study of obstacle problems associated to Kolmogorov operators in a more general and natural setting, i.e. by considering weak solutions to the obstacle problem related to

$$\mathscr{K}u(v,x,t) := \nabla_v \cdot (A(v,x,t)\nabla_v u(v,x,t)) + v \cdot \nabla_x u(v,x,t) - \partial_t u(v,x,t),$$

in the functional space $\mathcal{W}$ introduced in Sect. 2. Specifically, in a standard manner (see [51, Chapter 6]), in [9] it is assumed that obstacle ψ and boundary data g inherit the same regularity of the function u, namely $\psi \in \mathcal{W}(\Omega_v \times \Omega_{xt})$ and $g \in \mathcal{W}(\Omega_v \times \Omega_{xt})$, where $\Omega := \Omega_v \times \Omega_{xt}$ is a subset of $\mathbb{R}^{2n+1}$ satisfying the following assumption:

(D) $\Omega_v \subset \mathbb{R}^n$ is a bounded Lipschitz domain and $\Omega_{xt} \subset \mathbb{R}^{n+1}$ is a bounded domain with $C^{1,1}$-boundary, i.e. $C^{1,1}$-smooth with respect to the transport operator Y as well as t.

Thanks to this assumption, it is possible to introduce the outward-pointing unit normal N to Ω_{xt} and to classically define the Kolmogorov boundary of the set Ω as

$$\partial_K(\Omega_v \times \Omega_{xt}) := (\partial\Omega_v \times \Omega_{xt}) \cup \left\{(v, x, t) \in \overline{\Omega}_v \times \partial\Omega_{xt} \mid (v, -1) \cdot N_{xt} > 0\right\},$$

which serves in the context of the operator $\mathscr{L}$ as the natural hypoelliptic counterpart of the parabolic boundary considered in the context of Cauchy-Dirichlet problems for uniformly parabolic equations. In comparison with [30, 34, 64, 69], in [9] the authors weaken the regularity assumptions on the right-hand side by considering $f \in L^2(\Omega_{xt}, H^{-1}(\Omega_v))$. Furthermore, the following more general obstacle problem than (4.2) is considered

$$\begin{cases} \mathscr{K} u(v, x, t) = f(v, x, t) & (v, x, t) \in \Omega, \\ u(v, x, t) \geq \psi(v, x, t) & (v, x, t) \in \Omega, \\ u(v, x, t) = g & (v, x, t) \in \partial_K \Omega, \end{cases} \tag{4.3}$$

where the boundary condition needs to be considered as attained in the sense of traces, the obstacle condition holds in $\mathcal{W}(\Omega_v \times \Omega_{xt})$ and

$$\psi \leq g \quad \text{on } \partial_K(\Omega_v \times \Omega_{xt}) \quad \text{in } \mathcal{W}(\Omega_v \times \Omega_{xt}).$$

Finally, the condition $\mathscr{K} u(v, x, t) = f(v, x, t)$ for $(v, x, t) \in \Omega$ appearing in (4.3) needs to be interpreted as stating that

$$\begin{aligned} 0 = & \iiint\limits_{\Omega_v \times \Omega_{xt}} A(v, x, t) \nabla_v u \cdot \nabla_v \varphi \, \mathrm{d}v \, \mathrm{d}x \, \mathrm{d}t \\ & + \iint\limits_{\Omega_{xt}} \langle f(\cdot, x, t) - Yu(\cdot, x, t) | \varphi(\cdot, x, t) \rangle \, \mathrm{d}x \, \mathrm{d}t \end{aligned} \tag{4.4}$$

for every $\varphi \in L^2(\Omega_{xt}, H^1_c(\Omega_v))$ and where $\langle \cdot | \cdot \rangle$ is the standard duality pairing in $H^{-1}(\Omega_v)$. Then, the main result of [9] is the following.

Theorem 4.1 *Let us assume that the diffusion matrix A in* (4.1) *satisfies the ellipticity condition in **(H1)** for $m_0 = n$. Let $f \in L^2(\Omega_{xt}, H^{-1}(\Omega_v))$ and $g, \psi \in \mathcal{W}(\Omega_v \times \Omega_{xt})$, where Ω is a subset of $\mathbb{R}^{2n+1}$ satisfying assumption (D). Then there exists a unique weak solution $u \in \mathcal{W}(\Omega_v \times \Omega_{xt})$ in the sense of Eq.* (4.4) *to the obstacle problem* (4.3). *Moreover, there exists a constant C, which only depends on d and on $\Omega_v \times \Omega_{xt}$, such that*

$$\|u\|_{\mathcal{W}(\Omega_{xt}\times\Omega_v)} \leq C\left(\|g\|_{\mathcal{W}(\Omega_{xt}\times\Omega_v)} + \|f\|_{L^2(\Omega_{xt},H^{-1}(\Omega_v))}\right).$$

We eventually point out that with [9] the authors aim at initiating the study of the obstacle problem (4.3) in the framework of Calculus of Variations, by rewriting the problem of finding a solution to (4.3) as that of finding a null minimizer of the functional

$$\inf\Bigg\{ \iiint_{\Omega_v\times\Omega_{xt}} \frac{1}{2}\left(A\left(\nabla_v u - \mathfrak{J}\right)\right)\cdot\left(\nabla_v u - \mathfrak{J}\right)\, \mathrm{d}v\, \mathrm{d}x\, \mathrm{d}t \ :$$

$$\mathfrak{J} \in \left(L^2(\Omega_{xt}; L^2(\Omega_v))\right)^n \ s.t. \ \nabla_v\cdot(A\mathfrak{J}) = f - Yu\Bigg\}, \tag{4.5}$$

introduced for the first time in [2], and later on considered in [57]. It is clear that the infimum in (4.5) is non-negative and that, given a solution u to (4.3), if we choose $\mathfrak{J} = \nabla_v u$, then (4.5) vanishes at u. Moreover, it is easy to show that the functional in (4.5) is uniformly convex and attains its minimum at zero. Finally, we observe that the functional justifies the definition of functional kinetic space given in Sect. 2. However, it is still an open problem whether it is possible to employ classical tools from Calculus of Variations to study the variational problem associated to functional (4.5), see [Sections 4, 9].

4.2 *Relativistic Fokker-Planck Equation*

As pointed out in Example 1.1, the class of Kolmogorov operators (1.1) arises in many Physical applications. For this reason, we consider Eq. (1.15) in the framework of special relativity, namely

$$\mathscr{L}u(p,x,t) = \sqrt{|p|^2+1}\,\nabla_p\cdot\left(D\,\nabla_p u\right) - p\cdot\nabla_x u - \sqrt{|p|^2+1}\,\partial_t u = 0, \tag{4.6}$$

where $(p,x,t) \in \mathbb{R}^{2n+1}$ and D is the *relativistic diffusion matrix* given by

$$D = \frac{1}{\sqrt{|p|^2+1}}\left(\mathbb{I}_n + p\otimes p\right).$$

Here and in the following, $\mathbb{I}_n$ denotes the $n \times n$ identity matrix and

$$p \otimes p = \left(p_i p_j\right)_{i,j=1,\dots,n} .$$

In this context, a solution $u = u(p, x, t)$ to (4.6) denotes the density of particles in the phase space with momentum p and position x, at time t. Equation (4.6) is a generalization which agrees with special relativity of the frictionless kinetic Fokker-Planck equation (1.15), as the *relativistic velocity*

$$v = \frac{p}{\sqrt{|p|^2 + 1}},$$

clearly satisfies

$$\left| \frac{p}{\sqrt{|p|^2 + 1}} \right| < 1 \quad \text{for every } p \in \mathbb{R}^n,$$

in accordance with the relativity principles.[1]

Operator $\mathscr{L}$ in (4.6) serves as a suitable relativistic version of $\mathscr{K}_0$ in (1.15) as it preserves some relevant properties which hold in the non-relativistic setting (we refer the reader to [3] for a more rigorous derivation of Eq. (4.6)). In particular, operator $\mathscr{L}$ satisfies the relativistic analogue of property (1.17), i.e. it is invariant under Lorentz transformations. Taking advantage of this property, in [13] Polidoro and two of the authors costructed the invariance group of $\mathscr{L}$ by defining the composition law as follows

$$\begin{aligned}(p_0, x_0, t_0) \circ_{\mathcal{L}} (p, x, t) = \Big(& p\sqrt{|p_0|^2 + 1} + p_0\sqrt{|p|^2 + 1}, \\ & x_0 + x\sqrt{|p_0|^2 + 1} + p_0 t, t_0 + t\sqrt{|p_0|^2 + 1} + p_0 \cdot x\Big).\end{aligned} \tag{4.7}$$

We remark that for small velocities $\sqrt{1 + |p_0|^2} \approx 1$ and therefore (4.7) becomes precisely the non-relativistic composition law (1.16) for variables p and x. The introduction of the composition law (4.7) and consequently of an appropriate non-Euclidean structure on the space $\mathbb{R}^{2n+1}$ significantly simplifies the study of the regularity of operator $\mathscr{L}$.

As with the non-relativistic operator in (1.15), $\mathscr{L}$ is a strongly degenerate differential operator, since only second order derivatives with respect to the momentum variable $p \in \mathbb{R}^n$ appear. However, the first order part of $\mathscr{L}$ induces a strong regularizing property, namely $\mathscr{L}$ is hypoelliptic. Indeed (see [13, Appendix A])

[1] Here, we adopt a natural unit system with $c = 1$, where c is the speed of light.

we can write $\mathscr{L}$ as a *sum of squares plus a drift term*

$$\mathscr{L} := \sum_{j=1}^{n} X_j^2 + X_{n+1},$$

with

$$X_j = \sum_{k=1}^{n} \left(\delta_{jk} + \frac{p_j p_k}{1+\sqrt{|p|^2+1}}\right) \frac{\partial}{\partial p_k}, \quad j = 1, \dots, n, \text{ and } X_{n+1} = \sum_{k=1}^{n} c_k(p) X_k - Y,$$

where $c_1, \dots c_n$ are smooth functions and

$$Y = p \cdot \nabla_x + \sqrt{|p|^2 + 1}\, \frac{\partial}{\partial t}.$$

Moreover, as shown in [13, Appendix A], $\mathscr{L}$ satisfies Hörmander's rank condition (1.9) at every point $(p, x, t) \in \mathbb{R}^{2n+1}$.

It is then natural to consider the relativistic analogous of (1.6), namely

$$\begin{cases} P_s = p_0 + \sqrt{2} \int_0^s \sqrt{P_\tau^2 + 1}\, dW_\tau, \\ X_s = x_0 + \int_0^s P_\tau \mathrm{d}\tau, \\ T_s = t_0 + \int_0^s \sqrt{P_\tau^2 + 1}\, \mathrm{d}\tau, \end{cases} \tag{4.8}$$

where the third component is the time, which is not an absolute quantity in the relativistic setting. It is clear that (4.6) is the relativistic deterministic equation describing the density of the stochastic process (4.8). The main result of [13] provides us with a lower bound for such a density function and can be stated as follows for $(p, x, t) \in \mathbb{R}^3$.

Theorem 4.2 *Let Γ be the fundamental solution of $\mathscr{L}$ in* (4.6)*. Then for every $T > 0$ there exist three positive constants θ, c_T, C with $\theta < 1$, such that*

$$\Gamma(p_1, x_1, t_1; p_0, x_0, t_0) \geq \frac{c_T}{(t_1 - t_0)^2} \times \exp\left\{-C\,\Psi\left(p_1, x_1, y_1; p_0, x_0, \theta^2 t_0 + (1 - \theta^2) t_1\right)\right\}$$

for every $(p_0, x_0, t_0), (p_1, x_1, t_1) \in \mathbb{R}^3$ such that $0 < t_1 - t_0 < T$. The constants θ and C only depend on $\mathscr{L}$, while c_T also depends on T. Moreover, Ψ is the value function *of a suitable optimal control problem (see [13, Section 3]).*

We point out that Theorem 4.2 only constitutes a first step towards developing a systematic and more comprehensive study of $\mathscr{L}$ in its appropriate framework of PDEs theory. Indeed, the purpose of [13] is to propose an approach that may lead to various developments in Stochastic and Kinetic Theory, where the final aim is to extend the classical theory considered in [6] to the relativistic case.

4.3 Boltzmann Equation

Among the most classical applications of Kolmogorov operators to Physical modeling we find the Boltzmann operator, for which (1.5) represents a linearization. In $\mathbb{R}^{2n+1}$, the "*Boltzmann equation*" reads as

$$\partial_t u(v,x,t) + v \cdot \nabla_x u(v,x,t) = \mathcal{L}_K(u), \tag{4.9}$$

for $(v,x,t) \in B_1 \times B_1 \times (-1,0]$, where the function $u \equiv u(v,x,t)$ is defined for any $(v,x,t) \in \mathbb{R}^n \times B_1 \times (-1,0]$ and the nonlocal *collisional* operator $\mathcal{L}_K$ is given by

$$\mathcal{L}_K(u) := \iint_{\mathbb{R}^n \times \mathbb{S}^{n-1}} \big(u(v_*^{'})u(v^{'}) - u(v_*)u(v)\big)K$$
$$(|v - v_*|, \cos\theta)\ \mathrm{d}v_* \mathrm{d}\sigma, \tag{4.10}$$

where $v_*^{'}$ and $v^{'}$ are computed using v, v_* and σ via the following formulae

$$v' = \frac{v + v_*}{2} + \frac{|v - v_*|}{2}\sigma \quad \text{and} \quad v_*' = \frac{v + v_*}{2} - \frac{|v - v_*|}{2}\sigma,$$

and where θ is the angle measuring the deviation between v and v', whose cosinus is defined as

$$\cos\theta := \frac{v + v_*}{|v - v_*|}\sigma.$$

Equation (4.9) describes the dynamics of a dilute gas. Indeed, the transport term $\big(\partial_t + v \cdot \nabla_x\big)$ on the left-hand side of (4.9) describes the fact that particles travel in straight lines when no external force is applied, while the diffusion term $\mathcal{L}_K$ expresses the fluctuations in velocity arising from particles interactions. In this setting, the solution u represents the density of particles with position x and velocity v at time t, characterizing the state of the gas in a statistical way. Furthermore, the couples v, v_*, and $v^{'}$, $v_*^{'}$ represent pre and post collisional velocities of two distinct particles, respectively. The rate by which such particles switch from initial velocities v, v_* to $v^{'}$, $v_*^{'}$ after the collision is given by the kernel K, which is usually know as *collision kernel*. We refer the interested reader to [45, 46] and the refernces therein for some regularity results and related properties of solutions to (4.9) when the collision kernel does coincide with

$$K(r, \cos\theta) = r^{\alpha} b(\cos\theta), \qquad \text{with } b(\cos\theta) \approx |\sin(\theta/2)|^{-(n-1)-2s},$$

where $\alpha > -n$ and $s \in (0, 1)$.

5 Nonlocal Kolmogorov-Fokker-Planck Equations

In recent years an increasing interest has been focused on the study of fractional powers of nonlinear operators, not only because of their appearance in many concrete models in Physic, Biology and Finance, but also because of their challenging mathematical intrinsic nature, mainly due to the contemporary presence of a nonlinear and a nonlocal behaviour (see for example [21] and the references therein).

Naturally, presenting a comprehensive treatment of the related literature is beyond the scopes of the present overview, but we still mention [28, 29], where amongst other results the authors prove the De Giorgi-Nash-Moser theory for (elliptic) nonlinear fractional operators modeled on the fractional p-Laplacian; see also the survey article [67] and the references therein. Moreover, we recall that some notable results were established even in the parabolic setting, e.g. [31] for the De Giorgi-Nash-Moser theory in the superlinear case when $p \geq 2$, and the recent papers [1, 56] for Hölder regularity for any value of the exponent $p \in (1, \infty)$.

Hence, for the community it had been quite natural to begin considering ultraparabolic equations whose diffusion part is modeled on the fractional p-Laplacian, with the aim of extending the study of the regularity theory to this case. The reference model is now a kinetic Kolmogorov-Fokker-Planck equation whose diffusion part is made by an integro-differential operator

$$\partial_t u(v, x, t) + v \cdot \nabla_x u(v, x, t) + \mathcal{L}_K(u) = 0, \qquad \text{in } \mathbb{R}^{2n+1}. \tag{5.1}$$

The leading operator $\mathcal{L}_K$, representing the "diffusion" in the velocity variable, is an integro-differential operator of differentiability order $s \in (0, 1)$ and integrability exponent $p \in (1, \infty)$, whose explicit expression is given by

$$\mathcal{L}_K(u)(v, x, t) := \lim_{\varepsilon \to 0^+} \int_{\mathbb{R}^n \smallsetminus B_\varepsilon(v)} (u(v, x, t) - u(w, x, t)) K(v, w)\, dw, \tag{5.2}$$

The operator $\mathcal{L}_K$ is driven by its symmetric measurable kernel $K : \mathbb{R}^n \times \mathbb{R}^n \to [0, +\infty)$, which satisfies, for some $0 < \lambda \leq \Lambda$, the following bounds

$$\lambda |v - w|^{-n-2s} \leq K(v, w) \leq \Lambda |v - w|^{-n-2s}, \qquad \text{for a.e. } v, w \in \mathbb{R}^n. \tag{5.3}$$

Integro-differential kinetic equations such as (5.1) are strictly connected with concrete problems arising in Physics. In particular, one of the main area where kinetic equations with integro-differential diffusion find their application is the theory of anomalous diffusions, which derives the small-mean-free-path limit of fractional kinetic equations and shows that these equations describe the mesoscopic behaviour of fat-tailed equilibrium distribution, which arise in physical observations from astrophysics; see for instance [61, 62]. Furthermore, kinetic integro-differential equations such as (5.1) (when $p = 2$) do appear in Gas Dynamics when dealing with the Boltzmann equation (4.9) in the non-cutoff case. Indeed, for what concern the Boltzmann equation, if the macroscopical physical quantities involved in the statement of [46, Assumption 1.1], i.e.

$$M(x,t) := \int_{\mathbb{R}^n} u(v,x,t)\,\mathrm{d}v \qquad \text{(mass density)},$$

$$E(x,t) := \int_{\mathbb{R}^n} u(v,x,t)|v|^2\,\mathrm{d}v \qquad \text{(energy density)},$$

$$H(x,t) := \int_{\mathbb{R}^n} u \ln u(v,x,t)\,\mathrm{d}v \qquad \text{(entropy density)},$$

are uniformly bounded, then the Boltzmann collision operator (4.10) can be rewritten as the sum of a lower order term plus an integro-differential operator as (5.2) (taking $p = 2$ in there) driven by a kernel K satisfying some weaker integrability conditions than the uniformly elliptic constraints (5.3); see for instance [45, 46] and the references therein.

For what concerns the *fractional kinetic framework* many remarkable results are available, e. g. Hölder regularity [79], L^p-estimates [26, 43], hypoelliptic regularity [55], existence of weak solutions [82] and existence, uniqueness and regularity of solutions in the viscosity sense [44]. Furthermore, we recall the recent paper [58] where the author proves Hölder continuity together with some weak Harnack-type inequalities for nonnegative and apriori bounded weak solutions to integro-differential kinetic equations.

However, despite the recent interest in investigating qualitative and regularity properties of solutions to such kinetic integro equations, still the question of whether or not weak solutions are locally bounded remains open. This, seems linked to the intrinsic nonlocal nature of the underlying equations; see for example the clarifying explanation by the authors in [45, Pag. 548].

In order to overcome the aforementioned nonlocality issues mentioned in Theorem 5.1 we found a way to present a totally new *δ-interpolation L^∞-inequality with tail* for weak subsolutions to (5.1) which are not required to being nonnegative. Such result was established in the aforementioned [5] together with *the very Strong Harnack inequality for integro-differential kinetic equation*; see in particular Theorem 1.3 in [5]. We point out that the validity of Harnack type inequality are not for free even for the purely elliptic case of the fractional Laplacian $(-\Delta)^s$ (see for instance [47, 48]).

This section is organised as follow. We will firstly recall some helpful properties about the underlying geometrical and fractional functional setting suitable for the study of (5.1). Then, we recall the statement of the boundedness estimates for weak solutions to (5.1) for any value of the integrability exponent $p \in (1, +\infty)$ providing the reader with a comparison with existing results in literature and with related open problems and further developments.

5.1 Geometric and Functional Setting

As in the previous sections, we denote with $z = (v, x, t)$ points of $\mathbb{R}^{2n+1}$ and we define a family of anisotropic dilations $(\delta_r)_{r>0}$ on $\mathbb{R}^{2n+1}$ in the following way

$$\delta_r = \operatorname{diag}(r\mathbb{I}_n, r^{1+2s}\mathbb{I}_n, r^{2s}t), \quad \forall r > 0.$$

Firstly, we observe that when s tends to 1 we recover the family of dilations introduced in (1.12) for the linear Kolmogorov-Fokker-Planck operator. Moreover, analogously as in the linear case, Eq. (5.1) is homogeneous of degree 2 with respect to the dilation group $(\delta_r)_{r>0}$ just introduced. In addition, we endow $\mathbb{R}^{2n+1}$ with the product law

$$z_0 \circ z = (v_0 + v, x_0 + x + tv_0, t_0 + t), \qquad \forall z_0 = (v_0, x_0, t_0) \in \mathbb{R}^{2n+1}, \tag{5.4}$$

that is the same Galilean change of variables (1.16) considered in the linear Kolmogorov-Fokker-Planck case. This is due to the fact that, geometrically speaking, the family of translations is strongly connected to the shape of the transport operator, which in this case agrees with the one considered in (1.5), i.e. given by $\partial_t + v \cdot \nabla_x$. In this way, $\mathbb{K} := (\mathbb{R}^{2n+1}, \circ)$ is a Lie group with identity element $e := (0, 0, 0)$ and inverse given by

$$z^{-1} := (-v, -x + tv, -t), \qquad \forall z = (v, x, t) \in \mathbb{R}^{2n+1},$$

and, for sufficiently regular functions, Eq. (5.1) is invariant with respect to the Lie product " $\circ$".

Then, as it is done in the local framework, we introduce a famllies of fractional kinetic cylinders defined starting from the aforementioned dilations and product law

$$Q_r(z_0) := \left\{ z : \ |v - v_0| < r, \ |x - x_0 - (t - t_0)v_0| < r^{1+2s}, \ t_0 - r^{2s} < t < t_0 \right\}. \tag{5.5}$$

We now introduce the fractional functional setting. For any $s \in (0, 1)$ we define the fractional Sobolev space as

$$H^s(\mathbb{R}^n) := \left\{ g \in L^2(\mathbb{R}^n) : \iint_{\mathbb{R}^n \times \mathbb{R}^n} \frac{|g(v) - g(w)|^2}{|v - w|^{n+2s}} \, dv \, dw < \infty \right\},$$

and we endow it with the norm

$$\|g\|_{H^s(\mathbb{R}^n)} := \|g\|_{L^2(\mathbb{R}^n)} + \left(\iint_{\mathbb{R}^n \times \mathbb{R}^n} \frac{|g(v) - g(w)|^2}{|v - w|^{n+2s}} \, \mathrm{d}v \, \mathrm{d}w \right)^{\frac{1}{2}},$$

turning it into a Banach space. In a similar way one can define the fractional Sobolev space $H^s(\Omega_v)$ on an open and bounded set $\Omega_v \subset \mathbb{R}^n$. Finally, we denote with $H_0^s(\Omega_v)$ the closure with respect to $\|\cdot\|_{H^s(\Omega_v)}$ of $C_c^\infty(\Omega_v)$.

The difficulty in studying operators such as $\mathcal{L}_K$ lies in their own definition. Indeed, we have to deal with their nonlocal behaviour. Thus, as natural in the fractional framework, a tail-type contribution needs to be taken in consideration in order to carefully control the long-range interactions that naturally arise in the study of fractional problems. In the Kolmogorov-Fokker-Planck setting, the kinetic nonlocal tail mainly considers long-range contributions associated with the velocity variable appearing in the kernel.

Definition 5.1 Let $z_0 \in \Omega := \Omega_v \times \Omega_x \times (t_1, t_2) \subset \mathbb{R}^{2n+1}$ and $r > 0$ be such that $B_r(v_0) \times U_{2r}(x_0, t_0) \subset \Omega_v \times \Omega_x \times (t_1, t_2)$. Let u be a measurable function on $\mathbb{R}^n \times \Omega_x \times (t_1, t_2)$. Then for any $r > 0$ and any $p \in [1, \infty]$ we define the kinetic nonlocal tail of u with respect to z_0, and r as

$$\mathrm{Tail}_p(u; z_0, r) := \left(\fint_{U_{2r}(t_0, x_0)} \left(r^{2s} \int_{\mathbb{R}^n \setminus B_r(v_0)} \frac{|u(v, x, t)|}{|v - v_0|^{n+2s}} \, \mathrm{d}v \right)^p \mathrm{d}x \, \mathrm{d}t \right)^{\frac{1}{p}}.$$

The nonlocal tail of a function was firstly introduced in the aforementioned papers [28, 29] in the elliptic setting and subsequently used to derive fine properties of weak solutions to nonlocal operator; see [67] and the references therein. Various extensions in the parabolic setting can be found in [1, 31, 49, 50, 56, 80, 81]. Moreover, an analogous quantity has been defined in the sub-Riemannian framework of the n-dimensional Heisenberg group $\mathbb{H}^n$ in [60, 68] and, despite its young age, used to prove some regularity result for solutions to fractional (possibly nonlinear) equations; see [37, 74].

With this bit of notation it is possible to introduce the definition of weak solution to (5.1), that is a function belonging to the space

$$\mathcal{W} := \left\{ u \in L^2_{loc}(\mathbb{R}^{n+1}; H^s(\mathbb{R}^n)) \, : \, (\partial_t u + v \cdot \nabla_x u) \in L^2_{loc}(\mathbb{R}^{n+1}; H^{-s}(\mathbb{R}^n)) \right\}$$

and satisfying (5.1) in the following sense.

Definition 5.2 Let $\Omega := \Omega_v \times \Omega_x \times (t_1, t_2) \subset \mathbb{R}^{2n+1}$. A function $u \in \mathcal{W}$ is a weak subsolution (resp. supersolution) to (5.1) in Ω if

$$\int_{\Omega_x \times (t_1, t_2)} \iint_{\mathbb{R}^n \times \mathbb{R}^n} (u(v, x, t) - u(w, x, t)) \, (\varphi(v, x, t)$$

$$-\varphi(w,x,t))\,K(v,w)\,\mathrm{d}w\,\mathrm{d}v\,\mathrm{d}x\,\mathrm{d}t$$

$$\geq\ (\leq\,,\text{resp.})\int_{\mathbb{R}^n\times\Omega_x\times(t_1,t_2)}(u\,\partial_t\varphi+u\,v\cdot\nabla_x\varphi)\,(v,x,t)\,\mathrm{d}v\,\mathrm{d}x\,\mathrm{d}t,$$

for any nonnegative $\varphi\in W$, with $\operatorname{supp}(\varphi)\Subset\Omega$.

We say u is a weak solution to (5.1) in Ω if

$$\int_{\Omega_x\times(t_1,t_2)}\iint_{\mathbb{R}^n\times\mathbb{R}^n}u(v,x,t)-u(w,x,t)\,(\varphi(v,x,t)$$

$$-\varphi(w,x,t))\,K(v,w)\,\mathrm{d}w\,\mathrm{d}v\,\mathrm{d}x\,\mathrm{d}t$$

$$=\int_{\mathbb{R}^n\times\Omega_x\times(t_1,t_2)}(u\,\partial_t\varphi+u\,v\cdot\nabla_x\varphi)\,(v,x,t)\,\mathrm{d}v\,\mathrm{d}x\,\mathrm{d}t,$$

for any $\varphi\in W$ and $\operatorname{supp}(\varphi)\Subset\Omega$.

Note that, also in this framework, the transport derivative $\big(\partial_t+v\cdot\nabla_x\big)$ appearing in Definition 5.2 above needs to be intended in the duality sense, as suggested by our functional setting. Thus, our notation stands for the more formal one:

$$\int_{\mathbb{R}^n\times\Omega_x\times(t_1,t_2)}u\big(\partial_t\varphi+v\cdot\nabla_x\varphi\big)\,\mathrm{d}v\,\mathrm{d}x\,\mathrm{d}t=\int_{\Omega_x\times(t_1,t_2)}\langle\big(\partial_t u+v\cdot\nabla_x u\big)\mid\varphi\rangle\,\mathrm{d}x\,\mathrm{d}t,$$

where $\langle\cdot\mid\cdot\rangle$ denotes the standard duality pairing between $H^s(\mathbb{R}^n)$ and $H^{-s}(\mathbb{R}^n)$.

5.2 *Boundedness Estimate*

The boundedness estimate achieved in [5] reads as follows

Theorem 5.1 (δ-Interpolative $L^\infty-L^2$ Estimate) *For any $s\in(0,1)$, let $u\in\mathcal{W}$ be a weak subsolution to (5.2) in Ω and let $Q_1\equiv Q_1(\mathbf{0})\subset\Omega$. Then, for any $Q_{\frac{r}{2}}\equiv Q_{\frac{r}{2}}(\mathbf{0})$, any $\delta\in(0,\ 1]$, there exists $p^*=p^*(n,s)>2$ such that for any $p>p^*$, it holds*

$$\sup_{Q_{\frac{r}{2}}}u\leq c\left(\delta r^{4s}\right)^{-\frac{1}{2\epsilon_*}}\left(\int_{Q_r}u_+^2\,\mathrm{d}v\,\mathrm{d}x\,\mathrm{d}t\right)^{\frac{1}{2}}+\delta\mathrm{Tail}_p\left(u_+;\mathbf{0},\ r/2\right),$$

where $Q_r:=(-r^{2s},0]\ \times\ B_{r^{1+2s}}\times B_r$, and the quantity $\varepsilon_*>0$ depends only on the dimension n and the exponent s, whereas the positive constant c depends also on the kernel structural constant Λ.

Because of the particular choice of the involved kernel, we can not establish the desired estimate (5.1) by simply applying most of the methods used whenever $K \equiv |v - w|^{-n-2s}$ (e. g. velocity averaging and Fourier transform [19] or via a Dirichlet-to-Neumann's map [36]). For this reason, the proof of Theorem 5.1 rely on a precise energy estimate which essentially involve a Caccioppoli inequality with tail combined with a summability gain result for subsolutions—in turn based on the fundamental solution for the fractional Kolmogorov equation—as well as with a delicate iterative scheme taking into account both the Tail_p term and the desired interpolative effect.

5.3 *Further Developments*

Despite the increasing interest in nonlocal problems several questions still remains open. Below, we list just a few possible developments related to the results presented in this last section of the overview.

A subsequent further development of the estimate obtained in [5] would be trying to generalized them when a p-integrability is assumed in the diffusion operator $\mathcal{L}_K$.This seems rather a difficult task because of the involved competing nature of the main diffusion operator which is both nonlocal and nonlinear, together with the underlying "slanted" geometry. Moreover, such the quantitative approach may be feasible to treat even more general nonlocal equations. In this direction, one can consider more general nonlocal diffusions with non-standard growth conditions as done in [24, 25].

Also one could add a second integral-differential operator, $\mathcal{L}_{K;\alpha}$of differentiability exponent $t > s$ and summability growth $q > 1$, controlled by the zero set of a modulating coefficient $\alpha \equiv \alpha(x,t)$; that is, the so-called nonlocal double phase problem, in the same spirit of the elliptic case treated in [22, 33, 75].

Furthermore, one could investigate the regularity properties for solutions to a strictly related class of problems; that is, by adding in (5.1) a quasilinear operator modeled on the classical Laplacian. Mixed-type operators are a really recent fields of investigation but, regularity and related properties of weak solutions to (5.1) when the diffusion part in velocity $\mathcal{L}_K$ is a mixed type operator are almost unknown. Clearly, several results as e. g., in [15–17, 23, 32, 35, 65, 76], would be expected do still hold for mixed-type *kinetic* operators.

Acknowledgments The authors are members of the Gruppo Nazionale per l'Analisi Matematica, la Probabilità e le loro Applicazioni (GNAMPA) of the Istituto Nazionale di Alta Matematica (INdAM). The first and third authors are partially supported by the INdAM–GNAMPA project "Variational problems for Kolmogorov equations: long-time analysis and regularity estimates", CUP_E55F22000270001. The authors are grateful to the anonymous reviewer for the fruitful suggestions which helped improving the quality of the manuscript.

References

1. Adimurthi, K., Prasad, H., Tewary, V.: Local Hölder regularity for nonlocal parabolic p-Laplace equation (2022). arXiv2205.09695
2. Albritton, D., Armstrong, S., Mourrat, J.C., Novack, M.: Variational methods for the kinetic Fokker-Planck equation. Analysis & PDE, to appear
3. Alcántara, J.A., Calogero, S.: On a relativistic fokker-planck equation in kinetic theory. Kinet. Relat. Models **4**, 401–426 (2011)
4. Antonelli, F., Pascucci, A.: On the viscosity solutions of a stochastic differential utility problem. J. Differ. Equ. **186**, 69–87 (2002)
5. Anceschi, F., Palatucci, G., Piccinini, M.: Harnack inequalities for kinetic integral equations (2024). arXiv:2401.14182
6. Anceschi, F., Polidoro, S.: A survey on the classical theory for Kolmogorov equation. Matematiche **75**(1), 221–258 (2020)
7. Anceschi, F., Rebucci, A.: A note on the weak regularity theory for degenerate Kolmogorov equations. J. Differ. Equ. **341**, 538–88 (2022)
8. Anceschi, F., Rebucci, A.: On the fundamental solution for degenerate Kolmogorov equations with rough coefficients. J. Elliptic Parabol. Equ. **9**, 63–92 (2022)
9. Anceschi, F., Rebucci, A.: On the obstacle problem associated to the Kolmogorov-Fokker-Planck operator with rough coefficients. Ann. Mat. Pura Appl. (2024). https://doi.org/10.1007/s10231-024-01431-5
10. Anceschi, F., Eleuteri, M., Polidoro, S.: A geometric statement of the Harnack inequality for a degenerate Kolmogorov equation with rough coefficients. Commun. Contemp. Math. **21**(7), 1850057 (2019)
11. Anceschi, F., Polidoro, S., Ragusa, M.A.: Moser's estimates for degenerate Kolmogorov equations with non-negative divergence lower order coefficients. Nonlinear Anal. **189**, 111568 (2019)
12. Anceschi, F., Muzzioli, S., Polidoro, S.: Existence of a fundamental solution of partial differential equations associated to Asian options. Nonlinear Anal. Real World Appl. **62**, 1–29 (2021)
13. Anceschi, F., Polidoro, S., Rebucci, A.: Harnack inequality and asymptotic lower bounds for the relativistic Fokker-Planck operator (2022). arXiv:2211.05736
14. Barucci, E., Polidoro, S., Vespri, V.: Some results on partial differential equations and Asian options. Math. Models Methods Appl. Sci. **11**(03), 475–497 (2001)
15. Biagi, S., Dipierro, S., Valdinoci, E., Vecchi, E.: Semilinear elliptic equations involving mixed local and nonlocal operators. Proc. Roy. Soc. Edinburgh Sect. A **151**(5), 1611–1641 (2021)
16. Biagi, S., Dipierro, S., Valdinoci, E., Vecchi, E.: Mixed local and nonlocal elliptic operators: regularity and maximum principles. Commun. Partial Differ. Equ. **47**(3), 585–629 (2022)
17. Biswas, A., Modasiya, M., Sen, A.: Boundary regularity of mixed local-nonlocal operators and its application. Ann. Mat. Pura Appl. **202**(2), 679–710 (2023)
18. Bonfiglioli, A., Lanconelli, E., Uguzzoni, F.: Stratified Lie Groups and Potential Theory for Their Sub-Laplacians. Springer, Berlin (2007)
19. Bouchut, F.: Hypoelliptic regularity in kinetic equations. J. Math. Pures Appl. **81**(11), 1135–1159 (2002)
20. Brezis, H.: Functional Analysis, Sobolev Spaces and Partial Differential Equations. Universitext. Springer, New York (2011)
21. Bucur, C., Valdinoci, E.: Nonlocal Diffusion and Applications. Lecture Notes of the Unione Matematica Italiana, vol. 20. Springer, Berlin (2015)
22. Byun, S.S., Ok, J., Song, K.: Hölder regularity for weak solutions to nonlocal double phase problems. J. Math. Pures Appl. **168**, 110–142 (2022)
23. Byun, S.S., Lee, H.S., Song, K.: Regularity results for mixed local and nonlocal double phase functionals (2023). arXiv:2301.06234
24. Chaker, J., Minhyun, K., Weidner, M.: Regularity for nonlocal problems with non-standard growth. Calc. Var. Partial Differ. Equ. **61**(6), 227 (2022)

25. Chaker, J., Minhyun, K., Weidner, M.: Harnack inequality for nonlocal problems with non-standard growth. Math. Ann. 386, 533–550 (2023)
26. Chen, Z.Q., Zhang, X.: L^p-maximal hypoelliptic regularity of nonlocal kinetic Fokker-Planck operators. J. Math. Pures Appl. **116**(6), 52–87 (2018)
27. Cinti, C., Pascucci, A., Polidoro, S.: Pointwise estimates for a class of non-homogeneous Kolmogorov equations. Math. Ann. **340**(2), 237–264 (2008)
28. Di Castro, A., Kuusi, T., Palatucci, G.: Nonlocal Harnack inequalities. J. Funct. Anal. **267**(6), 1807–1836 (2014)
29. Di Castro, A., Kuusi, T., Palatucci, G.: Local behavior of fractional p-minimizers. Ann. Inst. H. Poincaré Anal. Non Linéaire **33**, 1279–1299 (2016)
30. Di Francesco, M., Pascucci, A., Polidoro, S.: The obstacle problem for a class of hypoelliptic ultraparabolic equations. Proc. R. Soc. Lond. Ser. A Math. Phys. Eng. Sci. **464**, 155–176 (2008)
31. Ding, M., Zhang, C., Zhou, S.: Local boundedness and Hölder continuity for the parabolic fractional p-Laplace equations. Calc. Var. Partial Differ. Equ. **60**(1), 1–21 (2021)
32. De Filippis, C., Mingione, G.: Gradient regularity in mixed local and nonlocal problems. Math. Ann. **388**, 261–328 (2024)
33. De Filippis, C., Palatucci, G.: Hölder regularity for nonlocal double phase equations. J. Differ. Equ. **267**(1), 547–586 (2019)
34. Frentz, M., Nyström, K., Pascucci, A., Polidoro, S.: Optimal regularity in the obstacle problem for Kolmogorov operators related to American Asian options. Math. Ann. **347**(4), 805–838 (2010)
35. Garain, P., Kinnunen, J.: On the regularity theory for mixed local and nonlocal quasilinear elliptic equations. Trans. Am. Math. Soc. **375**(8), 5393–5423 (2022)
36. Garofalo, N., Tralli, G.: A class of nonlocal hypoelliptic operators and their extensions. Ind. Univ. Math. J. **70**(5), 1717–1744 (2021)
37. Garofalo, N., Loiudice, A., Vassylev, D.: Optimal decay for solutions of nonlocal semilinear equations with critical exponent in homogeneous group (2022). arXiv:2210.16893
38. Golse, F., Imbert, C., Mouhot, C., Vasseur, A.F.: Harnack inequality for kinetic Fokker-Planck equations with rough coefficients and application to the Landau equation. Ann. Sc. Norm. Super. Pisa, Cl. Sci. **19**(1), 253–295 (2019)
39. Guerand, J., Imbert, C.: Log-transform and the weak Harnack inequality for kinetic Fokker-Planck equations. J. Inst. Math. Jussieu **2**, 1–26 (2022)
40. Guerand, J., Mouhot, C.: Quantitative De Giorgi methods in kinetic theory. J. Éc. Polytech. Math. **9**, 1159–1181 (2022)
41. Hobson, D.G., Rogers, L.C.G.: Complete models with stochastic volatility. Math. Finance **8**, 27–48 (1998)
42. Hörmander, L.: Hypoelliptic second order differential equations. Acta Math. **119**, 147–171 (1967)
43. Huang, L., Menozzi, S., Priola, E.: L^p estimates for degenerate non-local Kolmogorov operators. J. Math. Pures Appl. **9**(121), 162–215 (2019)
44. Imbert, C.: A non-local regularization of first order Hamilton-Jacobi equations. J. Differ. Equ. **211**(1), 218–246 (2005)
45. Imbert, C., Silvestre, L: The weak Harnack inequality for the Boltzmann equation without cut-off. J. Eur. Math. Soc. **22**(2), 507–592 (2020)
46. Imbert, C., Silvestre, L.: Global regularity estimates for the Boltzmann equation without cut-off. J. Am. Math. Soc. **3**, 625–703 (2022)
47. Kassmann, M.: The classical Harnack inequality fails for nonlocal operators (2007). https://citeseerx.ist.psu.edu/viewdoc/download?doi=10.1.1.454.223
48. Kassmann, M.: Harnack inequalities and Hölder regularity estimates for nonlocal operator revisited (2011). https://sfb701.math.uni-bielefeld.de/preprints/sfb11015.pdfhttps://sfb701.math.uni-bielefeld.de/preprints/sfb11015.pdf
49. Kassmann, M., Weidner, M.: The parabolic Harnack inequality for nonlocal equations. Duke Math. J., to appear
50. Kim, Y.C.: Nonlocal Harnack inequalities for nonlocal heat equations. J. Differ. Equ. **267**(11), 6691–6757 (2019)

51. Kinderlehrer, D., Stampacchia, G.: An Introduction to Variational Inequalities and Their Applications. Pure and Applied Mathematics, vol. 88. Academic Press, New York (1980). (A Subsidiary of Harcourt Brace Jovanovich, Publishers). XIV, 313
52. Kruzhkov, S.N.: A priori bounds and some properties of solutions of elliptic and parabolic equations. Mat. Sb. **65**(107), 522–570 (1964)
53. Lanconelli, E., Polidoro, S.: On a class of hypoelliptic evolution operators. Rend. Semin. Mat. Torino **52**(1), 29–63 (1994)
54. Lanconelli, A., Pascucci, A., Polidoro, S.: Gaussian lower bounds for non-homogeneous Kolmogorov equations with measurable coefficients. J. Evol. Equ. **20**(4), 1399–1417 (2020)
55. Li, W.X.: Global hypoelliptic estimates for fractional order kinetic equation. Math. Nachr. **287**(5–6), 610–637 (2014)
56. Liao, N.: Hölder regularity for parabolic fractional p-Laplacian. Calc. Var. **63**, 22 (2024)
57. Litsgard, M., Nyström, K.: The Dirichlet problem for Kolmogorov-Fokker-Planck type equations with rough coefficients. J. Funct. Anal. **281**(10), 1–39 (2021)
58. Loher, A.: Quantitative De Giorgi methods in kinetic theory for non-local operators. J. Funct. Anal. **286**(7) (2024)
59. Manfredini, M.: The Dirichlet problem for a class of ultraparabolic equations. Adv. Differ. Equ. **2**(5), 831–866 (1997)
60. Manfredini, M., Palatucci, G., Piccinini, M., Polidoro, S.: Hölder continuity and boundedness estimates for nonlinear fractional equations in the Heisenberg group. J. Geom. Anal. **33**, 77 (2023)
61. Mellet, A.: Fractional diffusion limit for collisional kinetic equations: a moments method. Ind. Univ. Math. J. **59**, 1333–1360 (2010)
62. Mellet, A., Mischler, S., Mouhot, C.: Fractional diffusion limit for collisional kinetic equations. Arch. Ration. Mech. Anal. **199**, 493–525 (2011)
63. Moser, J.: A rapidly convergent iteration method and non-linear differential equations. Ann. Sc. Norm. Super. Pisa Sci. Fis. Mat. III. Ser. **20**, 265–315, 499–535 (1966)
64. Nyström, K., Pascucci, A., Polidoro, S.: Regularity near the initial state in the obstacle problem for a class of hypoelliptic ultraparabolic operators. J. Differ. Equ. **249**, 2044–2060 (2010)
65. Oza, P., Tyagi, J.: Regularity of solutions to variable-exponent degenerate mixed fully nonlinear local and nonlocal equations (2023). arXiv:2302.06046
66. Pagliarani, S., Pascucci, A., Pignotti, M.: Intrinsic Taylor formula for Kolmogorov-type homogeneous groups. J. Math. Anal. Appl. **435**, 1054–1087 (2016)
67. Palatucci, G.: The Dirichlet problem for the p-fractional Laplace equation. Nonlinear Anal. **177**, 699–732 (2018)
68. Palatucci, G., Piccinini, M.: Nonlocal Harnack inequalities in the Heisenberg group. Calc. Var. Partial Differ. Equ. **61**, 185 (2022)
69. Pascucci, A.: Free boundary and optimal stopping problems for American Asian options. Finance Stoch. **12**, 21–41 (2008)
70. Pascucci, A.: PDE and Martingale Methods in Option Pricing. Bocconi & Springer Series 2, vol. xvii, 719pp. Springer, Milano (2011); Bocconi University Press, Milano. ISBN 978-88-470-1780-1/hbk; 978-88-470-1781-8/ebook
71. Pascucci, A., Pesce, A.: Sobolev embeddings for kinetic Fokker-Planck equations. J. Funct. Anal. **286**(7) (2024)
72. Pascucci, A., Polidoro, S.: The Moser's iterative method for a class of ultraparabolic equations. Commun. Contemp. Math. **6**(3), 395–417 (2004)
73. Pesce, A.: Approximation and Interpolation in Kolmogorov-type groups. J. Math. Anal. Appl. **535**(2) (2024)
74. Piccinini, M.: The obstacle problem and the Perron Method for nonlinear fractional equations in the Heisenberg group. Nonlinear Anal. **222**, 112966 (2022)
75. Prasad, H., Tewary, V.: Local boundedness of variational solutions to nonlocal double phase parabolic equations. J. Differ. Equ. **351**(5) (2023)
76. Shang, B., Zhang, C.: Harnack inequality for mixed local and nonlocal parabolic p-laplace equations. J. Geom. Anal. **33**(3), 124 (2023)

77. Silvestre, L.: Regularity estimates and open problems in kinetic equations. The IMA Volumes in Mathematics and its Applications (IMA, Volume 165) (2023)
78. Silvestre, L.: Hölder estimates for kinetic Fokker-Planck equations up to the boundary. ARS Inveniendi Analytica: Mathematics of Fluids, Gases and Plasmas **35** (2022)
79. Stokols, L.F.: Hölder continuity for a family of nonlocal hypoelliptic kinetic equations. SIAM J. Math. Anal. **51**(6), 4815–4847 (2019)
80. Strömqvist, M.: Local boundedness of solutions to non-local parabolic equations modeled on the fractional p-Laplacian. J. Differ. Equ. **266**(12), 7948–7979 (2019)
81. Strömqvist, M.: Harnack's inequality for parabolic nonlocal equations. Ann. Inst. H. Poincaré C Anal. Non Linéaire **36**(6), 1709–1745 (2019)
82. Wang, M., Duan, J.: Existence and regularity of a linear nonlocal Fokker-Planck equation with growing drift. J. Math. Anal. Appl. **449**(1), 228–243 (2017)
83. Wang, W., Zhang, L.: The C^α regularity of a class of non-homogeneous ultraparabolic equations. Sci. China Ser. A **52**(8), 1589–1606 (2009)
84. Wang, W., Zhang, L.: The C^α regularity of weak solutions of ultraparabolic equations. Discrete Contin. Dyn. Syst. **29**(3), 1261–1275 (2011)
85. Wang, W., Zhang, L.: C^α regularity of weak solutions of non-homogenous ultraparabolic equations with drift terms. Sci. China Math. **67**, 23–44 (2024)
86. Zhu, Y.: Regularity of kinetic Fokker-Planck equations in bounded domains (2022). arXiv:2206.04536

Schauder Estimates for Kolmogorov-Fokker-Planck Operators with Coefficients Measurable in Time and Hölder Continuous in Space

Stefano Biagi and Marco Bramanti

Abstract We consider degenerate Kolmogorov-Fokker-Planck operators

$$\mathcal{L}u = \sum_{i,j=1}^{q} a_{ij}(x,t)\partial^2_{x_i x_j} u + \sum_{k,j=1}^{N} b_{jk} x_k \partial_{x_j} u - \partial_t u, \ (x,t) \in \mathbb{R}^{N+1}, N \geq q \geq 1$$

such that the corresponding model operator having constant a_{ij} is hypoelliptic, translation invariant w.r.t. a Lie group operation in $\mathbb{R}^{N+1}$ and 2-homogeneous w.r.t. a family of nonisotropic dilations. The coefficients a_{ij} are bounded and Hölder continuous in space (w.r.t. some homogeneous norm $\|\cdot\|$ induced by $\mathcal{L}$ in $\mathbb{R}^N$) and only bounded measurable in time; the matrix $\{a_{ij}\}_{i,j=1}^q$ is symmetric and uniformly positive definite on $\mathbb{R}^q$. We prove "partial Schauder a priori estimates" of the kind

$$\sum_{i,j=1}^{q} \|\partial^2_{x_i x_j} u\|_{C^\alpha_x(S_T)} + \|Yu\|_{C^\alpha_x(S_T)} \leq c\left\{\|\mathcal{L}u\|_{C^\alpha_x(S_T)} + \|u\|_{C^0(S_T)}\right\}$$

for suitable functions u, where

$$\|f\|_{C^\alpha_x(S_T)} = \sup_{t \leq T} \sup_{x_1,x_2 \in \mathbb{R}^N, x_1 \neq x_2} \frac{|f(x_1,t)-f(x_2,t)|}{\|x_1-x_2\|^\alpha} + \|f\|_{L^\infty(S_T)}.$$

We also prove that the derivatives $\partial^2_{x_i x_j} u$ are locally Hölder continuous in space and time while $\partial_{x_i} u$ and u are globally Hölder continuous in space and time.

S. Biagi · M. Bramanti (✉)
Dipartimento di Matematica, Politecnico di Milano, Milano, Italy
e-mail: stefano.biagi@polimi.it; marco.bramanti@polimi.it

S. Menozzi et al. (eds.), *Kolmogorov Operators and Their Applications*, Springer INdAM Series 56, https://doi.org/10.1007/978-981-97-0225-1_4

1 Introduction and Main Results

In this note[1] we deal with Kolmogorov-Fokker-Planck (KFP, in short) operators of the form

$$\mathcal{L}u = \sum_{i,j=1}^{q} a_{ij}(x,t)\partial^2_{x_i x_j}u + \sum_{k,j=1}^{N} b_{jk}x_k\partial_{x_j}u - \partial_t u, \qquad (x,t) \in \mathbb{R}^{N+1} \tag{1.1}$$

where $N \geq q \geq 1$. The first-order part of the operator, also called *the drift term*, will be briefly denoted by

$$Yu = \sum_{k,j=1}^{N} b_{jk}x_k\partial_{x_j}u - \partial_t u. \tag{1.2}$$

We will make the following assumptions:

(H1) $A_0(x,t) = (a_{ij}(x,t))_{i,j=1}^{q}$ is a symmetric uniformly positive definite matrix on $\mathbb{R}^q$ of bounded coefficients defined in $\mathbb{R}^{N+1}$, so that

$$\nu|\xi|^2 \leq \sum_{i,j=1}^{q} a_{ij}(x,t)\xi_i\xi_j \leq \nu^{-1}|\xi|^2 \tag{1.3}$$

for some constant $\nu > 0$, every $\xi \in \mathbb{R}^q$, every $x \in \mathbb{R}^N$ and a.e. $t \in \mathbb{R}$. The regularity assumptions on the coefficients will be specified later (see Assumption (H3)).

(H2) The matrix $B = (b_{ij})_{i,j=1}^{N}$ satisfies the following condition: for $m_0 = q$ and suitable positive integers $m_1, \ldots, m_k$ such that

$$m_0 \geq m_1 \geq \ldots \geq m_k \geq 1 \quad \text{and} \quad m_0 + m_1 + \ldots + m_k = N, \tag{1.4}$$

we have

$$B = \begin{pmatrix} \mathbb{O} & \mathbb{O} & \ldots & \mathbb{O} & \mathbb{O} \\ B_1 & \mathbb{O} & \ldots & \ldots & \ldots \\ \mathbb{O} & B_2 & \ldots & \mathbb{O} & \mathbb{O} \\ \vdots & \vdots & \ddots & \vdots & \vdots \\ \mathbb{O} & \mathbb{O} & \ldots & B_k & \mathbb{O} \end{pmatrix} \tag{1.5}$$

[1] The present note contains the announcements of results proved in [1] and reflects the conference held by Bramanti at the INdAM Meeting "Kolmogorov Operators and their Applications"–Cortona—13–17 June 2022, to which this volume is devoted.

where every block B_j is an $m_j \times m_{j-1}$ matrix of rank m_j (for $j = 1, 2, \ldots, k$).

We explicitly note that, when $q < N$, the operator $\mathcal{L}$ is *ultraparabolic*; nevertheless, under assumptions (H1)–(H2), as proved by Lanconelli and Polidoro [19], the model operator $\overline{\mathcal{L}}$ corresponding to the case of *constant* a_{ij} is hypoelliptic and satisfies the following properties:

(a) $\overline{\mathcal{L}}$ is left-invariant on the *non-commutative Lie group* $\mathbb{G} = (\mathbb{R}^{N+1}, \circ)$, where

$$(y, s) \circ (x, t) = (x + E(t)y, t + s)$$
$$(y, s)^{-1} = (-E(-s)y, -s),$$

and $E(t) = \exp(-tB)$. Note that

$$(y, s)^{-1} \circ (x, t) = (x - E(t - s)y, t - s), \tag{1.6}$$

and that the Lebesgue measure is the Haar measure, which is also invariant with respect to the inversion.

(b) $\overline{\mathcal{L}}$ is homogeneous of degree 2 with respect to a nonisotropic family of *dilations* in $\mathbb{R}^{N+1}$, which are automorphisms of $\mathbb{G}$ and are defined by

$$D(\lambda)(x, t) \equiv (D_0(\lambda)(x), \lambda^2 t) = (\lambda^{q_1} x_1, \ldots, \lambda^{q_N} x_N, \lambda^2 t), \tag{1.7}$$

where the N-tuple $(q_1, \ldots, q_N)$ is given by

$$(q_1, \ldots, q_N) = (\underbrace{1, \ldots, 1}_{m_0}, \underbrace{3, \ldots, 3}_{m_1}, \ldots, \underbrace{2k+1, \ldots, 2k+1}_{m_k}).$$

The integer

$$Q = \textstyle\sum_{i=1}^{N} q_i > N \tag{1.8}$$

is called the *homogeneous dimension* of $\mathbb{R}^N$, while $Q + 2$ is the homogeneous dimension of $\mathbb{R}^{N+1}$.

Also, an explicit fundamental solution is exhibited for the operator $\overline{\mathcal{L}}$ in [19].

A general idea is that these properties of the model operator $\overline{\mathcal{L}}$ allow to prove that some good regularity properties still hold for the operator $\mathcal{L}$ with variable coefficients, under suitable regularity assumptions on the coefficients $a_{ij}(x, t)$. For instance, Schauder estimates have been investigated by Di Francesco-Polidoro in [9] on bounded domains, assuming the coefficients a_{ij} Hölder continuous with respect to the intrinsic distance induced in $\mathbb{R}^{N+1}$ by the vector fields $\partial_{x_1}, \ldots \partial_{x_q}, Y$. We point out also the papers by Lunardi [24], Priola [29], Imbert-Mouhot [15], Wang-Zhang [30], and the references therein, on related issues about Schauder estimates for KFP operators. Also, continuity estimates on $u_{x_i x_j}$ under a Dini continuity assumption on a_{ij} and $\mathcal{L}u$ have been proved by Polidoro, Rebucci, Stroffolini in [28].

Recent researches, especially in the field of stochastic differential equations (see e.g. [27]), which are the main motivation to study KFP operators, suggest the importance of developing a theory allowing the coefficients a_{ij} to be rough in t (say, L^∞), and Hölder continuous (in a suitable sense) only w.r.t. the space variables. The Schauder estimates that one can reasonably expect under this mild assumption consist in controlling the Hölder seminorms w.r.t. x of the derivatives involved in the equation, uniformly in time (the precise result will be stated later). These estimates are sometimes called "partial Schauder estimates". Similar results can be expected when Hölder continuity is replaced by Dini continuity. Results of this kind (in the Hölder case) are well-known for uniformly parabolic operators (see [6, 16], and more recent papers quoted in the references in [1]). Also, in the parabolic case, it is known that $\partial^2_{x_i x_j} u$ satisfies a continuity estimate *in time*, under the same assumptions of continuity in space of a_{ij} and $\mathcal{L}u$.

The aim of this note is to announce and briefly describe some results of this kind which have been recently proved by the Authors in [1]. In the forthcoming paper [2] similar results will be extended to the Dini continuity setting. Our main results will be stated in Theorem 1.4. A key feature of our results is that, like in the parabolic case, we are able to prove a time continuity estimate of $u_{x_i x_j}$ only assuming a suitable continuity property *in space* of a_{ij} and $\mathcal{L}u$.

Partial Schauder estimates for degenerate KFP operators have been proved also in the recent paper [7] by Chaudru de Raynal, Honoré, Menozzi, with different techniques and without getting the Hölder control in time of second order derivatives. We also quote some other papers on KFP operators with coefficients Hölder continuous in space and L^∞ in time: [22], by Lucertini, Pagliarani, Pascucci, dealing with the construction of a fundamental solution for these operators, with consequent results about the Cauchy problem; [12], by Henderson and Wang, containing partial Schauder estimates for a special class of KFP operators, with applications to the Landau equation; [11], by Dong and Yastrzhembskiy, proving partial Schauder estimates for KFP operators with kinetic drift. The results in [11, 12, 22] are independent from and do not contain our results. We also quote the more recent preprint [23], again by Lucertini, Pagliarani, Pascucci, which extends partial Schauder estimates to operators such that $\overline{\mathcal{L}}$ is not dilation invariant.

In order to state precisely our results, we need some preliminaries.

Points of $\mathbb{R}^{N+1}$ will be sometimes denoted by the compact notation

$$\xi = (x, t), \quad \eta = (y, s).$$

The vector fields

$$X_1 = \partial_{x_1}, \ldots, X_q = \partial_q, X_0 = Y$$

form a system of Hörmander vector fields in $\mathbb{R}^{N+1}$, left-invariant w.r.t. $\circ$; moreover, $X_i = \partial_{x_i}$ $(i = 1, \dots, q)$ are homogeneous of degree 1, while $X_0 = Y$ is homogeneous of degree 2 w.r.t. the dilations $D(\lambda)$. As every set of Hörmander vector fields with drift, this system induces a (weighted) *control distance* $d_{\mathbf{X}}(\xi, \eta)$ in $\mathbb{R}^{N+1}$. (For this and other basic notions which are not explicitly recalled here, see [4, Chaps. 1–3]).

The function

$$\rho(\xi) = \rho(x, t) := \|x\| + \sqrt{|t|} = \sum_{i=1}^{N} |x_i|^{1/q_i} + \sqrt{|t|} \tag{1.9}$$

is a *homogeneous norm* in $\mathbb{R}^{N+1}$ and the function

$$d(\xi, \eta) := \rho(\eta^{-1} \circ \xi)$$

is a left-invariant, 1-homogeneous *quasi-distance* on $\mathbb{R}^{N+1}$, which can be proved to be *globally equivalent* to the control distance $d_{\mathbf{X}}(\xi, \eta)$. Throughout the following we will systematically use this quasi-distance d and the associated balls

$$B_r(\xi) := \left\{\eta \in \mathbb{R}^{N+1} : d(\eta, \xi) < r\right\} \qquad (\text{for } \xi \in \mathbb{R}^{N+1} \text{ and } r > 0).$$

Saying that d is a quasi-distance means that there exists a positive constant κ such that

$$d(\xi, \eta) \le \kappa\big(d(\xi, \zeta) + d(\eta, \zeta)\big) \qquad \forall\, \xi, \eta, \zeta \in \mathbb{R}^{N+1}; \tag{1.10}$$

$$d(\xi, \eta) \le \kappa\, d(\eta, \xi) \qquad \forall\, \xi, \eta \in \mathbb{R}^{N+1}. \tag{1.11}$$

Note that d has the following explicit expression

$$d((x, t), (y, s)) = \|x - E(t - s)y\| + \sqrt{|t - s|}, \tag{1.12}$$

which, since $E(0) = \mathbb{O}$, implies

$$d((x, t), (y, t)) = \|x - y\| \qquad \text{for every } x, y \in \mathbb{R}^N \text{ and } t \in \mathbb{R}, \tag{1.13}$$

that is the quasi-distance d *is symmetric when applied to points with the same t-coordinate.*

Note that, instead,

$$d((x, t), (x, s)) = \|x - E(t - s)x\| + \sqrt{|t - s|} \neq \sqrt{|t - s|}.$$

Also, note that

$$|B_r(\xi)| = |B_r(0)| = \omega r^{Q+2} \tag{1.14}$$

where $\omega := |B_1(0)| > 0$.

The quasi-distance d allows us to define the Hölder spaces which will be used in the paper. We are interested both in Hölder norms which measure the *joint continuity in* (x, t) and in Hölder norms which measure the *continuity in x alone, for fixed t.*

Definition 1.1 For $-\infty \le \tau < T \le +\infty$, let $\Omega = \mathbb{R}^N \times (\tau, T)$, and let $f : \Omega \to \mathbb{R}$. Given any number $\alpha \in (0, 1)$, we introduce the notation:

$$|f|_{C^\alpha(\Omega)} = \sup\left\{\frac{|f(\xi) - f(\eta)|}{d(\xi,\eta)^\alpha} : \xi, \eta \in \Omega \text{ and } \xi \ne \eta\right\}$$

$$\begin{aligned}|f|_{C_x^\alpha(\Omega)} &= \underset{t\in(\tau,T)}{\text{supess}} \sup\left\{\frac{|f(x,t) - f(y,t)|}{d((x,t),(y,t))^\alpha} : x, y \in \mathbb{R}^N, x \ne y\right\} \\ &= \underset{t\in(\tau,T)}{\text{supess}} \sup\left\{\frac{|f(x,t) - f(y,t)|}{\|x-y\|^\alpha} : x, y \in \mathbb{R}^N, x \ne y\right\}\end{aligned}$$

(where the last equality holds by (1.13)). Accordingly, we define the spaces $C^\alpha(\Omega)$ and $C_x^\alpha(\Omega)$ as follows:

$$C^\alpha(\Omega) := \left\{f \in C(\Omega) \cap L^\infty(\Omega) : |f|_{C^\alpha(\Omega)} < \infty\right\} \tag{1.15}$$

$$C_x^\alpha(\Omega) := \left\{f \in L^\infty(\Omega) : |f|_{C_x^\alpha(\Omega)} < \infty\right\} \tag{1.16}$$

and endow these spaces with the following norms

$$\begin{aligned}\|f\|_{C^\alpha(\Omega)} &:= \|f\|_{L^\infty(\Omega)} + |f|_{C^\alpha(\Omega)} \\ \|f\|_{C_x^\alpha(\Omega)} &:= \|f\|_{L^\infty(\Omega)} + |f|_{C_x^\alpha(\Omega)}.\end{aligned}$$

We can now make precise our regularity assumption on the coefficients a_{ij}.

(H3) There exists $\alpha \in (0, 1)$ such that

$$a_{ij} \in C_x^\alpha\left(\mathbb{R}^{N+1}\right) \qquad \text{for every } 1 \le i, j \le q.$$

In all the estimates appearing in next sections, the number

$$\Lambda = \max_{i,j=1,\dots,q} \|a_{ij}\|_{C_x^\alpha(\mathbb{R}^{N+1})}, \tag{1.17}$$

together with the ellipticity constant ν in (1.3), will quantify the dependence of the constants on the coefficients a_{ij}.

We now turn to define the functions spaces to which our *solution* u will belong.

Definition 1.2 Throughout the following, given $T \in \mathbb{R}$ we set

$$S_T := \mathbb{R}^N \times (-\infty, T).$$

We then define $\mathcal{S}^0(S_T)$ as the space of all functions $u : \overline{S}_T \to \mathbb{R}$ such that

(i) $u \in C(\overline{S_T}) \cap L^\infty(S_T)$;
(ii) for every $1 \le i, j \le q$, the *distributional derivatives* $\partial_{x_i} u, \partial^2_{x_i x_j} u \in L^\infty(S_T)$;
(iii) the *distributional derivative* $Yu \in L^\infty(S_T)$.

Moreover, given any number $\alpha \in (0, 1)$, we define

$$\mathcal{S}^\alpha(S_T) := \{u \in \mathcal{S}^0(S_T) : \partial_{x_i} u, \partial_{x_i x_j} u, \ Yu \in C^\alpha_x(S_T) \text{ for } 1 \le i, j \le q\}.$$

Finally, given any $\tau \in \mathbb{R}$ with $\tau < T$, we define

$$\mathcal{S}^0(\tau; T) = \{u \in \mathcal{S}^0(S_T) : u(x, t) = 0 \text{ for every } t \le \tau\},$$
$$\mathcal{S}^\alpha(\tau; T) := \mathcal{S}^\alpha(S_T) \cap \mathcal{S}^0(\tau; T).$$

Remark 1.3 On account of assumption (H3), we immediately obtain the following facts.

(1) If $u \in \mathcal{S}^0(S_T)$, then $\mathcal{L}u \in L^\infty(S_T)$.
(2) If $u \in \mathcal{S}^\alpha(S_T)$, then $\mathcal{L}u \in C^\alpha_x(S_T)$.
(3) If $u \in \mathcal{S}^0(S_T)$ and $\partial^2_{x_i x_j} u, \ \mathcal{L}u \in C^\alpha_x(S_T)$ $(1 \le i, j \le q)$, then $Yu \in C^\alpha_x(S_T)$.

The following further regularity properties are not obvious, but are actually proved in [1]:

(4) if $u \in \mathcal{S}^0(S_T)$, then u and $\partial_{x_1} u, \dots, \partial_{x_q} u$ are locally Hölder-continuous in the joint variables (see, precisely, Proposition 3.3);
(5) if $u \in \mathcal{S}^\alpha(S_T)$ for some $\alpha \in (0, 1)$, then the distributional derivatives $\partial^2_{x_i x_j} u$ (for $1 \le i, j \le q$) are actually continuous (and locally Hölder continuous in a weaker sense) on S_T (see Theorem 1.4).

As a consequence, every function $u \in \mathcal{S}^\alpha(S_T)$ actually has *classical continuous derivatives* $\partial^2_{x_i} u, \partial^2_{x_i x_j} u$ (for $1 \le i, j \le q$); instead, the distributional derivative Yu is continuous in space for every fixed t, but it may be only L^∞ w.r.t. time.

We are finally in position to state the main results proved in [1]:

Theorem 1.4 (Schauder Estimates) *Let $\mathcal{L}$ be an operator as in* (1.1)*, and assume that* (H1), (H2), (H3) *hold, for some* $\alpha \in (0, 1)$.

Then, the following Schauder-type estimates hold true.

(1) *For every* $T > 0$ *there exists a constant* $c > 0$, *depending on* T, α, *the matrix* B *in* (1.5) *and the numbers* ν *and* Λ *in* (1.3)–(1.17), *respectively, such that*

$$\sum_{i,j=1}^{q} \|\partial^2_{x_i x_j} u\|_{C^\alpha_x(S_T)} + \|Yu\|_{C^\alpha_x(S_T)} + \sum_{i=1}^{q} \|\partial_{x_i} u\|_{C^\alpha(S_T)} + \|u\|_{C^\alpha(S_T)}$$
$$\le c\big(\|\mathcal{L}u\|_{C^\alpha_x(S_T)} + \|u\|_{C^0(S_T)}\big).$$

for every function $u \in \mathcal{S}^\alpha(S_T)$.

(2) *For every* $T > \tau > -\infty$ *and every compact set* $K \subset \mathbb{R}^N$ *there exists a constant* $c > 0$, *depending on* $K, \tau, T, \alpha, B, \nu, \Lambda$, *such that*

$$|\partial^2_{x_i x_j} u(\xi) - \partial^2_{x_i x_j} u(\eta)|$$
$$\le c\big(\|\mathcal{L}u\|_{C^\alpha_x(S_T)} + \|u\|_{C^\alpha(S_T)}\big)\big(d(\xi,\eta)^\alpha + |t-s|^{\alpha/q_N}\big)$$

for every $\xi = (x,t)$, $\eta = (y,s) \in K \times [\tau, T]$ *and every* $u \in \mathcal{S}^\alpha(S_T)$. *Here, the number* q_N *is the largest exponent in the dilations* $D(\lambda)$, *see* (1.7).

Remark 1.5

(i) As observed in Remark 1.3, the finiteness of the quantities

$$\textstyle\sum_{i=1}^{q} \|\partial_{x_i} u\|_{C^\alpha(S_T)}, \qquad \|u\|_{C^\alpha(S_T)},$$

as well as the finiteness of the space-time Hölder quotient in point (2) of the above theorem, are not obvious *a priori* for a function in $\mathcal{S}^\alpha(S_T)$, but can be actually proved.

(ii) While $\partial^2_{x_i x_j} u$ $(i, j = 1, 2, \dots, q)$ are locally Hölder continuous in space and time, note that a similar property cannot be assured, in general, for Yu. To see this, it is enough to consider an equation of the kind

$$\mathcal{L}u(x,t) = f(t)$$

with f bounded discontinuous function, and u independent of x.

(iii) Since, in the degenerate case, $q_N \ge 3$, the term $|t_1 - t_2|^{\alpha/q_N}$ in the right-hand side of (2.24) is larger than the 'expected'

$$|t_1 - t_2|^{\alpha/2},$$

(at least when $|t_1 - t_2| \le 1$). Also, the constant $c > 0$ depends on the fixed compact set $K \times [\tau, T] \subseteq S_T$. On the other hand, we observe that this mild t-continuity of $\partial^2_{x_i x_j} u$ is obtained *without any* t*-continuity assumption on* $\mathcal{L}u$. Moreover, from the proof of this result it is apparent that in the uniformly parabolic case ($B = 0$ and $q = N$) our argument would give exactly

$$|\partial^2_{x_i x_j} u(\xi) - \partial^2_{x_i x_j} u(\eta)| \le c\left\{\|\mathcal{L}u\|_{C^\alpha_x(S_T)} + \|u\|_{C^\alpha(S_T)}\right\}\left(|x-y| + |t-s|^{\alpha/2}\right).$$

The result in point 2 of Theorem 1.4 is therefore consistent with the classical result in [16] which holds for uniformly parabolic operators on bounded cylinders.

1.1 Structure of the Paper

We are now going to explain the strategy that we adopt to prove our a priori estimates. In the classical Schauder theory, the operator with variable coefficients is seen as a small local perturbation of the constant coefficient operator obtained freezing the coefficients a_{ij} at some $(\overline{x}, \overline{t})$. In the present situation, since the coefficients are not continuous in t, we can only see our operator as a small local perturbation of the operator with coefficients only depending on t, obtained freezing the $a_{ij}(\cdot, t)$ at some point $\overline{x}$. Therefore our model operator is the one with bounded measurable coefficients $a_{ij}(t)$:

$$\mathcal{L}u = \sum_{i,j=1}^{q} a_{ij}(t)\partial^2_{x_i x_j} u + \sum_{k,j=1}^{N} b_{jk} x_k \partial_{x_j} u - \partial_t u. \tag{1.18}$$

So, an important preliminary part of our job (see Sect. 2) is a careful study of the operator $\mathcal{L}$ with bounded measurable coefficients $a_{ij}(t)$. For this operator an explicit fundamental solution has been computed and studied by Bramanti and Polidoro in [5]. Here we need to establish representation formulas for u and $u_{x_i x_j}$ in terms of $\mathcal{L}u$, exploiting this fundamental solution. Then, by singular integral techniques, we can prove the desired a priori estimates for this model operator. In turn, implementing this singular integral technique requires sharp estimates on the space derivatives of every order of the fundamental solution, as well as the proof of suitable cancellation properties.

In the second part of the work (see Sect. 3) we then study the operator with coefficients $a_{ij}(x, t)$. Here we apply the classical "Korn's trick" of freezing the coefficients of $\mathcal{L}$, in our case only w.r.t. x, writing representation formulas for $u_{x_i x_j}$ and then regard the original operator as a small local perturbation of the frozen one. This allows to prove the desired a priori estimates for functions with small compact support. Removing this restriction requires the use of cutoff functions and interpolation inequalities for the derivatives of intermediate order.

2 Operators with Measurable Coefficients $a_{ij}(t)$

Throughout this section, we consider an operator $\mathcal{L}$ of the form (1.1) and satisfying (H1)–(H2), with bounded measurable coefficients a_{ij} only depending on t, that is,

$$\mathcal{L}u = \sum_{i,j=1}^{q} a_{ij}(t)\partial^2_{x_i x_j}u + \sum_{k,j=1}^{N} b_{jk}x_k\partial_{x_j}u - \partial_t u, \qquad (x,t) \in \mathbb{R}^{N+1}. \tag{2.1}$$

2.1 The Fundamental Solution of the Operator with Time Dependent Coefficients

In [5], an explicit fundamental solution for $\mathcal{L}$ is computed, and its properties are studied. The next theorem summarizes some results in [5] that we will need.

We point out that, since our assumption (H1) on the matrix B is stronger than the one made in [5] (here the model operator with constant a_{ij} is both left invariant and homogeneous, while in [5] it is only left invariant), here we specialize the formulas and results to our simpler situation.

Theorem 2.1 (Fundamental Solution for Operators with t-Variable Coefficients) *Under assumptions* (H1)–(H2) *above, let $C(t,s)$ be the $N \times N$ matrix defined as*

$$C(t,s) = \int_s^t E(t-\sigma) \cdot \begin{pmatrix} A_0(\sigma) & 0 \\ 0 & 0 \end{pmatrix} \cdot E(t-\sigma)^T \, d\sigma \quad (\text{with } t > s) \tag{2.2}$$

(*we recall that* $E(\sigma) = \exp(-\sigma B)$, *see* (1.5)). *Then, the matrix $C(t,s)$ is* symmetric and positive definite *for every $t > s$. Moreover, if we define*

$$\Gamma(x,t;y,s) = \frac{1}{(4\pi)^{N/2}\sqrt{\det C(t,s)}} e^{-\frac{1}{4}\langle C(t,s)^{-1}(x-E(t-s)y),\, x-E(t-s)y\rangle} \cdot \mathbf{1}_{\{t>s\}} \tag{2.3}$$

(*where $\mathbf{1}_A$ denotes the indicator function of a set A*), *then Γ enjoys the following properties, so that Γ is the* fundamental solution *for $\mathcal{L}$ with pole at (y,s).*

(1) *In the open set $\mathcal{O} := \{(x,t;y,s) \in \mathbb{R}^{2N+2} : (x,t) \neq (y,s)\}$, the function Γ is* jointly continuous *in $(x,t;y,s)$ and C^∞ with respect to x, y. Moreover, for every multi-indexes α, β the functions*

$$\partial_x^\alpha \partial_y^\beta \Gamma = \frac{\partial^{\alpha+\beta}\Gamma}{\partial x^\alpha \partial y^\beta}$$

are jointly continuous *in* $(x, t; y, s) \in \mathcal{O}$. *Finally,* Γ *and* $\partial_x^\alpha \partial_y^\beta \Gamma$ *are* Lipschitz continuous *with respect to* t, s *in any region* $\mathcal{R}$ *of the form*

$$\mathcal{R} = \{(x, t; y, s) \in \mathbb{R}^{2N+2} : H \leq s + \delta \leq t \leq K\},$$

where $H, K \in \mathbb{R}$ *and* $\delta > 0$ *are arbitrarily fixed.*

(2) *For every fixed* $y \in \mathbb{R}^N$ *and* $t > s$, *we have*

$$\lim_{|x| \to +\infty} \Gamma(x, t; y, s) = 0.$$

(3) *For every fixed* $(y, s) \in \mathbb{R}^{N+1}$, *we have*

$$\mathcal{L}\Gamma(\cdot; y, s)(x, t) = 0 \qquad \text{for every } x \in \mathbb{R}^N \text{ and a.e. } t > s.$$

(4) *For every fixed* $x \in \mathbb{R}^N$ *and every* $t > s$, *we have*

$$\int_{\mathbb{R}^N} \Gamma(x, t; y, s)\, dy = 1. \tag{2.4}$$

(5) *For every* $f \in C(\mathbb{R}^N) \cap L^\infty(\mathbb{R}^N)$ *and every* $s \in \mathbb{R}$, *the function*

$$u(x, t) = \int_{\mathbb{R}^N} \Gamma(x, t; y, s) f(y)\, dy$$

is the unique solution to the Cauchy problem

$$\begin{cases} \mathcal{L}u = 0 & \text{in } \mathbb{R}^N \times (s, \infty) \\ u(\cdot, s) = f \end{cases} \tag{2.5}$$

In particular, $u(\cdot, s) \to f$ *uniformly in* $\mathbb{R}^N$ *as* $t \to s^+$.

Finally, the function $\Gamma^*(x, t; y, s) := \Gamma(y, s; x, t)$ *satisfies dual properties of* (2)–(4) *with respect to the* formal adjoint *of* $\mathcal{L}$, *that is,*

$$\mathcal{L}^* = \textstyle\sum_{i,j=1}^q a_{ij}(s)\partial_{y_i y_j} - \sum_{k,j=1}^N b_{jk} y_k \partial_{y_i} + \partial_s,$$

and thus Γ^* *is the fundamental solution of* $\mathcal{L}^*$.

Let us now further specialize our class of operators to the model operators with *constant coefficients* a_{ij}. Keeping our assumption (H2) on the matrix B, let

$$\mathcal{L}_\alpha u = \alpha \sum_{i=1}^q \partial_{x_i x_i}^2 u + \sum_{k,j=1}^N b_{jk} x_k \partial_{x_j} u - \partial_t u \tag{2.6}$$

for some $\alpha > 0$. Then the results of the above theorem apply in a simpler form. Actually, the following facts are proved already in [19].

Theorem 2.2 (Fundamental Solution for Operators with Constant Coefficients) *Let $\alpha > 0$ be fixed, and let Γ_α be the fundamental solution of the operator $\mathcal{L}_\alpha$ in* (2.6)*, whose existence is guaranteed by Theorem 2.1. Then:*

(1) Γ_α *is a* kernel of convolution type, *that is,*

$$\begin{aligned}\Gamma_\alpha(x,t;y,s) &= \Gamma_\alpha\big(x - E(t-s)y, t-s; 0, 0\big) \\ &= \Gamma_\alpha\big((y,s)^{-1} \circ (x,t); 0, 0\big);\end{aligned} \tag{2.7}$$

(2) *The matrix $C(t,s)$ in* (2.2) *takes the simpler form*

$$C(t,s) = C_0(t-s), \tag{2.8}$$

where $C_0(\tau)$ is the $N \times N$ matrix defined as

$$C_0(\tau) = \alpha \int_0^\tau E(t-\sigma) \cdot \begin{pmatrix} I_q & 0 \\ 0 & 0 \end{pmatrix} \cdot E(t-\sigma)^T d\sigma \qquad (\tau > 0).$$

Furthermore, one has the 'homogeneity property'

$$C_0(\tau) = D_0(\sqrt{\tau}) C_0(1) D_0(\sqrt{\tau}) \qquad \forall\, \tau > 0. \tag{2.9}$$

In particular, by combining (2.3) *with* (2.8) *and* (2.9)*, we can write*

$$\begin{aligned}\Gamma_\alpha(x,t;0,0) &= \frac{1}{(4\pi\alpha)^{N/2}\sqrt{\det C_0(t)}} e^{-\frac{1}{4\alpha} x^T C_0(t)^{-1} x} \\ &= \frac{1}{(4\pi\alpha)^{N/2} t^{Q/2} \sqrt{\det C_0(1)}} e^{-\frac{1}{4\alpha}\langle C_0(1)^{-1}\left(D_0\left(\frac{1}{\sqrt{t}}\right)x\right),\, D_0\left(\frac{1}{\sqrt{t}}\right)x\rangle}.\end{aligned} \tag{2.10}$$

In [5, Thm.1.7], the next useful comparison result is proved.

Theorem 2.3 *Let Γ be as in Theorem 2.1, and let $\nu > 0$ be as in* (1.3)*. Then, for every $s, t \in \mathbb{R}$ with $s < t$, one has the following estimate*

$$\nu^N \Gamma_\nu(x,t;y,s) \le \Gamma(x,t;y,s) \le \frac{1}{\nu^N} \Gamma_{\nu^{-1}}(x,t;y,s), \tag{2.11}$$

where Γ_ν is the fundamental solution of the operator $\mathcal{L}_\nu$ in (2.6).

As already explained in the introduction, in order to prove representation formulas for $u_{x_i x_j}$ in terms of $\mathcal{L}u$ and to exploit them for proving a priori estimates for operators with coefficients a_{ij} only depending on t, we need to deepen the study of Γ started in [5] proving sharp upper bounds on Γ and its derivatives of every order. To state precisely the result we prove, we need to introduce some notation. For a multi-index

$$\boldsymbol{\alpha} = (\alpha_1, \ldots, \alpha_N) \in \mathbb{N}^N,$$

we define its *order*

$$\omega(\boldsymbol{\alpha}) := \sum_{i=1}^{N} \alpha_i q_i.$$

Also, for

$$\boldsymbol{\alpha} = (\boldsymbol{\alpha}_1, \boldsymbol{\alpha}_2)$$

with $\boldsymbol{\alpha}_1, \boldsymbol{\alpha}_2 \in \mathbb{N}^N$, we let

$$\omega(\boldsymbol{\alpha}) = \omega(\boldsymbol{\alpha}_1) + \omega(\boldsymbol{\alpha}_2).$$

Our sharp estimates are contained in the following:

Theorem 2.4 (See [1, Thm. 3.5]) *Let Γ be as in Theorem 2.1, and let $\nu > 0$ be as in* (1.3). *Moreover, let $\boldsymbol{\alpha} = (\boldsymbol{\alpha}_1, \boldsymbol{\alpha}_2) \in \mathbb{N}^{2N}$ be a fixed multi-index. Then, there exist $c = c(\nu, \boldsymbol{\alpha}) > 0$ and a constant $c_1 > 0$, independent of ν and $\boldsymbol{\alpha}$, such that*

$$\begin{aligned} \left| D^{\boldsymbol{\alpha}}_{(x,y)} \Gamma(\xi;\eta) \right| &= \left| D^{\boldsymbol{\alpha}_1}_x D^{\boldsymbol{\alpha}_2}_y \Gamma(\xi;\eta) \right| \\ &\leq \frac{c}{(t-s)^{\omega(\boldsymbol{\alpha})/2}} \Gamma_{c_1\nu^{-1}}(\xi;\eta) \\ &\leq \frac{c}{d(\xi,\eta)^{Q+\omega(\boldsymbol{\alpha})}} \end{aligned} \tag{2.12}$$

for every $\xi, \eta \in \mathbb{R}^N$ with $t \neq s$. The resulting inequality

$$\left| D^{\boldsymbol{\alpha}}_{(x,y)} \Gamma(\xi;\eta) \right| \leq \frac{c}{d(\xi,\eta)^{Q+\omega(\boldsymbol{\alpha})}}$$

actually holds for every $\xi, \eta \in \mathbb{R}^N$ with $\xi \neq \eta$.

The technique of the proof of the above theorem is the following. We "explicitly" compute (by suitable iterative formulas, and distinguishing several cases) the derivative $D^{\boldsymbol{\alpha}}_{(x,y)} \Gamma(\xi;\eta)$; then we rewrite the resulting expression in the form $\Gamma \cdot$ (something); after that, we exploit Thm. 2.3 (and some heavy computation) to get the desired uniform bound (first inequality in (2.12)). The second inequality in (2.12) is a more standard consequence of Gaussian-type estimates.

Remark 2.5 Let $(y, s) \in \mathbb{R}^{N+1}$ be fixed. Since we know from Theorem 2.1-(3) that $\mathcal{L}\Gamma(\cdot; y, s)(x, t) = 0$ for every $x \in \mathbb{R}^N$ and a.e. $t > s$, we can express

$$\partial_t (D^{\boldsymbol{\alpha}}_{(x,y)} \Gamma) \qquad \text{(for every } \boldsymbol{\alpha} \in \mathbb{N}^{2N})$$

as a combination of quantities that, by (2.12) and the exponential decay of the right-hand side of this inequality as $t \to s^+$ (with $x \neq y$), are locally essentially bounded in $\mathbb{R}^{N+1} \setminus \{(y, s)\}$. In particular, the same is true of $Y(D^{\boldsymbol{\alpha}}_{(x,y)} \Gamma)$.

A simple consequence of Theorem 2.4 and of (2.4) is the following:

Lemma 2.6 (See [1, Lemma 3.8]) *Let Γ be as in Theorem 2.4, and let $\boldsymbol{\alpha} = (\alpha_1, \ldots, \alpha_N) \in \mathbb{N}^N$ be a fixed non-zero multi-index. Then, we have*

$$\int_{\mathbb{R}^N} D^{\boldsymbol{\alpha}}_x \Gamma(x, t; y, s)\, dy = 0 \qquad \textit{for every } x \in \mathbb{R}^N \textit{ and every } s < t. \tag{2.13}$$

Combining the sharp estimates on the derivatives of the fundamental solution contained in Theorem 2.4 with a "Lagrange-type" theorem for vector fields, the following can also be proved:

Theorem 2.7 (Mean Value Inequality for Fractional and Singular Kernels) (See [1, Thm. 3.9]) *Let Γ be as in Theorem 2.1, and let $\eta = (y, s) \in \mathbb{R}^{N+1}$ be fixed. Moreover, let*

$$\boldsymbol{\alpha} = (\alpha_1, \ldots, \alpha_N)$$

be a fixed multi-index. Then, there exists a constant $c = c(\boldsymbol{\alpha}) > 0$ such that

$$|D^{\boldsymbol{\alpha}}_x \Gamma(\xi_1, \eta) - D^{\boldsymbol{\alpha}}_x \Gamma(\xi_2, \eta)| \leq c \frac{d(\xi_1, \xi_2)}{d(\xi_1, \eta)^{Q+\omega(\boldsymbol{\alpha})+1}}$$

for every $\xi_1 = (x_1, t_1), \xi_2 = (x_2, t_2) \in \mathbb{R}^{N+1}$ such that

$$d(\xi_1, \eta) \geq 4\kappa d(\xi_1, \xi_2) > 0,$$

where κ is the constant in (1.10) and (1.11).

The last important property of the singular kernel $\partial^2_{x_i x_j} \Gamma$ that we need to implement singular integral estimates is the following cancellation property:

Theorem 2.8 (Cancellation Property of the Singular Kernel) (See [1, Thm. 3.16]) *There exists a constant $c > 0$ such that, for every $1 \leq i, j \leq q$, one has* *the estimate*

$$I_{r,\tau}(x,t) := \int_\tau^t \left| \int_{\{y\in\mathbb{R}^N:\, d((x,t),(y,s))\geq r\}} \partial^2_{x_i x_j}\Gamma(x,t;y,s)\,dy \right| ds \leq c, \tag{2.14}$$

for every $x \in \mathbb{R}^N$, $\tau < t$ *and* $r > 0$.

The above cancellation inequality is nonstandard in its formulation, and has a quite delicate proof.

2.2 Representation Formulas for *u* and $\partial^2_{x_i x_j} u$ in terms of $\mathcal{L}u$

Here we are going to establish some representation formulas for u and for its derivatives in terms of $\mathcal{L}u$.

We start with the following proposition.

Proposition 2.9 (See [1, Prop. 3.10]) *Let* $T \in \mathbb{R}$ *be fixed, and let* $g : S_T \to \mathbb{R}$ *be* continuous and bounded. *For every* $\varepsilon > 0$, *we consider the function*

$$v_\varepsilon : S_T \to \mathbb{R}, \qquad v_\varepsilon(x,t) := \int_{\mathbb{R}^N} \Gamma(x,t;y,t-\varepsilon)\,g(y,t-\varepsilon)\,dy.$$

Then, $v_\varepsilon \to g$ pointwise in S_T *as* $\varepsilon \to 0^+$.

Thanks to Proposition 2.9, in [1] the next key result is proved. Throughout the sequel, when dealing with integral over strips we tacitly understand that

$$\int_{\mathbb{R}^N\times(a,b)} \cdots = -\int_{\mathbb{R}^N\times(b,a)} \{\cdots\} \quad \text{when } b < a.$$

Theorem 2.10 (See [1, Thm. 3.11]) *Let* $T \in \mathbb{R}$ *be fixed, and let* $\tau < T$. *Moreover, let* $u \in \mathcal{S}^0(\tau;T)$. *Then, we have the following* representation formula

$$u(x,t) = -\int_{\mathbb{R}^N\times(\tau,t)} \Gamma(x,t;y,s)\mathcal{L}u(y,s)\,dy\,ds, \tag{2.15}$$

for every point $(x,t) \in S_T$.

Starting from the representation formula (2.15), we easily obtain the following representation formula for the *first-order derivatives* of u.

Corollary 2.11 (See [1, Corollary 3.12]) *Let* $T \in \mathbb{R}$ *be fixed, and let* $\tau < T$. *Moreover, let* $u \in \mathcal{S}^0(\tau;T)$ *and let* $1 \leq i \leq q$ *be fixed. Then, we have the* *representation formula*

$$\partial_{x_i} u(x,t) = -\int_{\mathbb{R}^N \times (\tau,t)} \partial_{x_i} \Gamma(x,t;\cdot)\, \mathcal{L}u \, dy\, ds, \tag{2.16}$$

for every $(x,t) \in S_T$. *Moreover,*

$$\|\partial_{x_i} u\|_{L^\infty(S_T)} \leq c\, \|\mathcal{L}u\|_{L^\infty(S_T)} \cdot \sqrt{T-\tau}. \tag{2.17}$$

With the representation formula (2.16) at hand, the next step is to establish a representation formula for the *derivatives* $\partial_{x_i x_j} u$ of a function $u \in \mathcal{S}^0(\tau;T)$.

This is made possible by the following proposition, which relies on the sharp estimates on $\partial^2_{x_i x_j}\Gamma$ proved in Theorem 2.4 and the explicit expression of Γ_α:

Proposition 2.12 (See [1, Proposition 3.13]) *Let* $\alpha \in (0,1)$ *be fixed, and let* $1 \leq i, j \leq q$. *Then, there exists a constant* $c = c(\alpha) > 0$ *such that, for every* $x \in \mathbb{R}^N$ *and every* $\tau < t$, *one has*

$$\int_{\mathbb{R}^N \times (\tau,t)} |\partial^2_{x_i x_j}\Gamma(x,t;y,s)| \cdot \|E(s-t)x - y\|^\alpha \, dy\, ds \leq c(t-\tau)^{\alpha/2}. \tag{2.18}$$

As a consequence, we have

$$\begin{gathered}\int_{\mathbb{R}^N \times (t-\varepsilon,t)} |\partial^2_{x_i x_j}\Gamma(x,t;y,s)| \cdot \|E(s-t)x - y\|^\alpha \, dy\, ds \to 0 \\ \text{uniformly w.r.t. } (x,t) \in \mathbb{R}^{N+1} \text{ as } \varepsilon \to 0^+.\end{gathered} \tag{2.19}$$

With Proposition 2.12 at hand, we can now prove the following theorem.

Theorem 2.13 (See [1, Theorem 3.14]) *For* $T > \tau > -\infty$ *and* $\alpha \in (0,1)$, *let* $u \in \mathcal{S}^0(\tau;T)$ *be such that* $\mathcal{L}u \in C^\alpha_x(S_T)$. *Then, we have*

$$\partial^2_{x_i x_j} u(x,t) = \int_{\mathbb{R}^N \times (\tau,t)} \partial^2_{x_i x_j}\Gamma(x,t;y,s)\big[\mathcal{L}u(E(s-t)x,s) - \mathcal{L}u(y,s)\big]\, dy\, ds, \tag{2.20}$$

for every $(x,t) \in S_T$ *and every* $1 \leq i, j \leq q$.

Note that Proposition 2.12 immediately implies the convergence of the integral in (2.20) when $\mathcal{L}u \in C^\alpha_x(S_T)$. The proof of equality (2.20) starting from (2.16) and (2.13) also requires some approximation argument.

2.3 Schauder Estimates in Space

The next step in our strategy is the special case of Schauder a priori estimates which hold for operators with coefficients a_{ij} only depending on t. Namely, we can prove the following result:

Theorem 2.14 (Global Schauder Estimates in Space) (See [1, Theorem 3.15]) *Let $T > \tau > -\infty$ and $\alpha \in (0, 1)$. Then, there exists $c > 0$, only depending on $(T - \tau), \alpha, \nu, B$, such that*

$$\sum_{i,j=1}^{q} \|\partial^2_{x_i x_j} u\|_{C^\alpha_x(S_T)} \leq c\, |\mathcal{L}u|_{C^\alpha_x(S_T)} \tag{2.21}$$

$$\|Yu\|_{C^\alpha_x(S_T)} \leq c\, \|\mathcal{L}u\|_{C^\alpha_x(S_T)}, \tag{2.22}$$

for every $u \in \mathcal{S}^0(\tau; T)$ with $\mathcal{L}u \in C^\alpha_x(S_T)$.

This result, which will be later suitably generalized to operators with coefficients $a_{ij}(x, t)$, quite plainly follows by the representation formula in Theorem 2.13 and the next result:

Theorem 2.15 (Hölder Continuity of Singular Integrals) (See [1, Theorem 3.17]) *For $T > \tau > -\infty$ and $\alpha \in (0, 1)$, let us introduce the function space*

$$C^\alpha_x(\tau; T) := \{f \in C^\alpha_x(S_T) : \ f(x, t) = 0 \text{ for every } t \leq \tau\},$$

and define, on this space $C^\alpha_x(\tau; T)$, the linear operator

$$f \mapsto T_{ij} f(x, t) := \int_{\mathbb{R}^N \times (\tau, t)} \partial^2_{x_i x_j} \Gamma(x, t; y, s) \left[f(E(s - t)x, s) - f(y, s) \right] dy\, ds.$$

Then, there exists a constant $c > 0$, depending on $(T - \tau)$ and α, such that

$$\|T_{ij} f\|_{C^\alpha_x(S_T)} \leq c\, |f|_{C^\alpha_x(S_T)} \quad \text{for every } f \in C^\alpha_x(\tau; T). \tag{2.23}$$

The proof of the above result relies on the properties of the singular kernel $\partial^2_{x_i x_j} \Gamma$ stated in the previous section: the upper bound on $\partial^2_{x_i x_j} \Gamma$ (Theorem 2.4), the mean value inequality (Theorem 2.7) and the cancellation property (Theorem 2.8).

2.4 Schauder Estimates in Space and Time

Theorem 2.14 shows that, for the derivatives $\partial^2_{x_i x_j} u$ (with $1 \leq i, j \leq q$), we can bound the C^α_x-norm in terms of the quantity $|\mathcal{L}u|_{C^\alpha_x(S_T)}$. We now aim to show how to improve the previous result, giving a control on the Hölder norm of $\partial^2_{x_i x_j} u$ with respect to both space *and time*, without strengthening the assumptions on $\mathcal{L}u$:

Theorem 2.16 (Local Schauder Estimates in Space-Time) (See [1, Theorem 3.18]) *Let* $T > \tau > -\infty$, $\alpha \in (0, 1)$, *and let* $K \subseteq \mathbb{R}^N$ *be a* compact set.

Then, there exists a constant $c = c(K, \tau, T) > 0$ *such that, for every* $u \in \mathcal{S}^0(\tau; T)$ *such that* $\mathcal{L}u \in C^\alpha_x(S_T)$, *one has*

$$\begin{aligned} &|\partial^2_{x_i x_j} u(x_1, t_1) - \partial^2_{x_i x_j} u(x_2, t_2)| \\ &\qquad \leq c\, |\mathcal{L}u|_{C^\alpha_x(S_T)} \big(d((x_1, t_1), (x_2, t_2))^\alpha + |t_1 - t_2|^{\alpha/q_N} \big) \end{aligned} \tag{2.24}$$

for every $1 \leq i, j \leq q$ *and every* $(x_1, t_1), (x_2, t_2) \in K \times [\tau, T]$. *We recall that* $q_N \geq 3$ *is the largest exponent in the dilations* $D_0(\lambda)$, *see* (1.7).

To prove Theorem 2.16, we need the following technical lemma.

Lemma 2.17 (See [1, Lemma 3.19]) *Let* $K \subseteq \mathbb{R}^N$ *be a fixed compact set, and let* $T > \tau > -\infty$. *There exists a constant* $c = c(K, \tau, T) > 0$ *such that*

$$\|x - E(t-s)x\| \leq c\, |t-s|^{1/q_N} \quad \text{for every } x \in K \text{ and } t, s \in [\tau, T] \tag{2.25}$$

$$\|(E(t) - E(s))x\| \leq c\, |t-s|^{1/q_N} \quad \text{for every } x \in K \text{ and } t, s \in [\tau, T]. \tag{2.26}$$

The above lemma follows by inspection of the explicit definition of $E(t)$ as an exponential matrix. The proof of Theorem 2.16 then follows combining the estimate in Theorem 2.14 with a bound on the difference

$$\partial^2_{x_i x_j} u\,(x, t_1) - \partial^2_{x_i x_j} u\,(x, t_2)\,,$$

which can be rewritten by the representation formula of Theorem 2.13, revising the proof of Theorem 2.15. Doing so, we are forced to use the bounds in Lemma 2.17, which are responsible of the appearance of the term $|t_1 - t_2|^{\alpha/q_N}$ in (2.24) (see Remark 1.5 (iii)).

3 Schauder Estimates for Operators with Coefficients Depending on (x, t)

Throughout this section we study operators (1.1) with coefficients $a_{ij}\,(x, t)$ depending on both space and time, fulfilling assumptions (H1), (H2), (H3) stated in Sect. 1.

Here we will present our main result, Theorem 1.4, relying on all the results discussed so far.

3.1 Local Schauder Estimates in Space

Throughout this section we will consider metric balls $B_r(\xi)$ centered at points $\xi \in \mathbb{R}^N \times (0, T)$ but possibly overlapping the hyperplanes $t = 0$ and $t = T$ (since these balls will eventually build a covering of $\mathbb{R}^N \times (0, T)$). Our functions $u \in \mathcal{S}^\alpha(0; T)$, so that they are actually defined and jointly continuous in the whole ball $B_r(\xi) \cap S_T$; however, the derivative Yu is merely an L^∞ function of the joint variables.

Notation Throughout the following, we will set

$$B_\rho^T(\xi) := B_\rho(\xi) \cap S_T \qquad \text{for every } \xi \in \mathbb{R}^{N+1} \text{ and } \rho > 0.$$

Theorem 3.1 (See [1, Theorem 4.1]) *Let $\mathcal{L}$ be an operator of type* (1.1) *satisfying assumptions* (H1), (H2), (H3) *stated in Sect. 1, for some $\alpha \in (0, 1)$.*

Then, there exist constants $c, r_0 > 0$ depending on T, α, the matrix B in (1.5) *and the numbers ν and Λ in* (1.3) *and* (1.17)*, respectively, such that, for every point $\overline{\xi} \in S_T$, $r \leq r_0$ and $u \in \mathcal{S}^\alpha(S_T)$ with $\operatorname{supp}(u) \subseteq B_r(\overline{\xi}) \cap \overline{S_T}$, one has*

$$\|\partial^2_{x_k x_h} u\|_{C^\alpha_x(B_r^T(\overline{\xi}))} \leq c\, |\mathcal{L}u|_{C^\alpha_x(B_r^T(\overline{\xi}))}, \tag{3.1}$$

for every $1 \leq h, k \leq q$. We stress that the constant c in (3.1) *is independent of the ball $B_r(\overline{\xi})$.*

Theorem 3.1 is obtained by the classical "Korn's trick": for a fixed $\overline{\xi} = (\overline{x}, \overline{t})$, we consider the operator $\mathcal{L}_{\overline{x}}$ with coefficients $a_{ij}(\overline{x}, t)$ (frozen in space, variable in time). Let $\Gamma^{\overline{x}}$ be its fundamental solution, as described in Theorem 2.1. Let $u \in \mathcal{S}^\alpha(S_T)$ with $\operatorname{supp}(u) \subseteq B_r(\overline{\xi}) \cap \overline{S_T}$ where $r \leq 1$ will be chosen later. Then $\mathcal{L}_{\overline{x}} u \in C^\alpha_x(S_T)$ and, by Theorem 2.13, we can write

$$\begin{aligned}
&\partial^2_{x_k x_h} u(x, t) \\
&= \int_{\overline{t}-1}^{t} \left(\int_{\mathbb{R}^N} \partial^2_{x_k x_h} \Gamma^{\overline{x}}(x, t; y, s) \big[\mathcal{L}_{\overline{x}} u(E(s-t)x, s) - \mathcal{L}_{\overline{x}} u(y, s)\big] dy \right) ds,
\end{aligned}$$

for every $(x, t) \in B_r^T(\overline{\xi})$ (so that, in particular, $|t - \overline{t}| \leq r \leq 1$). Writing

$$\mathcal{L}_{\overline{x}} = \mathcal{L} + (\mathcal{L}_{\overline{x}} - \mathcal{L}),$$

we can now write

$$\partial^2_{x_k x_h} u = A + \sum_{i,j=1}^{q} B_{ij}$$

where

$$\|A\|_{C_x^\alpha(S_T)} \leq c\,|\mathcal{L}u|_{C_x^\alpha(S_T)} \tag{3.2}$$

by Theorem 2.15, and

$$B_{ij} = \int_{\bar{t}-1}^{t}\int_{\mathbb{R}^N} \partial^2_{x_kx_h}\Gamma^{\overline{x}}(x,t;y,s)\big[f_{ij}(E(s-t)x,s) - f_{ij}(y,s)\big]dyds \tag{3.3}$$

with

$$f_{ij}(y,s) = [a_{ij}(\overline{x},s) - a_{ij}(y,s)]\partial^2_{x_ix_j}u(y,s),$$

hence, again by Theorem 2.15,

$$\|B_{ij}\|_{C_x^\alpha(S_T)} \leq c\,|f_{ij}|_{C_x^\alpha(S_T)}.$$

A standard reasoning exploiting the compact support of u and the assumption $a_{ij} \in C_x^\alpha(\mathbb{R}^{N+1})$) allows then to conclude that

$$|\partial^2_{x_kx_h}u|_\alpha + \sup|\partial^2_{x_kx_h}u| \leq c\,|\mathcal{L}u|_\alpha + \sum_{i,j=1}^{q} c\,|a_{ij}|_\alpha r^\alpha |\partial^2_{x_ix_j}u|_\alpha,$$

which for r small enough gives the desired estimate.

3.2 Some Interpolation Inequalities

Interpolation inequalities are a typical tool to deduce global estimates starting with local estimates for compactly supported functions. We will need the following:

Theorem 3.2 (See [1, Theorem 4.2]) *For every $r > 0$ there exist $c > 0$ and $\gamma > 1$ such that for every $\varepsilon \in (0,1)$, $\overline{\xi} \in S_T$ and $u \in \mathcal{S}^0(S_T)$,*

$$\sum_{h=1}^{q}\|\partial_{x_h}u\|_{C^\alpha(B_r^T(\overline{\xi}))} + \|u\|_{C^\alpha(B_r^T(\overline{\xi}))}$$

$$\leq \varepsilon\left\{\sum_{h,k=1}^{q}\left\|\partial^2_{x_kx_h}u\right\|_{C^0(B_{4r}^T(\overline{\xi}))} + \|Yu\|_{C^0(B_{4r}^T(\overline{\xi}))}\right\} + \frac{c}{\varepsilon^\gamma}\|u\|_{C^0(B_{4r}^T(\overline{\xi}))}. \tag{3.4}$$

The proof of the above inequality in [1] is quite technical and proceeds in several steps. We just point out the first step, which is based on techniques of fractional integrals, and has an independent interest, since it contains a regularity result for functions in $\mathcal{S}^0(S_T)$.

Proposition 3.3 (See [1, Proposition 4.3])

(i) *Let* $\mathcal{L}_0$ *be the constant-coefficient operator*

$$\mathcal{L}_0 = \sum_{i=1}^{q} \partial_{x_i x_i}^2 + Y$$

and let $R > 0$ *be fixed. For every* $\alpha \in (0, 1)$ *there exist* $\gamma > 2$ *and* $c > 0$ *such that, for every* $\overline{\xi} \in S_T$, $u \in \mathcal{S}^0(S_T)$ *with* $\operatorname{supp}(u) \subseteq B_R(\overline{\xi}) \cap \overline{S_T}$ *and every* $\varepsilon \in (0, 1)$ *we have:*

$$\begin{aligned} &\left\| \partial_{x_k} u \right\|_{C^\alpha(B_R^T(\overline{\xi}))} + \|u\|_{C^\alpha(B_R^T(\overline{\xi}))} \\ &\quad \le \varepsilon \left\| \mathcal{L}_0 u \right\|_{C^0(B_R^T(\overline{\xi}))} + \frac{c}{\varepsilon^\gamma} \|u\|_{C^0(B_R^T(\overline{\xi}))} \qquad \text{for } k = 1, 2, \dots, q. \end{aligned}$$

(The constant c depends on R and α *but not on* $\overline{\xi}$, *u and* ε*).*

(ii) *Let* $u \in \mathcal{S}^0(S_T)$, $\overline{\xi} \in S_T$ *and* $R > 0$*. Then, we have*

$$u,\ \partial_{x_k} u \in C^\alpha(B_R^T(\overline{\xi})) \text{ for every } 1 \le k \le q.$$

3.3 Global Schauder Estimates in Space

Here we want to get global Schauder estimates on the strip S_T, starting with the local Schauder estimates proved in Theorem 3.1 for functions which are compactly supported on small balls. To this aim, we will basically make use of cutoff functions and the interpolation inequalities contained in the previous section.

We start with a brief discussion about how a control of $C_x^\alpha(S_T)$-norm can be obtained starting with the control of $C_x^\alpha(B_r(\xi_i))$-norms for a suitable family of balls $\{B_r(\xi_i)\}_i$.

Let us start defining, for some fixed small $r > 0$, the seminorms

$$|f|_{C^\alpha_{x,r}(S_T)} \equiv \sup_{\substack{(x_1,t),(x_2,t)\in S_T \\ 0<\|x_1-x_2\|\le r}} \frac{|f(x_1,t) - f(x_2,t)|}{\|x_1 - x_2\|^\alpha}$$

$$|f|_{C_r^\alpha(S_T)} \equiv \sup_{\substack{\xi_1,\xi_2\in S_T \\ 0<d(\xi_1,\xi_2)\le r}} \frac{|f(\xi_1) - f(\xi_2)|}{d(\xi_1,\xi_2)^\alpha}$$

and let

$$\|f\|_{C^\alpha_{x,r}(S_T)} = |f|_{C^\alpha_{x,r}(S_T)} + \|f\|_{C^0(S_T)}.$$

Then the following holds:

Proposition 3.4 (See [1, Proposition 4.6]) *Let $r > 0$ and $\alpha \in (0, 1)$ be fixed, then:*

(i) There exists $c > 0$, depending on α and r, such that

$$\|f\|_{C_x^\alpha(S_T)} \leq c \, \|f\|_{C_{x,r}^\alpha(S_T)} \,. \tag{3.5}$$

(ii) Moreover, let $\left\{B_r\left(\overline{\xi}_i\right)\right\}_{i=1}^\infty$ be a covering of S_T, then

$$|f|_{C_{x,r}^\alpha(S_T)} \leq \sup_i |f|_{C_x^\alpha(B_{\theta r}^T(\overline{\xi}_i))} \tag{3.6}$$

$$|f|_{C_r^\alpha(S_T)} \leq \sup_i |f|_{C^\alpha(B_{\theta r}^T(\overline{\xi}_i))} \tag{3.7}$$

where $\theta \geq 1$ is an absolute constant.

With the previous ingredients at hand, one can now get the first part of our main result, that is point 1 of Theorem 1.4. The proof of this result proceeds along a quite standard line, exploiting the local Schauder estimates for functions with small compact support (Theorem 3.1), the use of cutoff functions, and interpolation inequalities (Theorem 3.2).

Finally, one can prove point 2 of Theorem 1.4, that is Hölder estimate in space and time. The proof of this result basically follows applying the analogous result for operators with coefficients $a_{ij}(t)$ (Theorem 2.16) to the operator $\mathcal{L}_{\overline{x}}$ with coefficients frozen at some point $\overline{x}$, and then bounding $|\mathcal{L}_{\overline{x}}u|_{C_x^\alpha(S_T)}$ in terms of $\left|\partial_{x_i x_j}^2 u\right|_{C_x^\alpha(S_T)}$, which we can already bound by point 1 of Theorem 1.4.

References

1. Biagi, S., Bramanti, M.: Schauder estimates for Kolmogorov-Fokker-Planck operators with coefficients measurable in time and Hölder continuous in space. J. Math. Anal. Appl. **533**(1), 65 pp. (2024). Paper No. 127996
2. Biagi, S., Bramanti, M., Stroffolini, B.: Kolmogorov-Fokker-Planck operators with coefficients measurable in time and Dini continuous in space. J. Evol. Equ. (2023, to appear). https://arxiv.org/abs/2305.11641
3. Bramanti, M., Brandolini, M.: Schauder estimates for parabolic nondivergence operators of Hörmander type. J. Differ. Equ. **234**, 177–245 (2007)
4. Bramanti, M., Brandolini, M.: Hörmander Operators, pp. XXVIII+693. World Scientific, Singapore (2023)
5. Bramanti, M., Polidoro, S.: Fundamental solutions for Kolmogorov-Fokker-Planck operators with time-depending measurable coefficients. Math. Eng. **2**(4), 734–771 (2020)
6. Brandt, A.: Interior Schauder estimates for parabolic differential- (or difference-) equations via the maximum principle. Israel J. Math. **7**, 254–262 (1969)
7. Chaudru de Raynal, P.E., Honoré, I., Menozzi, S.: Sharp Schauder estimates for some degenerate Kolmogorov equations. Ann. Sc. Norm. Super. Pisa Cl. Sci. **22**(3), 989–1089 (2021)

8. Constantine, G.M., Savits, T.H.: A multivariate Faà di Bruno formula with applications. Trans. Am. Math. Soc. **348**(2), 503–520 (1996)
9. Di Francesco, M., Polidoro, S.: Schauder estimates, Harnack inequality and Gaussian lower bound for Kolmogorov-type operators in non-divergence form. Adv. Differ. Equ. **11**(11), 1261–1320 (2006)
10. Dong, H., Kim, S.: Partial Schauder estimates for second-order elliptic and parabolic equations: a revisit. Int. Math. Res. Not. IMRN **2019**(7), 2085–2136 (2019)
11. Dong, H., Yastrzhembskiy, T.: Global Schauder estimates for kinetic Kolmogorov-Fokker-Planck equations (2022, preprint). https://arxiv.org/abs/2209.00769
12. Henderson, C., Wang, W.: Kinetic Schauder estimates with time-irregular coefficients and uniqueness for the Landau equation. Discrete Contin. Dyn. Syst. **44**(4), 1026–1072 (2024)
13. Hörmander, L.: Pseudo-differential operators and hypoelliptic equations. In: Singular Integrals (Proceedings of Symposium Pure Mathematics, vol. X, Chicago, Ill, 1966), pp. 138–183. American Mathematical Society, Providence (1967)
14. Hörmander, L.: Hypoelliptic second order differential equations. Acta Math. **119**, 147–171 (1967)
15. Imbert, C., Mouhot, C.: The Schauder estimate in kinetic theory with application to a toy nonlinear model. Ann. H. Lebesgue **4**, 369–405 (2021)
16. Knerr, B.F.: Parabolic interior Schauder estimates by the maximum principle. Arch. Rational Mech. Anal. **75**(1), 51–58 (1980/1981)
17. Kolmogorov, A.N.: Zufällige Bewegungen (Zur Theorie der Brownschen Bewegung). Ann. Math. **35**(2), 116–117 (1934)
18. Krylov, N.V., Priola, E.: Elliptic and parabolic second-order PDEs with growing coefficients. Commun. Partial Differ. Equ. **35**(1), 1–22 (2010)
19. Lanconelli, E., Polidoro, S.: On a class of hypoelliptic evolution operators. Partial differential equations, II (Turin, 1993). Rend. Sem. Mat. Univ. Politec. Torino **52**(1), 29–63 (1994)
20. Lieberman, G.M.: Intermediate Schauder theory for second order parabolic equations. IV. Time irregularity and regularity. Differ. Integral Equ. **5**(6), 1219–1236 (1992)
21. Lorenzi, L.: Optimal Hölder regularity for nonautonomous Kolmogorov equations. Discrete Contin. Dyn. Syst. Ser. S **4**(1), 169–191 (2011)
22. Lucertini, G., Pagliarani, S., Pascucci, A.: Optimal regularity for degenerate Kolmogorov equations in non-divergence form with rough-in-time coefficients. J. Evol. Equ. **23**, 69 (2023)
23. Lucertini, G., Pagliarani, S., Pascucci, A.: Optimal Schauder estimates for kinetic Kolmogorov equations with time measurable coefficients (preprint 2023). https://arxiv.org/abs/2304.13392
24. Lunardi, A.: Schauder estimates for a class of degenerate elliptic and parabolic operators with unbounded coefficients in $\mathbb{R}^n$. Ann. Scuola Norm. Sup. Pisa Cl. Sci. **24**(1), 133–164 (1997)
25. Menozzi, S.: Martingale problems for some degenerate Kolmogorov equations. Stoch. Process. Appl. **128**(3), 756–802 (2018)
26. Nagel, A., Stein, E.M., Wainger, S.: Balls and metrics defined by vector fields. I. Basic properties. Acta Math. **155**, 103–147 (1985)
27. Pascucci, A., Pesce, A.: On stochastic Langevin and Fokker-Planck equations: the two-dimensional case. J. Differ. Equ. **310**, 443–483 (2022)
28. Polidoro, S., Rebucci, A., Stroffolini, B.: Schauder type estimates for degenerate Kolmogorov equations with Dini continuous coefficients. Commun. Pure Appl. Anal. **21**(4), 1385–1416 (2022)
29. Priola, E.: Global Schauder estimates for a class of degenerate Kolmogorov equations. Stud. Math. **194**(2), 117–153 (2009)
30. Wang, W.D., Zhang, L.Q.: The C^α regularity of a class of non-homogeneous ultraparabolic equations. Sci. China Ser. A **52**(8), 1589–1606 (2009)

A New Proof of the Geometric Sobolev Embedding for Generalised Kolmogorov Operators

Nicola Garofalo and Giulio Tralli

Abstract In this note we revisit nonlocal isoperimetric inequalities and the related embeddings for Besov spaces adapted to a class of Hörmander operators of Kolmogorov-type (Garofalo and Tralli, J Funct Anal 279:108591, 2020). We provide here a new proof which exploits a weak-type Sobolev embedding (Garofalo and Tralli, Math Ann 383: 1–38, 2022).

1 Introduction

Consider the Kolmogorov-Fokker-Planck operators in $\mathbb{R}^N$, $N \geq 2$, defined as follows

$$\mathscr{A} u = \operatorname{tr}(Q\nabla^2 u) + \langle BX, \nabla u\rangle \,, \tag{1.1}$$

where the $N \times N$ matrices Q and B have real, constant coefficients, $Q = Q^\star \geq 0$, and X stands for the generic point in $\mathbb{R}^N$. The operators $\mathscr{A}$ in (1.1) were introduced in [13], where Hörmander showed that they are hypoelliptic if and only if the covariance matrix

$$K(t) = \frac{1}{t}\int_0^t e^{sB} Q e^{sB^\star}\, ds \quad \text{is positive definite for every } t > 0. \tag{1.2}$$

This condition will be henceforth assumed throughout the whole paper. Equations such as (1.1) are of considerable interest in physics, probability and finance, and have been the subject of intense study during the past three decades (see the

N. Garofalo
Dipartimento d'Ingegneria Civile e Ambientale (DICEA), Università di Padova, Padova, Italy
e-mail: nicola.garofalo@unipd.it

G. Tralli (✉)
Dipartimento di Matematica e Informatica, Università di Ferrara, Ferrara, Italy
e-mail: giulio.tralli@unife.it

S. Menozzi et al. (eds.), *Kolmogorov Operators and Their Applications*,
Springer INdAM Series 56, https://doi.org/10.1007/978-981-97-0225-1_5

recent survey article [2], and the references therein). Besides the classical Laplace equation (which corresponds to the non-degenerate model $Q = I_N$, $B = O_N$), they encompass the Ornstein-Uhlenbeck operator (which is obtained by taking $Q = I_N$ and $B = -I_N$), as well as the degenerate operator of Kolmogorov in $\mathbb{R}^{2n}$

$$\mathscr{A}_0 u = \Delta_v u + < v, \nabla_x u >, \tag{1.3}$$

corresponding to the choice $N = 2n$, $Q = \begin{pmatrix} I_n & 0_n \\ 0_n & 0_n \end{pmatrix}$, and $B = \begin{pmatrix} 0_n & 0_n \\ I_n & 0_n \end{pmatrix}$. Such operator arises in the kinetic theory of gases and was first introduced in the seminal note [14] on Brownian motion. One should note that $\mathscr{A}_0$ fails to be elliptic since it is missing the diffusive term $\Delta_x u$. However, it does satisfy Hörmander's hypoellipticity condition since one easily checks that $K(t) = \begin{pmatrix} I_n & t/2\, I_n \\ t/2\, I_n & t^2/3\, I_n \end{pmatrix} > 0$ for every $t > 0$. In this respect, it should be noted that Kolmogorov himself had already shown the hypoellipticity of his operator since in [14] he constructed an explicit heat kernel for $\mathscr{A}_0$ which is C^∞ outside the diagonal. Kolmogorov's construction was generalised in [13], where it was shown that the heat kernel for $\mathscr{A}$ can be explicitly written as

$$p(X, Y, t) = \frac{\omega_N (4\pi)^{-\frac{N}{2}}}{V(t)} \exp\left(-\frac{m_t(X, Y)^2}{4t}\right). \tag{1.4}$$

In (1.4), for $X, Y \in \mathbb{R}^N$ we have let, for $t > 0$,

$$\begin{aligned} m_t(X, Y) &= \sqrt{\langle K(t)^{-1}(Y - e^{tB}X), Y - e^{tB}X \rangle}, \\ B_t(X, r) &= \{Y \in \mathbb{R}^N \mid m_t(X, Y) < r\} \\ V(t) &= \left| B_t(X, \sqrt{t}) \right| = \omega_N (\det(tK(t)))^{1/2}, \end{aligned} \tag{1.5}$$

where ω_N indicates the Lebesgue measure $|\cdot|$ of the unit ball in $\mathbb{R}^N$. If we indicate with

$$P_t f(X) = \int_{\mathbb{R}^N} p(X, Y, t) f(Y) dY$$

the Hörmander semigroup, then it is well-known that

$$P_t 1 \equiv 1 \qquad \text{and} \qquad P_t^* 1 \equiv e^{-t \operatorname{tr} B}.$$

As a consequence, under the assumption that the matrix B of the drift satisfies

$$\operatorname{tr} B \geq 0, \tag{1.6}$$

we obtain a non-symmetric semigroup which is contractive in $L^p := L^p(\mathbb{R}^N, dX)$, $1 \le p \le \infty$. Under the condition (1.6), in a series of papers [3, 8–11] we have developed some basic functional analytic aspects of the class (1.1). In particular, since the operators $\mathscr{A}$ do not possess a variational structure, one focus of our investigation has been a notion of *gradient*. Guiding by the idea that the heat kernel (1.4) should encapsulate the geometry underlying $\mathscr{A}$, we introduced the following class of Besov spaces naturally associated with the semigroup P_t.

Definition 1.1 For any $0 < s < 1$ and $1 \le p < \infty$, the Besov space $\mathfrak{B}^{\mathscr{A}}_{s,p}$ is the collection of all functions $f \in L^p$ such that

$$\mathscr{N}^{\mathscr{A}}_{s,p}(f) = \left(\int_0^\infty \frac{1}{t^{\frac{sp}{2}+1}} \int_{\mathbb{R}^N} P_t \left(|f - f(X)|^p \right)(X) dX dt \right)^{\frac{1}{p}} < \infty. \tag{1.7}$$

Under condition (1.6), we know that smooth functions with compact support are contained and dense in $\mathfrak{B}^{\mathscr{A}}_{s,p}$ (see Proposition 2.5 below). Denoting by $\mathbf{1}_E$ the indicator function of a set E, we have introduced the following.

Definition 1.2 Fix $0 < s < \frac{1}{2}$. We say that a measurable set $E \subset \mathbb{R}^N$ has finite *s-perimeter* if $\mathbf{1}_E \in \mathfrak{B}^{\mathscr{A}}_{2s,1}$ and we define the s-perimeter associated to $\mathscr{A}$ as

$$\mathfrak{P}_{\mathscr{A},s}(E) = \mathscr{N}^{\mathscr{A}}_{2s,1}(\mathbf{1}_E).$$

The reader should keep in mind that

$$\mathscr{N}^{\mathscr{A}}_{2s,1}(\mathbf{1}_E) = \int_0^\infty \frac{1}{t^{1+s}} \| P_t \mathbf{1}_E - \mathbf{1}_E \|_1 dt, \tag{1.8}$$

see in this respect [8, Corollary 3.6] and [9, Section 3]. Formula (1.8) underscores the role of the 1-parameter family of pseudo-distances $m_t(\cdot,\cdot)$ in (1.4) via the s-dependent average of $\| P_t \mathbf{1}_E - \mathbf{1}_E \|_1$.

When specialized to $\mathscr{A} = \Delta$, up to a renormalizing factor the seminorm $\mathscr{N}^{\mathscr{A}}_{s,p}(\cdot)$ coincides with the classical Aronszajn-Gagliardo-Slobedetzky seminorm: in such framework it is nowadays a common practice to call *nonlocal perimeter* the Aronszajn-Gagliardo-Slobedetzky seminorm of the indicator function, and we refer the reader to the influential work [4] where the structure of the critical points of nonlocal perimeters was first analyzed. The notion of nonlocal perimeter in the Euclidean setting was implicitly present in [1, 17], and the nonlocal counterpart of the classical De Giorgi's isoperimetric inequality [5] was established in [1, 7] with respect to such notion.

Concerning the general class of operators $\mathscr{A}$ in (1.1), we established in [9, Theorem 1.1 and Theorem 1.3] the following nonlocal isoperimetric inequalities and the related embeddings of the Besov spaces $\mathfrak{B}^{\mathscr{A}}_{2s,1}$ in the relevant L^q-space of functions with higher integrability properties.

Theorem 1.3 *Let $0 < s < \frac{1}{2}$, and assume* (1.6)*. Suppose there exist $D, \gamma_D > 0$ such that*

$$V(t) \geq \gamma_D \, t^{D/2} \quad \textit{for all } t > 0. \tag{1.9}$$

Then

$$\mathfrak{B}^{\mathscr{A}}_{2s,1} \hookrightarrow L^{\frac{D}{D-2s}}.$$

More precisely, there exists a positive constant c (depending on N, D, s, γ_D) such that for every $f \in \mathfrak{B}^{\mathscr{A}}_{2s,1}$ one has

$$||f||_{\frac{D}{D-2s}} \leq \frac{1}{c} \mathscr{N}^{\mathscr{A}}_{2s,1}(f). \tag{1.10}$$

In particular, for any measurable set $E \subset \mathbb{R}^N$ with $|E| < \infty$, one has

$$\mathfrak{P}_{\mathscr{A},s}(E) \geq c \, |E|^{(D-2s)/D}.$$

In Theorem 1.3 the assumption (1.9) on the growth of the volume function $V(t)$ is crucial as it allows to detect the dimensional parameter $q = \frac{D}{D-2s}$, and it deserves a detailed explanation. If we first take a look at the Kolmogorov example $\mathscr{A}_0$ in $\mathbb{R}^{2n}$ recalled in (1.3), the constant $D = 4n$ is determined by the direct computation $V(t) = \omega_{2n} 12^{-\frac{n}{2}} t^{2n}$. More generally, the fact that one can uniquely identify the constant D in (1.9) is shared by the subclass of (1.1) which possesses invariances with respect to a family of non-isotropic dilations since in this case $V(t) \equiv V(1) t^{\frac{D}{2}}$ (to fix the ideas, for $\mathscr{A}_0$ the dilations are defined by $\delta_\lambda(v, x) = (\lambda v, \lambda^3 x)$): such homogeneous class was introduced in [15] where Lanconelli and Polidoro provided an explicit characterization in terms of the matrices Q and B. Besides the homogeneous case, for the general class (1.1) one can always say

$$\exists D_0 \geq N \ \text{ such that } \ V(t) \cong t^{D_0/2} \ \text{ as } t \to 0^+, \tag{1.11}$$

see [11, Section 2.4]. We refer to the number D_0 as the intrinsic dimension of the semigroup $\{P_t\}_{t>0}$ at zero. With this perspective in mind, the validity of the assumption (1.9) readily implies that $D \geq D_0$ (which in particular ensures $D > 2s$ and $\frac{D}{D-2s} > 1$) and it underscores the importance of the behaviour of $V(t)$ for large t. For example, the operator

$$\Delta_v + \langle v, \nabla_v \rangle + \langle v, \nabla_x \rangle \qquad \text{in } \mathbb{R}^{2n}$$

fits the assumptions of Theorem 1.3 for every $D \geq 4n$: as a matter of fact one can check that in this case $V(t) = \omega_{2n} \left(2e^t - \frac{t}{2} - 1 + \frac{t}{2} e^{2t} - e^{2t}\right)^n$ which tells that $D_0 = 4n$ and $V(t)$ grows exponentially fast at $t \to \infty$. In [11, Definition 3.4] we introduced the notion of intrinsic dimension at infinity of the semigroup $\{P_t\}_{t>0}$ in order to handle the behaviour for large t of the volume function. We are going to recall such notion in Definition 2.1 below, and we denote by D_∞ such a constant. It turns out that $D_\infty \in [2, \infty]$ under assumption (1.6), and the validity of (1.9) forces $D_\infty \geq D_0$ and $D \in [D_0, D_\infty]$. This gives a way to understand the operators $\mathscr{A}$ which do not satisfy the volume growth condition (1.9). In fact, we can consider the following operator

$$\partial_v^2 + v\partial_x - x\partial_v \qquad \text{in } \mathbb{R}^2$$

for which $V(t) = \pi \left(\frac{t^2}{4} + \frac{1}{8}(\cos(2t) - 1) \right)^{\frac{1}{2}}$: in this situation we thus have $D_0 = 4 > D_\infty = 2$ and therefore Theorem 1.3 cannot apply to such case. In [9, Theorem 1.2 and Theorem 7.6] we established the following substitute result to treat the operators (1.1) with $D_0 > D_\infty$.

Theorem 1.4 *Let $0 < s < \frac{1}{2}$. Assume (1.6), and $D_0 > D_\infty$. Suppose there exists $\gamma > 0$ such that*

$$V(t) \geq \gamma \min\{t^{D_0/2}, t^{D_\infty/2}\}. \quad \textit{for all } t > 0. \tag{1.12}$$

Then

$$\mathfrak{B}^{\mathscr{A}}_{2s,1} \hookrightarrow L^{\frac{D_0}{D_0-2s}} + L^{\frac{D_\infty}{D_\infty-2s}}.$$

More precisely, there exists a positive constant c (depending on $N, D_0, D_\infty, s, \gamma$) such that for every $f \in \mathfrak{B}^{\mathscr{A}}_{2s,1}$ one has

$$||f||_{L^{\frac{D_0}{D_0-2s}} + L^{\frac{D_\infty}{D_\infty-2s}}} \leq \frac{1}{c} \mathscr{N}^{\mathscr{A}}_{2s,1}(f). \tag{1.13}$$

Moreover, there exists a positive constant $\tilde{c}$ (depending on $N, D_0, D_\infty, s, \gamma$) such that for any measurable set $E \subset \mathbb{R}^N$ with $|E| < \infty$, one has

$$\mathfrak{P}_{\mathscr{A},s}(E) \geq c \min\left\{ |E|^{\frac{D_0-2s}{D_0}}, |E|^{\frac{D_\infty-2s}{D_\infty}} \right\}.$$

We notice that, under the assumptions of Theorem 1.4, we have $D_0 > D_\infty > 2s$ and then $\frac{D_\infty}{D_\infty-2s} > \frac{D_0}{D_0-2s} > 1$. The proofs of Theorems 1.3 and 1.4 which can be found in [9] were inspired by the powerful and flexible semigroup approach to isoperimetric inequalities which is due to Ledoux [16] in the local case. We also refer the interested reader to [20, 21] for more insights on such a heat-kernel approach to perimeters and isoperimetric properties. The purpose of the present note is to provide a different proof of Theorems 1.3 and 1.4 which instead relies on an embedding in a weak L^q-space of a fractional ($p = 1$)-Sobolev space: these Sobolev spaces are tailored on the fractional powers of $\mathscr{A}$ and their relevant embeddings were established in [11]. For a proper historical perspective concerning the classical Sobolev spaces, we recall that from the representation formula $|f(X)| \leq C(N) \int_{\mathbb{R}^N} \frac{|\nabla f(Y)|}{|X-Y|^{N-1}} dY$, and the L^1-mapping properties of the Riesz potentials, one knows that $W^{1,1} \hookrightarrow L^{N/(N-1),\infty}$. A remarkable aspect of the end-point case $p = 1$ is that such weak Sobolev embedding in fact implies the classical isoperimetric inequality $P(E) \geq C_N |E|^{\frac{N-1}{N}}$. The latter, in turn, combined with the coarea formula, is equivalent to the strong embedding $W^{1,1} \hookrightarrow L^{N/(N-1)}$. This establishes the beautiful fact that, in the geometric case $p = 1$, the weak Sobolev embedding is equivalent to the strong one, and they are both equivalent to the isoperimetric inequality, see [18]. The main focus of this paper is a semigroup generalisation of this circle of ideas to the nonlocal degenerate setting of the operators $\mathscr{A}$.

2 Preliminaries

In this section we recall the main ingredients we shall need in the proofs of Theorems 1.3 and 1.4 that we present in Sect. 3. The main character in our analysis is the kernel in (1.4). As we mentioned it is well-known that $p(X, Y, t)$ is the heat kernel, i.e. that $p(\cdot, Y, \cdot)$ is a solution of the heat equation $\mathscr{A} u = \partial_t u$ in $\mathbb{R}^N \times (0, \infty)$ for any $Y \in \mathbb{R}^N$ and $p(X, \cdot, t)$ tends to the Dirac delta δ_X in the distributional sense as $t \to 0^+$ for any $X \in \mathbb{R}^N$. In particular we are going to exploit the following property concerning such an approximation of the identity

$$P_t\varphi(X) \underset{t\to 0^+}{\longrightarrow} \varphi(X) \qquad \text{for every } \varphi \in L^\infty \cap C\left(\mathbb{R}^N\right) \text{ and } X \in \mathbb{R}^N. \tag{2.1}$$

The limit in (2.1) can be verified directly using the Markovian condition $P_t 1 \equiv 1$ and the properties of the positive definite matrices $K(t)$ in (1.2) (see, e.g., [12, Proposition 2.1]; see also the analytic and probabilistic tools in [6, 19] for a treatment of a more general class of operators with varying coefficients).
It is very convenient for us to exploit the notations we adopted with the explicit expression (1.4), as in this way we can put the 1-parameter family of pseudoballs $B_t(X, \sqrt{t})$ at the center stage together with their volume function $V(t)$ in (1.5). We already stressed that, as a by product of the analysis in [15], the small-time behaviour of $V(t)$ is governed by a suitable infinitesimal homogeneous structure which we encode in the number D_0 defined via (1.11) (i.e. the intrinsic dimension of the semigroup $\{P_t\}_{t>0}$ at zero). On the other hand, the following definition allows us to handle the large-time behaviour of $V(t)$.

Definition 2.1 Consider the set

$$\Sigma_\infty = \left\{\alpha > 0 \Big| \int_1^\infty \frac{t^{\frac{\alpha}{2}-1}}{V(t)} dt < \infty\right\}.$$

We call the number

$$D_\infty = \sup \Sigma_\infty$$

the *intrinsic dimension at infinity* of the semigroup $\{P_t\}_{t>0}$.

Thanks to the study of the large-time behaviour of the eigenvalues of $tK(t)$ performed in [11, Section 3] (see also [3, Proposition 2.3]), we know that $2 \leq D_\infty \leq \infty$ under the assumption (1.6).

2.1 Fractional Powers of $\mathscr{A}$, and Their Sobolev Embeddings

In [10] we developed a fractional calculus for $\mathscr{A}$. On functions belonging to the Schwartz class $\mathscr{S}$ the nonlocal operator $(-\mathscr{A})^s$, for $0 < s < 1$, is defined through the following pointwise formula

$$(-\mathscr{A})^s f(X) = -\frac{s}{\Gamma(1-s)} \int_0^\infty t^{-(1+s)} [P_t f(X) - f(X)] dt, \qquad X \in \mathbb{R}^N. \tag{2.2}$$

If tr $B \geq 0$, the formula in (2.2) defines in fact an L^p-function for any $1 \leq p < \infty$. We can then extend the operator $(-\mathscr{A})^s$ to a closed operator on its domain in L^p endowed with the graph norm: this is precisely what we are doing with the following definition.

Definition 2.2 Let $1 \leq p < \infty$, $0 < s < 1$, and assume (1.6). We define the Sobolev space $\mathscr{L}^{2s,p}$ as

$$\mathscr{L}^{2s,p} = \{f \in L^p \mid (-\mathscr{A})^s f \in L^p\}$$

with

$$\|f\|_{\mathscr{L}^{2s,p}} = \|f\|_p + \|(-\mathscr{A})^s f\|_p.$$

Thanks to a density result the Banach space $\mathscr{L}^{2s,p}$ coincides with the completion of $\mathscr{S}$ with respect to $\|\cdot\|_{\mathscr{L}^{2s,p}}$, see in this respect [11, Definition 4.4 and Proposition 4.6] as well as [9, Proposition 2.13].

In [11, Theorem 7.5 and Theorem 7.7] we proved the following

Theorem 2.3 *Let* $0 < s < 1$, *and assume* (1.6).

(a) If (1.9) *hold, then we have* $\mathscr{L}^{2s,1} \hookrightarrow L^{\frac{D}{D-2s},\infty}$. *More precisely, there exists a constant* $S_{1,s} > 0$, *depending on* N, D, s, γ_D, *such that for any* $f \in \mathscr{L}^{2s,1}$ *one has*

$$\sup_{\lambda>0} \lambda |\{X \in \mathbb{R}^N \mid |f(X)| > \lambda\}|^{\frac{D-2s}{D}} \leq S_{1,s} \|(-\mathscr{A})^s f\|_1. \tag{2.3}$$

(b) If instead $D_0 > D_\infty$ *and* (1.12) *hold, we have* $\mathscr{L}^{2s,1} \hookrightarrow L^{\frac{D_0}{D_0-2s},\infty} + L^{\frac{D_\infty}{D_\infty-2s},\infty}$. *More precisely, there exists a constant* $S_{1,s} > 0$, *depending on* $N, D_\infty, D_0, s, \gamma$, *such that for any* $f \in \mathscr{L}^{2s,1}$ *one has*

$$\min\{\sup_{\lambda>0} \lambda \, |\{X \mid |f(X)| > \lambda\}|^{\frac{D_0-2s}{D_0}}, \sup_{\lambda>0} \lambda \, |\{X \mid |f(X)| > \lambda\}|^{\frac{D_\infty-2s}{D_\infty}}\} \leq S_{1,s} \|(-\mathscr{A})^s f\|_1. \tag{2.4}$$

We refer the reader to [11] for the case $p > 1$, where we established the strong embeddings

$$\mathscr{L}^{2s,p} \hookrightarrow L^{\frac{pD}{D-2sp}} \qquad \text{and} \qquad \mathscr{L}^{2s,p} \hookrightarrow L^{\frac{pD_0}{D_0-2sp}} + L^{\frac{pD_\infty}{D_\infty-2sp}} \tag{2.5}$$

under the respective assumptions (1.9) and (1.12) (with $p < \frac{D}{2s}$ and $p < \frac{D_\infty}{2s}$).
Let us spend some words on the proofs provided in [11] of Theorem 2.3 and (2.5), which were inspired by the works by Varopoulos in the framework of positive symmetric semigroups (see, e.g., [23]). With the aid of a crucial inversion formula for the fractional powers of $\mathscr{A}$ in terms of suitable Riesz-type potentials having the following semigroup representation

$$f \mapsto \frac{1}{\Gamma(s)} \int_0^\infty t^{s-1} P_t f dt,$$

we were able to deduce the proof of our Sobolev-type embeddings from the $L^p - L^q$ mapping properties of these Riesz potentials. The identification of such q is the point in the proof where the volume growth conditions (1.9) and (1.12) come into play. The key technical tool to show these mapping properties is the introduction of a maximal function related to $\mathscr{A}$, which we believe has interest in its own. Influenced by the powerful ideas by Stein in [22], we exploited the Poisson semigroup $e^{z\sqrt{-\mathscr{A}}}$ to define the maximal function

$$\mathscr{M}^\star f(X) = \sup_{z>0} \left| \frac{1}{\sqrt{4\pi}} \int_0^\infty \frac{z}{t^{3/2}} e^{-\frac{z^2}{4t}} P_t f(X) dt \right|, \qquad X \in \mathbb{R}^N.$$

The operator $\mathscr{M}^\star$ maps in fact continuously L^1 in $L^{1,\infty}$ and any L^p in itself for $p > 1$ (see [11, Theorem 5.5]), i.e. the following maximal theorem for the class (1.1) holds true.

Theorem 2.4 *Assume* (1.6).

(i) There exists a universal constant $A_1 > 0$ *such that, given* $f \in L^1$, *one has*

$$\sup_{\lambda>0} \lambda |\{X \in \mathbb{R}^N \mid \mathscr{M}^\star f(X) > \lambda\}| \le A_1 \|f\|_1;$$

(ii) If $1 < p \le \infty$, *there exists a universal constant* $A_p > 0$ *such that for any* $f \in L^p$ *one has*

$$\|\mathscr{M}^\star f\|_p \le A_p \|f\|_p.$$

2.2 Nonlocal Perimeters and Coarea Formulas

As we want to go back to the study of the Besov spaces $\mathfrak{B}^{\mathscr{A}}_{s,p}$ and their seminorm $\mathscr{N}^{\mathscr{A}}_{s,p}$ defined in (1.7), we start by recalling the relationship between the s-perimeter associated to $\mathscr{A}$ and the fractional power $(-\mathscr{A})^s$. Keeping in mind Definition 1.2 and (2.2), we have in fact

$$\mathfrak{P}_{\mathscr{A},s}(E) = \frac{\Gamma(1-s)}{s} \|(-\mathscr{A})^s \mathbf{1}_E\|_1 \qquad \text{if } \mathbf{1}_E \in \mathfrak{B}^{\mathscr{A}}_{2s,1}. \tag{2.6}$$

We refer to [8, Corollary 3.6] and [9, Lemma 3.3] for a proof of (2.6). Moreover, in [8, Proposition 3.3] we showed the boundedness of the map $(-\mathscr{A})^s : \mathfrak{B}^{\mathscr{A}}_{2s,1} \longrightarrow L^1$. This says that

$$\mathfrak{B}^{\mathscr{A}}_{2s,1} \hookrightarrow \mathscr{L}^{2s,1}. \tag{2.7}$$

Like the Sobolev-type spaces $\mathscr{L}^{2s,p}$, also the Besov spaces $\mathfrak{B}^{\mathscr{A}}_{s,p}$ enjoy useful density properties. In the next section it will play a role the following density result which was established in [3, Proposition 3.2] (see also [9, Lemma 7.3]).

Proposition 2.5 *Assume* (1.6). *For every* $0 < s < 1$ *and* $1 \le p < \infty$, *we have*

$$\overline{C_0^\infty}^{\mathfrak{B}^{\mathscr{A}}_{s,p}} = \mathfrak{B}^{\mathscr{A}}_{s,p}.$$

Another important tool for our purposes is the validity of a coarea formula which yields a further link between the seminorm $\mathscr{N}^{\mathscr{A}}_{2s,1}$ and the nonlocal perimeter $\mathfrak{P}_{\mathscr{A},s}$.

Proposition 2.6 *Let* $0 < s < \frac{1}{2}$, *and assume* (1.6). *For any* $f \in \mathfrak{B}^{\mathscr{A}}_{2s,1}$ *we have*

$$\mathscr{N}^{\mathscr{A}}_{2s,1}(f) = \int_{\mathbb{R}} \mathfrak{P}_{\mathscr{A},s}(\{f > \sigma\})\, d\sigma. \tag{2.8}$$

Proof It is a consequence of the results in [9, Proposition 7.4 and formula (7.4)], once we keep in mind (2.6) and (1.8). □

Remark 2.7 If one compares Definition 1.2 with the definition of s-perimeter associated to $\mathscr{A}$ given in [9, Section 4.1], one can notice a difference that we now want to comment on. Via a relaxation procedure, in [9] we denoted by $\mathfrak{P}^{\mathscr{A}}_s(E)$ the following

$$\inf\left\{\liminf_{k\to\infty} \|(-\mathscr{A})^s f_k\|_1 \ \mid\ \{f_k\}_{k\in\mathbb{N}} \in \mathscr{S} \text{ such that } f_k \to \mathbf{1}_E \text{ in } L^1\right\}$$

for $E \subset \mathbb{R}^N$ measurable and with finite volume. Thanks to Proposition 2.5 it is clear that, if $\mathbf{1}_E \in \mathfrak{B}^{\mathscr{A}}_{2s,1}$, then there exists $\{f_k\}_{k\in\mathbb{N}} \in C_0^\infty \subset \mathscr{S}$ converging to $\mathbf{1}_E$ in $\mathfrak{B}^{\mathscr{A}}_{2s,1}$. For such a sequence we then obtain

$$\lim_{k\to\infty} \|(-\mathscr{A})^s f_k\|_1 = \|(-\mathscr{A})^s \mathbf{1}_E\|_1 = \frac{s}{\Gamma(1-s)}\mathfrak{P}_{\mathscr{A},s}(E),$$

where we exploited (2.7) and (2.6). Hence we have

$$\mathfrak{P}^{\mathscr{A}}_s(E) \le \frac{s}{\Gamma(1-s)}\mathfrak{P}_{\mathscr{A},s}(E) \quad \text{in case } \mathbf{1}_E \in \mathfrak{B}^{\mathscr{A}}_{2s,1}.$$

Remark 2.8 In [9, Section 4.2] we introduced another notion of s-perimeter which we denoted by $\mathfrak{P}^{\mathscr{A},\star}_s(E)$. For bounded measurable sets $E \subset \mathbb{R}^N$, we let

$$\mathfrak{P}^{\mathscr{A},\star}_s(E) = \lim_{t\to 0^+} \|(-\mathscr{A})^s P_t \mathbf{1}_E\|_1.$$

The previous definition makes sense as $P_t\mathbf{1}_E$ belongs to $\mathscr{S}$ for any $t > 0$ and $t \mapsto \|(-\mathscr{A})^s P_t \mathbf{1}_E\|_1$ is monotone decreasing (see [9, pg. 21]). Keeping in mind (2.7), we have by [9, Corollary 3.5] and (2.6) that

$$\mathfrak{P}^{\mathscr{A},\star}_s(E) = \|(-\mathscr{A})^s \mathbf{1}_E\|_1 = \frac{s}{\Gamma(1-s)}\mathfrak{P}_{\mathscr{A},s}(E) \quad \text{in case } \mathbf{1}_E \in \mathfrak{B}^{\mathscr{A}}_{2s,1}.$$

From (2.7) and Theorem 2.3, we immediately deduce the validity of the embeddings

$$\mathfrak{B}^{\mathscr{A}}_{2s,1} \hookrightarrow L^{\frac{D}{D-2s},\infty} \qquad \text{and} \qquad \mathfrak{B}^{\mathscr{A}}_{2s,1} \hookrightarrow L^{\frac{D_0}{D_0-2s},\infty} + L^{\frac{D_\infty}{D_\infty-2s},\infty} \tag{2.9}$$

under the respective volume growth conditions (1.9) and (1.12). The objective of the next section is to replace $L^{q,\infty}$ with the strong spaces L^q in (2.9).

3 Proofs

The proof of Theorem 1.3 and Theorem 1.4 will result as a combination of Theorem 2.3 with the following lemma.

Lemma 3.1 *Consider a measurable set $E \subset \mathbb{R}^N$ with $|E| < \infty$. Then we have*

$$\liminf_{t\to 0^+} \left| \left\{ X \in \mathbb{R}^N \mid P_t \mathbf{1}_E(X) > \frac{1}{2} \right\} \right| \geq |\mathrm{int} E|$$

Proof We start by noticing that

$$\text{for every } X \in \mathrm{int} E \text{ we have } \lim_{t\to 0^+} P_t \mathbf{1}_E(X) = 1. \tag{3.1}$$

The statement in (3.1) is a consequence of (2.1). As a matter of fact, for any $X \in \mathrm{int}E$, we can pick $\rho_X > 0$ such that an open neighborhood of size ρ_X is contained in E. Thanks to this fact, it is easy to construct a continuous function φ_X such that

$$0 \leq \varphi_X \leq 1, \quad \varphi_X(X) = 1, \quad \varphi_X \equiv 0 \text{ in } \mathbb{R}^N \setminus E.$$

Hence we have

$$1 \geq P_t \mathbf{1}_E(X) \geq P_t \varphi_X(X) \underset{t\to 0^+}{\longrightarrow} \varphi_X(X) = 1,$$

which ensures the validity of (3.1). Therefore

$$\text{for every } X \in \mathrm{int} E \ \text{ there exists } t_X > 0 \ \text{ such that } P_t \mathbf{1}_E(X) > \frac{1}{2} \ \text{ for } 0 < t < t_X.$$

This fact implies that

$$\lim_{t\to 0^+} \mathbf{1}_{E_t}(X) = 1 \quad \text{ for every } X \in \mathrm{int} E$$

once we denote

$$E_t = \left\{ X \in \mathbb{R}^N \mid P_t \mathbf{1}_E(X) > \frac{1}{2} \right\}.$$

By Fatou's Lemma we then obtain

$$\liminf_{t\to 0^+}\int_{\mathbb{R}^N}\mathbf{1}_{E_t}(X)dX \geq \int_{\mathbb{R}^N}\left(\liminf_{t\to 0^+}\mathbf{1}_{E_t}(X)\right)dX \geq \int_{\mathrm{int}E} 1dX,$$

which completes the proof of the desired statement. □

We are thus ready to provide the proof of Theorem 1.3.

Proof of Theorem 1.3 For any $f \in \mathfrak{B}^{\mathscr{A}}_{2s,1}$, we denote

$$E_\sigma = \{X \in \mathbb{R}^N \mid |f(X)| > \sigma\}, \qquad \text{for } \sigma > 0.$$

Since by Proposition 2.6 we have

$$\int_0^\infty \mathfrak{P}_{\mathscr{A},s}(E_\sigma)\,d\sigma = \mathscr{N}^{\mathscr{A}}_{2s,1}(|f|) \leq \mathscr{N}^{\mathscr{A}}_{2s,1}(f) < \infty,$$

it is clear that for almost any σ one has $1_{E_\sigma} \in \mathfrak{B}^{\mathscr{A}}_{2s,1}$. As a consequence, for such σ, we obtain from [9, Corollary 3.5] and (2.6) that

$$\lim_{t\to 0^+}\|(-\mathscr{A})^s P_t 1_{E_\sigma}\|_1 = \|(-\mathscr{A})^s 1_{E_\sigma}\|_1 = \frac{s}{\Gamma(1-s)}\mathfrak{P}_{\mathscr{A},s}(E_\sigma). \tag{3.2}$$

The aim of the proof is to establish the following bound

$$||f||_{\frac{D}{D-2s}} \leq \frac{2s\,S_{1,s}}{\Gamma(1-s)}\mathscr{N}^{\mathscr{A}}_{2s,1}(f) \quad \text{for every } f \in C^\infty \cap \mathfrak{B}^{\mathscr{A}}_{2s,1}, \tag{3.3}$$

where $S_{1,s}$ is the positive constant appearing in (2.3). Thus, let us fix $f \in C^\infty \cap \mathfrak{B}^{\mathscr{A}}_{2s,1}$ and denote $G(\sigma) = |E_\sigma|$. Since E_σ has finite measure (by Chebyshev's inequality) and it is open (by the continuity of f), we are entitled to apply Lemma 3.1 and we obtain

$$G(\sigma) \leq \liminf_{t\to 0^+}\left|\left\{X \in \mathbb{R}^N \mid P_t\mathbf{1}_{E_\sigma}(X) > \frac{1}{2}\right\}\right|. \tag{3.4}$$

On the other hand, since for $t > 0$ the positive function $P_t 1_{E_\sigma}$ belongs to $\mathscr{L}^{2s,1}$ (this holds true for any σ such that $1_{E_\sigma} \in \mathfrak{B}^{\mathscr{A}}_{2s,1}$, see in this respect [9, Lemma 3.4]), Theorem 2.3 yields

$$\frac{1}{2}\left|\left\{X \in \mathbb{R}^N \mid P_t\mathbf{1}_{E_\sigma}(X) > \frac{1}{2}\right\}\right|^{\frac{D-2s}{D}} \leq S_{1,s}||(-\mathscr{A})^s P_t\mathbf{1}_{E_\sigma}||_1. \tag{3.5}$$

Therefore, from the combination of (3.4), (3.5), and (3.2), we obtain

$$G(\sigma)^{\frac{D-2s}{D}} \leq \frac{2s\,S_{1,s}}{\Gamma(1-s)} \mathfrak{P}_{\mathscr{A},s}(E_\sigma)$$

Since G is non-increasing and $D \geq D_0 \geq 2 > 2s$, we then have

$$\begin{aligned}
||f||_{\frac{D}{D-2s}} &= \left(\int_{\mathbb{R}^N} |f|^{\frac{D}{D-2s}}(X)dX\right)^{\frac{D-2s}{D}} = \left(\frac{D}{D-2s}\int_0^\infty \sigma^{\frac{2s}{D-2s}} G(\sigma)d\sigma\right)^{\frac{D-2s}{D}} \\
&\leq \int_0^\infty G(\sigma)^{\frac{D-2s}{D}} d\sigma \leq \frac{2s\,S_{1,s}}{\Gamma(1-s)} \int_0^\infty \mathfrak{P}_{\mathscr{A},s}(E_\sigma)\,d\sigma \\
&= \frac{2s\,S_{1,s}}{\Gamma(1-s)} \mathscr{N}^{\mathscr{A}}_{2s,1}(|f|) \\
&\leq \frac{2s\,S_{1,s}}{\Gamma(1-s)} \mathscr{N}^{\mathscr{A}}_{2s,1}(f).
\end{aligned}$$

The previous inequality proves the desired (3.3). Hence, the density of C^∞ in $\mathfrak{B}^{\mathscr{A}}_{2s,1}$ (which is a consequence of Proposition 2.5) implies the validity of (1.10) with the choice

$$c = \frac{\Gamma(1-s)}{2s\,S_{1,s}}.$$

In particular, if $E \subset \mathbb{R}^N$ has finite s-perimeter, we can plug $f = \mathbf{1}_E$ in (1.10) and we deduce

$$\mathfrak{P}_{\mathscr{A},s}(E) = \mathscr{N}^{\mathscr{A}}_{2s,1}(\mathbf{1}_E) \geq \frac{\Gamma(1-s)}{2s\,S_{1,s}} \|\mathbf{1}_E\|_{\frac{D}{D-2s}} = \frac{\Gamma(1-s)}{2s\,S_{1,s}} |E|^{\frac{D-2s}{D}},$$

which completes the proof of the theorem. □

We conclude the paper with the proof of Theorem 1.4. To this aim we fix some notation. We set

$$q_0 = \frac{D_0}{D_0 - 2s} \quad \text{and} \quad q_\infty = \frac{D_\infty}{D_\infty - 2s}.$$

We recall that, in the assumptions of the theorem, we have $D_0 > D_\infty \geq 2 > 2s$. This says in particular that

$$q_\infty > q_0 > 1. \tag{3.6}$$

Let us also recall that, when we write $L^{q_0} + L^{q_\infty}$, we mean the Banach space of functions f which can be written as $f = f_1 + f_2$ with $f_1 \in L^{q_0}$ and $f_2 \in L^{q_\infty}$, endowed with the norm

$$||f||_{L^{q_0}+L^{q_\infty}} = \inf_{f=f_1+f_2,\, f_1\in L^{q_0},\, f_2\in L^{q_\infty}} ||f_1||_{q_0} + ||f_2||_{q_\infty}.$$

Proof of Theorem 1.4 We want to argue as similar as possible to the proof of Theorem 1.3, from which we also borrow the notations for the sets E_σ and the non-increasing function $G(\sigma)$. We then fix an arbitrary function $f \in C^\infty \cap \mathfrak{B}^{\mathscr{A}}_{2s,1}$. From the combination of Lemma 3.1, Theorem 2.3, and (3.2), we obtain

$$\min\left\{G(\sigma)^{\frac{1}{q_0}}, G(\sigma)^{\frac{1}{q_\infty}}\right\} \le \frac{2s\,S_{1,s}}{\Gamma(1-s)}\mathfrak{P}_{\mathscr{A},s}(E_\sigma),$$

where $S_{1,s}$ is the positive constant appearing in (2.4). If we then exploit the coarea formula (2.8), we deduce that

$$\int_0^\infty \min\left\{G(\sigma)^{\frac{1}{q_0}}, G(\sigma)^{\frac{1}{q_\infty}}\right\} d\sigma \le \frac{2s\,S_{1,s}}{\Gamma(1-s)}\mathscr{N}^{\mathscr{A}}_{2s,1}(|f|) \le \frac{2s\,S_{1,s}}{\Gamma(1-s)}\mathscr{N}^{\mathscr{A}}_{2s,1}(f). \tag{3.7}$$

We now want to provide a lower bound for the left-hand side of (3.7) in terms of $||f||_{L^{q_0}+L^{q_\infty}}$. To this aim, we denote

$$\sigma_f = \sup\{\sigma > 0 \,:\, G(\sigma) > 1\}.$$

If $|E_\sigma| \le 1$ for all σ, we agree to let $\sigma_f = 0$. Since $f \in L^1$ we have that $\sigma_f \in [0,\infty)$. We consider

$$f_1(X) = f(X)\mathbf{1}_{E_{\sigma_f}}(X) \quad \text{and} \quad f_2(X) = f(X)(1-\mathbf{1}_{E_{\sigma_f}}(X)). \tag{3.8}$$

We notice that $f_1(X)$ and $f_2(X)$ cannot be both non-null for the same X, and in particular the following holds true

$$|f(X)| = |f_1(X) + f_2(X)| = |f_1(X)| + |f_2(X)|.$$

We also make use of the notation $E^i_\sigma = \{X \in \mathbb{R}^N \,:\, |f_i(X)| > \sigma\}$ for $i \in \{1,2\}$. One can check the following relations (see also [9, pg. 38])

$$E^1_\sigma = \begin{cases} E_\sigma & \text{if } \sigma > \sigma_f, \\ E_{\sigma_f} & \text{if } \sigma \le \sigma_f \end{cases} \qquad \text{and} \qquad E^2_\sigma = \begin{cases} \varnothing & \text{if } \sigma > \sigma_f, \\ E_\sigma \setminus E_{\sigma_f} & \text{if } \sigma \le \sigma_f. \end{cases}$$

Since $|E_\sigma^1| \le 1$ and $|E_\sigma^1| \le G(\sigma)$ for all σ, by (3.6) we obtain

$$\begin{aligned} ||f_1||_{q_0} &= \left(q_0 \int_0^\infty \sigma^{q_0 - 1} |E_\sigma^1| d\sigma \right)^{\frac{1}{q_0}} \le \int_0^\infty |E_\sigma^1|^{\frac{1}{q_0}} d\sigma \\ &= \int_0^\infty \min \left\{ |E_\sigma^1|^{\frac{1}{q_0}}, |E_\sigma^1|^{\frac{1}{q_\infty}} \right\} d\sigma \\ &\le \int_0^\infty \min \left\{ G(\sigma)^{\frac{1}{q_0}}, G(\sigma)^{\frac{1}{q_\infty}} \right\} d\sigma. \end{aligned} \tag{3.9}$$

On the other hand, since $G(\sigma)$ is bigger than 1 on the interval $(0, \sigma_f)$, by (3.6) we also have

$$\begin{aligned} ||f_2||_{q_\infty} &= \left(q_\infty \int_0^\infty \sigma^{q_\infty - 1} |E_\sigma^2| d\sigma \right)^{\frac{1}{q_\infty}} \le \int_0^\infty |E_\sigma^2|^{\frac{1}{q_\infty}} d\sigma = \int_0^{\sigma_f} |E_\sigma^2|^{\frac{D_\infty - 2s}{D_\infty}} d\sigma \\ &\le \int_0^{\sigma_f} G(\sigma)^{\frac{1}{q_\infty}} d\sigma \\ &= \int_0^{\sigma_f} \min \left\{ G(\sigma)^{\frac{1}{q_0}}, G(\sigma)^{\frac{1}{q_\infty}} \right\} d\sigma. \end{aligned} \tag{3.10}$$

The combination of (3.9) and (3.10) yields

$$||f_1||_{q_0} + ||f_2||_{q_\infty} \le 2 \int_0^\infty \min \left\{ G(\sigma)^{\frac{1}{q_0}}, G(\sigma)^{\frac{1}{q_\infty}} \right\} d\sigma,$$

Keeping in mind (3.7), we have just proved that

$$||f||_{L^{q_0} + L^{q_\infty}} \le ||f_1||_{q_0} + ||f_2||_{q_\infty} \le \frac{4s S_{1,s}}{\Gamma(1-s)} \mathscr{N}_{2s,1}^{\mathscr{A}}(f) \quad \text{for every } f \in C^\infty \cap \mathfrak{B}_{2s,1}^{\mathscr{A}}. \tag{3.11}$$

The density of C^∞ in $\mathfrak{B}_{2s,1}^{\mathscr{A}}$ implies then the validity of (1.13) with the choice

$$c = \frac{\Gamma(1-s)}{4s S_{1,s}}.$$

Finally, since it is not completely obvious to deduce from (1.13) the nonlocal isoperimetric inequality stated in Theorem 1.4, we provide here the details. Let us thus take a set $E \subset \mathbb{R}^N$ with finite s-perimeter. We can consider a sequence $\{f_k\}_{k \in \mathbb{N}}$

of Friedrichs' mollifiers for the function $\mathbf{1}_E$. We recall that

$$f_k \in C^\infty \cap \mathfrak{B}^{\mathscr{A}}_{2s,1}, \qquad\qquad 0 \le f_k \le 1.$$

If we let $(f_k)_1$ and $(f_k)_2$ the splitting of the function f_k according to (3.8), since $0 \le (f_k)_1, (f_k)_2 \le 1$ we have

$$\begin{aligned}
\min&\left\{\left(\int_{\mathbb{R}^N} |(f_k)_1(X)|^{q_0} dX + \int_{\mathbb{R}^N} |(f_k)_2(X)|^{q_\infty} dX\right)^{\frac{1}{q_0}},\right. \\
&\left.\left(\int_{\mathbb{R}^N} |(f_k)_1(X)|^{q_0} dX + \int_{\mathbb{R}^N} |(f_k)_2(X)|^{q_\infty} dX\right)^{\frac{1}{q_\infty}}\right\} \\
\ge \min&\left\{\left(\int_{\mathbb{R}^N} |(f_k)_1(X)|^{q_\infty} dX + \int_{\mathbb{R}^N} |(f_k)_2(X)|^{q_\infty} dX\right)^{\frac{1}{q_0}},\right. \\
&\left.\left(\int_{\mathbb{R}^N} |(f_k)_1(X)|^{q_\infty} dX + \int_{\mathbb{R}^N} |(f_k)_2(X)|^{q_\infty} dX\right)^{\frac{1}{q_\infty}}\right\} \\
= \min&\left\{\|f_k\|_{q_\infty}^{\frac{q_\infty}{q_0}}, \|f_k\|_{q_\infty}\right\},
\end{aligned} \tag{3.12}$$

where we used (3.6) and the fact that $(f_k)_1(f_k)_2 \equiv 0$. On the other hand, denoting $c(q_0, q_\infty) = \min\{x^{\frac{1}{q_0}} + y^{\frac{1}{q_\infty}} \ : \ x, y \ge 0, \ \text{and } x + y = 1\}$, we have the validity of the simple inequality

$$x^{\frac{1}{q_0}} + y^{\frac{1}{q_\infty}} \ge c(q_0, q_\infty) \min\left\{(x+y)^{\frac{1}{q_0}}, (x+y)^{\frac{1}{q_\infty}}\right\} \qquad \text{for all } x, y \ge 0,$$

which implies

$$\begin{aligned}
\|(f_k)_1\|_{q_0} + \|(f_k)_2\|_{q_\infty} &\ge c(q_0, q_\infty) \\
&\min\left\{\left(\int_{\mathbb{R}^N} |(f_k)_1(X)|^{q_0} dX + \int_{\mathbb{R}^N} |(f_k)_2(X)|^{q_\infty} dX\right)^{\frac{1}{q_0}},\right. \\
&\left.\left(\int_{\mathbb{R}^N} |(f_k)_1(X)|^{q_0} dX + \int_{\mathbb{R}^N} |(f_k)_2(X)|^{q_\infty} dX\right)^{\frac{1}{q_\infty}}\right\}.
\end{aligned} \tag{3.13}$$

Putting together (3.11) with (3.12)–(3.13), we obtain

$$\begin{aligned}\mathscr{N}^{\mathscr{A}}_{2s,1}(f_k) &\geq \frac{\Gamma(1-s)}{4s S_{1,s}} \left(\|(f_k)_1\|_{q_0} + \|(f_k)_2\|_{q_\infty}\right) \\ &\geq \frac{\Gamma(1-s)c(q_0,q_\infty)}{4s S_{1,s}} \min\left\{\|f_k\|_{q_\infty}^{\frac{q_\infty}{q_0}}, \|f_k\|_{q_\infty}\right\}\end{aligned}$$

for any $k \in \mathbb{N}$. By letting $k \to \infty$, since $f_k \to \mathbf{1}_E$ in every L^p-space and also in $\mathfrak{B}^{\mathscr{A}}_{2s,1}$ (see [3, Proposition 3.2, Step I]), we deduce the desired

$$\mathfrak{P}^{\mathscr{A}}_s(E) \geq \frac{\Gamma(1-s)c(q_0,q_\infty)}{4s S_{1,s}} \min\left\{|E|^{\frac{1}{q_0}}, |E|^{\frac{1}{q_\infty}}\right\}.$$

□

Acknowledgments We wish to thank Stéphane Menozzi, Andrea Pascucci, and Sergio Polidoro for the organization and the kind invitation to the conference "*Kolmogorov operators and their applications*" held in June 2022 in Cortona.

Both authors are supported in part by a Progetto SID: "Aspects of nonlocal operators via fine properties of heat kernels", University of Padova, 2022. The first author has also been partially supported by a Visiting Professorship at the Arizona State University. The second author has been partially supported by the Gruppo Nazionale per l'Analisi Matematica, la Probabilità e le loro Applicazioni (GNAMPA) of the Istituto Nazionale di Alta Matematica (INdAM).

References

1. Almgren, F.J. Jr., Lieb, E.H.: Symmetric decreasing rearrangement is sometimes continuous. J. Am. Math. Soc. **2**(4), 683–773 (1989)
2. Anceschi, F., Polidoro, S.: A survey on the classical theory for Kolmogorov equation. Matematiche (Catania) **75**, 221–258 (2020)
3. Buseghin, F., Garofalo, N., Tralli, G.: On the limiting behaviour of some nonlocal seminorms: a new phenomenon. Ann. Sc. Norm. Super. Pisa Cl. Sci. **23**, 837–875 (2022)
4. Caffarelli, L., Roquejoffre, J.M., Savin, O.: Nonlocal minimal surfaces. Commun. Pure Appl. Math. **63**(9), 1111–1144 (2010)
5. De Giorgi, E.: Su una teoria generale della misura (r-1)-dimensionale in uno spazio ad r dimensioni. Ann. Mat. Pura Appl. **36**, 191–213 (1954)
6. Di Francesco, M., Pascucci, A.: On a class of degenerate parabolic equations of Kolmogorov type. AMRX Appl. Math. Res. Express **3**, 77–116 (2005)
7. Frank, R.L., Seiringer, R.: Non-linear ground state representations and sharp Hardy inequalities. J. Funct. Anal. **255**(12), 3407–3430 (2008)
8. Garofalo, N., Tralli, G.: Functional inequalities for class of nonlocal hypoelliptic equations of Hörmander type. Nonlinear Anal. **193**, 111567 (2020). Special issue 'Nonlocal and Fractional Phenomena'
9. Garofalo, N., Tralli, G.: Nonlocal isoperimetric inequalities for Kolmogorov-Fokker-Planck operators. J. Funct. Anal. **279**, 108591 (2020)

10. Garofalo, N., Tralli, G.: A class of nonlocal hypoelliptic operators and their extensions. Ind. Univ. Math. J. **70**, 1717–1744 (2021)
11. Garofalo, N., Tralli, G.: Hardy-Littlewood-Sobolev inequalities for a class of non-symmetric and non-doubling hypoelliptic semigroups. Math. Ann. **383**, 1–38 (2022)
12. Garofalo, N., Tralli, G.: Heat kernels for a class of hybrid evolution equations. Potential Anal. **59**, 823–856 (2023)
13. Hörmander, L.: Hypoelliptic second order differential equations. Acta Math. **119**, 147–171 (1967)
14. Kolmogorov, A.N.: Zufällige Bewegungen (Zur Theorie der Brownschen Bewegung). Ann. Math. **35**, 116–117 (1934)
15. Lanconelli, E., Polidoro, S.: On a class of hypoelliptic evolution operators, Partial differential equations, II (Turin, 1993). Rend. Sem. Mat. Univ. Politec. Torino **52**(1), 29–63 (1994)
16. Ledoux, M.: Semigroup proofs of the isoperimetric inequality in Euclidean and Gauss space. Bull. Sci. Math. **118**(6), 485–510 (1994)
17. Maz'ya, V.: Lectures on isoperimetric and isocapacitary inequalities in the theory of Sobolev spaces. In: Heat Kernels and Analysis on Manifolds, Graphs, and Metric Spaces (Paris, 2002). Contemporary Mathematics, vol. 338, pp. 307–340. American Mathematical Society, Providence (2003)
18. Maz'ya, V.: Sobolev Spaces with Applications to Elliptic Partial Differential Equations. Second, revised and augmented edition. Grundlehren der Mathematischen Wissenschaften [Fundamental Principles of Mathematical Sciences], vol. 342. Springer, Heidelberg (2011)
19. Menozzi, S.: Parametrix techniques and martingale problems for some degenerate Kolmogorov equations. Electron. Commun. Probab. **16**, 234–250 (2011)
20. Miranda, M., Jr., Pallara, D., Paronetto, F., Preunkert, M.: Short-time heat flow and functions of bounded variation in $\mathbb{R}^N$. Ann. Fac. Sci. Toulouse Math. **16**(1), 125–145 (2007)
21. Preunkert, M.: A semigroup version of the isoperimetric inequality. Semigroup Forum **68**(2), 233–245 (2004)
22. Stein, E.M.: Topics in Harmonic Analysis Related to the Littlewood-Paley Theory. Princeton University Press, Princeton (1970)
23. Varopoulos, N.T.: Hardy-Littlewood theory for semigroups. J. Funct. Anal. **63**, 240–260 (1985)

Intrinsic Taylor Formula for Non-homogeneous Kolmogorov-Type Lie Groups

Stefano Pagliarani and Michele Pignotti

Abstract In this note we prove an intrinsic Taylor-like formula for a class of Lie groups arising in the study of some Kolmogorov operators of degenerate-type. The estimate of the remainder is in terms of the intrinsic norm induced by such operators. These results naturally extend those in Pagliarani et al. (J Math Anal Appl 435:1054–1087, 2016), where a full characterization of the intrinsic Hölder spaces and their Taylor polynomials were given under the additional assumption that the Lie group is homogeneous in the sense of Folland and Stein (Hardy Spaces on Homogeneous Groups. Mathematical Notes, vol. 28. Princeton University Press, Princeton, 1982). The intrinsic Taylor polynomial admits the same representation as in the homogeneous case.

1 Introduction

Consider the non-commutative group $\mathcal{G}_B = (\mathbb{R} \times \mathbb{R}^d, \circ)$ defined by

$$(t, x) \circ (s, \xi) = \left(t + s, e^{sB}x + \xi\right), \quad (t, x)^{-1} = \left(-t, -e^{-tB}x\right),$$
$$\mathrm{Id} = (0, 0), \quad (t, x), (s, \xi) \in \mathbb{R} \times \mathbb{R}^d,$$

S. Pagliarani (✉)
Dipartimento di Matematica, Università di Bologna, Bologna, Italy
e-mail: stefano.pagliarani9@unibo.it

M. Pignotti
CSE Consorzio Servizi Bancari, San Lazzaro di Savena, Italy

S. Menozzi et al. (eds.), *Kolmogorov Operators and Their Applications*,
Springer INdAM Series 56, https://doi.org/10.1007/978-981-97-0225-1_6

where B is a $(d \times d)$-square matrix such that

$$B = \begin{pmatrix} B_{0,0} & B_{0,1} & \cdots & B_{0,r-1} & B_{0,r} \\ B_{1,0} & B_{1,1} & \cdots & B_{1,r-1} & B_{1,r} \\ 0 & B_{2,1} & \cdots & B_{2,r-1} & B_{2,r} \\ \vdots & \vdots & \ddots & \vdots & \vdots \\ 0 & 0 & \cdots & B_{r,r-1} & B_{r,r} \end{pmatrix}, \qquad \begin{array}{l} \bullet\ B_{i,j} \in \mathcal{M}^{p_i \times p_j} \\ \bullet\ B_{j,j-1} \text{ have rank } p_j \\ \bullet\ p_0 \geq p_1 \geq \cdots \geq p_r \geq 1 \\ \bullet\ p_0 + p_1 + \cdots + p_r = d. \end{array} \tag{1}$$

Hereafter, $\mathcal{M}^{p\times q}$ denotes the spaces of the $(p \times q)$-matrices with real entries, while I_p will denote the $(p \times p)$ identity matrix. The group $\mathcal{G}_B$ plays a crucial role in the study of degenerate Kolmogorov operators

$$\mathcal{L} = \frac{1}{2}\sum_{i=1}^{p_0} A_{ij}(t,x)\partial_{x_i x_j} + \langle Bx, \nabla_x\rangle + \partial_t, \qquad (t,x) \in \mathbb{R}\times\mathbb{R}^d, \qquad 1 \leq p_0 \leq d,$$

with $A(t,x) \in \mathcal{M}^{p_0\times p_0}$, which is closely related to averaged-diffusion stochastic processes and whose applications include Mathematical Finance and Physics among others. In particular, $\mathcal{G}_B$ was first introduced in [10] as a group of left-invariant translations for the operator $\mathcal{L}_0$ obtained by $\mathcal{L}$ setting $A = I_{p_0}$. Defining

$$X_j := \partial_{x_j}, \quad j = 1, \ldots, p_0, \qquad \text{and} \qquad Y := \langle Bx, \nabla_x\rangle + \partial_t,$$

in [10] it was proved that (1) is equivalent to the Hörmander's condition, namely $\text{rank}\big(\text{Lie}(X_1, \ldots, X_{p_0}, Y)\big) = d + 1$, which in turn implies that $\mathcal{L}_0$ is hypoelliptic. Later (see [4, 19, 21] among others), several results for an $\mathcal{L}$-like operator with B as in (1), including regularity estimates and Gaussian upper bounds for its fundamental solution, were proved under various regularity assumptions and with $A(t,x)$ satisfying a uniform ellipticity condition on $\mathbb{R}^{p_0}$. In this stream of literature, the classical notions of regularity based on the Euclidean distance are replaced with intrinsic notions of regularity related to the B-semi-distance on $\mathbb{R}\times\mathbb{R}^d$ given by

$$\|\zeta^{-1}\circ z\|_B, \qquad z = (t,x), \zeta = (s,\xi) \in \mathbb{R}\times\mathbb{R}^d, \tag{2}$$

which was first introduced in [10], where the B-norm $\|\cdot\|_B$ is defined as

$$\|(t,x)\|_B := |t|^{1/2} + |x|_B, \qquad |x|_B := \sum_{j=0}^{r}\sum_{i=\bar{p}_{j-1}+1}^{\bar{p}_j} |x_i|^{2j+1}, \qquad \bar{p}_j := \sum_{k=0}^{j} p_k. \tag{3}$$

It is now standard practice to conduct the study of $\mathcal{L}$-like operators by taking into account the intrinsic geometry induced by the operator, in particular by adopting suitable intrinsic Hölder spaces related to the semi-distance (2). While customized versions of such spaces were employed in several works (see [1, 3, 5, 13–15] and [21] among others), a complete characterization of the intrinsic Hölder spaces at any order and a systematic study of the related intrinsic Taylor polynomials were first performed in [16], under the additional assumption that the blocks $(B_{i,j})_{i\leq j}$ in (1) are null. Taylor formulas are crucial to uncover the full regularizing effect of the fundamental solution of $\mathcal{L}$-like operators, and are relevant to derive regularity estimates of Schauder-type (see [8, 9, 20]), asymptotic expansions (see [17]) and Itô-type formulas (see [11]). In [16], the theory of intrinsic regularity was developed so as to mimic the general classical theory, namely: first one defines, recursively, the intrinsic Hölder space $C_B^{n,\alpha}$ by only specifying the regularity along the vector fields $X_1, \cdots, X_{p_0}, Y$, and then one proves an n-th order Taylor formula for a function $u \in C_B^{n,\alpha}$ with remainder expressed in terms of the intrinsic semi-distance (2). In this note we show how the program pursued in [16] can be extended to the case of a general matrix B satisfying (1); i.e.:

1. give a definition of intrinsic $C_B^{n,\alpha}$ regularity for $n \in \mathbb{N}_0$, $\alpha \in]0, 1]$; this appears to be the minimal regularity required in order for a Taylor formula to hold with an estimate of the remainder proportional to $\|u\|_{C_B^{n,\alpha}} \|\zeta^{-1} \circ z\|_B^{n+\alpha}$;
2. prove an explicit expression for the Taylor polynomial $T_n u(\zeta, \cdot)$ centered in $\zeta = (s, \xi) \in \mathbb{R} \times \mathbb{R}^d$ of a function $u \in C_B^{n,\alpha}$, namely

$$T_n u(\zeta, z) := \sum_{\substack{k\in\mathbb{N}_0,\, \beta\in\mathbb{N}_0^d \\ 0\leq 2k+|\beta|_B\leq n}} \frac{1}{k!\,\beta!}\big(Y^k \partial_\xi^\beta u(s, \xi)\big)(t-s)^k \big(x - e^{(t-s)B}\xi\big)^\beta,$$

$$z = (t, x) \in \mathbb{R} \times \mathbb{R}^d, \tag{4}$$

where $|\beta|_B := \sum_{j=0}^{r} \sum_{i=\bar{p}_{j-1}+1}^{\bar{p}_j} (2j+1)\beta_i$ is the B-length of β, $\partial_\xi^\beta := \partial_{\xi_1}^{\beta_1} \partial_{\xi_2}^{\beta_2} \cdots \partial_{\beta_d}^{\xi_d}$ is a Euclidean multi-derivative, whereas Y^k is meant as a k-th order Lie derivative.

We recall that, from the algebraic stand-point, the hypothesis $B_{i,j} = 0$, $i \leq j$, previously employed in [16] is equivalent to saying that $\mathcal{G}_B$ is homogeneous (in the sense of [6]) with respect to the family of automorphisms $(D(\lambda))_{\lambda>0}$, called dilations, defined as $D(\lambda) = \mathrm{diag}\big(\lambda^2, \lambda I_{p_0}, \lambda^3 I_{p_1}, \cdots, \lambda^{2r+1} I_{p_r}\big)$, with respect to which the norm $\|\cdot\|_B$ is homogeneous of degree one. Due to this reason the latter is sometimes referred to as *homogeneous* norm. Previous results about intrinsic Taylor polynomials on homogeneous Lie groups were proved in greater generality in [6], and also recently in [2]. However, in these references the Taylor polynomials were defined for functions that are differentiable in the Euclidean sense; consequently the constants appearing in the estimates of the remainder would depend on the Euclidean Hölder norms. Moreover, the number of terms appearing in the Taylor polynomials grew exponentially with n, while in (3) it grows only linearly. More importantly, those results strongly rely on the use of the dilations and thus on the homogeneity assumption on $\mathcal{G}_B$. By opposite, the technique adopted in [16] owes to the fact that the B-intrinsic norm in (3) is somehow well-behaved with respect to the *stratification*

$$\begin{aligned}\mathbb{R} \times \mathbb{R}^d &= \mathbb{R} \times \mathbb{R}^{p_0} \times \mathbb{R}^{p_1} \times \cdots \times \mathbb{R}^{p_r} \\ &\cong \underbrace{\text{span}(Y)|_{x=0}}_{=:W_{\text{time}}} \oplus \underbrace{\text{span}(X_1, \cdots, X_{p_0})}_{=:W_0} \oplus \underbrace{[W_0, W_{\text{time}}]|_{x=0}}_{=:W_1} \\ &\quad \oplus \cdots \oplus \underbrace{[W_{r-1}, W_{\text{time}}]|_{x=0}}_{=:W_r}\end{aligned}$$

equipped with the *formal degrees* $\mathfrak{d}(W_0)$ and $\mathfrak{d}(W_{\text{time}})$ equal to 1 and 2 respectively, and the formal degrees $\mathfrak{d}(W_i) = \mathfrak{d}(W_{i-1}) + \mathfrak{d}(W_{\text{time}})$ for $i = 1, \cdots, r$; the extension to the non-homogeneous case relies on the fact that $\|\cdot\|_B$ remains well-behaved w.r.t. to this stratification. Also note that, $T_n u(\zeta, \cdot)$ admits the same representation as in the homogeneous case, though it ceases to be a polynomial w.r.t. the time-variable as the matrix B is in general not nilpotent.

We point out that slightly weaker modifications of the intrinsic space $C_B^{2,\alpha}$ of the current note, and their related Taylor formulas, appear in [20] and [12]. Beyond the Hölder setting, in [23] (and in a series of related works) the authors consider anisotropic distributional Besov spaces in the spirit of [22]. These spaces, however, do not take into account the regularity along the vector field Y. Finally, recent results for intrinsic Sobolev spaces were derived in [18] and [7].

2 Intrinsic Hölder Spaces and Taylor Formula

We start by introducing the notions of B-intrinsic regularity in terms of suitable B-intrinsic Hölder spaces. For any $z = (t, x) \in \mathbb{R} \times \mathbb{R}^d$, we denote by

$$e^{\delta X_i}(t, x) = (t, x + \delta e_i), \quad i = 1, \cdots, p_0, \quad e^{\delta Y}(t, x) = (t + \delta, e^{\delta B} x), \quad \delta > 0, \tag{1}$$

the integral curves of the fields $X_1, \cdots, X_{p_0}, Y$ starting at z. Here e_i denotes the i-th element of the canonical basis of $\mathbb{R}^d$. Now, let Ω be a domain in $\mathbb{R} \times \mathbb{R}^d$. For any $z \in \Omega$ we set

$$\delta_z := \sup \left\{ \bar{\delta} \in]0, 1] \mid e^{\delta X_1}(z), \cdots, e^{\delta X_{p_0}}(z), e^{\delta Y}(z) \in \Omega \text{ for any } \delta \in [-\bar{\delta}, \bar{\delta}] \right\}.$$

If Ω_0 is a bounded domain with $\overline{\Omega_0} \subseteq \Omega$, we set $\delta_{\Omega_0} = \min_{z \in \overline{\Omega_0}} \delta_z$. Note that $\delta_{\Omega_0} \in]0, 1]$.

Definition 2.1 For $\alpha_1 \in]0, 1]$ and $\alpha_2 \in]0, 2]$, we say that $u \in C_{X_i}^{\alpha_1}(\Omega)$, $i = 1, \cdots, p_0$, and $v \in C_Y^{\alpha_2}(\Omega)$, if for any bounded domain Ω_0 with $\overline{\Omega_0} \subseteq \Omega$, the following semi-norms are finite:

$$\|u\|_{C^{\alpha_1}_{X_i}(\Omega_0)} := \sup_{\substack{z\in\Omega_0 \\ 0<|\delta|<\delta_{\Omega_0}}} \frac{\left|u\left(e^{\delta X_i}(z)\right) - u(z)\right|}{|\delta|^{\alpha_1}}, \qquad i = 1, \cdots, p_0,$$

$$\|v\|_{C^{\alpha_2}_{Y}(\Omega_0)} := \sup_{\substack{z\in\Omega_0 \\ 0<|\delta|<\delta_{\Omega_0}}} \frac{\left|v\left(e^{\delta Y}(z)\right) - v(z)\right|}{|\delta|^{\frac{\alpha_2}{2}}}.$$

We can now define the B-intrinsic Hölder spaces on $\mathcal{G}_B$ (B-Hölder spaces).

Definition 2.2 Let $\alpha \in]0, 1]$, then:

(i) $u \in C^{0,\alpha}_B(\Omega)$ if $u \in C^{\alpha}_Y(\Omega)$ and $u \in C^{\alpha}_{X_i}(\Omega)$ for any $i = 1, \ldots, p_0$;
(ii) $u \in C^{1,\alpha}_B(\Omega)$ if $u \in C^{1+\alpha}_Y(\Omega)$ and $\partial_{x_i} u \in C^{0,\alpha}_B(\Omega)$ for any $i = 1, \ldots, p_0$;
(iii) For $k \in \mathbb{N}$ with $k \geq 2$, $u \in C^{k,\alpha}_B(\Omega)$ if $Yu \in C^{k-2,\alpha}_B(\Omega)$ and $\partial_{x_i} u \in C^{k-1,\alpha}_B(\Omega)$ for any $i = 1, \ldots, p_0$.

Moreover, the space $C^{k,\alpha}_B(\Omega)$ is equipped, for any bounded domain Ω_0 compactly contained in Ω, with the Ω_0-seminorm

$$\|u\|_{C^{k,\alpha}_B(\Omega_0)} := \begin{cases} \|u\|_{C^{\alpha}_Y(\Omega_0)} + \sum_{i=1}^{p_0} \|u\|_{C^{\alpha}_{X_i}(\Omega_0)}, & k = 0 \\ \|u\|_{C^{\alpha+1}_Y(\Omega_0)} + \sum_{i=1}^{p_0} \|\partial_{x_i} u\|_{C^{0,\alpha}_B(\Omega_0)}, & k = 1 \\ \|Yu\|_{C^{k-2,\alpha}_B(\Omega_0)} + \sum_{i=1}^{p_0} \|\partial_{x_i} u\|_{C^{k-1,\alpha}_B(\Omega_0)}, & k \geq 2. \end{cases}$$

Recall that, in the previous definition as well as in the whole paper, Yu is meant as a Lie derivative.

Theorem 2.1 *Let Ω be a domain of $\mathbb{R} \times \mathbb{R}^d$, $\alpha \in]0, 1]$ and $n \in \mathbb{N}_0$. If $u \in C^{n,\alpha}_B(\Omega)$ then it holds:*

(1) there exist

$$Y^k \partial^{\beta}_x u \in C^{n-2k-|\beta|_B,\alpha}_B(\Omega), \qquad 0 \leq 2k + |\beta|_B \leq n;$$

(2) for any $\zeta_0 \in \Omega$, there exist two bounded domains U, V, such that $\zeta_0 \in U \subset V \subset \Omega$ and

$$|u(z) - T_n u(\zeta, z)| \leq c_{B,U} \|u\|_{C^{n,\alpha}_B(V)} \|\zeta^{-1} \circ z\|^{n+\alpha}_B, \qquad z, \zeta \in U, \tag{2}$$

where $c_{B,U}$ *is a positive constant and* $T_n u(\zeta, \cdot)$ *is the* n-th order B-intrinsic Taylor polynomial of u centered at ζ *as defined in* (4).

3 Proof of Theorem 2.1

For sake of brevity, we only prove the statement for $r = 1$, which is $B = (B_{i,j})_{i,j\in\{0,1\}}$ with $B_{i,j} \in \mathcal{M}^{p_i \times p_j}$ and $B_{1,0}$ has full rank. This case is complex enough to see the conceptual difficulties that arise from dropping the homogeneity assumption on $\mathcal{G}_B$. On the other hand, the proof for a general $r \geq 1$ is only a lengthy and technical extension.

Notation 3.1 Throughout this section we will use the notation $z = (t, x, y)$ or $\zeta = (s, \xi, \eta)$ to indicate a general element of $\mathbb{R} \times \mathbb{R}^{p_0} \times \mathbb{R}^{p_1}$. Moreover, we will denote by c any positive constant that depends on B and on the domain U in Theorem 2.1, at most.

The first task is connecting two points in $\mathbb{R} \times \mathbb{R}^{p_0} \times \mathbb{R}^{p_1}$ using integral curves. To obtain an increment in the x-variables it is enough to move along the integral curves of the fields $X_1, \cdots, X_{p_0}$, i.e.

$$\gamma^{(0)}_{v,\delta}(t, x, y) := (t, x + \delta v, y), \qquad v \in \mathbb{R}^{p_0},\ \delta \in \mathbb{R}.$$

To understand how to obtain an increment in the y-variables, it is useful to observe that, formally, we have

$$[v_1 X_1 + \cdots + v_{p_0} X_{p_0}, Y] - \langle \nabla_x, B_{0,0} v\rangle = \langle \nabla_y, B_{1,0} v\rangle, \qquad v \in \mathbb{R}^{p_0}. \tag{1}$$

It is thus reasonable to approximate the integral curves of the vector field on the right-hand side as

$$\begin{aligned}\gamma^{(1)}_{v,\delta}(t, x, y) &:= \gamma^{(0)}_{B_{0,0}v, -\delta^3}\left(e^{-\delta^2 Y}\left(\gamma^{(0)}_{v,-\delta}\left(e^{\delta^2 Y}\left(\gamma^{(0)}_{v,\delta}(t, x, y)\right)\right)\right)\right) \\ &= \left(t, x, y + \delta^3 B_{1,0} v\right) - \delta^5\left(0, \sum_{n=0}^{\infty} \frac{(-1)^n \delta^{2n}}{(n+2)!} B^{n+2}(v, 0)^\top\right), \\ &\qquad v \in \mathbb{R}^{p_0},\ \delta \in \mathbb{R}.\end{aligned} \tag{2}$$

The leading order increment is proportional to δ^3, along the y variable only. However, due to the non-homogeneous structure of B (the block $B_{0,0}$ is not null), the higher order increment affects both the components x and y. To correct this, we employ again the curve $\gamma^{(0)}$. Set

$$g_{v,\delta}(t, x, y) := \gamma^{(0)}_{v', \delta^5}\left(\gamma^{(1)}_{v,\delta}(t, x, y)\right), \qquad v \in \mathbb{R}^{p_0},\ \delta \in \mathbb{R}, \tag{3}$$

where

$$v' = v'(\delta, v) = \sum_{n=0}^{\infty} \frac{(-1)^n \delta^{2n}}{(n+2)!} B_{0,0}^{n+2} v \tag{4}$$

and $B_{0,0}^{n+2}$ is the top-left $(p_0 \times p_0)$-submatrix of B^{n+2}.

Lemma 3.2 *There exists $\varepsilon > 0$, only dependent on B, such that: for any $\eta \in \mathbb{R}^{p_1}$ with $|\eta| \le \varepsilon$, there exist $v \in \mathbb{R}^{p_0}$ with $|v| = 1$ and $\delta \ge 0$ such that*

$$g_{v,\delta}(t, x, y) = (t, x, y + \eta), \qquad \textit{and} \qquad |\delta| \le \frac{2}{\|B_{1,0}\|} |\eta|^{\frac{1}{3}}. \tag{5}$$

Proof By (3) and (2) we obtain

$$g_{v,\delta}(t, x, y) - (t, x, y) = \big(0, 0, \delta^3 R(\delta, v)\big), \quad R(\delta, v) := \sum_{n=0}^{\infty} \frac{(-1)^n \delta^{2n}}{(n+1)!} B_{1,0}^{n+1} v,$$

where $B_{1,0}^{n+1}$ is the bottom-left $(p_1 \times p_0)$-submatrix of B^{n+1}. Therefore, denoting by $\mathbb{S}^{p_0-1}$ the unitary sphere in $\mathbb{R}^{p_0}$, we have to find some $(\delta, v) \in [0, \infty[\times \mathbb{S}^{p_0-1}$ that solves the equation

$$\delta^3 R(\delta, v) = \eta. \tag{6}$$

Since $B_{1,0}$ has full rank it is not restrictive to assume $p_0 = p_1$, and thus $B_{1,0}$ invertible. In particular, $R(0, v) = B_{1,0} v$, which implies that $R(0, \cdot)$ is a bijective and linear function. Moreover, there exists $\bar{\delta} > 0$, only dependent on B, such that

$$|R(\delta, v)| > \|B_{1,0}\|/2, \qquad v \in \mathbb{S}^{p_0-1}, \ \delta \le \bar{\delta} \tag{7}$$

and $R(\delta, \cdot)$ is still a bijective linear function for any $\delta \le \bar{\delta}$. In particular, when restricted to $\mathbb{S}^{p_0-1}$, $(\delta^3 R(\delta, \cdot))_{0 \le \delta \le \bar{\delta}}$ is a continuous family of embeddings that collapses to zero at $\delta = 0$. Therefore, by employing Jordan-Brouwer's Theorem, it is possible to prove that Eq. (6) admits a solution $\big(\delta(\eta), v(\eta)\big) \in [0, \bar{\delta}] \times \mathbb{S}^{p_0-1}$ for any $|\eta| \le \varepsilon$, where $\varepsilon > 0$ only depends on B. Eventually, (6) and (7) imply the second part of (5). □

We are now in the position to prove Theorem 2.1.

Proof of Theorem 2.1 Like in the homogeneous setting, the cases $n = 0, 1, 2, 3$ have to be proved separately, while for $n > 3$ the proof can be done by induction on n. For sake of brevity, here we only provide a proof for $n = 0$ and $n = 3$, these being the most interesting and difficult steps. On the one hand, the proof for $n = 0$ allows to appreciate how the connection Lemma 3.2, together with the regularity along the fields, can be used, in a rather simple way, in order to obtain the most basic result: the Hölder-continuity with respect to the B-semi-distance (2). On the other hand, the case $n = 3$ exhibits the main difficulty of the whole proof, namely proving the existence of the first order partial derivative w.r.t. the y-variable. Note that the existence of the latter is not trivially ensured by the definition of $C_B^{3,\alpha}(\Omega)$, as the existence of $X_i Y u$ and $Y X_i u$, and thus the commutators $[X_i, Y]u$, are only meant in the sense of Lie derivatives. As for the steps $n = 1$, $n = 2$, these are just simpler versions of the case $n = 3$, whereas the inductive step for $n > 3$ is totally analogous to the homogeneous case.

Case $n = 0$: We only need to prove Part (2). Let $U \subset \Omega$ be a bounded domain suitably small so as to ensure that all the integral curves that are employed below to connect z and ζ are entirely contained in a bounded domain V, compactly contained in Ω. The first step is to bound the increment w.r.t. the time variable by employing the integral curve of Y in (1):

$$\begin{aligned} &|u(t,x,y) - u(s,\xi,\eta)| \\ &\leq \left|u(t,x,y) - u\big(e^{(t-s)Y}(s,\xi,\eta)\big)\right| + \left|u\big(e^{(t-s)Y}(s,\xi,\eta)\big) - u(s,\xi,\eta)\right| \\ &\leq \left|u(t,x,y) - u\big(t, e^{(t-s)B}(\xi,\eta)^\top\big)\right| + c\,\|u\|_{C_Y^\alpha(V)}|s-t|^{\frac{\alpha}{2}}, \end{aligned} \tag{8}$$

where we used triangular inequality in the first line and $u \in C_Y^\alpha(\Omega)$ in the second. Note that $\left|(x,y)^\top - e^{(t-s)B}(\xi,\eta)^\top\right|_B^\alpha \leq \|\zeta^{-1} \circ z\|_B^\alpha$ and thus we only need to prove

$$|u(t,\xi,\eta) - u(t,x,y)| \leq c\,\|u\|_{C_B^{0,\alpha}(V)}(|\xi - x| + |\eta - y|^{\frac{1}{3}})^\alpha. \tag{9}$$

We can use again triangular inequality and write

$$\begin{aligned} u(t,\xi,\eta) - u(t,x,y) &= \big(u(t,\xi,\eta) - u(t,x,\eta)\big) + \big(u(t,x,\eta) - u(t,x,y)\big) \\ &= \big(u(t,\xi,\eta) - u(t,x,\eta)\big) \\ &\quad + \Big(u\big(g_{\delta,v}(t,x,y)\big) - u(t,x,y)\Big) \quad \text{(by Lemma 3.2)} \end{aligned}$$

with $|v| = 1$ and $|\delta| \leq c|\eta - y|^{\frac{1}{3}}$. By using $u \in C_{X_i}^\alpha(\Omega)$, $i = 1, \ldots, p_0$, in order to bound the first term, together with $u \in C_Y^\alpha(\Omega)$ to bound the second, we obtain (9), which concludes the proof for $n = 0$.

Case $n = 3$**:** To shorten notation we only prove the case $p_0 = p_1 = 1$. The difficulty in considering multi-dimensional blocks is purely notational. We first prove Part 1). Fix an arbitrary bounded domain $\Omega_0 \subset \Omega$. Proceeding as in the homogeneous case (see the proof of [16, Proposition 5.25]), one obtains

$$\left|u\big(\gamma_{\delta,v}^{(i)}(z)\big) - \bar{T}_3 u\big(z, \gamma_{\delta,v}^{(i)}(z)\big)\right| \leq c\,\|u\|_{C_B^{3,\alpha}(\Omega_0)}|\delta|^{3+\alpha}, \qquad i = 0, 1, \tag{10}$$

for any $z \in \Omega_0$, and $v, \delta \in \mathbb{R}$ with $|v| = 1$ and $|\delta|$ suitably small, where we set

$$\bar{T}_3 u(z, \zeta) = \sum_{i=0}^{3} \frac{(\xi - x)^i}{i!} \partial_x^i u(z) + \frac{\eta - y}{B_{1,0}} \big([\partial_x, Y] - B_{0,0}\partial_x\big) u(z),$$
$$z = (t, x, y),\ \zeta = (t, \xi, \eta).$$

The last term in the right-hand side is inspired by (1) to mimic a partial derivative w.r.t. y and is well defined when applied to $u \in C_B^{3,\alpha}(\Omega)$. We now prove

$$\big|u\big(g_{\delta,v}(z)\big) - \bar{T}_3 u\big(z, g_{\delta,v}(z)\big)\big| \leq c\, \|u\|_{C_B^{3,\alpha}(\Omega_0)} |\delta|^{3+\alpha}, \tag{11}$$

where $g_{v,\delta}$ is as defined in (3)–(4). Setting $z' := \gamma_{\delta,v}^{(1)}(z)$ and $z'' = g_{v,\delta}(z)$ we have

$$u\big(z''\big) - \bar{T}_3 u\big(z, z''\big) = F_1 + F_2,$$
$$F_1 = \big(u(z'') - \bar{T}_3 u(z', z'')\big) + \big(u(z') - \bar{T}_3 u(z, z')\big),$$
$$F_2 = \bar{T}_3 u(z', z'') + \bar{T}_3 u(z, z') - u(z') - \bar{T}_3 u(z, z'').$$

Now, (10) and (3) yield $|F_1| \leq c\, \|u\|_{C_B^{3,\alpha}(\Omega_0)} |\delta|^{3+\alpha}$; as for F_2 we have:

$$|F_2| = \left|\sum_{i=1}^{3} \frac{1}{i!} \Big(\partial_x^i u(z') - \partial_x^i u(z)\Big) \delta^5 (v')^i\right| \leq c\, \|u\|_{C_B^{3,\alpha}(\Omega_0)} |\delta|^{3+\alpha},$$

where we used $\partial_x^i u \in C_B^{3-i,\alpha}(\Omega)$ and Theorem 2.1 for $n = 0, 1, 2$, together with (4). This proves (11) for any $z \in \Omega_0$, and $v, \delta \in \mathbb{R}$ with $|v| = 1$ and $|\delta|$ suitably small. We are now able to prove differentiability along the y direction. For any $z = (t, x, y) \in \Omega_0$ and $\eta \in \mathbb{R}$ with $|\eta|$ small enough, choosing v and δ as given by Lemma 3.2 yields

$$\begin{aligned}
&\big|u(t, x, y + \eta) - \bar{T}_3 u\big((t, x, y), (t, x, y + \eta)\big)\big| \\
&\quad = \big|u\big(g_{v,\delta}(t, x, y)\big) - \bar{T}_3 u\big((t, x, y), g_{\delta,v}(t, x, y)\big)\big| \\
&\quad \leq c \|u\|_{C_B^{3,\alpha}(\Omega_0)} |\delta|^{3+\alpha} \leq c \|u\|_{C_B^{3,\alpha}(\Omega_0)} |\eta|^{1+\frac{\alpha}{3}},
\end{aligned}$$

where we used (11) in to obtain the first inequality, and (5) to obtain the second. Thus $\partial_y u(z)$ exists and

$$\partial_y u(t, x, y) = \frac{1}{B_{1,0}} \big([\partial_x, Y] - B_{0,0}\partial_x\big) u(t, x, y).$$

Furthermore, $u \in C_B^{3,\alpha}(\Omega)$ implies $\partial_y u \in C_B^{0,\alpha}(\Omega)$, which is Part 1) of Theorem 2.1 for $n = 3$.

The proof of Part (2) is analogous to the homogeneous case treated in [16, Subsections 5.2.2, 5.2.3]. For brevity, we only give an account of the main idea. The first step is to prove (2) for z, ζ sharing the same time component, say $z = (t, x, y)$, $\zeta = (t, \xi, \eta)$. We define the intermediate point $z_1 := (t, \xi, y)$ and write

$$u(z) - T_3 u(\zeta, z) = (u(z) - T_3 u(z_1, z)) + (T_3 u(z_1, z) - T_3 u(\zeta, z)).$$

Now, an application of the mean value theorem along x in the first addend, and along y in in the second, yields the desired estimate. Note that this step strongly relies on Part 1), namely the differentiability w.r.t. the y variable. Finally, the case of general z, ζ can be reduced to the above one similarly to (8). □

References

1. Biagi, S., Bramanti, M.: Schauder estimates for Kolmogorov-Fokker-Planck operators with coefficients measurable in time and Hölder continuous in space (2022, preprint). arXiv:2205.10270
2. Bonfiglioli, A.: Taylor formula for homogeneous groups and applications. Math. Z. **262**, 255–279 (2009)
3. Chaudru de Raynal, P.E., Honoré, I., Menozzi, S.: Sharp Schauder estimates for some degenerate Kolmogorov equations. Ann. Sc. Norm. Super. Pisa Cl. Sci. **22**, 989–1089 (2021)
4. Di Francesco, M., Pascucci, A.: On a class of degenerate parabolic equations of Kolmogorov type. AMRX Appl. Math. Res. Exp. **2005**, 77–116 (2005)
5. Di Francesco, M., Polidoro, S.: Schauder estimates, Harnack inequality and Gaussian lower bound for Kolmogorov-type operators in non-divergence form. Adv. Differ. Equ. **11**, 1261–1320 (2006)
6. Folland, G.B., Stein, E.M.: Hardy Spaces on Homogeneous Groups. Mathematical Notes, vol. 28. Princeton University Press, Princeton (1982)
7. Garofalo, N., Tralli, G.: Hardy-Littlewood-Sobolev inequalities for a class of non-symmetric and non-doubling hypoelliptic semigroups. Math. Ann. **383**, 1–38 (2022)
8. Imbert, C., Mouhot, C.: The schauder estimate in kinetic theory with application to a toy nonlinear model. Ann. Henri Lebesgue **4**, 369–405 (2021)
9. Imbert, C., Silvestre, L.: The schauder estimate for kinetic integral equations. Anal. PDE **14**, 171–204 (2021)
10. Lanconelli, E., Polidoro, S.: On a class of hypoelliptic evolution operators. Rend. Sem. Mat. Univ. Politec. Torino **52**, 29–63 (1994). Partial differential equations, II (Turin, 1993)
11. Lanconelli, A., Pagliarani, S., Pascucci, A.: Local densities for a class of degenerate diffusions. Ann. Inst. Henri Poincaré Probab. Stat. **56**, 1440–1464 (2020)
12. Lucertini, G., Pagliarani, S., Pascucci, A.: Optimal Schauder estimates for kinetic Kolmogorov equations with time measurable coefficients (2023, preprint). arXiv:2304.13392
13. Lunardi, A.: Schauder estimates for a class of degenerate elliptic and parabolic operators with unbounded coefficients in $\mathbb{R}^n$. Ann. Scuola Norm. Sup. Pisa Cl. Sci. **24**, 133–164 (1997)
14. Manfredini, M.: The Dirichlet problem for a class of ultraparabolic equations. Adv. Differ. Equ. **2**, 831–866 (1997)

15. Nyström, K., Pascucci, A., Polidoro, S.: Regularity near the initial state in the obstacle problem for a class of hypoelliptic ultraparabolic operators, J. Differ. Equ. **249**, 2044–2060 (2010)
16. Pagliarani, S., Pascucci, A., & Pignotti, M.: Intrinsic Taylor formula for Kolmogorov-type homogeneous groups. J. Math. Anal. Appl. **435**, 1054–1087 (2016)
17. Pagliarani, S., Pascucci, A., Pignotti, M.: Intrinsic expansions for averaged diffusion processes. Stoch. Proces. Appl. **127**, 2560–2585 (2017)
18. Pascucci, A., Pesce, A.: Sobolev embeddings for kinetic Fokker-Planck equations (2022, preprint). arXiv:2209.05124
19. Polidoro, S.: On a class of ultraparabolic operators of Kolmogorov-Fokker-Planck type. Matematiche (Catania) **49**, 53–105 (1994/1995).
20. Polidoro, S., Rebucci, A., Stroffolini, B.: Schauder type estimates for degenerate Kolmogorov equations with Dini continuous coefficients. Commun. Pure Appl. Anal. **21**, 1385–1416 (2022)
21. Priola, E.: Global Schauder estimates for a class of degenerate Kolmogorov equations. Stud. Math. **194**, 117–153 (2009)
22. Triebel, H.: Theory of Function Spaces. II. Monographs in Mathematics, vol. 84. Birkhäuser Verlag, Basel (1992)
23. Zhang, X., Zhang, X.: Cauchy problem of stochastic kinetic equations(2021, preprint). arXiv: 2103.02267

Form-Boundedness and SDEs with Singular Drift

Damir Kinzebulatov

Abstract We survey and refine recent results on weak and strong well-posedness of stochastic differential equations with singular drift satisfying some minimal assumptions.

2010 Mathematics Subject Classification 60H10, 47D07 (primary), 35J75 (secondary)

1 Introduction

Various applications in physics and technology dictate the need to work with stochastic differential equations

$$dX_t = -b(X_t)dt + \sqrt{2}dW_t, \quad X_0 = x \in \mathbb{R}^d, \tag{1.1}$$

having an irregular, locally unbounded drift $b : \mathbb{R}^d \to \mathbb{R}^d$. Here $\{W_t\}_{t\geq 0}$ is a d-dimensional Brownian motion in $\mathbb{R}^d$ defined on some complete filtered probability space $(\Omega, \mathcal{F}_t, \mathcal{F}, \mathbf{P})$. This naturally leads to the problem of finding the least restrictive assumptions on b that ensure well-posedness of (1.1), in one sense or another. More specifically, one asks: what integral characteristics of b determine whether there exists a unique solution of (1.1)? The same question arises when one considers more general SDEs, also dictated by applications:

$$dX_t = -b(t, X_t)dt + \sqrt{2}dW_t, \tag{1.2}$$

The research of the author is supported by the NSERC (grant RGPIN-2017-05567).

D. Kinzebulatov (✉)
Département de mathématiques et de statistique, Université Laval, Québec, QC, Canada
e-mail: damir.kinzebulatov@mat.ulaval.ca

S. Menozzi et al. (eds.), *Kolmogorov Operators and Their Applications*,
Springer INdAM Series 56, https://doi.org/10.1007/978-981-97-0225-1_7

with drift $b : \mathbb{R} \times \mathbb{R}^d \to \mathbb{R}^d$ that can also be singular in time, and

$$dX_t = -b(t, X_t)dt + \sigma(t, X_t)dW_t \tag{1.3}$$

with diffusion coefficients $\sigma : \mathbb{R}_+ \times \mathbb{R}^d \to \mathbb{R}^{d\times d}$ that can be discontinuous. Regarding SDE (1.2), one illustrative example is the "passive tracer model" that describes the motion of a small particle in a turbulent flow, i.e. (1.2) with the velocity field b obtained by solving three-dimensional Navier-Stokes equations [73].

The paper deals with weak and strong well-posedness of SDEs (1.1)–(1.3), for every initial point $x \in \mathbb{R}^d$. Recall that a weak solution of (1.1)–(1.3) is a pair of continuous processes $\{(X_t, W_t)\}_{t\geq 0}$ defined on some complete probability space, such that $\{W_t\}$ is a Brownian motion and the identity in (1.1)–(1.3) holds a.s. for all $t \geq 0$. In turn, a strong solution of (1.1)–(1.3) is a continuous process X_t that is adapted to the natural filtration of the Brownian motion $\{W_t\}$, and such that the identity in (1.1)–(1.3) holds a.s. for $t \geq 0$. That is, X_t is a strong solution if it only depends on time and the driving process $\{W_s\}_{0\leq s\leq t}$.

The question of what local singularities of b are admissible, so that SDEs (1.1)–(1.3) are weakly or strongly well-posed, was thoroughly studied in the literature. Below we give a brief outline of the results on multidimensional SDEs with the focus on the singularities of the drift. We will keep the chronological order of appearance of preprints, where applicable. However, we will be somewhat loose with the terminology by including in "well-posedness" uniqueness results of varying strength (in general, within some large classes of solutions).

Veretennikov [96] was the first who established, using Zvonkin's method [113], strong well-posedness of (1.2) when $|b|$ is bounded measurable. Portenko [85] considered drift b in the sub-critical Ladyzhenskaya-Prodi-Serrin class

$$|b| \in L^l(\mathbb{R}, L^p(\mathbb{R}^d)), \quad p \geq d, l \geq 2, \quad \frac{d}{p} + \frac{2}{l} < 1 \tag{1.4}$$

and proved existence of weak solution to SDE (1.2) and its uniqueness in law. Krylov-Röckner [69] further established, using Yamada-Watanabe theorem, that for such b the SDE (1.2) is, in fact, strongly well-posed. A number of important results for SDEs with drift satisfying (1.4) were established next by X. Zhang [104–108]. In the period between the papers of Portenko and Krylov-Röckner, Bass-Chen [6] proved existence and uniqueness in law of weak solutions of (1.1) for time-homogeneous drift $b = b(x)$ in the Kato class of vector fields, with arbitrarily small δ, cf. (14.3). The Kato class of vector fields contains $\{|b| \in L^p(\mathbb{R}^d), p > d\}$ as well as some vector fields with entries not even in $L^{1+\varepsilon}_{\rm loc}(\mathbb{R}^d)$, $\varepsilon > 0$. However, Kato class does not contain $\{|b| \in L^d(\mathbb{R}^d)\}$.

Speaking of time-homogeneous drifts, the fact that $p = d$ is the optimal exponent on the scale of Lebesgue spaces can be seen already from rescaling the parabolic equation.

In [7], Beck-Flandoli-Gubinelli-Maurelli developed an approach to establishing strong well-posedness of (1.2) with drift b in the critical Ladyzhenskaya-Prodi-Serrin (LPS) class

$$|b| \in L^l(\mathbb{R}, L^p(\mathbb{R}^d)), \quad p \geq d, l \geq 2, \quad \frac{d}{p} + \frac{2}{l} \leq 1. \tag{1.5}$$

for a.e. initial point $x \in \mathbb{R}^d$, via stochastic transport and stochastic continuity equations. They also discussed the following attracting drift:

$$b(x) = \sqrt{\delta}\frac{d-2}{2}\mathbf{1}_{|x|<1}|x|^{-2}x \tag{1.6}$$

(note that $|b|$ misses $L^d(\mathbb{R}^d)$ by a logarithmic factor) and provided a detailed proof of the fact that for

$$\delta > 4(\frac{d}{d-2})^2, \tag{1.7}$$

i.e. when the attraction to the origin by the drift is strong enough, then SDE (1.1) with initial point $x = 0$ does not have a weak solution. In [46], Semënov and the author showed that the constructed in [37] Feller generator $-\Delta + b \cdot \nabla$ for b satisfying weak form-boundedness condition (1.11), see below, determines, for every initial point $x \in \mathbb{R}^d$, a weak solution to SDE (1.1) that is unique among weak solutions that can be constructed via approximation. To the best of the author's knowledge, this was the first result on weak well-posedness of (1.1) that included both $|b| \in L^d(\mathbb{R}^d)$ and the model vector field (1.6) with δ small. It also included the elliptic Morrey class $M_{1+\varepsilon}$, see below, and the Kato class considered by Bass-Chen. The construction of the Feller generator with such b used in an essential manner some inequalities for symmetric Markov generators, and hence required time-homogeneity of the drift. Returning to time-inhomogeneous drifts, we note that almost at the same time Jin [34] proved weak well-posedness of (1.2) with time-inhomogeneous Kato class drifts, Wei-Lv-Wu [97] and Nam [79] obtained results on weak well-posedness of (1.2), for every $x \in \mathbb{R}^d$, for time-inhomogeneous vector fields b that can be more singular than the ones in (1.4). Nevertheless, their results excluded $b = b(x)$ with $|b| \in L^d(\mathbb{R}^d)$. In [99], Xia-Xie-Zhang-Zhao established, among other results, weak well-posedness for every initial point of SDE (1.1) with b having entries in $C_b(\mathbb{R}, L^d(\mathbb{R}^d))$. In [62, 63, 65], Krylov proved weak and strong well-posedness of SDEs (1.1)–(1.3) with $|b|$ in $L^d(\mathbb{R}^d)$ and beyond, e.g. in a large Morrey class (in fact, below we use an argument from these papers to prove some gradient bounds). In [101], S. Yang-T. Zhang proved strong well-posedness of (1.2) for time-inhomogeneous drifts b with $|b|^2$ "almost" in the Kato class of potentials, cf. (3.9) (which, to make the comparison clear at least in the time-homogeneous case, is smaller than the Kato class of vector fields in [6]; still, the class considered by Yang-Zhang contains some drifts with $|b| \notin L^{2+\varepsilon}_{\rm loc}$, $\varepsilon > 0$). In [90], Röckner-Zhao established weak well-posedness of (1.1), for any initial point $x \in \mathbb{R}^d$, for drifts in $L^\infty(\mathbb{R}, L^{d,w}(\mathbb{R}^d))$, plus the drifts in the critical LPS class. Here $L^{d,w}(\mathbb{R}^d)$ denotes the weak L^d class that contains vector fields in $L^d(\mathbb{R}^d)$, as well as more singular vector fields, such as (1.6). In [91], the authors obtained strong well-posedness of (1.1), for any initial point, with b in the critical LPS class. In [43], Madou and the author established weak well-posedness of (1.2), for every initial point, for b in the class of time-inhomogeneous form-bounded drifts (containing (1.8) below). This class contains $L^\infty(\mathbb{R}, L^{d,w}(\mathbb{R}^d))$ and the critical LPS class, as well as some drifts that are not even in $L^\infty(\mathbb{R}, L^{2+\varepsilon}(\mathbb{R}^d))$ for a given $\varepsilon > 0$. In [51], Semënov and the author proved existence of a weak solution to SDE (1.2) with time-inhomogeneous form-bounded drift, see (1.8), covering the entire range of admissible form-bounds $\delta < 4$, cf. (1.7). The critical value $\delta = 4$ was recently attained at the PDE level in [42]. In [41], the author established weak well-posedness of SDE (1.2) and proved Feller property for time-inhomogeneous drifts in essentially the largest possible parabolic Morrey class, which contains the class of time-inhomogeneous form-bounded drifts, the time-homogeneous Morrey drifts in $M_{1+\varepsilon}$, and allows to include drifts b having critical behaviour both in time and in space, e.g.

$$|b(t,x)| \leq \frac{c_1}{|x|} + \frac{c_2}{\sqrt{t}}, \qquad t > 0, \quad x \in \mathbb{R}^d.$$

More recently, Krylov [66] established weak well-posedness of SDE (1.3) with discontinuous diffusion coefficients in the VMO class and time-inhomogeneous drifts in a large parabolic Morrey class; restricted to time-homogeneous drifts his assumptions read as $|b| \in M_q$, $q > \frac{d}{2}$. This result was further refined by Krylov in [67], see also [68] regarding regularity theory of parabolic equations with VMO diffusion coefficients and drift and potential in Morrey classes.

Below we survey and refine recent results on weak and strong well-posedness of SDEs (1.1)–(1.3) with $|b|$ satisfying some minimal constraints, as in [41, 43, 46, 51] mentioned above. For instance, regarding SDE (1.1), our assumption on the order of singularities of $|b|$ is basically that $-\Delta + b \cdot \nabla$ must generate a quasi contraction strongly continuous semigroup in L^2. That is, we will be assuming that $|b|$ is form-bounded:

$$|b|^2 \leq \delta(-\Delta) + c \quad \text{(in the sense of quadratic forms)} \tag{1.8}$$

for some constants δ and c. See Definition 3.1 below. This translates, by means of the Cauchy-Schwarz inequality, into the assumption of smallness of $b \cdot \nabla$ with respect to $-\Delta$. A broad elementary sufficient condition for (1.8) is the scaling-invariant Morrey class $M_{2+\varepsilon}$, i.e.

$$\|b\|_{M_{2+\varepsilon}} := \sup_{r>0, x\in\mathbb{R}^d} r\left(\frac{1}{|B_r|}\int_{B_r(x)} |b|^{2+\varepsilon} dx\right)^{\frac{1}{2+\varepsilon}} < \infty \tag{1.9}$$

where $B_r(x)$ is the ball of radius r centered at x. Here ε can be taken arbitrarily small. One has $\delta = C\|b\|_{M_{2+\varepsilon}}$ for appropriate constant C. The class $M_{2+\varepsilon}$ contains all $|b|$ in L^d or, more generally, in the weak L^d class (we recall its definition in Sect. 2).

Regarding the relationship between operator $-\Delta + b \cdot \nabla$ and SDE (1.1), one expects that for $b = b(x)$ the function

$$v(t, x) := \mathbf{E}_{X_0=x}[f(X_t)], \tag{1.10}$$

solves Cauchy problem

$$(\partial_t - \Delta + b \cdot \nabla)v = 0, \quad v|_{t=0} = f.$$

The intimate relationship between parabolic equations and SDEs is a consequence of the fact that both describe the same physical process of diffusion.

In Sect. 15, we discuss results on weak well-posedness of (1.1) under more general assumption on the drift than its form-boundedness. Namely, our hypothesis on the order of singularities of $|b|$ will be that $-\Delta + b \cdot \nabla$ generates a quasi contraction strongly continuous semigroup in the Bessel space $\mathcal{W}^{1/2,2}$, i.e. we will require

$$|b| \leq \delta(\lambda - \Delta)^{\frac{1}{2}} \quad \text{(in the sense of quadratic forms).} \tag{1.11}$$

Such vector fields b are called weakly form-bounded. This class of weakly form-bounded vector fields contains scaling-invariant Morrey class $M_{1+\varepsilon}$. It also contains the Kato class of vector fields.

One of our goals will be to bootstrap the semigroup in L^2 or in $\mathcal{W}^{\frac{1}{2},2}$ to a strongly continuous semigroup in C_∞, the space of continuous functions on $\mathbb{R}^d$ vanishing at infinity, endowed with the sup-norm (that is, to a Feller semigroup). This will come at the cost of restricting admissible values of constant δ (in terms of the Morrey class, this means that the Morrey norm cannot be too large). We emphasize that while the classes (1.8), (1.11) determine the order of singularities of drift b, the value of δ measures the magnitude of its singularities. We are particularly interested in the optimal assumptions on δ.

The requirement that there should be a properly defined operator behind (1.1) is reasonable, since it gives a reasonably complete solution theory of the corresponding parabolic equation. That being said, there are situations where one does not want to insist on a strong link between parabolic equations and stochastic processes in order to treat, in one sense or another, more singular drifts. See e.g. [98] which considers solutions of martingale problem with test functions cutting out the singular set of the drift, or [80, 103] which deal with elliptic or parabolic equations with supercritical (in the sense of scaling) drift where one no longer has Hölder continuity of solution. We are interested, on the contrary, in finding the maximal singularities of the drift that still give a more or less classical theory of parabolic equations and SDEs, including the possibility to consider SDEs with arbitrarily fixed initial point, e.g. in the singular set of the drift.

One natural question is: why not restrict attention to the Morrey class of drifts (1.9), a broad subclass of (1.8) defined in elementary terms (and which also allows to use e.g. harmonic-analytic arguments that are not available for form-bounded or weakly form-bounded drifts, cf. Remark 16.1)? First, there is a refinement of the Morrey class (1.9) due to Chang-Wilson-Wolff, which is still contained in the class of form-bounded vector fields (see Appendix B), and there is no reason to believe that their result itself cannot be refined, in elementary terms, even further. Second, and more importantly, by assuming e.g. form-boundedness of b (1.8) we impose a condition on b that is ultimate in some precise sense, i.e. the existence of quasi contraction strongly continuous semigroup in L^2. The latter is, arguably, an extra hypothesis on the diffusion process, but it is explicitly spelled out. Also, at least in the case of Schrödinger operators, see below, form-boundedness is a physical assumption on the potential.

We impose similar assumptions on the time-inhomogeneous drift in SDE (1.2). We will deal with SDE (1.3) with diffusion coefficients σ that can have some critical discontinuities, but only those that are allowed by the singularities of the drift b, which is our main focus.

In Sect. 17 we discuss weak well-posedness of SDE

$$X_t = x - \int_0^t b(X_s)ds + Z_t, \quad t \geq 0, \quad x \in \mathbb{R}^d, \tag{1.12}$$

driven by rotationally symmetric α-stable process Z_t, $1 < \alpha < 2$, with drift $b : \mathbb{R}^d \to \mathbb{R}^d$ satisfying

$$|b| \leq \delta(\lambda - \Delta)^{\frac{\alpha-1}{2}} \quad \text{(in the sense of quadratic forms)}, \tag{1.13}$$

e.g.

$$|b|^{\frac{1}{\alpha-1}} \in M_{1+\varepsilon}$$

for some $\varepsilon > 0$. There is a rich literature devoted to Eq. (1.12), which also covers the range $0 < \alpha \leq 1$ (see Remark 17.1). In the case $1 < \alpha < 2$, which allows to deal with locally unbounded drifts, earlier results on weak well-posedness of (1.12) include $|b| \in L^\infty$ due to Komatsu [59], $|b| \in L^p$, $p > \frac{d}{\alpha-1}$ due to Portenko and Podolynny-Portenko [86, 87], and, more generally, $|b|$ in appropriate Kato class of vector fields, see Chen-Kim-Song [18], Kim-Song [36], Chen-Wang [16]. All of these classes are contained in (1.13).

See also Priola [88] and X. Zhang [107] regarding strong well-posedness of (1.12) and its generalizations.

We discuss conditions on b stated in terms of $|b|$. The latter allows to include measure-valued drifts, see Remark 15.5. However, distributional drifts lie outside of the scope of this paper (regarding distributional drifts, see [14, 15, 25, 83, 110] and references therein). We also do not discuss here many interesting results that require additional structure of b such as existence of non-positive divergence $\operatorname{div} b$ or b of the form $b = \nabla V$ for appropriate potential V; we only refer to Bresch-Jabin-Wang [12], Fournier-Jourdain [26], Röckner-Zhao [90] and references therein.

The drifts that we consider in this paper in general destroy both the upper and the lower Gaussian bounds on the heat kernel of the corresponding to (1.1) and (1.2) parabolic equations.

In this paper we are interested in local singularities of the drift, although our drifts can still be unbounded at infinity along some subsets (e.g. $b(x) = \sum_{k=1}^{\infty} c_k|x - a_k|^{-2}(x - a_k)$, $x \in \mathbb{R}^d$, where $\sum_{k=1}^{\infty} |c_k|^{\frac{1}{2}} < \infty$ and $a_k \in \mathbb{R}^d$, $|a_k| \to \infty$).

Throughout the paper, dimension $d \geq 3$. Dimension $d = 2$ does not present an obstacle for our methods, however, in our opinion it deserves a separate study. The exposition of the results does not follow the chronological order of their appearance (on arXiv), but proceeds from weaker singularities of the drift to stronger singularities.

Structure of the Paper In Sect. 3 we introduce the class of time-homogeneous form-bounded vector fields.

In Sect. 4 we discuss the results on weak solvability of SDE (1.1) and on solution theory of the corresponding parabolic equation under essentially sharp assumption on the value of form-bound δ of the drift.

In Sect. 5 we describe two existing approaches to constructing a Feller semigroup associated with $-\Delta + b \cdot \nabla$, and introduce a new one.

In Sects. 6, 7 and 8 we prove a detailed weak well-posedness result for (1.1) with form-bounded drift, however, at expense of requiring smaller δ.

In Sects. 9 and 10 we discuss results on weak well-posedness of SDE (1.2) having time-inhomogeneous form-bounded drift. Their proofs use a different (iteration) technique.

In Sect. 11 we discuss an extension of the previous results to some discontinuous diffusion coefficients.

In Sects. 12 and 13 we discuss strong well-posedness of SDEs (1.1) and (1.2), via stochastic transport equation and via relative compactness criterion for random fields on the Wiener-Sobolev space (Röckner-Zhao's approach).

In Sects. 14 and 15 we substantially enlarge the class of admissible in SDE (1.1) time-homogeneous drifts (i.e. to weakly form-bounded drifts), but at expense of δ that now needs to be smaller than in Sect. 6.

In Sect. 16 we consider again time-inhomogeneous drifts and strengthen and simplify all aspects of the results from Sects. 9 and 10 except their assumptions on δ. We reach, in particular, critical singularities of drift in the time variable. Compared to Sect. 15 we, however, restrict somewhat the class of the spatial singularities of the drift to essentially the largest possible Morrey class.

In Sect. 17 we discuss analogues of the weak well-posedness results from Sect. 14 for the SDE (1.12) driven by symmetric α-stable process.

2 Notations

1. $\mathbb{R}_+ := [0, \infty[$. In what follows, $B_r(x)$ is the open ball of radius r centered at $x \in \mathbb{R}^d$. Put

$$\nabla_i f := \partial_{x_i} f,$$

where $f = f(x)$ or $f = f(t, x)$, $x = (x_1, \dots, x_d)$.

For $\alpha, \beta \in \mathbb{R}$, set

$$\alpha \vee \beta := \max\{\alpha, \beta\}, \quad \alpha \wedge \beta := \min\{\alpha, \beta\}.$$

Let $\mathcal{B}(X, Y)$ denote the space of bounded linear operators between Banach spaces $X \to Y$, endowed with the operator norm $\|\cdot\|_{X\to Y}$. $\mathcal{B}(X) := \mathcal{B}(X, X)$.

We write $T = s\text{-}X\text{-}\lim_n T_n$ for $T, T_n \in \mathcal{B}(X, Y)$ if

$$\lim_n \|Tf - T_n f\|_Y = 0 \quad \text{for every } f \in X.$$

Put

$$L^p = L^p(\mathbb{R}^d), \quad W^{1,p} = W^{1,p}(\mathbb{R}^d).$$

Set

$$\|\cdot\|_p := \|\cdot\|_{L^p}$$

and

$$\|\cdot\|_{p\to q} := \|\cdot\|_{L^p \to L^q}.$$

Let $\mathcal{W}^{\alpha,p}$, $\alpha > 0$, denote the Bessel potential space on $\mathbb{R}^d$ endowed with norm $\|u\|_{p,\alpha} := \|g\|_p$, $u = (1-\Delta)^{-\frac{\alpha}{2}} g$, $g \in L^p$. Let $\mathcal{W}^{-\alpha,p'}$, $p' = p/(p-1)$ denote the anti-dual of $\mathcal{W}^{\alpha,p}$.

For a given vector field b and $1 \leq p < \infty$, put

$$b^{\frac{1}{p}} := b|b|^{-1+\frac{1}{p}}.$$

Put

$$\langle f, g \rangle = \langle f\bar{g} \rangle := \int_{\mathbb{R}^d} f\bar{g}\, dx$$

(some of our functions will be complex-valued).

C_c (C_c^∞) denotes the space of continuous (smooth) functions on $\mathbb{R}^d$ having compact support.

$C_\infty := \{f \in C(\mathbb{R}^d) \mid \lim_{x\to\infty} f(x) = 0\}$ (with the sup-norm).

$\mathcal{S}$ is the L. Schwartz' space of test functions.

We denote by $\upharpoonright$ the restriction of an operator (or a function) to a subspace (a subset).

Given linear operators A, B, we write $B \supset A$ if B is an extension of A.

Let

$$\big[A \upharpoonright D(A) \cap L^p\big]^{\text{clos}}_{L^p \to L^p}$$

denote the closure of operator A as an operator $L^p \to L^p$ (if it exists).

2. Fix $0 < T < \infty$. Let $D([0, T], \mathbb{R}^d)$, the space of right-continuous functions having left limits, be endowed with the filtration $\mathcal{B}'_t = \sigma(\omega_r \mid 0 \le r \le t)$, where ω_t is the coordinate process on $D([0, T], \mathbb{R}^d)$.

We will also need the canonical space $(C([0, T], \mathbb{R}^d), \mathcal{B}_t = \sigma(\omega_r \mid 0 \le r \le t))$, where ω_t is the coordinate process on $C([0, T], \mathbb{R}^d)$.

Recall that a probability measure $\mathbb{P}_x$, $x \in \mathbb{R}^d$ on $(C([0, T], \mathbb{R}^d), \mathcal{B}_t)$ is called a martingale solution to SDE

$$dX_t = -b(t, X_t)dt + \sqrt{2}dW_t, \quad X_0 = x \tag{2.1}$$

if

(1) $\mathbb{P}_x[\omega_0 = x] = 1$;

(2)

$$\mathbb{E}_x \int_0^T |b(r, \omega_r)| dr < \infty;$$

(3) for every $f \in C_c^2(\mathbb{R}^d)$ the process

$$t \mapsto f(\omega_t) - f(x) + \int_0^t (-\Delta f + b \cdot \nabla f)(\omega_r) dt$$

is a $\mathcal{B}_t$-martingale under $\mathbb{P}_x$.

A martingale solution $\mathbb{P}_x$ of (2.1) is called a weak solution if, upon completing filtration $\mathcal{B}_t$ with respect to $\mathbb{P}_x$ (to, say, $\hat{\mathcal{B}}_t$), there exists a Brownian motion $\{W_t\}$ on $\big(C([0, T], \mathbb{R}^d), \hat{\mathcal{B}}_t, \mathbb{P}_x\big)$ such that

$$\omega_t = x - \int_0^t b(r, \omega_r) dr + \sqrt{2} W_t, \quad 0 \le t \le T \quad \mathbb{P}_x\text{-a.s.}$$

3. A function $h : \mathbb{R}^d \to \mathbb{R}$ is said to be in the weak L^d class (denoted by $L^{d,w}$) if

$$\|h\|_{d,w} := \sup_{s>0} s|\{x \in \mathbb{R}^d : |h(x)| > s\}|^{1/d} < \infty.$$

Clearly, $L^d \subset L^{d,w}$, but not vice versa, e.g. $h(x) = |x|^{-1}$ is in $L^{d,w}$ but not in L^d.

4. The De Giorgi mollifier $E_\varepsilon \equiv E_\varepsilon^d$ on $\mathbb{R}^d$:

$$E_\varepsilon f(x) := e^{\varepsilon\Delta} f(x), \quad x \in \mathbb{R}^d, \quad \varepsilon > 0,$$

where $f \in L^1_{\rm loc}$.

The Friedrichs mollifier $E_\varepsilon \equiv E_\varepsilon^d$ on $\mathbb{R}^d$:

$$E_\varepsilon f(x) := \eta_\varepsilon * f(x),$$

where $\eta_\varepsilon(y) := \frac{1}{\varepsilon^d}\eta\left(\frac{y}{\varepsilon}\right)$, $\varepsilon > 0$ and

$$\eta(y) := \begin{cases} c\exp\left(\frac{1}{|y|^2-1}\right) & \text{if } |y| < 1, \\ 0, & \text{if } |y| \geqslant 1, \end{cases}$$

with constant c adjusted to $\int_{\mathbb{R}^d} \eta(x)dx = 1$.

3 Form-Bounded Drifts. Semigroup in L^2

First, we discuss sufficient conditions for existence of an operator realization of $-\Delta + b \cdot \nabla$ generating a strongly continuous semigroup in L^2.

Assume that a Borel measurable vector field $b : \mathbb{R}^d \to \mathbb{R}^d$ with $|b| \in L^2_{\rm loc}$ satisfies inequality

$$\|b\varphi\|_2^2 \leq \delta\|\nabla\varphi\|_2^2 + c_\delta\|\varphi\|_2^2 \quad \text{for all } \varphi \in W^{1,2} \tag{3.1}$$

for finite constants $\delta > 0$ and $0 \leq c_\delta < \infty$.

Definition 3.1 Such vector fields b are called *form-bounded* (written as $b \in \mathbf{F}_\delta$).

Inequality (3.1) can be re-stated as an operator norm inequality

$$\|b(\lambda - \Delta)^{-\frac{1}{2}}\|_{2\to 2} \leq \sqrt{\delta} \tag{3.2}$$

with $\lambda \equiv \lambda_\delta = c_\delta/\delta$.

Using the quadratic (or Cauchy-Schwarz) inequality

$$|\langle b\cdot\nabla\varphi,\varphi\rangle| \le \varepsilon\|b\varphi\|_2^2 + \frac{1}{4\varepsilon}\|\nabla\varphi\|_2^2, \quad \varepsilon>0, \tag{3.3}$$

one can see that the form-boundedness condition (3.1) *with* $\delta<1$ is what is needed to verify conditions of the Lax-Milgram theorem for the bilinear form $\tau[\varphi,\psi] := \lambda\langle\varphi,\psi\rangle + \langle\nabla\varphi,\nabla\psi\rangle + \langle b\cdot\nabla\varphi,\psi\rangle$ defined on the real space $W^{1,2}$. That is, coercivity for all $\lambda \ge c_\delta/2\sqrt{\delta}$:

$$\begin{aligned}|\tau[\varphi,\varphi]| &\ge \lambda\|\varphi\|_2^2 + \|\nabla\varphi\|_2^2 - \varepsilon\|b\varphi\|_2^2 - \frac{1}{4\varepsilon}\|\nabla\varphi\|_2^2 \\ &\qquad \text{(we apply (3.1) and select } \varepsilon = \frac{1}{2\sqrt{\delta}}) \\ &= (\lambda - \frac{c_\delta}{2\sqrt{\delta}})\|\varphi\|_2^2 + (1-\sqrt{\delta})\|\nabla\varphi\|_2^2 \end{aligned} \tag{3.4}$$

and boundedness

$$|\tau[\varphi,\psi]| \le C\|\varphi\|_{W^{1,2}}\|\psi\|_{W^{1,2}}. \tag{3.5}$$

So, by the Lax-Milgram theorem, there exists a unique weak solution to elliptic equation

$$(\mu - \Delta + b\cdot\nabla)u = f, \quad f\in L^2.$$

Furthermore, form-boundedness (3.1) with $\delta<1$ ensures that the sesquilinear form of $\lambda-\Delta+b\cdot\nabla$ defined on the complex space $W^{1,2}$ is sectorial, and hence by the KLMN theorem[1] it determines a unique closed densely defined operator $\Lambda \equiv \Lambda_2(b)$,

$$\Lambda \supset (-\Delta + b\cdot\nabla) \upharpoonright C_c^\infty(\mathbb{R}^d), \tag{3.6}$$

generating a strongly continuous quasi contraction semigroup in L^2.

Examples Let us mention some elementary sufficient conditions for form-boundedness.

1. If $|b|\in L^d$ (or $|b|\in L^d+L^\infty$, i.e. the sum of two functions, one in L^d and the other one in L^∞), then $b\in\mathbf{F}_\delta$ with δ that can be chosen arbitrarily small (at expense of increasing c_δ, see Appendix B).

[1] Kato-Lions-Lax-Milgam-Nelson theorem, see [35, Ch. VI], [81, Ch.1].

2. There are form-bounded vector fields that have stronger singularities than the ones covered by the class L^d. For instance, by Hardy's inequality

$$\left(\frac{d-2}{2}\right)^2 \||x|^{-1}\varphi\|_2^2 \leq \|\nabla\varphi\|_2^2, \quad \varphi \in W^{1,2},$$

the vector field

$$b(x) = \pm\sqrt{\delta}\frac{d-2}{2}\frac{x}{|x|^2}, \tag{3.7}$$

is form-bounded with $c_\delta = 0$. The constant in Hardy's inequality is sharp, and the last vector field is not in $\mathbf{F}_{\delta'}$ for any $\delta' < \delta$ regardless of the value of $c_{\delta'}$.

As we explain below, the value of constant δ determines weak solvability of SDEs (see (4.1) below), and thus is a key characteristics of the vector field b. However, the dependence of the solution theory of SDEs on δ is not visible if one deals only with $|b| \in L^d$. In this sense, the vector fields b with entries in L^d are sub-critical.

3. More generally, if vector field b belongs to the scaling-invariant Morrey class $M_{2+\varepsilon}$ for some $\varepsilon > 0$ arbitrarily small, i.e.

$$|b| \in L^{2+\varepsilon}_{\rm loc} \quad \text{and} \quad \|b\|_{M_{2+\varepsilon}} := \sup_{r>0, x\in\mathbb{R}^d} r\left(\frac{1}{|B_r|}\int_{B_r(x)} |b|^{2+\varepsilon}dx\right)^{\frac{1}{2+\varepsilon}} < \infty, \tag{3.8}$$

then b is form-bounded with $\delta = c\|b\|_{M_{2+\varepsilon}}$ for appropriate constant $c = c(d,\varepsilon)$. See [22], see also [21]. Note that

$$\|\cdot\|_{M_q} \leq \|\cdot\|_{M_{q_1}} \quad \text{if } q < q_1,$$

so Morrey class becomes larger as q becomes smaller (and so we are interested in the class M_q with q close to 2). This class contains all $|b| \in L^d$ and $|b| \in L^{d,w}$ (see definition in Sect. 2). It also contains, for every $\epsilon > 0$, vector fields b such that $|b| \notin L^{2+\epsilon}_{\rm loc}$.

On the other hand, it is easy to show, by considering translates of a bump function, that if $b \in \mathbf{F}_\delta$ (say, with $c_\delta = 0$), then $|b| \in M_2$.

4. If $|b|^2$ is in the Kato class of potentials $\mathbf{K}^d_\delta$, then vector field b is form-bounded. Recall that $V \in \mathbf{K}^d_\delta$ if

$$V \in L^1_{\rm loc} \quad \text{and} \quad \lim_{\lambda\to+\infty} \|(\lambda-\Delta)^{-1}|V|\|_\infty \leq \delta. \tag{3.9}$$

The Kato class condition first appeared in a 1961 article by M. Birman as an elementary sufficient condition for the form-boundedness of a potential V [9, Sect. 2].

Some other sufficient conditions for the form-boundedness of b, including those refining condition $b \in M_{2+\varepsilon}$, are given in Appendix B.

Let us also note that the sum b_1+b_2 of form-bounded vector fields $b_1 \in \mathbf{F}_{\delta_1}, b_2 \in \mathbf{F}_{\delta_2}$ is form-bounded with form-bound $\delta = (\sqrt{\delta_1}+\sqrt{\delta_2})^2$ (cf. (3.2)). In particular, vector field

$$b(x) := \sum_{k=1}^{\infty} c_k \frac{x-a_k}{|x-a_k|^2}$$

with $\sum_{k=1}^{\infty} |c_k|^{\frac{1}{2}} < \infty$ and $\{a_k\}$ constituting e.g. a dense subset of $\mathbb{R}^d$ is form-bounded.

The form-boundedness condition is well known in the literature on spectral and scattering theory of Schrödinger operators, in particular, dealing with the questions of self-adjointness, estimates on the number of bound states, resolvent convergence, see e.g. [8, 74, 94].[2] Regarding Kolmogorov operator (3.6), one can show, with little additional effort, that the semigroup $e^{-t\Lambda}$ is a Markov semigroup in L^2 (the proof can be found e.g. [47]). In the probabilistic context, however, one of the basic problems is to construct a Markov process that inherits the essential properties of the Brownian motion. In particular, it is natural to expect that the constructed Markov semigroup would be strongly continuous on the space C_∞ of continuous functions vanishing at infinity endowed with the sup-norm, i.e. that it would be a Feller semigroup. However, to show this, one needs to verify the strong continuity in the norm of C_∞, which is, of course, a lot more rigid that the norm of L^2 where one defines the form-boundedness of the drift. In this regard, let us note that the strong continuity of this semigroup in L^p for any finite $p \geq 2$, on the contrary, presents no problem. Indeed, since $e^{-t\Lambda}$ is Markov, we have $\|e^{-t\Lambda} f\|_\infty \leq \|f\|_\infty$, $f \in L^2 \cap L^\infty$, so one can use an interpolation argument to define

$$e^{-t\Lambda_p} := \Big[e^{-t\Lambda} \upharpoonright L^2 \cap L^p\Big]^{\text{clos}}_{L^p \to L^p}. \tag{3.10}$$

The left-hand side is a quasi contraction strongly continuous semigroup on L^p (see the proof e.g. in [71]), its generator Λ_p is an appropriate operator realization of $-\Delta + b \cdot \nabla$ in L^p. So, the difficulty is in the strong continuity in C_∞. A major advancement came with the fundamental paper [60] of Kovalenko-Semënov that, among other results, presented a construction of a Feller realization of $-\Delta + b \cdot \nabla$ in C_∞ for form-bounded b using an $L^2 \to L^\infty$ iteration procedure. To the best of the author's knowledge, surprisingly, the first reaction to this result was [38], almost three decades later.

[2] If one is willing to ignore different roles played by the positive and the negative parts of potential V in the theory of Schrödinger operator $-\Delta + V$, then V is form-bounded if $\||V|^{\frac{1}{2}}\varphi\|_2^2 \leq \delta\|\nabla\varphi\|_2^2 + c_\delta\|\varphi\|_2^2$, $\varphi \in W^{1,2}$.

Apart from the results described in the present paper, we also refer to [17, 23] regarding form-boundedness appearing in probabilistic settings.

Form-boundedness and similar conditions, sometimes supplemented with other hypothesis on b, also appear in the regularity theory of elliptic and parabolic equations, which includes the Harnack inequality, Gaussian and non-Gaussian heat kernel bounds. See [3, 32, 49, 55, 72, 84, 93] and references therein.

Regarding the necessity of the form-boundedness condition (3.1) with $\delta < 1$ for the existence of L^2 semigroup theory of $-\Delta + b \cdot \nabla$, let us mention the following consequence of the result in [74]. Let b be a distributional vector field. The sesquilinear form $-\Delta + b \cdot \nabla$ is bounded, i.e.

$$|\langle \nabla\varphi, \nabla\psi\rangle + \langle b \cdot \nabla\varphi, \psi\rangle| \leq C\|\varphi\|_{W^{1,2}}\|\psi\|_{W^{1,2}} \tag{3.11}$$

for some constant C, for all $\varphi, \psi \in C_c^\infty$, if and only if b can be represented as $b = b^{(1)} + b^{(2)}$, where $b^{(1)} \in \mathbf{F}_\delta$ for some δ and $b^{(2)}$ is divergence-free and is in the class[3] BMO^{-1}. Thus, since in this paper we are interested in conditions on $|b|$, i.e. not assuming any additional structure of b such as zero divergence, our condition (3.1) is essentially necessary for (3.11) to hold. However, (3.11) is not synonymous with the existence of a realization of $-\Delta + b \cdot \nabla$ in L^2 generating a strongly continuous semigroup. Should we require or expect (3.11) to hold in order to have L^2 semigroup theory? As we explain below, for $-\Delta + b \cdot \nabla$ the answer is “no”: in the next section we will abandon (3.11) and will go beyond the class $\mathbf{F}_\delta$. However, we will have (3.11) when will be dealing with SDE (1.3) having discontinuous diffusion coefficients. See Sect. 11.

Concerning the difference between a popular condition[4] $|b| \in L^d$ and more general condition $b \in \mathbf{F}_\delta$, let us note the following: if v is a weak solution of the elliptic equation

$$(\lambda - \Delta + b \cdot \nabla)v = f, \quad \lambda > 0, \quad f \in C_c^\infty$$

[3] I.e. the components of $b^{(2)}$ satisfy

$$b_k^{(2)} = \sum_{i=1}^{d} \nabla_i F_{ik}, \quad 1 \leq k \leq d,$$

for a matrix F with entries $F_{ik} = -F_{ki} \in \mathrm{BMO}$. Recall that a function $f \in \mathrm{BMO}$ if $f \in L^1_{\mathrm{loc}}$ and $\|f\|_{\mathrm{BMO}} := \sup_Q \frac{1}{|Q|}\int_Q |f - (f)_Q| dx < \infty$, where the supremum taken over all cubes $Q \subset \mathbb{R}^d$ with sides parallel to the axes, and $(f)_Q$ is the average of f over Q.

[4] As is well known, on the Lebesgue scale, $|b| \in L^d$ is the best possible condition providing the solvability of (1.1).

with $|b| \in L^d$ and $v \in W^{1,r}$ for r large (which is valid by a classical result), then, by Hölder's inequality,

$$\Delta v \in L^{\frac{rd}{d+r}}_{\rm loc}.$$

However, for $b \in \mathbf{F}_\delta$, one can only say that

$$\Delta v \in L^{\frac{2d}{d+2}}_{\rm loc}$$

(one can in fact show that $v \in W^{2,2}$). Thus, in case $b \in \mathbf{F}_\delta$, any $W^{2,p}$ estimate on the solution v of the elliptic equation for p large is out of question.

Remark 3.1 That being said, if $\delta < 1$, then one has

$$v \in \mathcal{W}^{1+\frac{2}{q},p}, \quad p \in \Big[2, \frac{2}{\sqrt{\delta}}\Big[, \quad q > p$$

(see Theorem 6.1 below). After applying the Sobolev embedding theorem, one obtains $|\nabla v| \in L^\gamma$ for $\gamma < \frac{dp}{d-2}$ arbitrarily close to $\frac{dp}{d-2}$ (depending on how close q is to p). Thus, although p can be as large as one wants provided that δ is chosen sufficiently small, one never arrives at $|\nabla v| \in L^\infty$. For a form-bounded drift b, the gradient of v is in general unbounded.

4 Sharp Solvability

The constant c_δ in (3.1) controls the growth of the semigroup $e^{-t\Lambda}$ as $t \to +\infty$, see (6.13), and allows to include in the class $\mathbf{F}_\delta$ bounded vector fields. It is of secondary interest to us in this paper.

The constant δ in the assumption $b \in \mathbf{F}_\delta$, on the contrary, is very important to us since it determines weak solvability of SDE (1.1). Moreover, there is a quantitative dependence between δ and the regularity properties of solutions to the corresponding elliptic and parabolic equations, see Theorem 6.1 and other results below.

The following example, analysed in detail in [7], shows that δ cannot be too large. Consider SDE

$$X_t = -\sqrt{\delta}\frac{d-2}{2}\int_0^t |X_r|^{-2}X_r dr + \sqrt{2}W_t, \tag{4.1}$$

which corresponds to the choice of attracting drift

$$b(x) = \sqrt{\delta}\frac{d-2}{2}\frac{x}{|x|^2} \in \mathbf{F}_\delta \tag{4.2}$$

and the initial point $x = 0$ in SDE (1.1). Then, if

$$\delta > 4\Big(\frac{d}{d-2}\Big)^2, \tag{4.3}$$

the SDE (4.1) does not have a weak solution. Indeed, suppose that a weak solution exists. Then $X(t) = (X_t^1, \ldots, X_t^d)$ is a continuous semimartingale with cross-variation $[X^i, X^k]_t = 2\delta_{ik}t$. By Itô's formula,

$$|X_t|^2 = -2\int_0^t X_s b(X_s)ds + 2\sqrt{2}\int_0^t X_s dW_s + 2\int_0^t d[W, W]_s,$$

i.e.

$$|X_t|^2 = -2\sqrt{\delta}\frac{d-2}{2}\int_0^t \mathbf{1}_{X_s \neq 0} ds + 2\sqrt{2}\int_0^t X_s dW_s + 2td.$$

One has $\int_0^t \mathbf{1}_{X_s=0}ds = 0$ almost surely (see details in [7]), so

$$|X_t|^2 = 2\big(d - \sqrt{\delta}\frac{d-2}{2}\big)\int_0^t \mathbf{1}_{X_s \neq 0} ds + 2\sqrt{2}\int_0^t X_s dW_s \quad \text{a.s.}$$

Therefore, $X_t^2 \geq 0$ is a local supermartingale if $\sqrt{\delta}\frac{d-2}{2} > d$. Then a.s. $X_0 = 0 \Rightarrow X_t = 0$, which contradicts to $[X^1, X^1]_t = 2t$. This argument was used earlier in [20] in the one-dimensional setting. [7] furthermore showed if $\delta > 4$, then a trajectory started outside of the origin arrives at $x = 0$ in finite time with positive probability; in this regard, see also [98].

Although at the first sight this counterexample can be interpreted (and sometimes was) as a counterexample showing the optimality of the condition $|b| \in L^d$, we argued in [46] that this is a counterexample to admissible values of constant δ. In fact, we have the following theorem.

Theorem 4.1 *Let $b \in \mathbf{F}_\delta$ with*

$$\delta < 4. \tag{C_1}$$

Then, for every $x \in \mathbb{R}^d$, the SDE

$$X_t = x - \int_0^b b(X_s)ds + \sqrt{2}W_t, \quad t \geq 0, \tag{4.4}$$

has a martingale solution.

Theorem 4.1 was proved in [51]. (In fact, it was proved there for time-inhomogeneous form-bounded drifts b with $\delta < 4$, see Definition 9.1.)

Comparing "$\delta > 4(\frac{d}{d-2})^2$" in (4.3) and "$\delta < 4$", one sees that the result is essentially sharp in high dimensions.

Let us note that the well-posedness of SDEs and parabolic equations with δ reaching the critical value (up to the strict inequality) but for b having additional structure (e.g. $b = \nabla V$ for appropriate potential V such as $V(x) = c \log |x|$), was also addressed by Fournier-Jourdan [26] (in dimension 2), Bresch-Jabin-Wang [12], see also references therein. A crucial feature of Theorem 4.1 is that it attains the critical threshold $\delta = 4$, up to the strict inequality, for the entire class of form-bounded vector fields, i.e. without any assumptions on the structure of b.

Let us explain where does "$\delta < 4$" come from (and also how one can handle $1 \leq \delta < 4$ given that the KLMN theorem requires $\delta < 1$). Multiplying the corresponding to (4.4) parabolic equation $(\partial_t - \Delta + b \cdot \nabla)u = 0$ by $u|u|^{p-2}$, integrating by parts and using the quadratic inequality and form-boundedness, one obtains that the admissible p that give e.g. an energy inequality turn out to be $p > 2/(2 - \sqrt{\delta})$. In fact, it was proved in [60] that if $b = b(x)$ has form-bound $\delta < 4$, then there exists a realization of $-\Delta + b \cdot \nabla$ in L^p, $p > 2/(2-\sqrt{\delta})$ generating a quasi contraction strongly continuous semigroup there. The interval of contraction solvability can be extended to $[2/(2-\sqrt{\delta}), \infty[$ and is sharp, see [47]. Now, as $\delta \uparrow 4$, this interval disappears, and with it disappears the theory of operator $-\Delta + b \cdot \nabla$ (see, however, the next section).

The proof of Theorem 4.1 is based on the following analytic result, which allows to use the standard tightness argument (see [90]) to construct a martingale solution of (4.4). Namely, put

$$\rho(x) = (1 + \kappa|x|^{-2})^{-\beta}, \quad \beta > \frac{d}{4}, \tag{4.5}$$

with $\kappa > 0$ sufficiently small. Let u be the classical solution to Cauchy problem

$$(\partial_t - \Delta + b \cdot \nabla)u = |\mathsf{h}|f, \quad u(0) = 0, \tag{4.6}$$

where $f \in C_c$ and $b \in \mathbf{F}_\delta \cap C_c^\infty$, $\delta < 4$ and $\mathsf{h} \in \mathbf{F}_\nu \cap C_c^\infty$, $\nu < \infty$. Fix $T > 0$ and $1 < \theta < \frac{d}{d-1}$. For all $p > \frac{2}{2-\sqrt{\delta}}$, $p \geq 2$, there exists a constant C independent of smoothness of b and h such that

$$\|u\|_{L^\infty([0,T]\times\mathbb{R}^d)} \leq C \sup_{z\in\mathbb{Z}^d} \left(\int_0^T \left\langle \left(\mathbf{1}_{\{|\mathsf{h}|\geq 1\}} + \mathbf{1}_{\{|\mathsf{h}|<1\}}|\mathsf{h}|^p\right)^{\theta'} |f|^{p\theta'} \rho_z^2 \right\rangle \right)^{\frac{1}{p\theta'}} \tag{4.7}$$

where $\rho_z(x) := \rho(x - z)$.

Now, let b_n be smooth vector fields having compact support, e.g. defined by (6.2) below, approximating b in the sense of (6.3), (6.4). Fix $x \in \mathbb{R}^d$. By a classical result, there exist strong solutions X^n to SDEs

$$X_t^n = x - \int_0^t b_n(X_s^n)ds + \sqrt{2}W_t, \quad n = 1, 2, \dots,$$

where $\{W_t\}_{t\geq 0}$ is a Brownian motion in $\mathbb{R}^d$ on a fixed complete probability space $(\Omega, \mathcal{F}, \mathcal{F}_t, \mathbf{P})$. Then (4.7) yields

$$\left|\mathbf{E}\int_{t_0}^{t_1} |\mathsf{h}(X_s^n)| f(X_s^n)ds\right|$$
$$\leq C \sup_{z\in\mathbb{Z}^d} \left(\int_{t_0}^{t_1} \left\langle \left(\mathbf{1}_{\{|\mathsf{h}|\geq 1\}} + \mathbf{1}_{\{|\mathsf{h}|<1\}}|\mathsf{h}|^p\right)^{\theta'} |f|^{p\theta'} \rho_z^2 \right\rangle \right)^{\frac{1}{p\theta'}}, \tag{4.8}$$

where $0 \leq t_0 < t_1 \leq T$. We now apply (4.8) with $\mathsf{h} = b_n$ and $f \equiv 1$ (here $f \in C_c$ $\Rightarrow f \equiv 1$ using Fatou's Lemma):

$$\mathbf{E}\int_{t_0}^{t_1} |b_n(X_s^n)|ds \leq C \sup_{z\in\mathbb{Z}^d} \left(\int_{t_0}^{t_1} \left\langle \left(\mathbf{1}_{\{|b_n|\geq 1\}} + \mathbf{1}_{\{|b_n|<1\}}|b_n|^p\right)^{\theta'} \rho_z^2 \right\rangle \right)^{\frac{1}{p\theta'}}$$
$$\leq C_0(t_1 - t_0)^{\mu} \quad \text{for generic } \mu > 0 \text{ and } C_0. \tag{4.9}$$

The latter allows to verify tightness of probability measures

$$\mathbb{P}_x^n := (\mathbf{P} \circ X^n)^{-1}$$

on $\big(C([0, T], \mathbb{R}^d), \mathcal{B}_t\big)$, so for every $x \in \mathbb{R}^d$ there exists a subsequence $\{\mathbb{P}_x^{n_k}\}$ and a probability measure $\mathbb{P}_x$ such that

$$\mathbb{P}_x^{n_k} \to \mathbb{P}_x \text{ weakly.} \tag{4.10}$$

Another application of (4.8) but with $\mathsf{h} = b_{n_1} - b_{n_2}$ allows to conclude that $\mathbb{P}_x$ is a martingale solution of (4.4), see [51] for details.

Estimate (4.7) is proved using De Giorgi's iterations in L^p for $p > 2/(2 - \sqrt{\delta})$. Thus, $p > 2$ if $1 < \delta < 4$. In this regard, let us note that passing to L^p right away, using the fact that $u^{\frac{p}{2}}$ is a sub-solution, and then applying to $u^{\frac{p}{2}}$ the standard De Giorgi's iterations in L^2, does not allow to treat $1 \leq \delta < 4$. We have to follow the iteration procedure from the very beginning and adjust it accordingly.

Earlier, De Giorgi's method in L^2 was used by Zhang-Zhao [109], Zhao [111], Röckner-Zhao [90] to prove, among other results, weak well-posedness of (4.4) with b having not too singular divergence and satisfying

$$|b| \in L^q_{\rm loc}(\mathbb{R}_+, L^r + L^\infty), \quad \frac{d}{r} + \frac{2}{q} < 2.$$

Remark 4.1 Looking at the counterexample (4.1)–(4.3) and Theorem 4.1, one can draw an analogy with the celebrated result of Baras-Goldstein for Schroödiner operator

$$-\Delta - V_0, \quad V_0(x) = \delta\frac{(d-2)^2}{4}|x|^{-2}$$

on $\mathbb{R}^d$, $d \geq 3$. This V_0 is a form-bounded potential, i.e. $\langle V_0\varphi, \varphi\rangle \leq \delta\langle\nabla\varphi, \nabla\varphi\rangle$ for all $\varphi \in W^{1,2}$ (this is Hardy's inequality). If $0 < \delta < 1$, then the self-adjoint operator realization H of $-\Delta - V_0$ on L^2, defined e.g. via the KLMN theorem, satisfies

$$e^{-tH} = s\text{-}L^2\text{-}\lim_{\varepsilon\downarrow 0} e^{-tH(V_\varepsilon)},$$

where $V_\varepsilon(x) = \delta\frac{(d-2)^2}{4}|x|_\varepsilon^{-2}$, $|x|_\varepsilon^2 := |x|^2 + \varepsilon$, $\varepsilon > 0$. For $\delta > 1$, however, by the result in [5] (see also [28]),

$$\lim_{\varepsilon\downarrow 0} e^{-tH(V_\varepsilon)} f(x) = \infty, \quad t > 0, \quad x \in \mathbb{R}^d, \quad f \geq 0, \ f \not\equiv 0,$$

i.e. all positive solutions of the corresponding parabolic equation explode instantly at any point. This phenomenon is not observable on any $V_0 \in L^{\frac{d}{2}}$ since any such potential has arbitrarily small form-bound (similarly to how any b with $|b| \in L^d$ has arbitrarily small form-bound δ).

4.1 *Critical Magnitude of Drift Singularities*

It turns out that one still has a strong[5] solution (i.e. semigroup) theory of parabolic equation $(\partial_t - \Delta + b\cdot\nabla)u = 0$ with $b \in \mathbf{F}_\delta$ in the critical borderline case $\delta = 4$, although not in L^p, as in the previous section, but in an Orlicz space. Orlicz spaces are known to appear in various borderline situations in analysis (e.g. Trudinger's theorem).

We will work over the d-dimensional torus Π^d obtained as the quotient of $[-\frac{1}{2}, \frac{1}{2}]^d$. This is not a technical assumption since the volume of the torus enters the estimates. Still, since we are interested in local singularities of b, working on the torus is sufficient for our purposes. The definition of form-bounded vector fields does not change, except that now we integrate over the torus. In this section, $\langle\cdot,\cdot\rangle$ denotes the integration over torus Π^d.

[5] Here "strong" refers to the differentiability of the solution in time.

Put

$$\Phi(y) := \cosh(y) - 1, \quad y \in \mathbb{R}$$

Expanding cosh in the Taylor series, one sees right away that $\Phi(y) = \Phi(|y|)$. This function is convex on $\mathbb{R}$, $\Phi(y) = 0$ if and only if $y = 0$, $\Phi(y)/y \to 0$ if $y \to 0$ and $\Phi(y)/y \to \infty$ if $y \to \infty$. Therefore, the space $\mathcal{L}_\Phi = \mathcal{L}_\Phi(\mathbb{R}^d)$ of real-valued measurable functions f on Π^d satisfying

$$\|f\|_\Phi := \inf\left\{c > 0 \mid \langle\Phi\left(\frac{f}{c}\right)\rangle \leq 1\right\} < \infty, \tag{4.11}$$

is a Banach space with respect to Orlicz norm $\|\cdot\|_\Phi$. See e.g. [2, Ch. 8]. Let L_Φ denote the closure $\mathcal{L}_\Phi$ of the subspace of smooth functions on Π^d. This is our Orlicz space.

Note that

$$\|\cdot\|_\Phi \geq \frac{1}{(2p)!}\|\cdot\|_{2p}, \quad p = 1, 2, \ldots, \tag{4.12}$$

i.e. we are dealing with an Orlicz norm that is stronger that any L^p norm.

Let u_n be the classical solution to Cauchy problem

$$\begin{cases} (\partial_t - \Delta + b_n \cdot \nabla)u_n = 0 \text{ on } [0, \infty[\times\Pi^d, \\ \qquad\qquad u_n(0, \cdot) = f(\cdot) \in C^\infty(\Pi^d), \end{cases}$$

where b_n are bounded smooth vector fields such that $b_n \in \mathbf{F}_\delta$ with the same c_δ and $b_n \to b$ in $L^2(\Pi^d)$ (e.g. obtained upon applying to b the De Giorgi mollifier on Π^d). Let $e^{-t\Lambda(b_n)}$, $\Lambda(b_n) := -\Delta + b_n \cdot \nabla$ denote the corresponding semigroup, i.e.

$$e^{-t\Lambda(b_n)}f := u_n(t).$$

On the smooth initial functions, $[0, \infty[\ni t \mapsto e^{-t\Lambda(b_n)}f$ is strongly continuous in the norm of L_Φ since it is strongly continuous in the norm of L^∞.

Theorem 4.2 *Let $b \in \mathbf{F}_\delta$, $0 < \delta \leq 4$. The following are true:*

(i) *For all $n \geq 1$, $f \in C^\infty(\Pi^d)$,*

$$\|e^{-t\Lambda(b_n)}f\|_\Phi \leq e^{2\frac{c_\delta}{\sqrt{\delta}}t}\|f\|_\Phi, \quad t \geq 0.$$

(ii) *There exists a strongly continuous quasi contraction semigroup $e^{-t\Lambda(b)}$ on L_Φ such that, for every $f \in C^\infty(\Pi^d)$,*

$$\|e^{-t\Lambda(b)}f - e^{-t\Lambda(b_n)}f\|_\Phi \to 0 \quad \textit{as } n \to \infty \textit{ loc. uniformly in } t \geq 0.$$

It follows that $e^{-t\Lambda(b)}$ is a positivity preserving L^∞ contraction. Its generator $\Lambda(b)$ is the appropriate operator realization of the formal operator $-\Delta+b\cdot\nabla$ in L_Φ.

This semigroup is unique in the sense that it does not depend on the choice of smooth vector fields $\{b_n\}$, $b_n \to b$ in $L^2(\Pi^d)$, as long as they do not increase constants δ, c_δ.

(iii) The following energy inequality holds for $u = e^{-t\Lambda(b_n)}f$:

$$\frac{1}{2}\sup_{s\in[0,t]}\langle e^{u^p(s)}\rangle + 4\frac{(p-1)}{p}\langle(\nabla u^{\frac{p}{2}})^2 e^{u^p}\rangle \le \langle e^{f^p}\rangle, \quad p = 2, 4, \dots$$

provided $\frac{c_\delta}{\sqrt{\delta}}t < \frac{1}{2}$; the last constraint can be removed using the semigroup property.

The last assertion is noteworthy, since, at the first sight, it seems like one can reach $\delta = 4$ only at the cost of killing off the dispersion term.

Theorem 4.2 was proved in [42]. The following calculation illustrates the main observation behind this result. Below we are looking for integral bounds on u_n that can depend on δ and c_δ, but not on the smoothness or boundedness of b_n (i.e. they should not depend on n). Put $u = u_n$, $b = b_n$. Replacing v by $v = e^{-\lambda t}u$, $\lambda \ge 0$, we will deal with Cauchy problem

$$(\lambda + \partial_t - \Delta + b\cdot\nabla)v = 0, \quad v(0) = f.$$

We multiply the equation by e^v and integrate:

$$\lambda\langle v, e^v\rangle + \langle\partial_t(e^v - 1)\rangle + 4\langle(\nabla e^{\frac{v}{2}})^2\rangle + 2\langle be^{\frac{v}{2}}, \nabla e^{\frac{v}{2}}\rangle = 0.$$

By quadratic inequality,

$$\lambda\langle v, e^v\rangle + \langle\partial_t(e^v - 1)\rangle + 4\langle(\nabla e^{\frac{v}{2}})^2\rangle \le \alpha\langle b^2 e^v\rangle + \frac{1}{\alpha}\langle(\nabla e^{\frac{v}{2}})^2\rangle.$$

Now, selecting $\alpha = \frac{1}{\sqrt{\delta}}$ and using $b \in \mathbf{F}_\delta$, we obtain

$$\lambda\langle v, e^v\rangle + \langle\partial_t(e^v - 1)\rangle + (4 - 2\sqrt{\delta})\langle(\nabla e^{\frac{v}{2}})^2\rangle \le \frac{c_\delta}{\sqrt{\delta}}\langle e^v\rangle.$$

Using $\delta \le 4$, we obtain after integrating in time from 0 to t,

$$\lambda\int_0^t\langle v, e^v\rangle ds + \langle e^{v(t)} - 1\rangle \le \langle e^f - 1\rangle + \frac{c_\delta}{\sqrt{\delta}}\int_0^t\langle e^v\rangle ds.$$

Replacing in the last inequality u by $-u$ and adding up the resulting inequalities, we obtain

$$\lambda \int_0^t \langle v \sinh(v)\rangle ds + \langle \cosh(v(t)) - 1\rangle \leq \langle \cosh(f) - 1\rangle + \frac{c_\delta}{\sqrt{\delta}} \int_0^t \langle \cosh(v)\rangle ds.$$

Finally, applying $v \sinh(v) \geq \cosh(v) - 1$, we arrive at

$$(\lambda - \frac{c_\delta}{\sqrt{\delta}}) \int_0^t \langle \cosh(v) - 1\rangle ds + \langle \cosh(v(t)) - 1\rangle \leq \langle \cosh(f) - 1\rangle + \frac{c_\delta}{\sqrt{\delta}} t, \quad (4.13)$$

where at the last step we have used the fact that volume $|\Pi^d| = 1$. Let $\lambda \geq \frac{c_\delta}{\sqrt{\delta}}$. Estimate (4.13) suggests that one should work in the topology determined by the "norm" $\langle \cosh(v) - 1\rangle$ or, better, in the corresponding Orlicz space L_Φ.

5 Three Approaches to Constructing Feller Semigroup for $-\Delta + b \cdot \nabla$

We have at our disposal the following approaches to constructing Feller semigroup associated with the Kolmogorov operator $-\Delta + b \cdot \nabla$ with $b \in \mathbf{F}_\delta$. These approaches require $\delta < 1$ ($\ll 1$ in high dimensions).

(1) By using the iteration procedure of [60] for solutions u_n to elliptic equations

$$(\mu - \Delta + b_n \cdot \nabla) u_n = f, \quad f \in C_c,$$

with bounded smooth drifts b_n approximating $b = b(x)$ (in the sense of (6.2), (6.3) below). It yields inequality

$$\|u_n - u_m\|_\infty \leq B \|u_n - u_m\|_2^\gamma \quad \text{for some } \gamma > 0 \text{ independent of } n, m.$$

The latter, in turn, transfers the verification of the Cauchy criterion in C_∞ to a much easier to deal with[6] space L^2. The convergence of the iteration procedure depends on the uniform in n gradient bound

$$\|\nabla u_n\|_{\frac{qd}{d-2}} \leq C \|f\|_q, \quad q > d - 2,$$

[6] Because it has much weaker topology than C_∞ and because form-boundedness is an L^2 assumption on $|b|$.

which was also established in [60], a pioneer work on the elliptic regularity theory of $-\Delta + b \cdot \nabla$ with form-bounded b, for $q > 2 \vee (d-2)$. This bound requires $\delta < 1 \vee (\frac{2}{d-2})^2$.

We describe this approach, or rather its relatively recent counterpart for time-inhomogeneous form-bounded drift $b = b(t,x)$, in Sect. 9. In the time-inhomogeneous case, one obtains a Feller evolution family (see definition in Sect. 9), and the convergence of the iteration procedure depends on the uniform in n gradient bound for a $q > d$:

$$\sup_{t\in[s,T]} \|\nabla u_n(t)\|_q^q + c\int_s^T \|\nabla u_n\|^q_{\frac{qd}{d-2}}\,dt \leq C\|\nabla f\|_q^q, \tag{5.1}$$

where u_n is the solution to parabolic equation $(\partial_t - \Delta + b_n(t,x)\cdot\nabla u_n) = 0$, $u_n(0) = f \in C_c^1$, and positive constants c, C are independent of n. See (9.12). The time-inhomogeneous form-bounded vector fields are defined as: $|b| \in L^2_{\text{loc}}(\mathbb{R}\times\mathbb{R}^d)$ and there exist a constant $\delta > 0$ and a function $0 \leq g \in L^1_{\text{loc}}(\mathbb{R})$ such that for a.e. $t \in \mathbb{R}$

$$\|b(t,\cdot)\varphi\|_2^2 \leq \delta\|\nabla\varphi\|_2^2 + g(t)\|\varphi\|_2^2 \tag{5.2}$$

for all $\varphi \in W^{1,2}$ (Definition 9.1).

(2) By constructing the Feller resolvent (Theorem 6.1):

$$(\mu - \Delta + b\cdot\nabla)^{-1} f := (\mu-\Delta)^{-1} f - (\mu-\Delta)^{-\frac{1}{2}-\frac{1}{q}} Q_p(1+T_p)^{-1} G_p(\mu-\Delta)^{-\frac{1}{2}+\frac{1}{r}} f,$$

where $\mu > 0$, on functions $f \in C_\infty \cap L^p$ for $r < p < q$. The operators Q_p, T_p, G_p are bounded on L^p and, moreover, $\|T_p\|_{p\to p} < 1$ under appropriate assumptions on δ, so the geometric series converges. With p chosen larger than $2\vee(d-2)$, we select q sufficiently close to p so that, by the Sobolev embedding theorem, the free Bessel potential $(\mu-\Delta)^{-1/2-1/q}$ will take us from L^p to C_∞. The difficulty in this approach is ensuring boundedness of Q_p, T_p, G_p in L^p given that form-boundedness is an L^2 assumption on $|b|$.

We note that $\|T_p\|_{p\to p} < 1$ for $p > 2\vee(d-2)$ provided that $\delta < 1\vee(\frac{2}{d-2})^2$, i.e. we arrive at the same constraint on δ as in the previous approach.

This approach was developed later in [37, 40]. It is arguably simpler than (1), due to the use of the linear structure of $-\Delta + b\cdot\nabla$. It also gives an explicit representation for the Feller resolvent of $-\Delta + b\cdot\nabla$. Throughout most of this paper, we pursue approach (2).

In the case of time-inhomogeneous form-bounded $b = b(t,x)$, one constructs solution to the inhomogeneous parabolic equation $(\mu+\partial_t-\Delta+b\cdot\nabla)u = f$ on $\mathbb{R}\times\mathbb{R}^d$ as

$$\begin{aligned}(\mu+\partial_t-\Delta+b\cdot\nabla)^{-1}f &:= (\mu+\partial_t-\Delta)^{-1}\\ &-(\mu+\partial_t-\Delta)^{-\frac{1}{2}-\frac{1}{q}}Q_p(1+T_p)^{-1}G_p(\mu+\partial_t-\Delta)^{-\frac{1}{2}+\frac{1}{r}}f,\end{aligned}$$

for appropriately defined *parabolic* operators Q_p, T_p, G_p. Armed with this result, one can obtain the sought Feller evolution family for $\partial_t-\Delta+b\cdot\nabla$, also in an explicit form. This approach is developed in Sect. 16.

The two methods of constructing Feller semigroup, (1) and (2), are quite different, but they give results of more or less the same strength (see Remark 6.1). However, for the larger class of *weakly form-bounded vector fields*, described in Sects. 14–15, only approach (2) is available. That being said, the iteration procedure in (1) has degrees of flexibility that have not been fully explored yet.

Let us now present the third approach to constructing Feller semigroup (Feller evolution family) associated with $-\Delta+b\cdot\nabla$:

(3) Using De Giorgi's method. Let us show that $\{u_n\}$, solutions to parabolic equations $(\partial_t-\Delta+b_n(t,x)\cdot\nabla u_n)=0$, $u_n(0)=f\in C_c^1$, constitute a Cauchy sequence in $L^\infty([0,T]\times\mathbb{R}^d)$. For simplicity, let us assume that b has compact support (this is not necessary, see Remark 5.1 below).

Set $g := u_n-u_m$. We have

$$\partial_t g-\Delta g+b_n\cdot\nabla g = -(b_n-b_m)\cdot\nabla u_m,\quad g(0)=0.$$

This is a Cauchy problem for an inhomogeneous parabolic equation of the same form as (4.6) (with $\mathsf{h}=b_n-b_m$, $f=\nabla u_m$; the fact that these are vector-valued functions is not an obstacle). Therefore, by (4.7),

$$\begin{aligned}&\|g\|_{L^\infty([0,T]\times\mathbb{R}^d)}\\ &\leq C\sup_{z\in\mathbb{Z}^d}\biggl(\int_0^T\Bigl\langle\bigl(\mathbf{1}_{\{|b_n-b_m|\geq 1\}}+\mathbf{1}_{\{|b_n-b_m|<1\}}|b_n-b_m|^p\bigr)^{\theta'}\\ &\quad\times|\nabla u_m|^{p\theta'}\rho_z^2\Bigr\rangle\biggr)^{\frac{1}{p\theta'}},\end{aligned}\tag{5.3}$$

where $\theta < \frac{d}{d-1}$ is chosen close to $\frac{d}{d-1}$, so that $\theta' > d$ is close to d. Using Hölder's inequality, we obtain that

$$\|g\|_{L^\infty([0,T]\times\mathbb{R}^d)}$$
$$\leq C \sup_{z\in\mathbb{Z}^d} \left(\int_0^T \left\langle \left(\mathbf{1}_{\{|b_n-b_m|\geq 1\}} + \mathbf{1}_{\{|b_n-b_m|<1\}}|b_n - b_m|^p\right)^{\frac{s'}{p}} \rho_z^2 \right\rangle\right)^{\frac{1}{s'}}$$
$$\left(\int_0^T \langle|\nabla u_m|^s \rho_z^2\rangle\right)^{\frac{1}{s}}$$

for some $s > p\theta' > pd$ is close to pd (upon selecting θ' close to d). Since b has compact support, by the Dominated convergence theorem the first multiple in the RHS tends to zero as $n, m \to \infty$. Therefore,

$$\|g\|_{L^\infty([0,T]\times\mathbb{R}^d)} = \|u_n - u_m\|_{L^\infty([0,T]\times\mathbb{R}^d)} \to 0 \quad \text{as } n, m \to \infty \tag{5.4}$$

if

$$\sup_m \|\nabla u_m\|_{L^s([0,T]\times\mathbb{R}^d)} < \infty \tag{5.5}$$

for $s > pd$ is close to pd. By the gradient estimate (5.1), after applying the interpolation inequality, we have

$$\sup_m \|\nabla u_m\|_{L^{\frac{qd}{d-2+\frac{4}{d+2}}}([0,T]\times\mathbb{R}^d)} \leq C\|\nabla f\|_q, \tag{5.6}$$

so (5.5) holds with $s = \frac{qd}{d-2+\frac{4}{d+2}}$, hence "$s > pd$" gives us constraint

$$\frac{qd}{d-2+\frac{4}{d+2}} > dp, \quad p > \frac{2}{2-\sqrt{\delta}}. \tag{5.7}$$

Additionally, the gradient estimate (5.1) imposes its own constraint on q, see Remark 9.2. These two constraints on q (which has to be as small as possible for admissible δ to be as large as possible) ensure (5.4) and hence the existence of the Feller semigroup (evolution family).

The proof of (4.7) in [51], which gives us (5.3), also uses the assumption $p \geq 2$, so q in (5.7) is not as small as one hopes. Hence, we obtain more restrictive assumption on δ than we have in approaches (1) and (2) (see Sects. 6 and 9). It is possible, however, to weaken "$p \geq 2$" by modifying (5.3), which we will address elsewhere.

Remark 5.1 To exclude the assumption that b has compact support we estimate

$$
\begin{aligned}
&\|g\|_{L^\infty([0,T]\times\mathbb{R}^d)} \\
&\leq C \sup_{z\in\mathbb{Z}^d,|z|\leq R} \bigg(\int_0^T \Big\langle\big(\mathbf{1}_{\{|b_n-b_m|\geq 1\}} + \mathbf{1}_{\{|b_n-b_m|<1\}}|b_n - b_m|^p\big)^{\frac{s'}{p}} \rho_z^2\Big\rangle\bigg)^{\frac{1}{s'}} \\
&\bigg(\int_0^T \langle|\nabla u_m|^s \rho_z^2\rangle\bigg)^{\frac{1}{s}} \\
&+ \sup_{z\in\mathbb{Z}^d,|z|>R} \bigg(\int_0^T \Big\langle\big(\mathbf{1}_{\{|b_n-b_m|\geq 1\}} + \mathbf{1}_{\{|b_n-b_m|<1\}}|b_n - b_m|^p\big)^{\frac{s'}{p}} \rho_z^2\Big\rangle\bigg)^{\frac{1}{s'}} \\
&\bigg(\int_0^T \langle|\nabla u_m|^s \rho_z^2\rangle\bigg)^{\frac{1}{s}},
\end{aligned}
$$

where the second sum can be made as small as needed by selecting R sufficiently large. To show this, one needs to use instead of (5.6) the estimate

$$
\sup_m \int_0^T \langle|\nabla u_m|^s \rho_z^2\rangle^{\frac{1}{s}} \leq C\langle|\nabla f|^q \rho_z^2\rangle^{\frac{1}{q}}, \quad s = \frac{qd}{d-2+\frac{4}{d+2}}
$$

with C independent of m and z (this is a consequence of (10.3)). Here $\langle|\nabla f|^q \rho_z^2\rangle$ is small if $|z| > R$ for R large compared to the support of f. Now, for R fixed sufficiently large, we can treat the first sum in the same way as in the case of a compact support b.

It should be noted that the methods of constructing a Feller realization of $-\Delta + b\cdot\nabla$ actually yield other regularity results. This is discussed in the rest of this paper.

6 Basic Result on Weak Well-Posedness of SDEs with Singular Drift

1. The theory of SDE

$$
dX_t = -b(X_t)dt + \sqrt{2}dW_t, \quad X_0 = x \in \mathbb{R}^d, \tag{6.1}
$$

with $b \in \mathbf{F}_\delta$, becomes more detailed as form-bound δ gets smaller. Namely, if $\delta < 1 \wedge (\frac{2}{d-2})^2$, then there is a realization of $-\Delta + b \cdot \nabla$ in C_∞ that generates a Feller semigroup. The latter, in turn, determines weak solutions of (1.1). For every $x \in \mathbb{R}^d$, the constructed weak solution is unique in some large classes (e.g. in the class of weak solutions satisfying Krylov-type estimates, or in the class of weak solutions that can be obtained via a "reasonable approximation procedure"). See Theorems 6.1 and 6.2 below.

Set $b^{\frac{2}{p}} := b|b|^{-1+\frac{2}{p}}$. For given $p \in [2, \infty[$ and $1 \leq r < p < q < \infty$, define operators ($\mu > 0$)

$$G_p(r) := b^{\frac{2}{p}} \cdot \nabla(\mu - \Delta)^{-\frac{1}{2}-\frac{1}{r}},$$

$$Q_p(q) \upharpoonright \mathcal{E} := (\mu - \Delta)^{-\frac{1}{2}+\frac{1}{q}} |b|^{1-\frac{2}{p}},$$

$$T_p \upharpoonright \mathcal{E} := b^{\frac{2}{p}} \cdot \nabla(\mu - \Delta)^{-1} |b|^{1-\frac{2}{p}}.$$

We define the last two operators on $\mathcal{E} := \bigcup_{\varepsilon>0} e^{-\varepsilon|b|} L^p$, a dense subspace of L^p, to remove any question regarding the summability of $|b|^{1-\frac{2}{p}} f$, $f \in L^p$, on which we act with the Bessel potential.

Set

$$c_{\delta,p} := \Big(\frac{p}{2}\delta + \frac{p-2}{2}\sqrt{\delta}\Big)^{\frac{1}{p}} \Big(p - 1 - (p-1)\frac{p-2}{2}\sqrt{\delta} - \frac{p(p-2)}{4}\delta\Big)^{-\frac{1}{p}}.$$

Lemma 6.1 ([40]) *Let $b \in \mathbf{F}_\delta$. Then for every $p \in [2, \infty[$, there exists $\mu_0 = \mu_0(d, p, q)$ such that the following is true for all $\mu \geq \mu_0$:*

(i) $T_p \upharpoonright \mathcal{E}$ admits extension by continuity to L^p, denoted by T_p. One has

$$\|T_p\|_{p\to p} \leq c_{\delta,p}.$$

In particular, if $\delta < 1$, then

$$\|T_p\|_{p\to p} < 1 \quad \textit{for every } p \in [2, \frac{2}{\sqrt{\delta}}].$$

(ii) $Q_p(q) \upharpoonright \mathcal{E}$ admits extension by continuity to L^p, denoted by $Q_p(q)$.
(iii) $G_p(r)$ is bounded on L^p.

Lemma 6.1 is a key result needed to prove Theorem 6.1. Its proof, which uses only elementary arguments, is included in Appendix A.

Let us fix the following approximation of b by smooth vector fields having compact supports:

$$b_n := c_n \eta_{\varepsilon_n} * (\mathbf{1}_n b), \tag{6.2}$$

where $\mathbf{1}_n$ is the indicator of $\{x \mid |x| \leq n, |b(x)| \leq n\}$ and η_{ε_n} is the Friedrichs mollifier (Sect. 2), and we choose $\varepsilon_n \downarrow 0$ sufficiently rapidly so that, for appropriate $c_n \uparrow 1$, one has

$$b_n \to b \quad \text{in } L^2_{\text{loc}}(\mathbb{R}^d, \mathbb{R}^d) \tag{6.3}$$

and

$$b_n \in \mathbf{F}_\delta \quad \text{with some } c_\delta \text{ independent of } n = 1, 2, \dots \tag{6.4}$$

see Appendix C.1. Actually, in the next theorem any bounded smooth b_n satisfying (6.3), (6.4) will do, not necessarily the ones defined by (6.2), but at the moment we save ourselves some efforts by considering (6.2) since these are, essentially, cutoffs of b. Later, however, we will consider any bounded smooth $\{b_n\}$ satisfying (6.3), (6.4). This is important because it will give us a uniqueness result on its own: the constructed semigroups or weak solutions to SDEs will not depend on the choice of approximating $\{b_n\}$, as long as they satisfy (6.3), (6.4). In this regard, it is worth introducing the following definition:

Definition 6.1 We say that vector fields b_n satisfying (6.4) are uniformly (in n) form-bounded.

Theorem 6.1 *Let $b \in \mathbf{F}_\delta$, $\delta < 1$. There exists $\mu_0 > 0$ such that the following is true:*

(*i*) *For every $p \in [2, \frac{2}{\sqrt{\delta}}[$, for all $\mu \geq \mu_0$ the function*[7]

$$u = (\mu - \Delta)^{-1} f - (\mu - \Delta)^{-\frac{1}{2}-\frac{1}{q}} Q_p(q)(1 + T_p)^{-1} G_p(r)(\mu - \Delta)^{-\frac{1}{2}+\frac{1}{r}} f, \quad f \in L^p \tag{6.5}$$

is a weak solution to equation

$$(\mu - \Delta + b \cdot \nabla)u = f \tag{6.6}$$

i.e.

$$\mu\langle u, \psi\rangle + \langle \nabla u, \nabla \psi\rangle + \langle b \cdot \nabla u, \psi\rangle = \langle f, \psi\rangle \quad \text{for all } \psi \in C_c^\infty.$$

Moreover, if $f \in L^p \cap L^2$, then u is the unique in $W^{1,2}$ weak solution to (6.6).

(*ii*) *It follows from* (6.5) *that*

$$u \in \mathcal{W}^{1+\frac{2}{q},p} \text{ (Bessel potential space)}, \quad q > p. \tag{6.7}$$

In particular, if $\delta < 1 \wedge \left(\frac{2}{d-2}\right)^2$, then in the interval $p \in [2, \frac{2}{\sqrt{\delta}}]$ we can select $p > d - 2$, and then select q sufficiently close to p, so that by the Sobolev embedding theorem u is Hölder continuous (possibly after a modification on a

[7] Note that after expanding $(1 + T_p)^{-1}$ in (6.5) in the geometric series, we obtain the formal Neumann series representation for u.

measure zero set), with the Hölder continuity exponent less than but arbitrarily close to $1 - \frac{d-2}{p}$.

(*iii*) *The operator-valued function in* (6.5)

$$\Theta_p(\mu, b) := (\mu-\Delta)^{-1} - (\mu-\Delta)^{-\frac{1}{2}-\frac{1}{q}} Q_p(q)(1+T_p)^{-1} G_p(r)(\mu-\Delta)^{-\frac{1}{2}+\frac{1}{r}},$$

defined on $\{\mu \geq \mu_0\}$, *takes values in* $\mathcal{B}(\mathcal{W}^{-1+\frac{2}{r},p}, \mathcal{W}^{1+\frac{2}{q},p})$.

(*iv*) *Let* $\delta < 1 \wedge \left(\frac{2}{d-2}\right)^2$. *Fix* $p \in]d-2, \frac{2}{\sqrt{\delta}}[$ *if* $d \geq 4$, *or* $p \in [2, \frac{2}{\sqrt{\delta}}[$ *if* $d = 3$. *Then*

$$(\mu + \Lambda_{C_\infty}(b))^{-1} := \left(\Theta_p(\mu, b) \upharpoonright L^p \cap C_\infty\right)^{\text{clos}}_{C_\infty \to C_\infty}, \quad \mu \geq \mu_0, \tag{6.8}$$

determines the resolvent of a Feller generator on C_∞. *This semigroup satisfies*

$$e^{-t\Lambda_{C_\infty}(b)} = s\text{-}C_\infty\text{-}\lim_n e^{-t\Lambda_{C_\infty}(b_n)} \quad \textit{locally uniformly in } t \geq 0,$$

where bounded smooth b_n *are defined by* (6.2) *and the approximating operators* $\Lambda_{C_\infty}(b_n) := -\Delta + b_n \cdot \nabla$ *with domain* $D(\Lambda_{C_\infty}(b_n)) := (1-\Delta)^{-1} C_\infty$ *are, by a classical result, Feller generators.*

(*v*) *Feller semigroup* $e^{-t\Lambda_{C_\infty}(b)}$ *is conservative, i.e. its integral kernel* $e^{-t\Lambda_{C_\infty}}(x, \cdot)$ *satisfies*

$$\langle e^{-t\Lambda_{C_\infty}(b)}(x, \cdot)\rangle = 1 \quad \textit{for all } x \in \mathbb{R}^d, t > 0. \tag{6.9}$$

Excluding the cases where b is sufficiently regular, one has little information about the domain $D(\Lambda_{C_\infty}(b))$ of the Feller generator $\Lambda_{C_\infty}(b)$. But, it is easily seen, one can be certain that $C_c^2 \not\subset D(\Lambda_{C_\infty}(b))$ already if $|b| \in L^\infty - C_b$.

Remark 6.1 The construction of the Feller semigroup via the iteration procedure of [60] requires gradient bounds on solutions u_n of $(\mu - \Delta + b_n \cdot \nabla)u_n = f$, $f \in C_c^\infty$. Namely, it is established in [60, proof of Lemma 5] that if $\delta < 1 \wedge (\frac{2}{d-2})^2$, then for $p > 2 \vee (d-2)$ close to $2 \vee (d-2)$

$$\|\nabla |\nabla u_n|^{\frac{p}{2}}\|_2^2 \leq K \|f\|_p^p, \tag{6.10}$$

for a constant K independent of n. (We discuss a parabolic analogue of (6.10) below.) Then, applying the Sobolev embedding theorem twice, one obtains that u_n has Hölder continuity exponent $1 - \frac{d-2}{p}$ (independent of n). Taking into account that p satisfies a strict inequality, one thus obtains the same Hölder continuity result as in Theorem 6.1(*ii*). However, there is substantial difference between gradient bounds (6.7) and (6.10): (6.7) allows to control order > 1 derivatives of u_n, while (6.10) does not need additional strict inequalities such as "$q > p$" in (6.7) (clearly, if in

(6.7) one could take $q = p$, then it would make the regularity result stronger, but one cannot do this).

Above we mentioned that the approach of Theorem 6.2 is somewhat simpler than [60]. One reason is that it uses to the full extent the linear structure of the equation. There is another reason: in the iteration procedure of [60] one shows that the solutions of the approximating equations constitute a Cauchy sequence, while in the proof of Theorem 6.2 one already has a candidate for the limit of this sequence, cf. (6.8).

The free Bessel potential $(\mu - \Delta)^{-\frac{1}{2}-\frac{1}{q}}$ in (6.5) provides a "trampoline" from L^p to C_∞. One advantage of working in L^p with p large is that it simplifies the proof of approximation results as e.g. in assertion (*iv*) of Theorem 6.1, since it is easier to prove convergence in L^p than in C_∞. In fact, this "trampoline" was used in [59] who considered bounded drifts of α-stable process. There, however, working in L^p is a matter of convenience (one can also stay entirely in C_∞ while dealing with unbounded drifts, see [6]). Our goal, however, is to transition from an L^2 assumptions on the drift (i.e. the form-boundedness), via L^p, to a semigroup in C_∞. It is the transition from L^2 to L^p with p large that is the most difficult one.

The operator-valued function $\mu \mapsto \Theta_p(\mu, b) \in \mathcal{B}(L^p)$ in Theorem 6.1 is itself the resolvent of the generator of a quasi contraction semigroup in L^p. In fact, this semigroup coincides with the semigroup $e^{-t\Lambda_p(b)}$ constructed via (3.10), and satisfies

$$e^{-t\Lambda_p(b)} = s\text{-}C_\infty\text{-}\lim_n e^{-t\Lambda_p(b_n)} \quad \text{locally uniformly in } t \geq 0, \tag{6.11}$$

where $\Lambda_p(b_n) = -\Delta + b_n \cdot \nabla$ having domain $\mathcal{W}^{2,p}$. See [40] for details.

By the way, it is easy to write a similar to $\Theta_p(\mu, b)$ operator-valued function representation for the resolvent of $-\Delta - \nabla \cdot b$ with $b \in \mathbf{F}_\delta$, and to modify the proofs in [40] to work for this operator.

Combining (6.11) with Theorem 6.1, one of course obtains

$$e^{-t\Lambda_{C_\infty}(b)} \upharpoonright L^2 \cap C_\infty = e^{-t\Lambda_p(b)} \upharpoonright L^2 \cap C_\infty = e^{-t\Lambda(b)} \upharpoonright L^2 \cap C_\infty, \tag{6.12}$$

where the last semigroup is provided by the KLMN theorem.

The resolvent representations of the type $\Theta_p(\mu, b)$ were considered earlier for Schrödinger operators with form-bounded potentials, to obtain information about Sobolev regularity of their domains and hence regularity of their eigenfunctions, see [8, 71]. Let us note, however, that a direct comparison between Feller theories of Kolmogorov and Schrödinger operators is not possible, see Remark 14.2.

The semigroup $e^{-t\Lambda_p(b)}$ can, in fact, be defined via the limit (6.11) for all $\delta < 4$, $p > \frac{2}{2-\sqrt{\delta}}$ [60], and satisfies

$$\|e^{-t\Lambda_p(b)} f\|_q \leq c e^{t\omega_p} t^{-\frac{d}{2}(\frac{1}{p}-\frac{1}{q})} \|f\|_p, \quad t > 0, \quad \omega_p := \frac{c_\delta}{2(p-1)}, \tag{6.13}$$

for all $f \in L^q \cap L^p$. In view of (6.12), the same estimate on $e^{-t\Lambda_{C_\infty}(b)}f$, $f \in L^p \cap C_\infty$ is valid (of course, under the assumptions on δ and p of Theorem 6.1(*iv*)). Regarding the properties of $e^{-t\Lambda_p(b)}$ (in particular, regarding extending the interval $p \in]\frac{2}{2-\sqrt{\delta}}, \infty[$ to a larger interval), see [47].

Nevertheless, both the upper and the lower Gaussian bounds on the heat kernel of $-\Delta + b \cdot \nabla$ are easily destroyed by form-bounded drifts. (The heat kernel is defined, up to a modification on a measure zero set, as the integral kernel $e^{-t\Lambda(b)}(x, y)$ of $e^{-t\Lambda_p(b)}$. The heat kernel does not depend on p.) In fact, it suffices to consider drifts

$$b(x) = \pm\sqrt{\delta}\frac{d-2}{2}|x|^{-2}x, \tag{6.14}$$

which are form-bounded. The singularity of (6.14) is so strong that it introduces an extra factor $\varphi_t(y)$ in the Gaussian bounds,

$$c_1 t^{-\frac{d}{2}} e^{-\frac{|x-y|^2}{c_2 t}} \varphi_t(y) \le e^{-t\Lambda(b)}(x, y) \le c_3 t^{-\frac{d}{2}} e^{-\frac{|x-y|^2}{c_4 t}} \varphi_t(y),$$

where $\varphi_t(t)$ either explodes or vanishes at the origin depending on the sign in front of $\sqrt{\delta}$ in (6.14) (moreover, the rate of explosion or vanishing is an explicit function of δ) [76]. Of course, if one considers a sum (or a series) of such drifts with singularities at different points (which is still form-bounded), the situation at the level of heat kernel bounds becomes even more complicated. A detailed discussion on this subject can be found in [49] where the authors prove Gaussian lower and/or upper bound on the heat kernel of $-\nabla \cdot a \cdot \nabla + b \cdot \nabla$ with measurable uniformly elliptic matrix a and drift b that is form-bounded or even more singular, under additional constraints on $\operatorname{div} b$.

Let us also note that the two-sided Gaussian bound on the heat kernel of $-\Delta + b \cdot \nabla$ or of $-\nabla \cdot a \cdot \nabla + b \cdot \nabla$ hold, without any assumptions on $\operatorname{div} b$, when b is in the Kato class of vector fields[8] [102], or in the Nash class[9] [50, 92], respectively. Moreover, one can go beyond the Kato class and prove Gaussian bounds for distributional drifts, see [83, 110]. Both the Kato class and the Nash class neither contain the class $\mathbf{F}_\delta$ nor are contained in it (but the Kato class of vector fields is contained in the class of *weakly form-bounded vector fields*, considered in the next section).

Remark 6.2 It is clear from (6.5) that we do not have (and cannot have) information about L^p summability of the second derivatives of u for $p > 2$ large (in this

---

[8] See definition in Sect. 14.

[9] $b = b(x)$ is in the Nash class $\mathbf{N}_\delta$ if $|b| \in L^2_{\text{loc}}$ and

$$\inf_{h>0} \sup_{x \in \mathbb{R}^d} \int_0^h \sqrt{e^{t\Delta}|b|^2(x)} \frac{dt}{\sqrt{t}} \le \delta.$$

It contains b with $|b| \in L^p$, $p > d$, and it also contains vector fields with $|b| \notin L^{2+\varepsilon}_{\text{loc}}$, $\varepsilon > 0$.

regard, see discussion before Theorem 4.1). However, we have weighted estimates on the second derivatives. Assume for simplicity that b is bounded and smooth, so we are looking for estimates with constants that do not depend on smoothness of b or its boundedness. It follows from (6.5) that

$$(|b|+1)^{-1+\frac{2}{p}}(\mu-\Delta)u$$
$$=(|b|+1)^{-1+\frac{2}{p}}f-\left(\frac{|b|}{|b|+1}\right)^{1-\frac{2}{p}}(1+T_p)^{-1}T_p'\,(|b|+1)^{-1+\frac{2}{p}}f$$

where $T_p'=b^{\frac{2}{p}}\cdot\nabla(\mu-\Delta)^{-1}(|b|+1)^{1-\frac{2}{p}}$ is bounded on L^p, just like T_p, and so

$$\|(|b|+1)^{-1+\frac{2}{p}}(\mu-\Delta)u\|_p\leq(1+C)\|(|b|+1)^{-1+\frac{2}{p}}f\|_p,\quad C<\infty.$$

Thus, if $p>2$, then at the points where $|b|$ is infinite the factor $(|b|+1)^{-1+\frac{2}{p}}$ vanishes, and so around the singular set of b the information about L^p summability of $(\mu-\Delta)u$ disappears, but in a controlled way.

Finally, let us add that in Theorem 6.1 we could also write

$$u=(\mu-\Delta)^{-1}f-(\mu-\Delta)^{-\frac{1}{2}-\frac{1}{q}}\hat{Q}_p(q)(1+\hat{T}_p)^{-1}\hat{G}_p(\mu-\Delta)^{-\frac{1}{2}+\frac{1}{r}}f,$$

where operators $\hat{Q}_p$, $\hat{T}_p$, $\hat{R}_p(r)$ have "classical" form but are bounded in the weighted L^p space with weight $|b|_1^2:=(|b|+1)^2$:

$$\hat{Q}_p(q)=(\mu-\Delta)^{-\frac{1}{2}+\frac{1}{q}}b\cdot\quad\text{is in }\mathcal{B}\big(\big[L^p(\mathbb{R}^d,|b|_1^2dx)\big]^d,L^p(\mathbb{R}^d)\big),$$

$$\hat{T}_p=\nabla(\mu-\Delta)^{-1}b\cdot\quad\text{is in }\mathcal{B}\big(\big[L^p(\mathbb{R}^d,|b|_1^2dx)\big]^d\big),$$

$$\hat{G}_p(r)=\nabla(\mu-\Delta)^{-\frac{1}{2}-\frac{1}{r}}\quad\text{is in }\mathcal{B}\big(L^p(\mathbb{R}^d),\big[L^p(\mathbb{R}^d,|b|_1^2dx)\big]^d\big)$$

where $[\cdot]^d$ denotes vector with d components. Their boundedness follows from the boundedness of Q_p, T_p, G_p on $L^p=L^p(\mathbb{R}^d,dx)$.

2. We are now in position to prove the following result on weak well-posedness of SDE

$$X_t=x-\int_0^t b(X_r)dr+\sqrt{2}W_t,\quad 0\leq t\leq T \tag{6.15}$$

with $x\in\mathbb{R}^d$ fixed.

Theorem 6.2 *Let $b\in\mathbf{F}_\delta$ with $\delta<1\wedge\left(\frac{2}{d-2}\right)^2$. Let $e^{-t\Lambda_{C_\infty}(b)}$ be the Feller semigroup constructed in Theorem 6.1. Fix $T>0$. The following is true:*

(*i*) *There exist probability measures* $\{\mathbb{P}_x\}_{x\in\mathbb{R}^d}$ *on the canonical space* $(C[0,T],\mathcal{B}_t)$ *such that*

$$\mathbb{E}_{\mathbb{P}_x}[f(X_t)] = (e^{-t\Lambda_{C_\infty}(b)}f)(x), \quad f \in C_\infty, \quad x \in \mathbb{R}^d.$$

For every $x \in \mathbb{R}^d$ *the measure* $\mathbb{P}_x$ *is a weak solution to SDE* (6.15).

(*ii*) *If* δ *is sufficiently small and, additionally,* $|b| \in L^{\frac{d}{2}+\varepsilon}$ *for some* $\varepsilon > 0$, *then the constructed in (i) weak solution* $\mathbb{P}_x$ *belongs to and is unique in the class of weak solutions satisfying the following Krylov-type estimate for* $q > \frac{d}{2}$ *sufficiently close to* $\frac{d}{2}$ *(depending on how small* ε *is):*

$$\mathbb{E}_{\mathbb{P}_x}\int_0^T |h(t,\omega_t)|dt \le c\|h\|_{L^q([0,T]\times\mathbb{R}^d)}. \tag{6.16}$$

for all $h \in C_c(\mathbb{R}^{d+1})$, *for generic constant* c.

Remark 6.3 Arguing as in [46], one can also prove the following "approximation uniqueness" result. If $\{\mathbb{Q}_x\}_{x\in\mathbb{R}^d}$ is another weak solution to (6.15) such that

$$\mathbb{Q}_x = w\text{-}\lim_n \mathbb{P}_x(\tilde{b}_n) \quad \text{for every } x \in \mathbb{R}^d,$$

for some $\{\tilde{b}_n\} \subset \mathbf{F}_{\delta_1}$ with $\delta < 1 \wedge \left(\frac{2}{d-2}\right)^2$ and c_δ independent of n, then $\{\mathbb{Q}_x\}_{x\in\mathbb{R}^d} = \{\mathbb{P}_x\}_{x\in\mathbb{R}^d}$. In other words, the constructed weak solutions $\mathbb{P}_x$ of (6.15) are unique among those that can be obtained via an approximation procedure. Note that we do not require here any convergence of $\tilde{b}_n$ to b.

In Sect. 15 we will discuss analogues of Theorems 6.1, 6.2 for drifts that can be more singular than the form-bounded drifts. See Theorems 15.1, 15.2. This, however, will come at the cost of imposing more restrictive assumption on δ, and losing the possibility to include discontinuous diffusion coefficients, as we do for the form-bounded drifts in end of this section. Also, while the proof of the analogue of Lemma 6.1 in Sect. 15 (i.e. Lemma 15.1 there) relies on some operator inequalities for fractional powers of the Laplacian, the proof of Lemma 6.1 uses only elementary arguments.

Remark 6.4 The proof of Theorem 6.2 and the construction of the Feller semigroup in Theorem 6.1(*iv*)–(*v*) rely on the elliptic regularity result of Theorem 6.1(*i*)–(*iii*). In the next sections we will be working directly in the parabolic setting, thus avoiding the use of the Trotter approximation theorem and, generally speaking, arriving at shorter proofs. However, by working with resolvents and using the semigroup theory (i.e. Trotter's theorem), we can construct a Feller semigroup departing from L^p for a smaller p, which leads to less restrictive assumptions on δ. For instance, one can compare Theorem 6.1(*iv*) where p is chosen to be strictly greater than $2 \vee (d-2)$, and we require $\delta < 1 \wedge (\frac{2}{d-2})^2$, and Theorem 9.1 where p

has to be strictly greater than d, and $\delta < \frac{1}{d^2}$; or, better, δ satisfies (C_5) below, which is still more restrictive than $\delta < 1 \wedge (\frac{2}{d-2})^2$.

7 Proof of Theorem 6.1

Assertion (*iii*) follows right away from Lemma 6.1. The first part of assertion (*i*), and assertion (*ii*), follow from (*iii*). The proof of the second part of (*i*), i.e. the characterization of u as the unique weak solution to the elliptic equation, is standard and we will attend to it in the end.

(*iv*) For every $n = 1, 2, \ldots$, the operator-valued function $\Theta_p(\mu, b_n)$ is a pseudo-resolvent, i.e. it satisfies

$$\Theta_p(\mu, b_n) - \Theta_p(\eta, b_n) = (\nu - \mu)\Theta_p(\mu, b_n)\Theta_p(\nu, b_n), \quad \mu, \nu \geq \mu_0, \tag{7.1}$$

where μ_0 is from Lemma 6.1. Identity (7.1) is verified via direct calculation. See [37, proof of Prop. 2] for details.[10]

By the classical theory, for every $n = 1, 2, \ldots$, the resolvent of the approximating operator $(\mu + \Lambda_{C_\infty}(b_n))^{-1}$ is defined on $\{\mu \geq \mu_n\}$, where μ_n depends e.g. on $\|b_n\|_\infty$. Our first observation is that we can replace μ_n by a μ_0 independent of n by establishing a link between $(\mu + \Lambda_{C_\infty}(b_n))^{-1}$ and the operator-valued function $\Theta_p(\mu, b_n)$. That is,

$$(\mu + \Lambda_{C_\infty}(b_n))^{-1} \upharpoonright \mathcal{S} = \Theta_p(\mu, b_n) \upharpoonright \mathcal{S} \quad \text{for all } \mu \geq \mu_0, \tag{7.2}$$

for some μ_0 independent of n. Indeed, we have

$$\Theta_p(\mu_n, b_n) \upharpoonright \mathcal{S} = (\mu_n + \Lambda_{C_\infty}(b_n))^{-1} \upharpoonright \mathcal{S}$$

for all sufficiently large μ_n. By $\Theta_p(\mu, b_n)\mathcal{S} \subset \mathcal{S}$, the previous identity and the resolvent identity (7.1),

$$\Theta_p(\mu, b_n) \upharpoonright \mathcal{S} = (\mu_n + \Lambda_{C_\infty}(b_n))^{-1}\big(1 + (\mu_n - \mu)\Theta_p(\mu, b_n)\big) \upharpoonright \mathcal{S}, \quad \mu \geq \mu_0,$$

so $\Theta_p(\mu, b_n) \upharpoonright \mathcal{S}$ is the right inverse of $\mu + \Lambda_{C_\infty}(b_n) \upharpoonright \mathcal{S}$ on $\mu \geq \mu_0$. Similarly, it is seen that $\Theta_p(\mu, b_n)|_{\mathcal{S}}$ is the left inverse of $\mu + \Lambda_{C_\infty}(b_n) \upharpoonright \mathcal{S}$ on $\mu \geq \mu_0$. This gives (7.2).

---

[10] Alternatively, one can verify, using the KLMN theorem, that $\Theta_2(\mu, b_n)$, $\mu \geq \mu_0$, for a $\mu_0 > 0$ independent of n, is the resolvent of $-\Delta + b_n \cdot \nabla$ in L^2, and so (7.1) holds on $L^2 \cap L^p$ and hence on L^p. But we do not need the L^2 theory of $-\Delta + b \cdot \nabla$ in the proof of Theorem 6.1.

Second, let us show that for every $\mu \geq \mu_0$

$$\Theta_p(\mu, b)\mathcal{S} \subset L^p \cap C_\infty \quad \text{(after a modification on a measure zero set)}, \tag{7.3}$$

and

$$\Theta_p(\mu, b_n) \xrightarrow{s} \Theta_p(\mu, b) \text{ in } L^p \cap C_\infty \quad (n \to \infty). \tag{7.4}$$

The inclusion into C_∞ in (7.3) is immediate due to the factor $(\mu - \Delta)^{-\frac{1}{2}-\frac{1}{q}}$ in the definition of $\Theta_p(\mu, b_n)$, upon applying the Sobolev embedding theorem (here we use the assumption $p > 2 \vee (d-2)$ and the choice of $q > p$ close to p). The second assertion (7.3) is proved using again the Sobolev embedding theorem and the convergence

$$Q_p(q, b_n) \xrightarrow{s} Q_p(q, b), \quad T_p(b_n) \xrightarrow{s} T_p(b), \quad G_p(r, b_n) \xrightarrow{s} G_p(r, b).$$

The latter, in turn, follow from the Dominated convergence theorem since b_n defined by (6.2) are, essentially, cutoffs of b (for details, if needed, see the proof of [37, Prop. 7]).

Third, we have

$$\sup_n \|\mu(\mu + \Lambda_{C_\infty}(b_n))^{-1}\|_{\infty \to \infty} \leq 1 \quad \text{for all } \mu \geq \mu_0. \tag{7.5}$$

Indeed, for every $n = 1, 2, \dots$, the semigroup $e^{-t\Lambda_{C_\infty}(b_n)}$ is an L^∞ contraction, so, integrating $\|e^{-\mu t}e^{-t\Lambda_{C_\infty}(b_n)}f\|_\infty \leq e^{-\mu t}\|f\|_\infty$ in t from 0 to ∞ we arrive at (7.5).

Fourth, we note that

$$\mu\Theta_p(\mu, b_n) \xrightarrow{s} 1 \text{ in } C_\infty \text{ as } \mu \uparrow \infty \text{ uniformly in } n. \tag{7.6}$$

Indeed, in view of (7.5), since $\mathcal{S}$ is dense in C_∞, it suffices to prove that $\mu\Theta_p(\mu, b_n)f \to f$ in C_∞ as $\mu \uparrow \infty$ for all $f \in \mathcal{S}$. In turn, since $\lim_{\mu\to\infty} \|\mu(\mu - \Delta)^{-1}f - f\|_\infty = 0$, it suffices to show that $\sup_n \|\mu\Theta_p(\mu, b_n)f - \mu(\mu - \Delta)^{-1}f\|_\infty \to 0$ as $\mu \uparrow \infty$. We have

$$\begin{aligned} &\Theta_p(\mu, b_n)f - (\mu - \Delta)^{-1}f \\ &\quad = -(\mu - \Delta)^{-\frac{1}{2}-\frac{1}{q}} Q_p(q)\big(1 + T_p\big)^{-1} b_n^{\frac{2}{p}} \cdot \nabla(\lambda - \Delta)^{-1}(\mu - \Delta)^{-1}(\lambda - \Delta)f, \end{aligned}$$

with $q > p$ where λ is sufficiently large but fixed. Note, that

$$\|(\mu - \Delta)^{-\frac{1}{2}-\frac{1}{q}}\|_{p \to \infty} \leq c\mu^{-\frac{1}{2}+\frac{d}{2p}-\frac{1}{q}},$$

and

$$\|b_n^{\frac{2}{p}} \cdot \nabla(\lambda - \Delta)^{-1}\|_{p \to p} \leq c_1$$

with c_1 independent of n and μ (since $\|G_p(r)\|_{p \to p} \equiv \|b_n^{\frac{2}{p}} \cdot \nabla(\lambda - \Delta)^{-\frac{1}{2}-\frac{1}{r}}\|_{p \to p}$ is uniformly bounded in n). Thus

$$\|\Theta_p(\mu, b_n)f - (\mu - \Delta)^{-1}f\|_\infty \leq C\mu^{-\frac{1}{2}+\frac{d}{2p}-\frac{1}{q}}\mu^{-1}\|(\lambda - \Delta)f\|_p.$$

Since $p > 2 \vee (d-2)$, we select q sufficiently close to p so that $-\frac{1}{2} + \frac{d}{2p} - \frac{1}{q} - 1 < -1$, and hence $\sup_n \|\mu\Theta_p(\mu, b_n)f - \mu(\mu - \Delta)^{-1}f\|_\infty \to 0$ as $\mu \uparrow \infty$, which yields (7.6).

We now prove assertion (*iv*) of the theorem using the Trotter approximation theorem (Appendix D). Its conditions

$$\sup_n \|(\mu + \Lambda_{C_\infty}(b_n))^{-1}\|_{\infty \to \infty} \leq \frac{1}{\mu} \quad \text{for all } \mu \geq \mu_0,$$

$$\text{there exists } s\text{-}C_\infty\text{-}\lim_n(\mu + \Lambda_{C_\infty}(b_n))^{-1} \quad \text{for some } \mu \geq \mu_0,$$

$$\mu(\mu + \Lambda_{C_\infty}(b_n))^{-1} \to 1 \quad \text{in } C_\infty \text{ as } \mu \uparrow \infty \text{ uniformly in } n$$

are verified in (7.2)–(7.6). So, Trotter's theorem yields (*iv*) including the strong convergence of semigroups in C_∞.

(*v*) The proof of (6.9) uses the localized estimate

$$\left\|\rho(\mu + \Lambda_{C_\infty}(b_n))^{-1}h\right\|_\infty \leq C_3\|\rho h\|_p, \tag{7.7}$$

where

$$\rho(x) := (1 + \kappa|x|^2)^{-\nu}$$

with $\nu > \frac{d}{2p}$ fixed (then $\rho \in L^p$) and $\kappa > 0$ to be chosen sufficiently small. We comment on the proof of (7.7) below. Estimate (7.7) yields: for every fixed $x \in \mathbb{R}^d$ there is C such that

$$|(\mu + \Lambda_{C_\infty}(b_n))^{-1}h(x)| \leq C\|\rho h\|_p.$$

By considering an increasing sequence $h \uparrow 1 - \mathbf{1}_{B_R(0)}$ we obtain

$$\langle(\mu + \Lambda_{C_\infty}(b_n))^{-1}(x, \cdot)(1 - \mathbf{1}_{B_R(0)}(\cdot))\rangle \leq C\|\rho(1 - \mathbf{1}_{B_R(0)}\|_p,$$

where, as is evident from the definition of ρ, the right-hand side can be made smaller than any ε uniformly in n by selecting radius $R > 0$ sufficiently large.

Since $\langle (\mu + \Lambda_{C_\infty}(b_n))^{-1}(x, \cdot) \rangle = \mu^{-1}$, $n = 1, 2, \dots$, we have

$$\mu^{-1} - \varepsilon \le \langle (\mu + \Lambda_{C_\infty}(b_n))^{-1}(x, \cdot) \mathbf{1}_{B_R(0)}(\cdot) \rangle \le \mu^{-1}.$$

By passing to the limit in n we obtain $\mu^{-1} - \varepsilon \le \langle (\mu + \Lambda_{C_\infty}(b))^{-1}(x, \cdot) \mathbf{1}_{B_R(0)}(\cdot) \rangle \le \mu$. Finally, sending $R \to \infty$, and then $\varepsilon \downarrow 0$, we arrive at

$$\langle (\mu + \Lambda_{C_\infty}(b))^{-1}(x, \cdot) \rangle = \mu^{-1}$$

which gives us (6.9).

Regarding the proof of (7.7), we can either commute ρ with the operators that constitute $\Theta_p(\mu, b_n) \equiv (\mu + \Lambda_{C_\infty}(b_n))^{-1}$ (see [46] for details), or we can note that the equation for u_n, i.e. $(\mu - \Delta + b_n \cdot \nabla) u_n = h$, can be rewritten as

$$\mu \rho u_n - \Delta(\rho u_n) + \tilde{b}_n \cdot \nabla(\rho u_n) = \rho h + K,$$

where

$$\tilde{b}_n := b_n + 2\frac{\nabla \rho}{\rho}, \qquad K = 2\frac{(\nabla \rho)^2}{\rho} u_n + (-\Delta \rho) u_n + b_n u_n \cdot \nabla \rho.$$

Now, we apply bounds

$$|\nabla \rho(x)| \le 2\nu\sqrt{\kappa}\rho(x), \qquad \frac{|\nabla \rho(x)|^2}{\rho(x)} \le 4\nu^2 \kappa \rho(x),$$
$$|\Delta \rho(x)| \le \big(4\nu^2 + (4 + 2d)\nu\big)\kappa \rho(x), \quad x \in \mathbb{R}^d.$$

By selecting κ sufficiently small, one can make the form-bound of $\tilde{b}$ as close to the form-bound δ of b as needed. Furthermore, the first two terms in K can be absorbed by $\mu \rho u_n$ (at the expense of replacing μ with $\mu - \mu_1$ for appropriate $\mu_1 = \mu_1(\kappa) > 0$). We have

$$\begin{aligned} \rho u_n &= (\mu - \mu_1 + \Lambda_{C_\infty}(\tilde{b}_n))^{-1}(\rho h + b_n u_n \cdot \nabla \rho) \\ &= \Theta_p(\mu - \mu_1, \tilde{b}_n)(\rho h + b_n u_n \cdot \nabla \rho), \quad \mu > \mu_1, \end{aligned}$$

so we can apply Lemma 6.1 to obtain (7.7). See details in [41].

Returning to the proof of (*i*), we note that u_n coincides with the classical solution to $(\mu - \Delta + b_n \cdot \nabla) u_n = f$. Now, using convergence

$$Q_p(q, b_n) \xrightarrow{s} Q_p(q, b), \quad T_p(b_n) \xrightarrow{s} T_p(b), \quad G_p(r, b_n) \xrightarrow{s} G_p(r, b),$$

(see the prof of (7.4)) it is easy to pass to the limit in n in

$$\mu\langle u_n, \psi\rangle + \langle \nabla u_n, \nabla \psi\rangle + \langle b_n \cdot \nabla u_n, \psi\rangle = \langle f, \psi\rangle \quad \text{for all } \psi \in C_c^\infty,$$

which shows that u is a weak solution. A standard argument (i.e. the Lax-Milgram theorem) yields that u is the unique weak solution. □

8 Proof of Theorem 6.2

(*i*) The following estimate will be needed: for all $h \in C_c$ and all $\mu \geq \mu_0$

$$\big\|(\mu + \Lambda_{C_\infty}(b))^{-1}|b_m|h\big\|_\infty \leq C_1\||b_m|^{\frac{2}{p}}h\|_p, \tag{8.1}$$

$$\|(\mu + \Lambda_{C_\infty}(b))^{-1}|b_m - b_n|h\|_\infty \leq C_2\big\||b_m - b_n|^{\frac{2}{p}}h\big\|_p. \tag{8.2}$$

for appropriate constants $C_i = C_i(\delta, p)$, $i = 1, 2$, where $p > 2 \vee (d-2)$ is fixed. These estimates follow right away from the construction of $(\mu + \Lambda_{C_\infty}(b))^{-1}$ via $\Theta_p(\mu, b)$ in Theorem 6.1(*iv*). Namely, since $|b_m|h \in C_c$,

$$\begin{aligned}(\mu + \Lambda_{C_\infty}(b))^{-1}|b_m|h &= \Theta_p(\mu, b)|b_m|h \\ &= (\mu - \Delta)^{-\frac{1}{2}-\frac{1}{q}}G_p(q, b_m)|b_m|^{\frac{2}{p}}h \\ &\quad - (\mu - \Delta)^{-\frac{1}{2}-\frac{1}{q}}Q_p(q, b)(1 + T_p(b))^{-1}T_p(b, b_m)|b_m|^{\frac{2}{p}}h.\end{aligned} \tag{8.3}$$

The operator $T_p(b, b_m) := b^{\frac{2}{p}} \cdot \nabla(\mu - \Delta)|b_m|^{1-\frac{2}{p}}$ is "almost $T_p(b)$". In fact, repeating the proof of Lemma 6.1(*i*), we obtain $\|T_p(b, b_m)\|_{p\to p} \leq c_\delta'$ with constant c_δ' independent of m. Now, applying Lemma 6.1 in (8.3) and using the Sobolev embedding theorem (recall $p > 2 \vee (d-2)$ and $q > p$ is close to p), we obtain (8.1). The same argument gives (8.2).

By a standard result (see e.g. [10, Sect. I.9]), given a conservative Feller semigroup $e^{-t\Lambda_{C_\infty}(b)}$, there exist probability measures $\mathbb{P}_x$ ($x \in \mathbb{R}^d$) on $(D(\mathbb{R}_+, \mathbb{R}^d), \mathcal{B}_t' = \sigma(\omega_r, 0 \leq r \leq t))$, where $D([0, T], \mathbb{R}^d)$ is the space of right-continuous functions having left limits, and ω_t is the coordinate process on $D(\mathbb{R}_+, \mathbb{R}^d)$, such that

$$\mathbb{E}_x[f(\omega_t)] = e^{-t\Lambda_{C_\infty}(b)}f(x), \quad f \in C_\infty, \quad t > 0.$$

Here and below, $\mathbb{E}_x := \mathbb{E}_{\mathbb{P}_x}$. We will show that $\mathbb{P}_x$ are actually concentrated on $(C(\mathbb{R}_+, \mathbb{R}^d), \mathcal{B}_t)$.

For every $n = 1, 2, \dots$, let $X_t^n = X_{t,x}^n$ denote the strong solution to the approximating SDE

$$X_t^n = x - \int_0^t b_n(s, X_s^n)ds + \sqrt{2}W_t, \quad x \in \mathbb{R}^d,$$

on a complete probability space $(\Omega, \mathcal{F}, \mathcal{F}_t, \mathbf{P})$. Put $\mathbb{P}_x^n := (\mathbf{P}X^n)^{-1}$, $n = 1, 2, \dots$, and set $\mathbb{E}_x^n := \mathbb{E}_{\mathbb{P}_x^n}$.

Fix $\mu \geq \mu_0$. In what follows, $0 < t \leq T < \infty$. For every $g \in C_c^2$, the following is true:

(a) $\mathbb{E}_x \int_0^t \big|b \cdot \nabla g\big|(\omega_s)ds < \infty$.
Indeed, since $b_n \to b$ everywhere outside of a measure zero set, we have by Fatou's lemma[11]

$$\begin{aligned}
&\mathbb{E}_x \int_0^t \big|b \cdot \nabla g\big|(\omega_s)ds \\
&\leq \liminf_n \mathbb{E}_x \int_0^t \big|b_n \cdot \nabla g\big|(\omega_s)ds = \liminf_n \int_0^t e^{-s\Lambda_{C_\infty}(b)}\big|b_n \cdot \nabla g\big|(x)ds \\
&\leq e^{\mu T} \liminf_n (\mu + \Lambda_{C_\infty}(b))^{-1}|b_n||\nabla g|(x)
\end{aligned}$$

Now, applying (8.1) with $h = |\nabla g|$, we obtain

$$\begin{aligned}
\mathbb{E}_x \int_0^t \big|b \cdot \nabla g\big|(\omega_s)ds &\leq C_1 e^{\mu T} \liminf_n \langle |b_n|^2 |\nabla g|^p\rangle^{\frac{2}{p}} \\
&= C_1 e^{\mu T} \langle |b|^2 |\nabla g|^p\rangle^{\frac{2}{p}} < \infty \quad (\text{by } |b| \in L^2_{\text{loc}}).
\end{aligned}$$

(b)

$$\mathbb{E}_x \int_0^t (b_n \cdot \nabla g)(\omega_s)ds - \mathbb{E}_x^n \int_0^t (b_n \cdot \nabla g)(\omega_s)ds \to 0 \quad \text{as } n \to \infty.$$

---

[11] When applying Fatou's lemma, we use

$$\mathbb{E}_x \int_0^t \big|b \cdot \nabla g\big|(\omega_s)ds = \mathbb{E}_x \int_0^t \liminf_n \big|b_n \cdot \nabla g\big|(\omega_s)ds.$$

Indeed, $\xi := \big|b \cdot \nabla g\big| - \liminf_n \big|b_n \cdot \nabla g\big| = 0$ a.e. on $\mathbb{R}^d$, but $|\mathbb{E}_x \int_0^t \xi(\omega_s)ds| = 0$ (as follows e.g. by representing $\{\xi \neq 0\} = \cap_k U_k$ for a decreasing sequence of open sets U_k such that $|U_k| \downarrow 0$, smoothing out $\mathbf{1}_{U_k}$ by replacing it with $e^{\varepsilon_k \Delta}\mathbf{1}_{U_k}$ with $\varepsilon_k \downarrow 0$ rapidly, and then applying $\mathbb{E}_x \int_0^t e^{\varepsilon_k\Delta}\mathbf{1}_{U_k}(\omega_s)ds \leq e^{\mu T}(\mu + \Lambda_{C_\infty}(b))^{-1} e^{\varepsilon_k \Delta}\mathbf{1}_{U_k}(x) \leq C e^{\mu T}\|e^{\varepsilon_k\Delta}\mathbf{1}_{U_k}\|_p \downarrow 0$ as $k \to \infty$. The last inequality follows from the construction of $(\mu + \Lambda_{C_\infty}(b))^{-1}$ via the operator-valued function $\Theta_p(\mu, b)$.

Indeed, we have:

$$\begin{aligned}
&\mathbb{E}_x \int_0^t (b_n \cdot \nabla g)(\omega_s)ds - \mathbb{E}_x^n \int_0^t (b_n \cdot \nabla g)(\omega_s)ds \\
&= \int_0^t \left(e^{-s\Lambda_{C_\infty}(b)} - e^{-s\Lambda_{C_\infty}(b_n)}\right)(b_n \cdot \nabla g)(x)ds \\
&= \int_0^t \left(e^{-s\Lambda_{C_\infty}(b)} - e^{-s\Lambda_{C_\infty}(b_n)}\right)((b_n - b_m) \cdot \nabla g)(x)ds \\
&+ \int_0^t \left(e^{-s\Lambda_{C_\infty}(b)} - e^{-s\Lambda_{C_\infty}(b_n)}\right)(b_m \cdot \nabla g)(x)ds \\
&=: S_1 + S_2,
\end{aligned}$$

where m is to be chosen. Reducing the estimates on the expectations of time integrals to the estimates on resolvents as in the proof of (**a**), we obtain:

$$\begin{aligned}
S_1(x) \leq e^{\mu T}(\mu + \Lambda_{C_\infty}(b))^{-1}|(b_n - b_m) \cdot \nabla g|(x) \\
+ e^{\mu T}(\mu + \Lambda_{C_\infty}(b_n))^{-1}|(b_n - b_m) \cdot \nabla g|(x).
\end{aligned}$$

Using (8.2) and the convergence $b_n - b_m \to 0$ in L^2_{loc} as $n, m \uparrow \infty$, we obtain $S_1 \to 0$ as $n, m \uparrow \infty$. Now, let us fix a sufficiently large m. Since $e^{-s\Lambda_{C_\infty}(b)} = s\text{-}C_\infty\text{-}\lim_n e^{-s\Lambda_{C_\infty}(b_n)}$ uniformly in $0 \leq s \leq T$ (i.e. assertion (*iv*) of Theorem 6.1), we have $S_2 \to 0$ as $n \uparrow \infty$. The proof of (**b**) is completed.

(**c**)

$$\mathbb{E}_x^n[g(\omega_t)] \to \mathbb{E}_x[g(\omega_t)]$$

and

$$\mathbb{E}_x \int_0^t (b \cdot \nabla g)(\omega_s)ds - \mathbb{E}_x^n \int_0^t (b_n \cdot \nabla g)(\omega_s)ds \to 0 \quad \text{as } n \to \infty.$$

The first convergence is direct a consequence of $e^{-s\Lambda_{C_\infty}(b)} = s\text{-}C_\infty\text{-}\lim_n e^{-s\Lambda_{C_\infty}(b_n)}$ uniformly in $0 \leq s \leq T$. The second convergence is a consequence of (**b**) and $\mathbb{E}_x \int_0^t ((b_n - b) \cdot \nabla g)(\omega_s)ds \to 0$ as $n \to \infty$, as follows from (8.2) upon applying Fatou's lemma in m there.

Now, since

$$M^g_{r,m} := g(\omega_r) - g(x) + \int_0^r (-\Delta g + b_m \cdot \nabla g)(\omega_t)dt$$

is a $\mathcal{B}'_r$-martingale under $\mathbb{P}_x^m$,

$$x \mapsto \mathbb{E}_x^m[g(\omega_r)] - g(x) + \mathbb{E}_x^m \int_0^r (-\Delta g + b_m \cdot \nabla g)(\omega_t)dt \quad \text{is identically zero on } \mathbb{R}^d,$$

and so by (**c**)

$$x \mapsto \mathbb{E}_x[g(\omega_r)] - g(x) + \mathbb{E}_x \int_0^r (-\Delta g + b \cdot \nabla g)(\omega_t)dt \quad \text{is identically zero in } \mathbb{R}^d.$$

Since $\{\mathbb{P}_x\}_{x \in \mathbb{R}^d}$ are determined by a Feller semigroup, and thus constitute a Markov process, we can conclude (see e.g. the proof of [62, Lemma 2.2]) that

$$M_r^g := g(\omega_r) - g(x) + \int_0^r (-\Delta g + b \cdot \nabla g)(\omega_t)dt,$$

is a $\mathcal{B}'_r$-martingale under $\mathbb{P}_x$.

Let us show now that $\{\mathbb{P}_x\}_{x \in \mathbb{R}^d}$ are concentrated on $(C([0, T], \mathbb{R}^d), \mathcal{B}_t)$. Since ω_t is a semimartingale under $\mathbb{P}_x$, Itô's formula yields, for every $g \in C_c^\infty(\mathbb{R}^d)$, that

$$g(\omega_t) - g(x) = \sum_{s \leq t} \big(g(\omega_s) - g(\omega_{s-})\big) + S_t, \tag{8.4}$$

where S_t is defined in terms of some integrals and sums of $(\partial_{x_i} g)(\omega_{s-})$ and $(\partial_{x_i} \partial_{x_j} g)(\omega_{s-})$ in s, see [18, Sect. 2] for details. Now, let A, B be arbitrary compact sets in $\mathbb{R}^d$ such that $\operatorname{dist}(A, B) > 0$. Fix $g \in C_c^\infty(\mathbb{R}^d)$ that separates A, B, say, $g = 0$ on A, $g = 1$ on B. Set

$$K_t^g := \int_0^t \mathbf{1}_A(\omega_{s-})dM_s.$$

In view of (8.4), when evaluating K_t^g one needs to integrate $\mathbf{1}_A(\omega_{s-})$ with respect to S_t, however, one obtains zero since $(\partial_{x_i} g)(\omega_{s-}) = (\partial_{x_i} \partial_{x_j} g)(\omega_{s-}) = 0$ if $\omega_{s-} \in A$. Thus,

$$\begin{aligned} K_t^g &= \sum_{s \leq t} \mathbf{1}_A\,(\omega_{s-})\, g(\omega_s) + \int_0^t \mathbf{1}_A(\omega_{s-})\big(- \Delta g + b \cdot \nabla g\big)(\omega_s)ds \\ &= \sum_{s \leq t} \mathbf{1}_A\,(\omega_{s-})\, g(\omega_s). \end{aligned}$$

Since M_t^g is a martingale, so is K_t^g. Thus, $\mathbb{E}_x\big[\sum_{s\leq t}\mathbf{1}_A(\omega_{s-})g(\omega_s)\big]=0$. Using the Dominated convergence theorem, we further obtain $\mathbb{E}_x\big[\sum_{s\leq t}\mathbf{1}_A(\omega_{s-})\mathbf{1}_B(\omega_s)\big]=0$, which yields the required. By the way, this construction, in a more general form, was used to control the jumps of stable process perturbed by a drift, see [18].

We denote the restriction of $\mathbb{P}_x$ from $(D([0,T],\mathbb{R}^d),\mathcal{B}'_t)$ to $(C([0,T],\mathbb{R}^d),\mathcal{B}_t)$ again by $\mathbb{P}_x$, and thus obtain that for every $x\in\mathbb{R}^d$ and all $g\in C_c^2(\mathbb{R}^d)$

$$M_r^g=g(\omega_r)-g(x)+\int_0^r(-\Delta g+b\cdot\nabla g)(\omega_t)dt,\quad \omega\in C([0,T],\mathbb{R}^d),$$

is a $\mathcal{B}_r$-martingale under $\mathbb{P}_x$. Thus, $\mathbb{P}_x$ is a martingale solution to (6.15).

To show that $\mathbb{P}_x$ is a weak solution it suffices to show that M_r^g is also a martingale for $g(x)=x_i$ and $g(x)=x_ix_j$ (proving along the way that $\mathbb{E}_x\int_0^t|b|(X(s))ds<\infty$), which can be done by following closely [46, proof of Lemma 6].

(*ii*) is obtained via a simple modification of the proof of the uniqueness result of Theorem 16.2(*iv*) below. □

9 Time-Inhomogeneous Form-Bounded Drifts and Feller Theory via Iterations

1. The following is the time-inhomogeneous counterpart of Definition 3.1.

Definition 9.1 A Borel measurable vector field $b:\mathbb{R}_+\times\mathbb{R}^d\to\mathbb{R}^d$ is said to be form-bounded if

$$|b|\in L^2_{\rm loc}(\mathbb{R}_+\times\mathbb{R}^d)$$

and there exist a constant $\delta>0$ and a function $0\leq g\in L^1_{\rm loc}(\mathbb{R}_+)$ such that for a.e. $t\in\mathbb{R}_+$

$$\|b(t,\cdot)\varphi\|_2^2\leq\delta\|\nabla\varphi\|_2^2+g(t)\|\varphi\|_2^2 \tag{9.1}$$

for all $\varphi\in W^{1,2}$.

This will be written as $b\in L^\infty\mathbf{F}_\delta+L^2_{\rm loc}(\mathbb{R}_+)$.

An equivalent form of the a.e. inequality (9.1) is: for every $0<T<\infty$,

$$\int_0^T\|b(t)\psi(t)\|_2^2dt\leq\delta\int_0^T\|\nabla\psi(t)\|_2^2dt+\int_0^T g(t)\|\psi(t)\|_2^2dt$$

for all $\psi\in L^\infty(\mathbb{R}_+,W^{1,2})$.

Examples The class of time-inhomogeneous form-bounded vector fields includes e.g. the critical Ladyzhenskaya-Prodi-Serrin class

$$|b| \in L^q_{\rm loc}(\mathbb{R}_+, L^r + L^\infty), \quad \frac{d}{r} + \frac{2}{q} \leq 1, \quad 2 \leq q \leq \infty,$$

as well as vector fields having stronger spatial singularities, see Appendix B.

We fix an approximation of $b \in L^\infty \mathbf{F}_\delta + L^2_{\rm loc}$ by smooth bounded vector fields b_m that preserve the form-bound δ and have functions g_n locally uniformly bounded in $L^1(\mathbb{R}_+)$, i.e.

$$b_n \to b \quad \text{in } L^2_{\rm loc}(\mathbb{R}_+ \times \mathbb{R}^d, \mathbb{R}^d) \tag{9.2}$$

and for all $t \geq 0$

$$\|b_n(t)\varphi\|_2^2 \leq \delta \|\nabla \varphi\|_2^2 + g_n(t)\|\varphi\|_2^2 \tag{9.3}$$

with g_n such that

$$\sup_n \int_0^T g_n(s)ds < \infty \qquad \text{for any } 0 < T < \infty. \tag{9.4}$$

Examples It is easy to show that the following b_n, with $\varepsilon_n \downarrow 0$ sufficiently rapidly and $c_n \uparrow 1$ sufficiently slow, satisfy (9.2)–(9.4).

$$b_n := c_n E^{1+d}_{\varepsilon_n}(\mathbf{1}_n b),$$

where $\mathbf{1}_n$ is the indicator of $\{(t,x) \mid |b(t,x)| \leq n, |x| \leq n, |t| \leq n\}$ (say, b is extended by 0 to $t < 0$), E^{1+d}_ε is the De Giorgi or the Friedrichs mollifier on $\mathbb{R} \times \mathbb{R}^d$. See details in Appendix C.1. Note that, by selecting $\varepsilon_n \downarrow 0$ rapidly, one can treat b_n as basically a cutoff of b times constant c_n.

Moreover, with some additional effort, one can simplify this approximation to

$$b_\varepsilon := E^1_\varepsilon E^d_\varepsilon b, \quad \varepsilon \downarrow 0,$$

where E^1_ε in the Friedrichs mollifier on $\mathbb{R}$, and E^d_ε is the De Giorgi or the Friedrichs mollifier on $\mathbb{R}^d$. See Appendix C.3.

The last approximation is important if one needs e.g. to transfer the form-boundedness assumption on "potential" $\operatorname{div} b$ to the uniform form-boundedness of $\operatorname{div} b_\varepsilon$ since then one commute div and the mollifiers, although we are not concerned with this here.

2. Our first goal is to construct the corresponding to $(\partial_t - \Delta + b \cdot \nabla)u = 0$, $b \in L^\infty \mathbf{F}_\delta + L^2_{\rm loc}$, Feller evolution family on $D_T = \{(s,t) \mid 0 \leq s \leq t \leq T\}$ for $T > 0$ fixed, i.e. a family of operators $\{U^{t,s}\}_{(s,t)\in D_T}$ that are bounded on C_∞, and

(1) $U^{t,r}U^{r,s} = U^{t,s}$, $r \in [s,t]$, and $U^{s,s} = I$,
(2) $\|U^{t,s} f\|_\infty \leq \|f\|_\infty$, $U^{t,s}[C_\infty^+] \subset C_\infty^+$,
(3)

$$U^{r,s} = s\text{-}C_\infty\text{-}\lim_{t\downarrow r} U^{t,s}, \quad r \geq s,$$

and which will, additionally, satisfy: $u(t) := U^{t,s} f$ is the unique weak solution of $(\partial_t - \Delta + b \cdot \nabla)u = 0$, $u(s) = f \in C_\infty \cap L^2$.

The sought Feller evolution family is produced as the limit $L^\infty(D_T, L^\infty)$ of

$$U_n^{t,s} f(\cdot) := u_n(t,\cdot), \quad (s,t) \in D_T$$

and u_n is the classical solution to initial problem

$$\big(\partial_t - \Delta + b_n \cdot \nabla\big)u_n = 0, \quad u_n(s,\cdot) = f(\cdot) \in C_c^\infty. \tag{9.5}$$

We will prove the uniform convergence of the functions $\{(t,s,x) \mapsto U_n^{t,s} f(x)\}$ on $D_T \times \mathbb{R}^d$ by showing that they constitute a Cauchy sequence in $L^\infty(D_T, L^\infty)$. To that end, we will employ a parabolic variant of the iteration procedure of [60]. This parabolic variant first appeared in [38] and was recently refined in [54].

Namely, subtracting the equations for u_m, u_n and setting

$$h := u_m - u_n,$$

one obtains

$$\partial_t h - \Delta h + b_m \cdot \nabla h + (b_m - b_n) \cdot \nabla u_n = 0, \quad h(s,\cdot) = 0 \tag{9.6}$$

Multiplying the last equation by $h|h|^{r-2}$, $r > \frac{2}{2-\sqrt{\delta}}$, integrating over $[s,T] \times \mathbb{R}^d$ and applying the Sobolev embedding theorem, we arrive at the inequality

$$c_d r^k \|h\|^r_{L^\infty([s,T],L^r)} + \|h\|^r_{L^r([s,T],L^{\frac{rd}{d-2}})}$$
$$\leq C r^{2k} e^{C_T} \sup_{\tau\in[s,T]} \|\nabla u_n(\tau)\|_q^2 \int_s^T \|h(\tau)\|^{r-2}_{\frac{q}{q-2}(r-2)} d\tau,$$

where for a fixed $q > d$, for constants C, C_T that are independent of m, n. Hence, applying the interpolation inequality in the left-hand side and setting $K = Ce^{C_T}$, one obtains

$$\|h\|_{\frac{r}{1-\beta},\frac{rd}{d-2+2\beta}} \leq K^{\frac{1}{r}}(r^{\frac{1}{r}})^{2k}\Big(\sup_{\tau\in[s,T]}\|\nabla u_n(\tau)\|_q\Big)^{\frac{2}{r}}\|h\|^{1-\frac{2}{r}}_{L^{r-2}([s,T],L^{\frac{q}{q-2}(r-2)})}.$$

(we only need $\delta < 4$ to prove this inequality). Now, with appropriate choice of β, one can iterate this inequality in essentially the same way as it was done [60] *provided that one has uniform in n bound on* $\sup_{\tau\in[s,T]}\|\nabla u(\tau)\|_q$ (see below), arriving at

$$\|u_m - u_n\|_{L^\infty([s,T],L^\infty)} \leq C_1\|u_m - u_n\|^\gamma_{L^{r_0}([s,T],L^{r_0})}$$

for $r_0 > \frac{2}{2-\sqrt{\delta}}$ and some $\gamma > 0$ (this strict inequality is the main concern of the iteration procedure). Now, a standard argument yields

$$\|u_m - u_n\|^\gamma_{L^{r_0}([s,T],L^{r_0})} \to 0 \quad \text{as } m, n \to \infty,$$

see e.g. [38, 54], and so we have our Cauchy sequence:

$$\|u_m - u_n\|_{L^\infty([s,T],L^\infty)} \to 0 \quad \text{as } m, n \to \infty,$$

moreover, the convergence is uniform in $s \in [0, T]$. So, we can define the sought Feller evolution family by

$$U^{t,s}f := s\text{-}C_\infty\text{-}\lim_n u_n(t), \quad (s,t) \in D_T,$$

for $f \in C_c^\infty$, as was assumed above, and then extend operators $U^{t,s}$ to all $f \in C_\infty$ by continuity using the fact that $U^{t,s}$ inherits the L^∞ contraction property from $U_n^{t,s}$. Let us emphasize that the a priori assumption $f \in C_c^\infty$ is needed for the uniform in n bound on $\sup_{\tau\in[s,T]}\|\nabla u(\tau)\|_p$, $p > d$.

Remark 9.1 Working in the elliptic setting (i.e. as in the proof of Theorem 6.1 or in [60]), after showing that solutions to the approximating elliptic equation converge in C_∞, one needs to verify the other conditions of the Trotter approximation theorem. This is not needed when one is working directly with the parabolic equation, so we arrive at shorter proofs even if $b = b(x)$, however, at expense of requiring smaller δ. We discussed this effect in Remark 6.4. It is fair to say that there is a fundamental difference between time-homogeneous and time-inhomogeneous cases when one is dealing with singular drifts.

3. To make the iteration procedure converge, one needs gradient bound

$$\sup_n \sup_{\tau\in[s,T]}\|\nabla u_n(\tau)\|_p < \infty, \quad \text{for some } p > d \tag{9.7}$$

for f in a dense subset of C_∞ (e.g. for $f \in C_c^\infty$). To obtain such a bound, one can differentiate the initial problem (9.5). Namely, writing for brevity

$$u := u_n, \quad b := b_n$$

and

$$w := \nabla u, \quad w_i := \nabla_i u,$$

we obtain

$$\partial_t w_j - \Delta w_j + b \cdot \nabla w_j + (\nabla_j b) \cdot w = 0, \quad w_j(0) = \nabla_j f, \quad 1 \leq j \leq d. \tag{9.8}$$

Now, one needs to "wrap up" this system and, additionally, get rid of the derivative $\nabla_j b$. For instance, one can consider all products $w_{i_1} \dots w_{i_m}$ for m fixed and then sum them up, as was done in [7] for solutions of stochastic transport equation and, after them, in [57]. However, we are interested in arguments that give less restrictive assumptions on δ. Another argument was used in [63], although in a different situation dealing with a more sophisticated system of parabolic equations. This argument still imposes more restrictive assumption on δ than the arguments employed [38] and [54] (compare (C_3) with (C_4), (C_5)). However, it is quite nice and simple (and, again, works for other equations), so we describe it here. Assume that that the form-bound δ of b satisfies

$$\sqrt{\delta} < \frac{d-1}{d(d+1)}. \tag{C_3}$$

Put

$$w_\eta := \eta \cdot w, \quad \eta = (\eta_j)_{j=1}^d \in \mathbb{R}^d.$$

We differentiate (9.5) in the direction η, i.e. multiply (9.8) by η_j and add the resulting identities in $j = 1, \dots, d$ to obtain

$$\partial_t w_\eta - \Delta_x w_\eta - b \cdot \nabla_x w_\eta - \sum_{i=1}^d (\eta \cdot \nabla b_i) \nabla_{\eta_i} w_\eta = 0,$$

where in the last term we have used $w_i = \nabla_{\eta_i} w_\eta$. Given a function $g(x, \eta)$, we denote by $\langle g \rangle_x$, $\langle g \rangle_x$ the integral of g over $\mathbb{R}^d$ in x and in η, respectively. Let $\langle g \rangle_{x,\eta}$ denote the corresponding repeated integral over $\mathbb{R}^d \times \mathbb{R}^d$. Set

$$h(\eta) := (1 + \kappa |\eta|^2)^{-\theta},$$

where $\kappa > 0$ is fixed arbitrarily, and $\theta > \frac{d+q}{2}$ so that $\langle|\eta|^q h\rangle_\eta < \infty$. Let $q \geq 2$ (in the iteration procedure we need $q > d$). Also, without loss of generality, q is rational with odd denominator, so we, if needed, we can raise negative numbers of power q. Multiply the previous identity by hw_η^{q-1} and integrate in $(x,\eta) \in \mathbb{R}^{2d}$ to obtain

$$\frac{1}{q}\partial_t\langle hw_\eta^q\rangle_{x,\eta} + \frac{4(q-1)}{q^2}\langle h|\nabla w_\eta^{\frac{q}{2}}|^2\rangle_{x,\eta} - \frac{2}{q}\langle b\cdot\nabla w_\eta^{\frac{q}{2}}, hw_\eta^{\frac{q}{2}}\rangle_{x,\eta} - \sum_{i=1}^d\langle(\eta\cdot\nabla b_i)\partial_{\eta_i}w_\eta, hw_\eta^{q-1}\rangle_{x,\eta} = 0, \tag{9.9}$$

where b_i are the components of b. The last term in the left-hand side is dealt with as follows:

$$\begin{aligned}
-\sum_{i=1}^d\langle(\eta\cdot\nabla b_i)\partial_{\eta_i}w_\eta, hw_\eta^{q-1}\rangle_{x,\eta} &= \sum_{i=1}^d\langle\eta b_i\nabla\partial_{\eta_i}w_\eta, hw_\eta^{q-1}\rangle_{x,\eta} \\
&\quad + (q-1)\sum_{i=1}^d\langle\eta b_i\partial_{\eta_i}w_\eta, hw_\eta^{q-2}\nabla w_\eta\rangle_{x,\eta} \\
&\quad \text{(now we integrate by parts in } \eta_i \text{ in the first term)} \\
&= -\langle b\cdot\nabla w_\eta, hw_\eta^{q-1}\rangle_{x,\eta} \\
&\quad - \sum_{i=1}^d\langle\eta b_i\nabla w_\eta, (\partial_{\eta_i}h)w_\eta^{q-1}\rangle_{x,\eta}.
\end{aligned}$$

Hence, (9.9) becomes

$$\frac{1}{q}\partial_t\langle hw_\eta^q\rangle_{x,\eta} + \frac{4(q-1)}{q^2}\langle h|\nabla w_\eta^{\frac{q}{2}}|^2\rangle_{x,\eta} - \frac{4}{q}\langle b\cdot\nabla w_\eta^{\frac{q}{2}}, hw_\eta^{\frac{q}{2}}\rangle_{x,\eta} - \frac{2}{q}\sum_{i=1}^d\langle\eta b_i\nabla w_\eta^{\frac{q}{2}}, (\partial_{\eta_i}h)w_\eta^{\frac{q}{2}}\rangle_{x,\eta} = 0,$$

so

$$\partial_t\langle hw_\eta^q\rangle_{x,\eta} + \frac{4(q-1)}{q}\langle h|\nabla w_\eta^{\frac{q}{2}}|^2\rangle_{x,\eta} - 4\langle b\cdot\nabla w_\eta^{\frac{q}{2}}, hw_\eta^{\frac{q}{2}}\rangle_{x,\eta} + 4\theta\sum_{i=1}^d\langle b_i\nabla w_\eta^{\frac{q}{2}}, \frac{\kappa\eta_i\eta}{1+\kappa|\eta|^2}hw_\eta^{\frac{q}{2}}\rangle_{x,\eta} = 0.$$

Hence

$$\partial_t \langle h w_\eta^q \rangle_{x,\eta} + \frac{4(q-1)}{q} \langle h |\nabla w_\eta^{\frac{q}{2}}|^2 \rangle_{x,\eta} - (4+4\theta) \langle |b| |\nabla w_\eta|^{\frac{q}{2}}, h w_\eta^{\frac{q}{2}} \rangle_{x,\eta} \leq 0. \tag{9.10}$$

We estimate

$$\begin{aligned} \langle |b| |\nabla w_\eta|^{\frac{q}{2}}, h w_\eta^{\frac{q}{2}} \rangle_{x,\eta} &\leq \varepsilon \langle |b|^2, h w_\eta^q \rangle_{x,\eta} + \frac{1}{4\varepsilon} \langle h |\nabla w_\eta|^q \rangle_{x,\eta} \\ &\leq \varepsilon \Big(\delta \langle h |\nabla w_\eta^{\frac{q}{2}}|^2 \rangle + g(t) \langle w_\eta^q \rangle \Big) \\ &\quad + \frac{1}{4\varepsilon} \langle h |\nabla w_\eta|^q \rangle_{x,\eta} \qquad \varepsilon := \frac{1}{2\sqrt{\delta}}. \end{aligned}$$

Thus, integrating (9.10) from s to t, one obtains

$$\begin{aligned} &\langle h w_\eta^q(t) \rangle_{x,\eta} + 4\Big(\frac{q-1}{q} - \sqrt{\delta}(1+\theta) \Big) \int_s^t \langle h |\nabla w_\eta^{\frac{q}{2}}(\tau)|^2 \rangle_{x,\eta} d\tau \\ &\leq \frac{1}{2\sqrt{\delta}} \int_s^t g(\tau) \langle w_\eta^q(\tau) \rangle d\tau + \langle h (\nabla f \cdot \eta)^q \rangle, \end{aligned}$$

which allows to conclude

$$\sup_{t \in [s,T]} \langle h w_\eta^q(t) \rangle_{x,\eta} + \int_s^t \langle h |\nabla w_\eta^{\frac{q}{2}}(\tau)|^2 \rangle_{x,\eta} d\tau \leq C \|\nabla f\|_q^q \tag{9.11}$$

for some for some $q > d$, for constant C independent of n, provided that

$$\frac{q-1}{q} - \sqrt{\delta}(1+\theta) > 0 \text{ for some } \theta > \frac{d+q}{2} \quad \Leftrightarrow \quad \sqrt{\delta} < \frac{d-1}{d(d+1)}.$$

It remains to derive (9.7) from (9.11). Put

$$A_{t,x} := \{ \eta \in \mathbb{R}^d \mid |\eta - \frac{w(t,x)}{|w(t,x)|}| < \frac{1}{2} \}, \quad x \in \mathbb{R}^d$$

(if $w(t,x) = 0$, fix $\eta = (1, 0, \dots, 0)$). Thus, $A_{t,x}$ is a ball of radius $\frac{1}{2}$ with centre placed at distance 1 from the origin. The angle between $w(t,x)$ and $\eta \in A_{t,x}$ is bounded from above by a generic constant, hence $|\eta \cdot w(t,x)| \geq c|w(t,x)|$ for some $c > 0$ independent of (t, x), for all $\eta \in A_{t,x}$. Therefore, for all $t \in [s, T]$

$$\begin{aligned}\langle h w_\eta^q \rangle_{x,\eta} &= \langle \langle h(\eta) |\eta \cdot w(x)|^q \rangle_\eta \rangle_x \\ &\geq \langle \langle h(\eta) |\eta \cdot w(x)|^q \mathbf{1}_{A_{t,x}}(\eta) \rangle_\eta \rangle_x \\ &\geq \langle c^q |w(x)|^q \langle h(\eta) \mathbf{1}_{A_{t,x}}(\eta) \rangle_\eta \rangle_x = C \langle |w|^q \rangle_x, \quad C > 0.\end{aligned}$$

Thus, we obtain from (9.11)

$$\sup_{t \in [s,T]} \|w(t)\|_q^q + c \int_s^T \|\nabla |w|^{\frac{q}{2}} \|_2^2 dt \leq C_T \|\nabla f\|_q^q, \quad c > 0. \tag{9.12}$$

Hence, one can run the iteration procedure under the assumption (C_3).

4. The proofs of (9.12) in [38, 54] choose a specific direction of the differentiation (following [60] which, by the way, appeared earlier than the other papers cited above):

$$\eta = \frac{w}{|w|},$$

which maximizes the directional derivative $w_\eta = w \cdot \eta$. Put differently, one multiplies the parabolic equation in (9.5) by the test function

$$\varphi = -\nabla \cdot (\frac{w}{|w|} |w|^{q-1}) \tag{9.13}$$

and then integrates by parts (the same test function is used in the proof of Lemma 6.1). This choice of the test function (or direction) leads to better assumptions on δ than (C_3). Indeed, the identity

$$\langle \partial_t u, \varphi \rangle + \langle -\Delta u, \varphi \rangle + \langle b_m \cdot w, \varphi \rangle = 0,$$

yields

$$\frac{1}{q} \partial_t \|w\|_q^q + I_q + (q-2) J_q = \langle b_m \cdot w, \nabla \cdot (w |w|^{q-2}) \rangle \tag{9.14}$$

where

$$I_q = \sum_{i=1}^d \langle |\nabla w_i|^2, |w|^{q-2} \rangle, \quad J_q = \langle |\nabla |w||^2, |w|^{q-2} \rangle$$

are the "good" terms, i.e. the right-hand side of (9.14) will be estimated in terms of I_q, J_q multiplied by coefficients that, thus, cannot be too large, hence our assumptions on δ. Namely, we represent

$$\begin{aligned}\langle b_m \cdot w, \nabla \cdot (w|w|^{q-2})\rangle &= \langle b_m \cdot w, \Delta u|w|^{q-2}\rangle + (q-2)\langle b_m \cdot w, |w|^{q-3} w \cdot \nabla |w|\rangle \\ &:= S_1 + S_2.\end{aligned}$$

Put $B_q = \langle (b_m \cdot w)^2, |w|^{q-2}\rangle$, then

$$S_2 \leq (q-2) B_q^{\frac{1}{2}} J_q^{\frac{1}{2}}, \tag{9.15}$$

where B_q is estimated using (9.3), i.e. the form-boundedness of b_m:

$$B_q \leq \frac{q\sqrt{\delta}}{2} J_q + g_m \|w\|_q^q. \tag{9.16}$$

Thus, S_2 is estimated in terms of J_q, and one can apply the resulting bound on S_2 in (9.14).

In [38], after estimating S_1 as

$$|S_1| \leq B_q^{\frac{1}{2}} \langle |\Delta u|^2, |w|^{q-2}\rangle^{\frac{1}{2}}, \tag{9.17}$$

the factor $\langle |\Delta u|^2, |w|^{q-2}\rangle$ was bounded by I_q and J_q without appealing to the equation, by representing $|\Delta u|^2 = (\nabla \cdot w)^2$ and integrating by parts twice:

$$\langle |w|^{q-2}|\Delta u|^2\rangle = -\langle w \cdot \nabla |w|^{q-2}, \Delta u\rangle + \sum_{r=1}^{d} \Big\langle w \cdot \nabla w_r, \nabla_r |w|^{q-2}\Big\rangle + I_q,$$

where

$$|\langle w \cdot \nabla |w|^{q-2}, \Delta u\rangle| \leqslant (q-2)\left(\frac{1}{4\varkappa}\langle |w|^{q-2}|\Delta u|^2\rangle + \varkappa J_q\right), \quad \kappa > 0,$$

and

$$|\sum_{r=1}^{d} \Big\langle w \cdot \nabla w_r, \nabla_r |w|^{q-2}\Big\rangle| \leqslant (q-2)\left(\frac{1}{2} I_q + \frac{1}{2} J_q\right).$$

Hence

$$\left(1 - \frac{q-2}{4\varkappa}\right)\langle |w|^{q-2}|\Delta u|^2\rangle \leqslant I_q + (q-2)\left(\varkappa J_q + \frac{1}{2} I_q + \frac{1}{2} J_q\right), \quad \varkappa > \frac{q-2}{4}. \tag{9.18}$$

The resulting from (9.17), (9.18) bound on $|S_1|$, combined with (9.15), (9.16), led in [38] to the gradient estimate (9.12) for a $q > d$ close to d provided that

$$\sqrt{\delta} < \frac{1}{d}. \tag{C_4}$$

One important advantage of working with the test function (9.13) is that one can "evaluate" it by representing $\Delta u = \partial_t u + b_m \cdot \nabla u$, thus using the equation one more time. This is what [60] did. The same can be done in the parabolic setting, and it leads to better assumptions on δ than (C_4) [54]. Specifically, in dimensions $3 \leq d \leq 6$, one abandons estimate (9.17) and represents $\Delta u = \partial_t u + b_m \cdot \nabla u$ to evaluate

$$S_1 = -\frac{1}{q}\frac{d}{dt}\|w\|_q^q + B_q - \langle |\partial_t u|^2, |w|^{q-2}\rangle - (q-2)\langle |w|^{q-3} w \cdot \nabla |w|, \partial_t u\rangle,$$

which, upon applying the quadratic inequality, gives

$$S_1 \leq -\frac{1}{q}\partial_t \|w\|_q^q + B_q + \frac{(q-2)^2}{4} J_q.$$

This bound, (9.15) and (9.16) applied in (9.14) give the desired gradient bound (9.12) for $\sqrt{\delta} \leq \big(\sqrt{q-1} - \frac{q-2}{2}\big)\frac{2}{q}$. In dimensions $d = 3$ and $d = 4$ this gives significantly less restrictive assumption on δ than (C_4), see (C_5) below.

In dimensions $d \geq 5$, one starts with two representations for S_1:

$$\begin{aligned} S_1 &= \langle -\partial_t u + \Delta u, |w|^{q-2}\Delta u\rangle \\ &= \langle |\Delta u|^2, |w|^{q-2}\rangle - \mathrm{Re}\langle \partial_t u, |w|^{q-2}\Delta u\rangle, \end{aligned}$$

$$\begin{aligned} S_1 &= \langle b_n \cdot w, |w|^{q-2}(\partial_t u + b_n \cdot w)\rangle \\ &= B_q + \langle b_n \cdot w, |w|^{q-2}\partial_t u\rangle. \end{aligned}$$

Equating the right-hand sides, one obtains

$$\begin{aligned} \langle |\Delta u|^2, |w|^{q-2}\rangle &= B_q + \langle \partial_t u, |w|^{q-2}(b_n \cdot w + \Delta u)\rangle \\ &= B_q + \langle \partial_t u, |w|^{q-2}(-\partial_t u + 2\Delta u)\rangle \\ &= B_q - \langle |\partial_t u|^2, |w|^{q-2}\rangle + 2\langle \partial_t u, |w|^{q-2}\Delta u\rangle \\ &= B_q - \langle |\partial_t u|^2, |w|^{q-2}\rangle \end{aligned}$$

$$
\begin{aligned}
&-\frac{2}{q}\frac{d}{dt}\|w\|_q^q - 2(q-2)\langle \partial_t u, |w|^{q-3} w \cdot \nabla |w|\rangle \\
&\leq B_q - \langle |\partial_t u|^2, |w|^{q-2}\rangle - \frac{2}{q}\frac{d}{dt}\|w\|_q^q + (q-2)^2 J_q \\
&\quad + \langle |\partial_t u|^2, |w|^{q-2}\rangle \\
&= B_q - \frac{2}{q}\partial_t \|w\|_q^q + (q-2)^2 J_q.
\end{aligned}
$$

This estimate on $\langle |\Delta u|^2, |w|^{q-2}\rangle$, which is more efficient than (9.18), when applied in (9.17) leads, together with (9.15), (9.16), to the following. *If form-bound δ satisfies*

$$
\begin{aligned}
&d \geq 5 \ \sqrt{\delta} \text{ satisfies } \tfrac{d\sqrt{\delta}}{2}\left(\sqrt{\tfrac{d^2\delta}{4} + (d-2)^2} + d - 2\right) < d - 1, \\
&d = 4 \ \sqrt{\delta} < \tfrac{2(\sqrt{3}-1)}{d} \approx 0.36602, \qquad\qquad (C_5) \\
&d = 3 \ \sqrt{\delta} < \tfrac{2\sqrt{2}-1}{d} \approx 0.60947,
\end{aligned}
$$

then gradient estimate (9.12) *holds for a $q > d$ close to d*. See [54] for the proof.

The assumption (C_5) is less restrictive than (C_4). In fact, if one assumes $\sqrt{\delta} = \frac{1}{d}$, then (9.12) holds even for $q = d + 1$.

Remark 9.2 The gradient bounds in [38, 54, 60] are proved not only for q close to d, but for the entire range of (δ, q) satisfying some algebraic inequalities. In particular, in [54],

$$
q - 1 - \frac{q\sqrt{\delta}}{2}\left(\sqrt{\frac{q^2\delta}{4} + (q-2)^2} + q - 2\right) > 0 \quad \text{if } d \geq 5
$$

and, as was mentioned above, for $d = 3, 4$, $\sqrt{\delta} \leq \left(\sqrt{q-1} - \frac{q-2}{2}\right)\frac{2}{q}$.

Thus, we have the following result.

Theorem 9.1 *Assume that $b \in L^\infty \mathbf{F}_\delta + L^2_{\mathrm{loc}}(\mathbb{R}_+)$ with δ that satisfying (C_4) or, better, (C_5). Then the following is true:*

(*i*) *The limit*

$$
U^{t,s} f := s\text{-}C_\infty\text{-}\lim_n U_n^{t,s} f \quad \textit{uniformly in } (s,t) \in D_T
$$

exists for all $f \in C_c^\infty$ and satisfies $\|U^{t,s} f\|_\infty \leq \|f\|_\infty$. Upon extending operators $U^{t,s}$ by continuity to all $f \in C_\infty$, one obtains a Feller evolution family.

(*ii*) *The Feller evolution family* $\{U^{t,s}\}_{(s,t)\in D_T}$ *is unique in the sense that it does not depend on the choice of the approximation vector fields* $\{b_n\}$, *as long as they satisfy* (9.2), (9.3), (9.4).

(*iii*) *If* $f \in C_\infty \cap L^2$, *then* $u(t) := U^{t,s} f$ *is the unique weak solution of* $(\partial_t - \Delta + b \cdot \nabla)u = 0$, $u(s) = f$, *in* L^2.

10 SDEs with Time-Inhomogeneous Form-Bounded Drifts

We return to the discussion of weak well-posedness of SDE

$$X_t = x - \int_0^b b(s, X_s)ds + \sqrt{2}W_t, \quad t \geq 0, \tag{10.1}$$

where $x \in \mathbb{R}^d$ is fixed and $b \in L^\infty \mathbf{F}_\delta + L^2_{\rm loc}(\mathbb{R}_+)$, where, we assume for simplicity, (C_4) holds.

We need to supplement Theorem 9.1 with a localized analogue of (9.12) for inhomogeneous parabolic equations, proved in [43]. Let $\mathsf{f} \in L^\infty \mathbf{F}_\nu + L^2_{\rm loc}(\mathbb{R}_+)$, $\nu < \infty$, define f_k similarly to b_m in (9.2)–(9.4). Let $h \in C([s, T], \mathcal{S})$, $g \in C_c^\infty(\mathbb{R}^d)$. Fix $T > s$. Let $u = u_{m,k}$ be the solution to Cauchy problem on $[s, T]$

$$(\partial_t - \Delta + b_m \cdot \nabla)u = |\mathsf{f}_k|h, \quad u(s, \cdot) = g. \tag{10.2}$$

Then, for every $q \in]d, \delta^{-\frac{1}{2}}[$, there exist constants C and κ such that, for all $0 \leq s \leq r \leq T$,

$$\|u\|^q_{L^\infty([s,r],L^q_\rho)} + \|\nabla u\|^q_{L^\infty([s,r],L^q_\rho)} + \|\nabla|\nabla u|^{\frac{q}{2}}\|^2_{L^2([s,r],L^2_\rho)} \\ \leq C\big(\|\mathsf{f}|h|^{\frac{q}{2}}\|^2_{L^2([s,r],L^2_\rho)} + \|\nabla g\|^q_{L^q_\rho} + \|g\|^q_{L^q_\rho}\big). \tag{10.3}$$

Here $\rho(x) := (1 + \kappa|x|^2)^{-\theta}$ $(x \in \mathbb{R}^d)$, where $\theta > \frac{d}{2}$ is fixed, and $L^2_\rho := L^2(\mathbb{R}^d, \rho dx)$.

Define backward Feller evolution family $(0 \leq t \leq r \leq T)$

$$P^{t,r}(b) = U^{T-t,T-r}(\tilde{b}), \quad \tilde{b}(t, x) = b(T - t, x),$$

where $U^{t,s}$ is the Feller evolution family from Theorem 9.1. Using (10.3) with $\mathsf{h} = 0$ and arguing essentially as in the proof of Theorem 6.1(v), one obtains that $\{P^{t,r}(b)\}_{0\leq t\leq r\leq T}$ is conservative, i.e. for all $x \in \mathbb{R}^d$ $\langle P^{t,r}(x, \cdot)\rangle = 1$. Now, by a standard result (see e.g. [31, Ch. 2]), given a conservative backward Feller evolution family, there exist probability measures $\mathbb{P}_x$ $(x \in \mathbb{R}^d)$ on $(D([0, T], \mathbb{R}^d), \mathcal{B}'_t)$, such that

$$\mathbb{E}_x[f(\omega_r)] = P^{0,r} f(x), \quad 0 \leq r \leq T.$$

Here and below, $\mathbb{E}_x := \mathbb{E}_{\mathbb{P}_x}$.

Let X_t^m ($m = 1, 2, \dots$) be the strong solution of

$$X_t^m = x - \int_0^t b_m(r, X_r^m) dr + \sqrt{2} W_t, \quad x \in \mathbb{R}^d$$

defined on some complete probability space $(\Omega, \mathcal{F}, \mathcal{F}_t, \mathbf{P})$.

We will require the following estimate: there exists a constant $C > 0$ independent of m, k such that

$$\sup_m \sup_{x \in \mathbb{R}^d} \mathbf{E} \int_s^r |b_k(t, X_t^m)| dt \leq C F(r - s) \tag{10.4}$$

for $0 \leq s \leq r \leq T$, where $F(h) := h + \sup_{s \in [0, T-h]} \int_s^{s+h} g(t) dt$. Here we assume, without loss of generality, that $b_k \in L^\infty \mathbf{F}_\delta + L^2_{\text{loc}}(\mathbb{R}_+)$ with the same function g as b (if not, then we can increase g, cf. (9.4)).

Indeed, let $v = v_{m,k}$ be the solution to the terminal-value problem

$$(\partial_t + \Delta - b_m \cdot \nabla) v = -|b_k|, \quad v(r, \cdot) = 0, \quad t \leq r. \tag{10.5}$$

By Itô's formula,

$$v(r, X_r^m) = v(s, X_s^m) + \int_s^r (\partial_t v + \Delta v - b_m \cdot \nabla v)(t, X_t^m) dt + \sqrt{2} \int_s^r \nabla v(t, X_t^m) dW_t.$$

Taking expectation, we obtain

$$\mathbb{E} \int_s^r |b_k(t, X_t^m)| dt = \mathbb{E} v(s, X_s^m).$$

Now, (10.3) applied to Eq. (10.5) (so, we reverse the direction of time and take $|\mathsf{f}| := |b_k|, h \equiv 1$ and $g = 0$) yields, upon applying the Sobolev embedding theorem,

$$\|v\|_{L^\infty([s,r] \times B_1(0))} \leq C_1 \|b_k \sqrt{\rho}\|^2_{L^2([s,r], L^2)},$$

where 0 is the "centre" of the weight ρ. Thus, considering its translates $\rho_z := \rho(x - z)$, we obtain

$$\|v\|_{L^\infty([s,r] \times \mathbb{R}^d)} \leq C_2 \sup_{z \in \mathbb{Z}^d} \|b_k \sqrt{\rho_z}\|_{L^2([s,r], L^2)}.$$

Since $\mathbb{E}v(s, X_s^m) \leq \|v(s,\cdot)\|_\infty$, we obtain

$$\mathbb{E}\int_s^r |b_k(t, X_t^m)|dt \leq C_2 \sup_{z\in\mathbb{Z}^d} \|b_k\sqrt{\rho_z}\|_{L^2([s,r],L^2)}.$$

Since $b_k \in L^\infty\mathbf{F}_\delta + L^2_{\rm loc}$, we have

$$\begin{aligned}\|b_k\sqrt{\rho_z}\|^2_{L^2([s,r],L^2)} &\leq \frac{\delta}{4}\int_s^r \langle\frac{|\nabla\rho_z|^2}{\rho_z}\rangle dt + \int_s^r g(t)\langle\rho_z\rangle dt \\ &\quad \text{(we are using } |\nabla\rho| \leq \theta\sqrt{\kappa}\rho \text{ and } \|\sqrt{\rho}\|_2 < \infty) \\ &\leq CF(r-s)\end{aligned}$$

for $0 \leq s \leq r \leq T$, so (10.4) follows.

Now, define probability measures $\mathbb{P}_x^n := (\mathbf{P}\circ X^n)^{-1}$ on $\big(C([0,T],\mathbb{R}^d),\mathcal{B}_t\big)$, so (10.4) takes form

$$\sup_m \sup_{x\in\mathbb{R}^d} \mathbb{E}_{\mathbb{P}_x^m}\int_s^r |b_k(t,\omega_t)|dt \leq CF(r-s),$$

where ω_t is the coordinate process. We apply in (10.4) the convergence result of Theorem 9.1 (in m) and then Fatou's lemma (in k) to obtain

$$\mathbb{E}\int_s^r |b(t,\omega_t)|dt \leq CF(r-s) < \infty$$

(which is one of the requirements in the definition of a martingale solution). Arguing similarly, we obtain, for every $f \in C_c^2$,

$$\begin{aligned}\mathbb{E}_{\mathbb{P}_x^m}\left|\int_0^r \big((b_m-b_n)\cdot\nabla f\big)(t,\omega_t)dr\right| &\leq C\|(b_m-b_n)|\nabla f|^{\frac{q}{2}}\|_{L^2([0,r],L^2)} \\ &\to 0 \quad (m,n\to\infty)\end{aligned}$$

since $b_m \to b$ in $L^2_{\rm loc}(\mathbb{R}_+\times\mathbb{R}^d)$ and f has compact support. This, and the convergence result of Theorem 9.1, allow to pass to the limit in the martingale problem for b_m in essentially the same way as in the proof of Theorem 6.2 to show that $\mathbb{P}_x$ is a martingale solution of (10.1) but on $(D([0,T],\mathbb{R}^d),\mathcal{B}_t')$. The latter allows to prove, arguing again as in the proof of Theorem 6.2, that $\mathbb{P}_x$ are actually concentrated on continuous trajectories. We arrive at the following result.

Theorem 10.1 *Under the assumptions of Theorem 9.1, let us also assume*[12] (C_4). *The following is true:*

(*i*) *For every* $x \in \mathbb{R}^d$, *the probability measure* $\mathbb{P}_x$ *is a weak solution to SDE* (10.1).

(*ii*) $\mathbb{P}_x$ *satisfies, for all* $\mathsf{f} \in L^\infty \mathbf{F}_\nu + L^2_{\rm loc}(\mathbb{R}_+)$, $\nu < \infty$, $h \in C([0,T], \mathcal{S})$, *for all* $q \in]d, \delta^{-\frac{1}{2}}[$, *the estimate*

$$\mathbb{E}_{\mathbb{P}_x} \int_0^T |\mathsf{f}(r,\omega_r)h(r,\omega_r)|dr \leq c \| |\mathsf{f}|h|^{\frac{q}{2}} \|^{\frac{2}{q}}_{L^2([0,T]\times\mathbb{R}^d)}. \tag{10.6}$$

On the other hand, if, for some $x \in \mathbb{R}^d$, $\mathbb{P}'_x$ *is a martingale solution of* (10.1) *that satisfies* (10.6) *for some* $q \in]d, \delta^{-\frac{1}{2}}[$ *with* $\mathsf{f} = b$, *then it coincides with* $\mathbb{P}_x$.

(*iii*) $\mathbb{P}_x$ *satisfies, for a given* $\nu > \frac{d+2}{2}$, *for all* $h \in C([0,T], \mathcal{S})$ *the following Krylov-type bound:*

$$\mathbb{E}_{\mathbb{P}_x} \int_0^T |h(r,\omega_r)|dr \leq c\|h\|_{L^\nu([0,T]\times\mathbb{R}^d)}. \tag{10.7}$$

On the other hand, if additionally $|b| \in L^{\frac{d+2}{2}+\varepsilon}_{\rm loc}(\mathbb{R}_+ \times \mathbb{R}^d)$ *for some* $\varepsilon > 0$ *and* δ *is sufficiently small, then any martingale solution* $\mathbb{P}'_x$ *of* (10.1) *that satisfies* (10.7) *for some* $\nu > \frac{d+2}{2}$ *sufficiently close to* $\frac{d+2}{2}$ *(depending on how small* ε *is) coincides with* $\mathbb{P}_x$.

The first two assertions of Theorem 10.1 were proved in [43], the last assertion will be proved in the next section, in fact, for a substantially larger than $L^\infty \mathbf{F}_\delta + L^2_{\rm loc}(\mathbb{R}_+)$ class of drifts.

Remark 10.1 One advantage of the uniqueness class in (*ii*), i.e.

$$\mathbb{E}_{\mathbb{P}_x} \int_0^T |b(r,\omega_r)h(r,\omega_r)|dr \leq c\| |b|h|^{\frac{q}{2}} \|^{\frac{2}{q}}_{L^2([0,T]\times\mathbb{R}^d)} \tag{10.8}$$

for some $q \in]d, \delta^{-\frac{1}{2}}[$ is that it senses the value of δ. Namely, as δ becomes smaller, one can take q larger, and so the verification of (10.8), in principle, becomes easier (e.g. $|b|$ is bounded, then δ can be arbitrarily small and hence q can be arbitrarily large).

Regarding the proof of Theorem 10.1(*i*), let us note that we can alternatively use the tightness argument, also employed in the proof of Theorem 4.1, and then apply the convergence result of Theorem 9.1. See details in [43]. The uniqueness results in assertions (*ii*), (*iii*), however, require gradient bounds (10.3).

[12] We can assume (C_5), but then we need to adjust interval $q \in]d, \delta^{-\frac{1}{2}}[$. For simplicity, we will not do this here.

11 "Form-Bounded" Diffusion Coefficients

1. The results of the previous two sections can be extended to Itô and Stratonovich SDEs

$$X_t = x - \int_0^t b(s, X_s)ds + \sqrt{2}\int_0^t \sigma(s, X_s)dW_s, \quad \in \mathbb{R}^d \tag{11.1}$$

$$X_t = x - \int_0^t b(s, X_s)ds + \sqrt{2}\int_0^t \sigma(s, X_s) \circ dW_s, \tag{11.2}$$

where the drift $b : \mathbb{R}^d \to \mathbb{R}^d$ is form-bounded and the diffusion coefficient $\sigma : \mathbb{R}^d \to \mathbb{R}^{d\times d}$ are bounded, uniformly elliptic and can be discontinuous. For time-homogeneous b and σ such extension was carried out in [48]. Namely, let $a := \sigma\sigma^\top$ satisfy,

$$\sigma I \leq a \leq \xi I \quad \text{a.e. on } \mathbb{R}^d \tag{$H_{\sigma,\xi}$}$$

for some $0 < \sigma \leq \xi < \infty$, and assume that the entries a_{ij} of a have form-bounded derivatives, that is,

$$(\nabla_r a_{ij})_{i=1}^d \in \mathbf{F}_{\delta_{rj}} \tag{11.3}$$

for some $\delta_{rj} > 0$. Equivalently, since the entries of σ are bounded, we can replace (11.3) with $(\nabla_r \sigma_{ij})_{i=1}^d \in \mathbf{F}_{\delta'_{rj}}$ for appropriate δ'_{rj}.

Examples

1. If $a \in W^{1,d}(\mathbb{R}^d, \mathbb{R}^{d\times d})$, then (11.3) holds with δ_{rj} that can be chosen arbitrarily small (Appendix B). More generally, if the derivatives of a_{ij} are in the Morrey class $M_{2+\varepsilon}$, then (11.3) holds.
2. Here is a concrete example of matrix a satisfying (11.3) and having a critical discontinuity at the origin:

$$a(x) = I + c\frac{x \otimes x}{|x|^2}, \quad \text{the constant } c > -1. \tag{11.4}$$

Indeed, $\nabla_r a_{ij} = c\mathbf{1}_{r=i}\frac{x_j}{|x|^2} + c\mathbf{1}_{r=j}\frac{x_i}{|x|^2} + cx_ix_j\frac{2x_r}{|x|^4}$, so

$$|(\nabla_r a_{ij})_{i=1}^d| \leq 2|c||x|^{-1} \quad \Rightarrow \quad (\nabla_r a_{ij})_{i=1}^d \in \mathbf{F}_{\delta_{rj}}, \quad \delta_{rj} = (4c)^2/(d-2)^2$$

by the Hardy inequality. Another example is

$$a(x) = I + c(\sin\log(|x|))^2 e \otimes e, \quad e \in \mathbb{R}^d, |e| = 1$$

(indeed, $\nabla_r a_{ij} = 2c(\sin\log|x|)(\cos\log|x|)|x|^{-2}x_r\, e_i e_j$, so using that the Hardy vector field (3.7) is form-bounded one obtains the required).

More generally, (11.3) holds for a that is an infinite sum of such matrices (properly normalized so that the series converges) with their points of discontinuity constituting e.g. a dense subset of $\mathbb{R}^d$.

Without loss of generality, in $(H_{\sigma,\xi})$ $\sigma = 1$. In [48], assuming that $b \in \mathbf{F}_\delta$ and

$$a \text{ satisfies } (H_{\sigma,\xi}) \text{ and } (11.3), \quad \nabla a \in \mathbf{F}_{\delta_a}, \tag{11.5}$$

where $(\nabla a)_k := \sum_{i=1}^d (\nabla_i a_{ik})$, with δ, δ_a and δ_{rn} satisfying, for some $q > 2 \vee (d-2)$,

$$1 - \frac{q}{4}(\sqrt{\gamma} + \|a - I\|_\infty \sqrt{\delta + \delta_a}) > 0, \tag{11.6}$$

where $\gamma := \sum_{r,n=1}^d \delta_{rn}$, and

$$(q-1)\big(1 - \frac{q\sqrt{\gamma}}{2}\big) - (\sqrt{\delta + \delta_a}\sqrt{\delta_a} + \delta + \delta_a)\frac{q^2}{4} \\ - (q-2)\frac{q\sqrt{\delta + \delta_a}}{2} - \|a - I\|_\infty \frac{q\sqrt{\delta + \delta_a}}{2} > 0, \tag{11.7}$$

the authors constructed a Feller semigroup and proved an analogue of Theorem 6.2(i), including the "approximation uniqueness" result in Remark 6.3, for the Itô SDE

$$X_t = x - \int_0^t b(X_s)ds + \sqrt{2}\int_0^t \sigma(X_s)dW_s. \tag{11.8}$$

The result for the Stratonovich SDE

$$X_t = x - \int_0^t b(X_s)ds + \sqrt{2}\int_0^t \sigma(X_s) \circ dW_s, \tag{11.9}$$

in [48] is valid under assumption (11.6), (11.7) but with δ replaced by $\delta + \delta_a + \delta_c$, where δ_c is the form-bound of

$$c := (c^i)_{i=1}^d, \quad \text{where } c^i := \frac{1}{\sqrt{2}} \sum_{r,j=1}^d (\nabla_r \sigma_{ij})\sigma_{rj}.$$

The assumptions (11.6), (11.7) imply that δ, δ_a and δ_{rj} cannot be too large. It is also easily seen that if $a = I$, then these assumptions reduce to $\delta < 1 \wedge (\frac{d}{d-2})^2$, i.e. then there exists $q > 2 \vee (d-2)$ such that (11.6), (11.7) hold.

The assumptions on diffusion coefficients σ of the form (11.3) go back to Veretennikov [96] who proved strong well-posedness of (11.1) for bounded measurable b and $\nabla_r\sigma_{ij} \in L^{2d}_{\rm loc}$. There are many other papers that consider assumptions of this type, see e.g. [106] and [63, 65] who considered $\nabla_r\sigma_{ij} \in L^p_{\rm loc}$ $(p > d)$ and $\nabla_r\sigma_{ij} \in L^d_{\rm loc}$.

The construction of the Feller semigroup in [48] is based on an extension of the iteration procedure described in Sect. 9 (in the elliptic setting) to solutions u_n of divergence-form equations

$$(\mu + \Lambda(a_n, b_n))u_n = f, \quad f \in C_c^\infty, \quad \mu \geq \mu_0, \tag{11.10}$$

where $\Lambda(a_n, b_n) = -\nabla \cdot a_n \cdot \nabla + b_n \cdot \nabla$, and uses the gradient bound

$$\|\nabla u_n\|_{\frac{qd}{d-2}} \leq K\|f\|_q, \quad K \text{ is independent of } n. \tag{11.11}$$

Here a_n, b_n are bounded and smooth, $\{b_n\}$ is uniformly (in n) form-bounded, and $\{a_n\}$ satisfy the same assumptions as a above (thus, with constants independent on n). This iteration procedure and (11.11) yield Feller semigroup $e^{-t\Lambda_{C_\infty}(a,b)}$. Then $e^{-t\Lambda_{C_\infty}(a,\nabla a+b)}$ is the sought Feller semigroup that produces weak solution to Itô SDE (11.8), where we used the identity

$$-a_n \cdot \nabla^2 + b_n \cdot \nabla = -\nabla \cdot a_n \cdot \nabla + (\nabla a_n + b_n) \cdot \nabla. \tag{11.12}$$

For Stratonovich SDE (11.9) one needs Feller semigroup $e^{-t\Lambda_{C_\infty}(a,\nabla a+b-c)}$.

Remark 11.1 For instance, the approximating vector fields b_n can be defined via (6.2), so they are uniformly (in n) in $\mathbf{F}_\delta$. The approximating matrices a_n can be defined via

$$a_n = \eta_{\varepsilon_n} * a,$$

where η_{ε_n} is the Friedrichs mollifier, and $\varepsilon_n \downarrow 0$. To see that a_n are such that ∇a_n are indeed uniformly in $\mathbf{F}_{\delta_a}$ and satisfy (11.3) with the same δ_{rn}, also uniformly in n, one can apply the result of Appendix B.3. In [48], there was an additional cutoff function under the mollifier, which is not necessary.

Let us add that already the task of proving the uniform in n gradient bound (11.11) for solutions u_n to the divergence-form equation (11.10) (which was the original interest of the authors of [48]) leads to condition (11.3).

Theorem 9.1 and the existence part of Theorem 10.1 can be extended to SDEs (11.1) and (11.2) with time-inhomogeneous $b \in L^\infty\mathbf{F}_\delta + L^2_{\rm loc}(\mathbb{R}_+)$ and time-inhomogeneous bounded σ such that $a = \sigma\sigma^\top$ is uniformly elliptic and satisfies

$$\nabla a \in L^\infty\mathbf{F}_{\delta_a} + L^2_{\rm loc}(\mathbb{R}_+), \quad (\nabla_r a_{ij})_{i=1}^d \in L^\infty\mathbf{F}_{\delta_{rj}} + L^2_{\rm loc}(\mathbb{R}_+)$$

for all $1 \leq r, j \leq d$, for appropriate δ_a and δ_{rj}. A direct extension of the weak uniqueness part of Theorem 10.1 to a_{ij} as above is problematic: one has to control second-order derivatives of u_n, but these are destroyed by form-bounded b. Thus, one needs extra assumptions both on b and a, such as Morrey class in [66], see below.

2. Below we assume, for simplicity, that a, b are time-homogeneous, although most of the results cited below are valid for time-inhomogeneous coefficients.

The condition (11.3) puts a in the class VMO,[13] see [65]. Recall that matrix a is in VMO if

$$\sup_{B_r} \frac{1}{|B_r|} \int_{B_r} \big|a - (a)_{B_r}\big| dx \to 0 \quad \text{as } \rho \downarrow 0,$$

where the supremum is taken over all balls of radius $\leq \rho$. Here $(a)_{B_r}$ denotes the average of a on B_r.

There is a very rich literature on well-posedness of parabolic equations and SDEs with VMO diffusion matrix and singular drift b satisfying more restrictive assumptions than the form-boundedness. The strongest result on weak well-posedness of SDE (11.1) with VMO diffusion coefficients is a very recent result of Krylov [66] who proved that there exist positive constants θ (sufficiently small) and ρ_a such that if a in the BMO class with norm $\leq \theta$, i.e.

$$\sup_{B_r,\, r \leq \rho} \frac{1}{|B_r|} \int_{B_r} \big|a - (a)_{B_r}\big| dx \leq \theta \quad \text{for all } \rho \leq \rho_a, \tag{11.13}$$

(e.g. if $a \in$ VMO) and

$$|b| \in M_{\frac{d}{2}+\varepsilon}, \quad \varepsilon > 0, \tag{11.14}$$

with sufficiently small norm, then (11.1) is weakly well-posed, i.e. the solution exists and is unique in a class similar to the one in Theorem 10.1.

In the case of constant diffusion coefficients σ one can prove weak well-posedness of (11.1) for substantially larger than $\mathbf{F}_\delta$ class of *weakly form-bounded drifts*, discussed in Sects. 14–15. This class contains e.g.

$$|b| \in M_{1+\varepsilon} \tag{11.15}$$

with arbitrarily small $\varepsilon > 0$.

If we were to exploit the relationship between non-divergence and divergence form operators in the case the matrix a is sufficiently discontinuous, cf. (11.12),

---

[13] In [48] there is an incorrect statement that there are matrices a satisfying (11.3) and not contained in the VMO class.

then the sesquilinear form of the divergence-form operator will have to satisfy

$$|\langle a \cdot \nabla\varphi, \nabla\psi\rangle + \langle b \cdot \nabla\varphi, \psi\rangle| \leq C\|\varphi\|_{W^{1,2}}\|\psi\|_{W^{1,2}}, \quad \varphi, \psi \in W^{1,2}, \tag{11.16}$$

which, by [74], would make $\nabla a + b$ form-bounded (modulo a divergence-free component in BMO^{-1}, see the discussion around (3.11)). See also Remarks 14.1 and 15.6 below.

12 Stochastic Transport Equation and Strong Solutions to SDEs

In [7], Beck-Flandoli-Gubinelli-Maurelli presented, among many results, an approach to proving strong well-posedness of SDE

$$X_t = x - \int_0^t b(r, X_r)dr + \sigma W_t, \tag{12.1}$$

for a.e. $x \in \mathbb{R}^d$, where $\{W_t\}_{t\geq 0}$ is a d-dimensional Brownian motion in $\mathbb{R}^d$ defined on a complete filtered probability space $(\Omega, \mathcal{F}_t, \mathcal{F}, \mathbf{P})$, with b in the critical Ladyzhenskaya-Prodi-Serrin class. Below we will discuss the time-homogeneous case $b : \mathbb{R}^d \to \mathbb{R}^d$, so their assumption on b reads as $|b| \in L^d + L^\infty$. Their approach is based on a detailed regularity theory of the stochastic transport equation (STE)

$$\begin{aligned} du + b \cdot \nabla u dt + \sigma \nabla u \circ dW_t = 0 \quad &\text{on } (0, \infty) \times \mathbb{R}^d, \\ u|_{t=0} = f, \quad & \end{aligned} \tag{12.2}$$

where $u(t, x)$ is a scalar random field, $\sigma \neq 0$ is a constant, f is in L^p or $W^{1,p}$, $\circ$ is the Statonovich multiplication.

Speaking of the STE (12.2), let us mention that the Cauchy problem for the deterministic transport equation $\partial_t u + b \cdot \nabla u = 0$ is in general not well posed already for a bounded but discontinuous b. Moreover, in that case, even if the initial function f is regular, one cannot hope that the corresponding solution u will be regular immediately after $t = 0$. This, however, changes if one adds the noise term $\sigma \nabla u \circ dW_t$, $\sigma > 0$. For the stochastic STE (12.2), a unique weak solution exists and is regular for some discontinuous b. This effect of regularization and well-posedness by noise, demonstrated by the STE, attracted considerable interest in the past few years, as a part of the more general program of establishing well-posedness by noise for SPDEs whose deterministic counterparts arising in fluid dynamics are not well-posed, see Flandoli-Gubinelli-Priola [24], Gess-Maurielli [27] for detailed discussions and references.

Let us make a few preliminary remarks regarding STE (12.2) in the case the drift is smooth.

1. Let $b \in C_c^\infty(\mathbb{R}^d, \mathbb{R}^d)$ and $f \in C_c^\infty$. Then there exists (see [70, Theorem 6.1.9]) a unique adapted strong solution to

$$u(t) - f + \int_0^t b \cdot \nabla u ds + \sigma \int_0^t \nabla u \circ dW_s = 0 \text{ a.s.}, \quad t \in [0, T], \tag{12.3}$$

given by

$$u(t) := f(\Psi_t^{-1}), \quad t \geqslant 0, \tag{12.4}$$

where $\Psi_t : \mathbb{R}^d \times \Omega \to \mathbb{R}^d$ is the stochastic flow for the SDE (12.1). The latter means that there exists $\Omega_0 \subset \Omega$, $\mathbb{P}(\Omega_0) = 1$, such that, for all $\omega \in \Omega_0$,

$$\Psi_t(\cdot, \omega)\Psi_s(\cdot, \omega) = \Psi_{t+s}(\cdot, \omega), \quad \Psi_0(x, \omega) = x,$$

for every $x \in \mathbb{R}^d$, the process $t \mapsto \Psi_t(x, \omega)$ is a strong solution to (12.1), and $\Psi_t(x, \omega)$ is continuous in (t, x), $\Psi_t(\cdot, \omega) : \mathbb{R}^d \to \mathbb{R}^d$ are homeomorphisms and $\Psi_t(\cdot, \omega)$, $\Psi_t^{-1}(\cdot, \omega) \in C^\infty(\mathbb{R}^d, \mathbb{R}^d)$.

2. Applying Itô's formula, one easily obtain that for every $\mu \geq 0$,

$$u(t) = e^{\mu t} f(\Psi_t^{-1}), \quad t \geqslant 0$$

solves

$$u(t) - f + \mu \int_0^t u ds + \int_0^t b \cdot \nabla u ds + \sigma \int_0^t \nabla u \circ dW_s = 0 \text{ a.s.}, \quad t \in [0, T]. \tag{12.5}$$

Thus, solutions of the Cauchy problems (12.3) and (12.5) differ by the factor $e^{-\mu t}$.

3. One can rewrite the equation in (12.5), using the identity relating Stratonovich and Itô integrals

$$\int_0^t \nabla u \circ dW_s = \int_0^t \nabla u dW_s - \frac{1}{2}\sum_{k=1}^d [\partial_{x_k} u, W^k]_t, \qquad W_t = (W_t^k)_{k=1}^d, \tag{12.6}$$

as

$$du + \mu u dt + b \cdot \nabla u dt + \sigma \nabla u dW_t - \frac{\sigma^2}{2} \Delta u dt = 0 \tag{12.7}$$

(the Itô form of the STE). Now, taking expectation, one obtains that $v := \mathbf{E}[u]$ solves Cauchy problem for the deterministic parabolic equation

$$\partial_t v + \mu v + b \cdot \nabla v - \frac{\sigma^2}{2}\Delta v = 0, \quad v|_{t=0} = f.$$

Let now b be discontinuous. The authors of [7], in a sense, reversed (12.4), i.e. given a $|b| \in L^d$ (in the time-homogeneous case) they used their Sobolev regularity theory of (12.2) to prove strong well-posedness of SDE (12.1) for a.e. initial point $x \in \mathbb{R}^d$.

In [57], the authors extended the approach of [7] to (time-homogeneous) form-bounded drifts b. We describe these results below.

Set

$$\rho(x) \equiv \rho_{\kappa,\theta}(x) := (1+\kappa|x|^2)^{-\theta}, \quad \kappa > 0, \quad \theta > \frac{d}{2}, \quad x \in \mathbb{R}^d.$$

Let $L^p_\rho \equiv L^p(\mathbb{R}^d, \rho dx)$. Denote by $\|\cdot\|_{p,\rho}$ the norm in L^p_ρ, and by $\langle \cdot,\cdot\rangle_\rho$ the inner product in L^2_ρ. Set

$$W^{1,2}_\rho := \{g \in W^{1,2}_{\mathrm{loc}} \mid \|g\|_{W^{1,2}_\rho} := \|g\|_{2,\rho} + \|\nabla g\|_{2,\rho} < \infty\}.$$

Fix $T > 0$ and put

$$\beta_{2q} := 1 + 4qd, \quad q = 1, 2, \ldots$$

Theorem 12.1 *Let $b \in \mathbf{F}_\delta$ with $\sqrt{\delta} < \frac{\sigma^2}{2\beta_2}$. Let $p \geq 2$. Provided that κ is chosen sufficiently small, there are generic constants $\mu_1 \geq 0$, $C_1 > 0, C_2 > 0$ (i.e. they depend only on δ, c_δ, p and T) such that for any $\mu \geq \mu_1$, for every $f \in L^{2p}$ there exists a function $u \in L^\infty([0,T], L^2(\Omega, L^2_\rho))$ for which the following are true:*

(i) For a.e. $\omega \in \Omega$,

$$\nabla \int_0^T u(s,\cdot,\omega)ds \in L^2_{\mathrm{loc}}(\mathbb{R}^d, \mathbb{R}^d),$$

so

$$b \cdot \nabla \int_0^T u(s,\cdot,\omega)ds \in L^1_{\mathrm{loc}},$$

and for every test function $\varphi \in C_c^\infty$, *we have a.s. for all* $t \in [0, T]$,

$$\begin{aligned}&\langle u(t), \varphi\rangle - \langle f, \varphi\rangle \\ &+ \mu\langle \int_0^t u ds, \varphi\rangle + \langle b \cdot \nabla \int_0^t u ds, \varphi\rangle - \sigma\langle \int_0^t u dW_s, \nabla\varphi\rangle \\ &+ \frac{\sigma^2}{2}\langle \nabla \int_0^t u ds, \nabla\varphi\rangle = 0. \qquad (12.8)\end{aligned}$$

(ii) *For any sequence of smooth vector fields* $b_m \in C_c^\infty(\mathbb{R}^d, \mathbb{R}^d)$, $m = 1, 2, \ldots$, *that are uniformly form-bounded in the sense that* $b_m \in \mathbf{F}_\delta$ *with* c_δ *independent of* m, *and are such that*

$$b_m \to b \text{ in } L^2_{\mathrm{loc}}(\mathbb{R}^d, \mathbb{R}^d) \text{ as } m \to \infty,$$

we have for any initial function $f \in C_c^\infty$,

$$u_m(t) \to u(t) \quad \text{in } L^2(\Omega, L^2_\rho) \quad \text{uniformly in } t \in [0, T],$$

where u_m *is the unique strong solution to* (12.7) *(with* $b = b_m$*) with initial condition* $u_m|_{t=0} = f$.

The last property implies that u does not depend on the choice of the approximating sequence $\{b_m\}$ as long as it preserves the class of form-bounded vector fields. This can be viewed as a uniqueness result on its own.

The next theorem establishes Sobolev regularity of u up to the initial time $t = 0$.

Theorem 12.2 *Let* $b \in \mathbf{F}_\delta$ *with* $\sqrt{\delta} < \frac{\sigma^2}{2\beta_2}$ *and* $f \in W^{1,4}$. *Let* κ *be sufficiently small and* μ_1 *be the constant in Theorem 12.1 with* $p = 2$. *For* $\mu \geq \mu_1$, *let* u *be the process constructed in Theorem 12.1. There exists generic constant* $\mu_2 \geq \mu_1$ *such that for* $\mu \geq \mu_2$, *the following are true:*

(a) $\mathbf{E}u^2, \mathbf{E}|\nabla u|^2 \in L^\infty([0, T], L^2)$, *so* $u \in L^\infty([0, T], L^2(\Omega, W^{1,2}_\rho))$.
(b) *For any test function* $\varphi \in C_c^\infty$, *the process* $t \mapsto \langle u(t), \varphi\rangle$ *is* $(\mathcal{F}_t)$*-progressively measurable and has a continuous* $(\mathcal{F}_t)$*-semi-martingale modification that satisfies a.s. for every* $t \in [0, T]$,

$$\begin{aligned}&\langle u(t), \varphi\rangle - \langle f, \varphi\rangle \\ &+ \mu \int_0^t \langle u, \varphi\rangle ds + \int_0^t \langle b \cdot \nabla u, \varphi\rangle ds - \sigma \int_0^t \langle u, \nabla\varphi\rangle dW_s \\ &+ \frac{\sigma^2}{2} \int_0^t \langle u, \Delta\varphi\rangle ds = 0. \qquad (12.9)\end{aligned}$$

Moreover, if $\sqrt{\delta} < \frac{\sigma^2}{2\beta_{2q}}$ *for some* $q = 1, 2, \ldots$, *then there exist generic constants* $\mu_2(q) \geq \mu_1$ *(with* $\mu_2(1)$ *equal to the* μ_2 *above) and* $C_1 > 0$ *such that when* $\mu \geq \mu_2(q)$ *and* $f \in W^{1,4q}$, *we have*

$$\sup_{0\leq\alpha\leq 1} \left\| \mathbf{E}|\nabla u|^{2q} \right\|_{L^{\frac{2}{1-\alpha}}([0,T], L^{\frac{2d}{d-2+2\alpha}})} \leq C_1 \|\nabla f\|_{4q}^{2q}. \tag{12.10}$$

In particular, there exists generic $C_2 > 0$ *such that*

$$\sup_{t\in[0,T]} \mathbf{E}\langle \rho |\nabla u|^{2q} \rangle \leq C_2 \|\nabla f\|_{4q}^{2q}.$$

If $2q > d$, *then for a.e.* $\omega \in \Omega$, $t \in [0, T]$, *the function* $x \mapsto u(t, x, \omega)$ *is Hölder continuous, possibly after modification on a set of measure zero in* $\mathbb{R}^d$ *(in general, depending on* ω*).*

The estimate (12.10) can be viewed as a counterpart of (9.12).

A function satisfying (a), (b) of Theorem 12.2 will be called a weak solution of Cauchy problem

$$\begin{aligned} du + \mu u dt + b \cdot \nabla u dt + \sigma \nabla u \circ dW_t = 0 \quad &\text{on } (0, \infty) \times \mathbb{R}^d, \\ u|_{t=0} = f \in L^p, \quad &p \geq 2. \end{aligned} \tag{12.11}$$

This definition of weak solution is close to [7, Definition 2.13].

Theorem 12.3 *Let* $b \in \mathbf{F}_\delta$ *with* $\sqrt{\delta} < \frac{\sigma^2}{2\beta_2}$ *and* $f \in W^{1,4}$. *Provided* κ *is sufficiently small, there exists generic* $\mu_3 \geq 0$ *such that for* $\mu \geq \mu_3$, *the Cauchy problem* (12.11) *has a unique weak solution in the class of functions satisfying (a), (b) of Theorem 12.2.*

Theorems 12.1–12.3 were proved in [57]. Theorem 12.2 extends a similar result in [7] for (in the time-homogeneous case) $|b| \in L^d$. The proof of the uniqueness result in Theorem 12.3 adopts the method of [7, Sect. 3].

It should be noted that the authors in [7] prove their uniqueness result in a larger class of weak solutions (not requiring any differentiability, see [7, Definition 3.3]) but under additional assumptions on b. Specialized to the time-dependent case, they assume that b satisfies

$$\operatorname{div} b \in L^d + L^\infty \tag{12.12}$$

in addition to $b \in L^d + L^\infty$. The latter is needed to establish (12.10) for solutions of the adjoint equation to the STE, i.e. the stochastic continuity equation (which allows to prove an even stronger result: the uniqueness of weak solution to the corresponding random transport equation), see [7, Sect. 3].

Armed with Theorems 12.1, 12.2, one can repeat the argument in [7, Sect. 4] to prove the following result. *Assuming that* $b \in \mathbf{F}_\delta$ *with* δ *sufficiently small, there exists a stochastic Lagrangian flow for SDE* (12.1)*, i.e. a measurable map* $\Phi : [0, T] \times \mathbb{R}^d \times \Omega \to \mathbb{R}^d$ *such that, for a.e.* $x \in \mathbb{R}^d$*, the process* $t \mapsto \Phi_t(x, \omega)$ *is a strong solution of the SDE* (12.1)*:*

$$\Phi_t(x, \omega) = x - \int_0^t b(s, \Phi_r(x, \omega))dr + \sigma W_t(\omega), \quad \text{a.s.}, \quad t \in [0, T],$$

and $\Phi_t(x, \cdot)$ *is* $\mathcal{F}_t$*-progressively measurable. If also* $\sqrt{\delta} < \frac{\sigma^2}{2\beta_{2q}}$, $q = 1, 2, \dots$, *then* $\Phi_t(\cdot, \omega) \in W^{1,2q}_{\mathrm{loc}}$ $(t \in [0, T])$ *for a.e.* $\omega \in \Omega$*. Moreover,* Φ_t *is unique, i.e. any two such stochastic flows coincide a.s. for every* $t > 0$ *for a.e.* x.

The restriction "for a.e. initial point" in the above strong existence result for SDE (12.1) can be removed using a different method discussed in the next section.

13 Strong Well-Posedness via Röckner-Zhao's Approach

In [91], Röckner and Zhao proved strong existence and uniqueness in a large class of strong solutions satisfying Krylov estimate (cf. (13.6) below) for SDE

$$X_t^x = x + \int_0^t b(s, X_s^x)ds + W_t, \quad 0 \leq t \leq T, \tag{13.1}$$

for every initial point $x \in \mathbb{R}^d$, provided that drift $b : \mathbb{R}^{d+1} \to \mathbb{R}^d$ satisfies the critical Ladyzhenskaya-Prodi-Serrin condition

$$b \in L^p(\mathbb{R}, L^q(\mathbb{R}^d)), \quad \frac{d}{q} + \frac{2}{p} \leq 1, \quad p > 2, \quad q \geq d. \tag{13.2}$$

Above $\{W_t\}_{0 \leq t \leq T}$ denotes, as before, a Brownian motion on a complete filtered probability space $(\Omega, \{\mathcal{F}_t\}_{0 \leq t \leq T}, \mathcal{F}, \mathbf{P})$.

The method of Röckner-Zhao is different from the other methods used in the literature on strong well-posedness of (13.1) (cf. [7] and [62–65]). Their proof of strong existence uses a relative compactness criterion for random fields on the Wiener-Sobolev space. Their proof of uniqueness uses their weak uniqueness result from [90] and Cherny's theorem [19] (briefly, strong existence and weak uniqueness $\Rightarrow$ strong uniqueness). The method of [91] is a far-reaching strengthening of the methods of Meyer-Brandis and Proske [77], Mohammed-Nilsen-Proske [78] (for $b \in L^\infty(\mathbb{R} \times \mathbb{R}^d)$) and Rezakhanlou [89] (for b in the sub-critical Ladyzhenskaya-Prodi-Serrin class).

Let us give a brief outline of their method. For a given vector field $Y = (Y_i)_{i=1}^d : \mathbb{R}^k \to \mathbb{R}^m$, denote

$$\nabla Y = \nabla_x Y(x) := \begin{pmatrix} \nabla_1 Y_1 & \nabla_2 Y_1 & \dots & \nabla_k Y_1 \\ & & \dots & \\ \nabla_1 Y_m & \nabla_2 Y_m & \dots & \nabla_k Y_m \end{pmatrix}. \tag{13.3}$$

The proof of strong existence in [91] is based on the following estimates. Let b be additionally bounded and smooth. For every $r \geq 1$, there exist constants K_1, K_2 independent of smoothness or boundedness of b such that

(*i*) $\|\nabla X_t^x - I\|_{L^{2r}(\mathbb{R}^d, L^r(\Omega))} \leq K_1 t^{\frac{1}{2r}}$ for all $0 \leq t \leq T$;

(*ii*) $\|D_s X_t^x - I\|_{L^{2r}(\mathbb{R}^d, L^r(\Omega))} \leq K_1 (t-s)^{\frac{1}{4r}}$ for a.e. $s \in [0, T]$ and $0 \leq s \leq t \leq T$, where $D_s X_t^x$ denotes the Malliavin derivative;

(*iii*) $\|D_s X_t^x - D_{s'} X_t^x\|_{L^{2r}(\mathbb{R}^d, L^r(\Omega))} \leq K_2 |s - s'|^{\frac{1}{4r}}$ for a.e. $s, s' \in [0, T]$ and $0 \leq s, s' \leq t \leq T$,

These estimates allow [91] to apply the relative compactness criterion on the Wiener-Sobolev space in order to construct a strong solution to (13.1).

The first step in the proof of (*i*)–(*iii*) is to differentiate SDE (13.1), e.g. for (*i*)

$$\nabla X_t^x - I = \int_0^t \nabla b(s, X_s^x) \nabla X_s^x ds,$$

with the goal of iterating this identity and, in the end, obtaining an expression for the left-hand side that one can control. This goal is achieved using the following bound (for (*i*)): there exist positive generic constants C_0, K such that, for every $n \geq 1$,

$$\int_{\mathbb{R}^d} \left| \mathbf{E} \int_{\Delta_n(T_0, T_1)} \prod_{i=1}^n \nabla_{\alpha_i} f_i(t_i, X_{t_i}^x) dt_1 \dots dt_n \right|^2 dx \leq C_0 K^n (T_1 - T_0), \tag{13.4}$$

where $1 \leq \alpha_i \leq d$ $(i \geq 1)$, $\nabla_i := \partial_{x_i}$, $\Delta_n(T_0, T_1) := \{(t_1, \dots, t_n) \mid T_0 \leq t_1 \leq \dots \leq t_n \leq T_1\}$, and functions f_i are taken to be the components of drift b. In [91], (13.4) is proved using Sobolev regularity estimates for solutions of parabolic equations with distributional right-hand side. These estimates are quite strong and are interesting on their own.

In [45], the authors noticed that (13.4) can be proved for time-inhomogeneous form-bounded drifts

$$b \in L^\infty \mathbf{F}_\delta + L^{2+\varepsilon}_{\text{loc}}(\mathbb{R}), \quad \text{(i.e. with function } g_\delta \in L^{1+\varepsilon/2}_{\text{loc}}(\mathbb{R})), \quad \varepsilon > 0,$$

that can have stronger spatial singularities than the drifts in (13.2), using a different argument which applies repeatedly integration by parts, quadratic inequality and the form-boundedness of b. The rest of the proof of strong existence essentially repeats [91]. This, combined with the weak uniqueness results from [43] and [41] via Cherny's theorem (see Sects. 10 and 16), yields the following result.

Theorem 13.1 ([45]) *Assume that $b \in L^\infty \mathbf{F}_\delta + L^{2+\varepsilon}_{\rm loc}(\mathbb{R})$ for some $\varepsilon > 0$. Also, assume that b has compact support. Then, provided that form-bound δ is sufficiently small, for every $x \in \mathbb{R}^d$, SDE* (13.1) *has a strong solution X_t^x. This strong solution satisfies the following Krylov-type bounds:*

(1) For a given $q \in]d, \delta^{-\frac{1}{2}}[$ and any vector field $\mathsf{g} \in L^\infty \mathbf{F}_{\delta_1} + L^{2+\varepsilon}_{\rm loc}(\mathbb{R})$, $\delta_1 < \infty$,

$$\mathbf{E}\int_0^T |\mathsf{g}h|(\tau, X^x_{0,\tau})d\tau \leq c\| |\mathsf{g}|h|^{\frac{q}{2}} \|^{\frac{2}{q}}_{L^2([0,T]\times\mathbb{R}^d)} \quad \text{for all } h \in C_c([0,T]\times\mathbb{R}^d). \tag{13.5}$$

(2) For a given $\mu > \frac{d+2}{2}$, there exists constant C such that

$$\mathbf{E}\bigg[\int_0^T |h(\tau, X^x_{0,\tau})|d\tau\bigg] \leq C\|h\|_{L^\mu([0,T]\times\mathbb{R}^d)} \quad \text{for all } h \in C_c([0,T]\times\mathbb{R}^d). \tag{13.6}$$

Solution X_t^x is unique among strong solutions to (13.1) *that satisfy* (13.5) *for some $q \in]d, \delta^{-\frac{1}{2}}[$ with $\mathsf{g} = 1$ and with $\mathsf{g} = b$. If, in addition to the hypothesis on b, one has $|b| \in L^{\frac{d+2}{2}+\epsilon}$ for some $\epsilon > 0$, then X_t^x is unique among strong solutions to* (13.1) *that satisfy* (13.6).

The assumption that b has compact support can be removed with an additional effort (using weight (4.5)).

Speaking of "$\varepsilon > 0$" in $b \in L^\infty \mathbf{F}_\delta + L^{2+\varepsilon}_{\rm loc}(\mathbb{R})$, this assumption does not allow us to include completely the critical Ladyzhenskaya-Prodi-Serrin class even with $p > 2$, see above, as is assumed in [91]. It does include, however, the case that interests us the most: $p = \infty$, $q = d$. It also includes with case $p > 2$, $q = \infty$.

The weak well-posedness of SDE (13.1) is known to hold for larger classes of singular drifts than class $L^\infty \mathbf{F}_\delta + L^2_{\rm loc}(\mathbb{R})$ discussed in Theorem 13.1. This is the subject of the next three sections.

14 More Singular than Form-Bounded. Semigroup in $\mathcal{W}^{\frac{1}{2},2}$

In Sects. 14, 15 and 16 we expand the classes of singular drifts considered in Theorems 6.1, 6.2 and 9.1, 10.1, although at expense of requiring smaller δ.

Definition 14.1 A Borel measurable vector field $b : \mathbb{R}^d \to \mathbb{R}^d$ with $|b| \in L^1_{\text{loc}}$ is said to be weakly form-bounded if there exists a constant $\delta > 0$ such that

$$\||b|^{\frac{1}{2}}(\lambda - \Delta)^{-\frac{1}{4}}\|_{2\to 2} \leq \sqrt{\delta}$$

for some $\lambda = \lambda_\delta \geq 0$. This is written as $b \in \mathbf{F}_\delta^{\frac{1}{2}}$.

There is an important difference between form-bounded vector fields and weakly form-bounded vector fields. Namely, when we dealt with $b \in \mathbf{F}_\delta$, we controlled the gradient term $b \cdot \nabla$ in the Kolmogorov operator $-\Delta + b \cdot \nabla$ using the quadratic (Cauchy-Schwarz) inequality

$$|\langle b \cdot \nabla\varphi, \varphi\rangle| \leq \varepsilon\|b\varphi\|_2^2 + \frac{1}{4\varepsilon}\|\nabla\varphi\|_2^2, \quad \varepsilon > 0 \tag{14.1}$$

(e.g. (3.3) in the verification of conditions of the Lax-Milgram theorem and the KLMN theorem in L^2, in the proof of Lemma 6.1, in the iteration procedure and gradient bound (9.7), etc.). *We can no longer do this when dealing with $b \in \mathbf{F}_\delta^{1/2}$ if only because $|b|$ is in general no longer locally in L^2.*

The form-bounded vector fields are weakly form-bounded. To show this, let us recall that the condition $b \in \mathbf{F}_\delta$ can be stated as an operator-norm inequality $\||b|(\lambda - \Delta)^{-\frac{1}{2}}\|_{2\to 2} \leq \sqrt{\delta}$. The Heinz-Kato inequality [33] allows us to take square roots in the operators that constitute the left-hand side, so we arrive at $\||b|^{\frac{1}{2}}(\lambda - \Delta)^{-\frac{1}{4}}\|_{2\to 2} \leq \delta^{\frac{1}{4}}$, and hence

$$b \in \mathbf{F}_\delta \quad \Rightarrow \quad b \in \mathbf{F}_{\sqrt{\delta}}^{\frac{1}{2}}.$$

The opposite inclusion is invalid: there are weakly form-bounded vector fields that are not form-bounded:

Examples

1. If $|b|$ belongs to scaling-invariant Morrey class $M_{1+\varepsilon}$, for arbitrarily fixed small $\varepsilon > 0$, i.e.

$$|b| \in L^{1+\varepsilon}_{\text{loc}} \quad \text{and} \quad \|b\|_{M_{1+\varepsilon}} := \sup_{r>0, x\in\mathbb{R}^d} r\left(\frac{1}{|B_r|}\int_{B_r(x)} |b|^{1+\varepsilon}dx\right)^{\frac{1}{1+\varepsilon}} < \infty, \tag{14.2}$$

then $b \in \mathbf{F}_\delta^{1/2}$. The proof of this inclusion follows right away from [1, Theorem 7.3].

Recalling that the class of form-bounded vector fields $\mathbf{F}_\delta$ satisfies $M_{2+\varepsilon} \subset \mathbf{F}_\delta \subset M_2$ (say, with $c_\delta = 0$), one can see that we gain quite a lot in admissible singularities of b by working with $\mathbf{F}_\delta^{1/2}$. In particular, we gain all b with $|b| \in M_{1+\varepsilon} - M_2$.

2. Recall that a vector field $b : \mathbb{R}^d \to \mathbb{R}^d$ is said to belong to the Kato class if $|b| \in L^1_{\rm loc}$ and

$$\|(\lambda - \Delta)^{-\frac{1}{2}}|b|\|_\infty \leq \sqrt{\delta} \tag{14.3}$$

for some $\delta > 0$ and $\lambda = \lambda_\delta \geq 0$. This is written as $b \in \mathbf{K}^{d+1}_\delta$. The Kato class vector fields are weakly form-bounded. Indeed, if $b \in \mathbf{K}^{d+1}_\delta$, then by duality one has

$$\||b|(\lambda - \Delta)^{-\frac{1}{2}}\|_{1\to 1} \leq \sqrt{\delta}. \tag{14.4}$$

Applying Stein's interpolation between (14.3) and (14.4), one has $\||b|^{\frac{1}{2}}(\lambda - \Delta)^{-\frac{1}{2}}|b|^{\frac{1}{2}}\|_{2\to 2} \leq \sqrt{\delta}$; in the left-hand side of the last inequality one has the norm of the product of $|b|^{\frac{1}{2}}(\lambda - \Delta)^{-\frac{1}{4}}$ and its adjoint. By a standard result this yields $\||b|^{\frac{1}{2}}(\lambda - \Delta)^{-\frac{1}{4}}\|_{2\to 2} \leq \delta^{\frac{1}{4}}$. Thus,

$$b \in \mathbf{K}^{d+1}_\delta \quad \Rightarrow \quad b \in \mathbf{F}^{1/2}_{\sqrt{\delta}}.$$

As was mentioned earlier, the Kato class $b \in \mathbf{K}^{d+1}_\delta$ with δ sufficiently small provides two-sided Gaussian bounds on the heat kernel of Kolmogorov operator $-\Delta + b \cdot \nabla$ [102]. The Kato class also provides uniqueness in law for SDE (15.1), see [6]. There the authors required δ to be arbitrarily small. In fact, they show that under the Kato class assumption on b the gradient of solution of elliptic equation $(\mu - \Delta + b \cdot \nabla)v = f$ is bounded. The reader can compare this with Remark 3.1 concerning the gradient of v for a form-bounded b.

The Kato class $\mathbf{K}^{d+1}_\delta$ does not contain a popular class $|b| \in L^d$ (if only because there are vector fields b with $|b| \in L^d$ that destroy two-sided Gaussian bounds on $-\Delta + b \cdot \nabla$), and so it does not contain $\mathbf{F}_\delta$. On the other hand, the Kato class is not contained in $\mathbf{F}_\delta$. However, both form-bounded and Kato class vector fields are contained in $\mathbf{F}^{1/2}_\delta$.

Let us demonstrate one way to arrive at the condition $b \in \mathbf{F}^{1/2}_\delta$, $\delta < 1$. First, let $b \in \mathbf{F}_\delta$, $\delta < 1$, and, for brevity, assume that $c_\delta = 0$. Also, let b be bounded and smooth so that all manipulations with the equation are justified, but the constants in the estimates below will not depend on the smoothness or boundedness of b. One can prove the following two L^2 regularity results for Cauchy problem $(\partial_t - \Delta + b \cdot \nabla)u = 0$, $u(0) = f$ with such b:

- Multiplying $(\partial_t - \Delta + b \cdot \nabla)u = 0$ by u, integrating over $[0, t] \times \mathbb{R}^d$ and applying $b \in \mathbf{F}_\delta$ and quadratic inequality (14.1), we obtain energy inequality

$$\frac{1}{2}\|u(t)\|_2^2 + (1 - \sqrt{\delta}) \int_0^t \|\nabla u\|_2^2 dr \leq \frac{1}{2}\|f\|_2^2. \tag{14.5}$$

– Multiplying $(\partial_t - \Delta + b \cdot \nabla)u = 0$ by $-\Delta u$, integrating over $[0,t] \times \mathbb{R}^d$, we obtain

$$\frac{1}{2}\partial_t \|\nabla u\|_2^2 + \|\Delta u\|_2^2 + \langle b \cdot \nabla u, -\Delta u\rangle = 0,$$

where we further estimate, using $b \in \mathbf{F}_\delta$ with $c_\delta = 0$,

$$\begin{aligned}
|\langle b \cdot \nabla u, -\Delta u\rangle| &\le \varepsilon \||b|\nabla u|\|_2 + \frac{1}{4\varepsilon}\|\Delta u\|_2^2 \\
&\le \varepsilon\delta \|(-\Delta)^{\frac{1}{2}}|\nabla u|\|_2^2 + \frac{1}{4\varepsilon}\|\Delta u\|_2^2 \\
&\quad \text{(use Beurling-Deny-type inequality } \|(-\Delta)^{\frac{1}{2}}|\nabla u|\|_2^2 \\
&\quad \le \|(-\Delta)^{\frac{1}{2}}\nabla u\|_2^2 \equiv \|\Delta u\|_2^2 \\
&\quad \text{and select } \varepsilon = \frac{1}{2\sqrt{\delta}}) \\
&\le \sqrt{\delta}\|\Delta u\|_2^2.
\end{aligned}$$

Thus, we obtain another "energy inequality":

$$\frac{1}{2}\|\nabla u(t)\|_2^2 + (1-\sqrt{\delta})\int_0^t \|\Delta u\|_2^2 dr \le \frac{1}{2}\|\nabla f\|_2^2. \tag{14.6}$$

– Now, one can ask what happens if we multiply $(\partial_t - \Delta + b \cdot \nabla)u = 0$ by an intermediate test function $(-\Delta)^s u$, $0 < s < 1$. One obtains an intermediate result between (14.5) and (14.6), but for a larger class of vector fields b, which becomes maximal if one multiplies by $(-\Delta)^{\frac{1}{2}}u$:

$$\begin{aligned}
|\langle b \cdot \nabla u, (-\Delta)^{\frac{1}{2}}u\rangle| &= |\langle b^{\frac{1}{2}}(-\Delta)^{-\frac{1}{4}}\nabla(-\Delta)^{\frac{1}{4}}u, |b|^{\frac{1}{2}}(-\Delta)^{-\frac{1}{4}}(-\Delta)^{\frac{3}{4}}u\rangle| \\
&\qquad b^{\frac{1}{2}} := b|b|^{-\frac{1}{2}} \\
&\le \||b|^{\frac{1}{2}}(-\Delta)^{-\frac{1}{4}}\|_{2\to 2}\|\nabla(-\Delta)^{\frac{1}{4}}u\|_2 \\
&\quad \||b|^{\frac{1}{2}}(-\Delta)^{-\frac{1}{4}}\|_{2\to 2}\|(-\Delta)^{\frac{3}{4}}u\|_2 \\
&= \||b|^{\frac{1}{2}}(-\Delta)^{-\frac{1}{4}}\|_{2\to 2}^2\|(-\Delta)^{\frac{3}{4}}u\|_2^2.
\end{aligned}$$

Thus, requiring

$$\||b|^{\frac{1}{2}}(-\Delta)^{-\frac{1}{4}}\|_{2\to 2} \le \sqrt{\delta} \quad (\text{i.e. } b \in \mathbf{F}_\delta^{1/2} \text{ with } \lambda = 0), \quad \delta < 1,$$

one obtains the following "energy inequality":

$$\frac{1}{2}\|(-\Delta)^{\frac{1}{4}}u(t)\|_2^2 + (1-\delta)\int_0^t \|(-\Delta)^{\frac{3}{4}}u\|_2^2 dr \le \frac{1}{2}\|(-\Delta)^{\frac{1}{4}}f\|_2^2. \tag{14.7}$$

From the look of (14.7), it is seen that one needs to work with the chain of Bessel spaces

$$\mathcal{W}^{\frac{3}{2},2} \subset \mathcal{W}^{\frac{1}{2},2} \subset \mathcal{W}^{-\frac{1}{2},2} \tag{14.8}$$

rather than the standard Sobolev triple $W^{1,2} \subset L^2 \subset W^{-1,2}$ (above $\mathcal{W}^{-\frac{1}{2},2}$ is the dual of $\mathcal{W}^{\frac{3}{2},2}$ with respect to the inner product in $\mathcal{W}^{\frac{1}{2},2}$) . Of course, by doing that, one sacrifices (3.11) and loses the possibility to consider general operator $-\nabla \cdot a \cdot \nabla + b \cdot \nabla$ unless uniformly elliptic matrix a satisfies additional regularity assumptions that make $-\nabla \cdot a \cdot \nabla$ a bounded operator from $\mathcal{W}^{\frac{3}{2},2}$ to $\mathcal{W}^{-\frac{1}{2},2}$, see Remark 14.1. In fact, the following result is true:

Proposition 14.1 *Let $b \in \mathbf{F}_\delta^{\frac{1}{2}}$, $\delta < 1$. Then for every $f \in \mathcal{W}^{\frac{1}{2},2}$ there exists a unique weak solution to Cauchy problem*

$$(\partial_t + \lambda - \Delta + b \cdot \nabla)u = 0, \quad u(0+) = f \tag{14.9}$$

where λ is from the condition $b \in \mathbf{F}_\delta^{1/2}$, i.e. a unique in $L^\infty_{\rm loc}(]0,\infty[, \mathcal{W}^{\frac{1}{2},2}) \cap L^2_{\rm loc}(]0,\infty[, \mathcal{W}^{\frac{3}{2},2})$ function u satisfying

$$\int_0^\infty \langle(\lambda-\Delta)^{\frac{1}{4}}u, \partial_t(\lambda-\Delta)^{\frac{1}{4}}\varphi\rangle dt = \int_0^\infty \langle(\lambda-\Delta)^{\frac{3}{4}}u, (\lambda-\Delta)^{\frac{3}{4}}\varphi\rangle dt + \int_0^\infty \langle b\cdot\nabla u, (\lambda-\Delta)^{\frac{1}{2}}\varphi\rangle$$

for all $\varphi \in C_c^\infty(]0,\infty[, \mathcal{S})$ and w-$\mathcal{W}^{\frac{1}{2},2}$-$\lim_{t\downarrow 0} u(t) = f$. One has $u \in C(\mathbb{R}_+, \mathcal{W}^{\frac{1}{2},2})$, the following energy inequality holds:

$$\|u(t)\|^2_{\mathcal{W}^{\frac{1}{2},2}} + (1-\delta)\int_0^t \|u\|^2_{\mathcal{W}^{\frac{3}{2},2}} dr \le \|f\|^2_{\mathcal{W}^{\frac{1}{2},2}}, \quad t \ge 0,$$

and $T^t f(\cdot) := u(t,\cdot)$ is a contraction strongly continuous in $\mathcal{W}^{\frac{1}{2},2}$ Markov semigroup. If $\{b_\varepsilon\}_{\varepsilon>0}$ is a family of bounded smooth vector fields such that $b_\varepsilon \in \mathbf{F}_\delta^{1/2}$ with the same λ as b, $b_\varepsilon \to b$ in $[L^1_{\rm loc}]^d$ as $\varepsilon \to 0$, and if u_ε denotes the solution to Cauchy problem (14.9) *with the vector field b_ε, then*

$$u_\varepsilon \to u \quad \text{weakly in } L^2_{\rm loc}(\mathbb{R}_+, \mathcal{W}^{\frac{3}{2},2}) \text{ as } \varepsilon \to 0.$$

The proof uses the standard J.-L. Lions approach in the scale (14.8), and was carried out in [53], in fact, in greater generality: for time-inhomogeneous $b \in L^\infty \mathbf{F}_\delta^{1/2}$, that is, satisfying for a.e. $t \in \mathbb{R}_+$ the operator inequality

$$\| |b(t)|^{\frac{1}{2}} (\lambda - \Delta)^{-\frac{1}{4}} \|_{2 \to 2} \leq \sqrt{\delta}$$

for some fixed $\lambda = \lambda_\delta$.

The above argument leading to the energy inequality (14.7) is not how the class $\mathbf{F}_\delta^{1/2}$ first appeared in the literature. The (L^p, L^q) estimate

$$\|e^{-t\Lambda_p(b)} f\|_q \leq c e^{t\omega_p} t^{-\frac{d}{2}(\frac{1}{p}-\frac{1}{q})} \|f\|_p, \quad t > 0, \quad \omega_p := \frac{c_\delta}{2(p-1)}, \tag{14.10}$$

can be proved separately for $b \in \mathbf{F}_\delta$, see (6.13), and for $b \in \mathbf{K}_\delta^{d+1}$ (as was mentioned above, for the Kato class one even has two-sided Gaussian bounds on the integral kernel $e^{-t\Lambda(b)}(x, y)$ of $e^{-t\Lambda_p(b)}$). It was noticed in [93] that the validity of estimate (14.10) depends, in fact, only on the weaker condition $\| |b|^{\frac{1}{2}} (\lambda - \Delta)^{-\frac{1}{4}} \|_{2 \to 2} \leq \sqrt{\delta}$ with $\delta < 1$, which led to the introduction of the class $\mathbf{F}_\delta^{1/2}$. Also, [93] proposed a way to construct a *quasi bounded* semigroup in L^2 associated with $-\Delta + b \cdot \nabla$ with weakly form-bounded $b \in \mathbf{F}_\delta^{1/2}$ by constructing its resolvent as the operator-valued function

$$\Phi(\zeta, b) := (\zeta - \Delta)^{-\frac{3}{4}} (1 + HS)^{-1} (\zeta - \Delta)^{-\frac{1}{4}}, \qquad \operatorname{Re} \zeta \geq \lambda_\delta,$$

where, by $b \in \mathbf{F}_\delta^{1/2}$, operators

$$H := (\zeta - \Delta)^{-\frac{1}{4}} |b|^{\frac{1}{2}}, \quad S := b^{\frac{1}{2}} \cdot \nabla (\zeta - \Delta)^{-\frac{3}{4}} \tag{14.11}$$

are bounded on L^2 with norm $\sqrt{\delta}$ each (for operator S, taking into account that $\nabla(\zeta - \Delta)^{-\frac{1}{2}}$ is bounded on L^2 with norm 1), and so $\Phi(\zeta, b)$ is bounded on L^2. Then, it is easily seen, $\Phi(\zeta, b)$ satisfies

$$\|\Phi(\zeta, b)\|_{2 \to 2} \leq (1 - \delta)^{-1} |\zeta|^{-1}, \qquad \text{on } \{\operatorname{Re} \zeta \geq \lambda_\delta\}. \tag{14.12}$$

The proof that $\Phi(\zeta, b)$ is indeed the resolvent of the generator Λ of a quasi bounded strongly continuous semigroup[14] $e^{-t\Lambda}$ on L^2 uses the general Trotter approximation theorem. The latter, in practice, requires the uniform in n estimate (14.12) for $\Phi(\zeta, b_n)$, where b_n are approximating vector fields for b (cf. (15.6), (15.7)). In other words, it is essential for the construction that one is working with a holomorphic semigroup. We refer to [47] for detailed discussion.

[14] But not a quasi contraction semigroup, as in Sect. 6.

Note also that one no longer has (3.6), i.e.

$$\Lambda \not\supset -\Delta + b\cdot\nabla \upharpoonright C_c^\infty.$$

The reason is that for a weakly form-bounded b its norm $|b|$ is in general not in $L^2_{\rm loc}$.

If b is form-bounded or in the Kato class of vector fields – two standard assumptions – then one can construct a realization of $-\Delta + b\cdot\nabla$ as the generator of a strongly continuous semigroup *in some* L^p by invoking the KLMN theorem in L^2 or the Miyadera theorem in L^1, respectively (see e.g. [47]). However these two theorems (and, generally speaking, the standard perturbation-theoretic tools) are inapplicable to $-\Delta + b\cdot\nabla$ in any L^p if b is in the class of weakly form-bounded drifts $\mathbf{F}_\delta^{1/2}$.

Remark 14.1 Let us comment on the assumptions on a measurable uniformly elliptic matrix a (i.e. $\sigma I \leq a \leq \xi I$ a.e. on $\mathbb{R}^d$ for $0 < \sigma \leq \xi < \infty$) that would allow to extend Proposition 14.1 to operator $-\nabla\cdot a\cdot\nabla + b\cdot\nabla$. If $b \in \mathbf{F}_\delta^{1/2}$, then it is easily seen that

$$b\cdot\nabla \in \mathcal{B}(\mathcal{W}^{\frac{3}{2},2}, \mathcal{W}^{-\frac{1}{2},2}).$$

The matrix a has to be such that

$$-\nabla\cdot a\cdot\nabla \in \mathcal{B}(\mathcal{W}^{\frac{3}{2},2}, \mathcal{W}^{-\frac{1}{2},2}) \tag{14.13}$$

(of course, if a is only measurable uniformly elliptic, then one only has $-\nabla\cdot a\cdot\nabla \in \mathcal{B}(W^{1,2}, W^{-1,2})$). Let us mention one elementary sufficient condition for (14.13). For simplicity we will stay at the a priori level (i.e. the matrix is smooth but the norm of the operator in (14.13) does not depend on smoothness of a). Also, assume that $a = I + a^0$ where a^0 has entries a^0_{ij} in $\mathcal{S}$. For given $\varphi \in \mathcal{W}^{\frac{3}{2},2}$, $\psi \in \mathcal{W}^{\frac{1}{2},2}$, we have

$$\begin{aligned}|\langle -\nabla\cdot a^0\cdot\nabla\varphi, \psi\rangle| &= |\langle (1-\Delta)^{\frac{1}{4}} a^0 (1-\Delta)^{-\frac{1}{4}}\cdot(1-\Delta)^{\frac{1}{4}}\nabla\varphi, (1-\Delta)^{-\frac{1}{4}}\nabla\psi\rangle| \\ &\leq \|(1-\Delta)^{\frac{1}{4}} a^0 (1-\Delta)^{-\frac{1}{4}}\|_{2\to 2}\|\varphi\|_{\mathcal{W}^{\frac{3}{2},2}}\|\psi\|_{\mathcal{W}^{\frac{1}{2},2}},\end{aligned}$$

where, in turn, by the Kato-Ponce inequality ($\equiv$ fractional Leibnitz rule) [30], for all $1 \leq i, j \leq d$,

$$\begin{aligned}\|(1-\Delta)^{\frac{1}{4}} a^0_{ij} (1-\Delta)^{-\frac{1}{4}} f\|_2 \leq \|(1-\Delta)^{\frac{1}{4}} a^0_{ij}\|_{2d}\|(1-\Delta)^{-\frac{1}{4}} f\|_{\frac{2d}{d-1}} \\ + \|a_{ij}\|_\infty\|f\|_2.\end{aligned}$$

Thus, if

$$\|a^0_{ij}\|_{\mathcal{W}^{\frac{1}{2},2d}} \leq c < \infty$$

for all i, j, for a generic constant c (i.e. a constant that does not depend on the smoothness of a^0_{ij}), then, using $\|(1-\Delta)^{-\frac{1}{4}} f\|_{\frac{2d}{d-1}} \leq C\|f\|_2$, we obtain $\|(1-\Delta)^{\frac{1}{4}} a^0 (1-\Delta)^{-\frac{1}{4}}\|_{2\to 2} \leq c'$ for a generic c', and hence (14.13) with the operator norm bounded by a generic constant.

Remark 14.2 At the level of Feller semigroups, one cannot really draw a parallel between the Kato class of potentials $\mathbf{K}^d_\delta$ (see (3.9)) and the Kato class of drifts $\mathbf{K}^{d+1}_\delta$. Indeed, in view of the results [82], for the Schrödinger operator $-\Delta + V$ condition $V \in \mathbf{K}^d_\delta$ is, basically, necessary and sufficient for the Feller property ($\equiv$ strong continuity of the semigroup on C_∞) to hold. For the Kolmogorov operator $-\Delta + b \cdot \nabla$, condition $b \in \mathbf{K}^{d+1}_\delta$ is only sufficient for the Feller property; as Theorem 15.1 below shows, one can go much farther, to weakly form-bounded drifts.

15 Weakly Form-Bounded Drifts and SDEs

In this section we construct the Feller semigroup for $-\Delta + b \cdot \nabla$ and prove weak well-posedness of SDE

$$X_t = x - \int_0^t b(X_r)dr + \sqrt{2}W_t, \quad t \geq 0, \tag{15.1}$$

for a fixed $x \in \mathbb{R}^d$, with $b : \mathbb{R}^d \to \mathbb{R}^d$ in the class of *weakly form-bounded vector fields* $\mathbf{F}^{1/2}_\delta$.

1. We construct the sought Feller generator by arguing essentially as in the proof of Theorem 6.1 (so, in particular, we do not use L^2 theory of $-\Delta + b \cdot \nabla$, $b \in \mathbf{F}^{1/2}_\delta$). First, we prove an analogue of Lemma 6.1 for weakly form-bounded drifts. Namely, for given $p \in]1, \infty[$, $1 \leq r < p < q < \infty$ and $\mu > 0$, define operators

$$\begin{aligned} G_p(r) &:= b^{\frac{1}{p}} \cdot \nabla(\mu - \Delta)^{-\frac{1}{2} - \frac{1}{2r}}, \\ Q_p(q) \upharpoonright \mathcal{E} &:= (\mu - \Delta)^{-\frac{1}{2} + \frac{1}{2q}} |b|^{1 - \frac{1}{p}}, \\ T_p \upharpoonright \mathcal{E} &:= b^{\frac{1}{p}} \cdot \nabla(\mu - \Delta)^{-1} |b|^{1 - \frac{1}{p}}. \end{aligned}$$

where, recall, $\mathcal{E} := \bigcup_{\varepsilon > 0} e^{-\varepsilon |b|} L^p$ is a dense subspace of L^p. Notice that the power $\frac{2}{p}$ in the definition of operators $G_p(r)$, T_p, $Q_p(q)$ in Lemma 6.1 is now replaced with $\frac{1}{p}$. Set

$$m_d := \pi^{\frac{1}{2}} (2e)^{-\frac{1}{2}} d^{\frac{d}{2}} (d-1)^{\frac{1-d}{2}}, \qquad c_p := pp'/4.$$

Lemma 15.1 *Let $b \in \mathbf{F}_\delta^{1/2}$. For every $p \in]1,\infty[$, the following is true for all $\mu \geq \kappa_d \lambda_\delta$,*

(i) $T_p \upharpoonright \mathcal{E}$ admits extension by continuity to L^p, denoted by T_p. One has

$$\|T\|_{p\to p} \leq m_d c_p \delta.$$

In particular, if δ satisfies $m_d\delta < 1$, then for every $p \in I_\delta :=]\frac{2}{1+\sqrt{1-m_d\delta}}, \frac{2}{1-\sqrt{1-m_d\delta}}[$ one has $\|T_p\|_{p\to p} < 1$.

(ii) $Q_p(q) \upharpoonright \mathcal{E}$ admits extension by continuity to L^p, denoted by $Q_p(q)$.

(iii) $G_p(r)$ is bounded on L^p.

Lemma 15.1 was proved in [37]. Let us demonstrate the proof of (*i*) to make it easier to compare Lemma 15.1 with Lemma 6.1 (proved in Appendix A). Define in L^2 operator $A = (\mu - \Delta)^{\frac{1}{2}}$, $D(A) = W^{1,2}$. This is a *symmetric* Markov generator. Therefore, we have for $p \in]1,\infty[$:

$$0 \leqslant u \in D(A_p) \quad \Rightarrow \quad u^{\frac{p}{2}} \in D(A^{\frac{1}{2}})$$

and the following inequality (sometimes called the Stroock-Varopoulos inequality) is valid:

$$c_p^{-1}\|A^{\frac{1}{2}}u^{\frac{p}{2}}\|_2^2 \leqslant \langle A_p u, u^{p-1}\rangle, \quad c_p := \frac{pp'}{4}, \quad p' = \frac{p}{p-1} \tag{15.2}$$

(see [71, Theorem 2.1], see also [52] for a useful vector-valued analogue of these inequalities). Here A_p is the generator of strongly continuous semigroup $e^{-tA_p} := [e^{-tA} \upharpoonright L^2 \cap L^p]^{\text{clos}}_{L^p \to L^p}$, cf. discussion in the beginning of the previous section. Now, let u be the solution of equation[15] $A_p u = |b|^{1/p'}|f|$, $f \in \mathcal{E}$. The condition $b \in \mathbf{F}_\delta^{1/2}$ yields, provided $\mu \geq \lambda_\delta$,

$$\||b|^{\frac{1}{2}}u^{\frac{p}{2}}\|_2^2 \leq \delta\|A^{\frac{1}{2}}u^{\frac{p}{2}}\|_2^2.$$

Hence, by (15.2),

$$(c_p\delta)^{-1}\||b|^{\frac{1}{2}}u^{\frac{p}{2}}\|_2^2 \leqslant \langle A_p u, u^{p-1}\rangle.$$

Now, noting that $\||b|^{\frac{1}{2}}u^{\frac{p}{2}}\|_2^2 = \||b|^{\frac{1}{p}}u\|_p^p$ and using $A_p u = |b|^{\frac{1}{p'}}|f|$, we obtain

$$\||b|^{\frac{1}{p}}u\|_p^p \leqslant c_p\delta\|f\|_p\||b|^{\frac{1}{p}}u\|_p^{p-1}.$$

---

[15] We can also carry out the proof of Lemma 15.1 for bounded b_n as e.g. defined below, and then pass to the limit in n.

Thus, $\||b|^{\frac{1}{p}}u\|_p \leqslant c_p\delta\|f\|_p$, so we arrive at the estimate

$$\||b|^{\frac{1}{p}}A^{-1}|b|^{\frac{1}{p'}}|f|\|_p \leq c_p\delta\|f\|_p. \tag{15.3}$$

To end the proof of (i), it remains to apply in the definition of T_p the pointwise inequality (this is where the constant m_d comes from)

$$|\nabla_x(\mu - \Delta)^{-1}(x, y)| \leqslant m_d(\kappa_d^{-1}\mu - \Delta)^{-\frac{1}{2}}(x, y), \quad x, y \in \mathbb{R}^d,\ x \neq y, \tag{15.4}$$

where

$$\kappa_d := \frac{d}{d-1},$$

and then apply (15.3) to the result.

Remark 15.1 Similar estimates, without the gradient, were used earlier in [8, 71] in the study of Schrödinger operators with form-bounded potentials.

Remark 15.2 By applying (15.4) in the definition of T_p we kill the gradient from the gradient term $b \cdot \nabla$. This allows us to apply to what is left the inequalities for symmetric Markov generators. By the way, this is why the interval I_δ of admissible p in Theorem 15.1 is symmetric, despite the fact that the operator $-\Delta+b\cdot\nabla$ is non-symmetric. In the proof of Lemma 6.1, when dealing with the condition $b \in \mathbf{F}_\delta$, we control the gradient in a more efficient way, which allows to impose less restrictive assumptions on δ than in Lemma 15.1. It is not clear at the moment how to prove Lemma 15.1 without either resorting to (15.4) or restricting the class $\mathbf{F}_\delta^{1/2}$ to Morrey class $M_{1+\varepsilon}$, $\varepsilon > 0$ (cf. the proof of Lemma 16.1 below).

The interval I_δ expands to $]1, \infty[$ as $\delta \downarrow 0$. In particular, if δ is sufficiently small, I_δ contains $p > d - 1$, which is what will be needed to construct the resolvent of a Feller generator in terms of $Q_p(q)$, T_p, $G_p(r)$ and some "free" Bessel potentials using the Sobolev embedding theorem. This is what is done in Theorem 15.1 below.

Set

$$b_n := c_n\eta_{\varepsilon_n} * (\mathbf{1}_n b), \tag{15.5}$$

where $\mathbf{1}_n$ is the indicator of $\{x \mid |x| \leq n, |b(x)| \leq n\}$, η_{ε_n} is the Friedrichs mollifier, and we choose $\varepsilon_n \downarrow 0$ (sufficiently rapidly) so that, for appropriate $c_n \uparrow 1$ (sufficiently slow), one has

$$b_n \to b \quad \text{in } L^1_{\text{loc}}(\mathbb{R}^d, \mathbb{R}^d) \tag{15.6}$$

and

$$b_n \in \mathbf{F}_\delta^{1/2} \quad \text{with some } \lambda_\delta \text{ independent of } n = 1, 2, \dots \tag{15.7}$$

see Appendix C.1. The following theorem was proved in [38].

Theorem 15.1 *Let $b \in \mathbf{F}_\delta^{\frac{1}{2}}$, $m_d\delta < 1$. The following is true for all $\mu \geq \kappa_d \lambda_\delta$.*

(i) *For every $p \in I_\delta =]\frac{2}{1+\sqrt{1-m_d\delta}}, \frac{2}{1-\sqrt{1-m_d\delta}}[$, the function*

$$\begin{aligned} u &= (\mu - \Delta)^{-1} f \\ &\quad - (\mu - \Delta)^{-\frac{1}{2}-\frac{1}{2q}} Q_p(q)(1+T_p)^{-1} G_p(r)(\mu - \Delta)^{-\frac{1}{2}+\frac{1}{2r}} f, \quad f \in L^p \end{aligned} \tag{15.8}$$

is a weak solution to the elliptic equation

$$(\lambda - \Delta + b \cdot \nabla) u = f, \tag{15.9}$$

i.e. $b \cdot \nabla u \in L^1_{\mathrm{loc}}$

$$\mu\langle u, \psi\rangle + \langle \nabla u, \nabla \psi\rangle + \langle b \cdot \nabla u, \psi \rangle = \langle f, \psi\rangle \quad \text{for all } \psi \in \mathcal{S}.$$

Moreover, if $f \in L^p \cap L^2$, then u is the unique in $\mathcal{W}^{\frac{3}{2},2}$ weak solution to (15.9).

(ii) *It follows from* (15.8) *that*

$$u \in \mathcal{W}^{1+\frac{1}{q},p}, \quad q > p.$$

In particular, if $m_d\delta < \frac{4(d-2)}{(d-1)^2}$, then in the interval $p \in I_\delta$ we can select $p > d-1$, and then select q sufficiently close to p, so that by the Sobolev embedding theorem u is Hölder continuous.

(iii) *The operator-valued function in* (15.8)

$$\Theta_p(\mu, b) := (\mu - \Delta)^{-1} - (\mu - \Delta)^{-\frac{1}{2}-\frac{1}{2q}} Q_p(q)(1+T_p)^{-1} G_p(r)(\mu - \Delta)^{-\frac{1}{2}+\frac{1}{2r}}$$

on $\{\mu > \mu_0\}$ takes values in $\mathcal{B}(\mathcal{W}^{-1+\frac{1}{r},p}, \mathcal{W}^{1+\frac{1}{q},p})$.

(iv) *Let δ satisfy $m_d\delta < \frac{4(d-2)}{(d-1)^2}$. Fix $p \in I_\delta$ such that $p > d-1$. Then*

$$(\mu + \Lambda_{C_\infty}(b))^{-1} := \big(\Theta_p(\mu, b) \upharpoonright L^p \cap C_\infty\big)^{\mathrm{clos}}_{C_\infty \to C_\infty}, \quad \mu \geq \kappa_d \lambda,$$

determines the resolvent of a Feller generator on C_∞. This semigroup satisfies

$$e^{-t\Lambda_{C_\infty}(b)} = s\text{-}C_\infty\text{-}\lim_n e^{-t\Lambda_{C_\infty}(b_n)} \quad \text{locally uniformly in } t \geq 0,$$

where b_n are defined by (15.5)*, and operators $\Lambda_{C_\infty}(b_n) := -\Delta + b_n \cdot \nabla$ with domain $D(\Lambda_{C_\infty}(b_n)) := (1-\Delta)^{-1}C_\infty$ are Feller generators.*

(v) *Feller semigroup $e^{-t\Lambda_{C_\infty}(b)}$ is conservative, i.e. its integral kernel $e^{-t\Lambda_{C_\infty}}(x,\cdot)$ satisfies*

$$\langle e^{-t\Lambda_{C_\infty}(b)}(x,\cdot)\rangle = 1 \quad \textit{for all } x \in \mathbb{R}^d,\ t>0.$$

For the proof, except for the part that concerns the weak solution to the elliptic equation, one can repeat the proof of Theorem 6.1 using Lemma 15.1 instead of Lemma 6.1.

Remark 15.3 The fact that u is a weak solution was proved in [38]. Moreover, since $f \in L^2 \cap L^p$, we have

$$u = \Theta_2(\mu, b)f = \Phi_2(\mu, b)f,$$

so $u \in \mathcal{W}^{\frac{3}{2},2}$. (We could also prove that if in Theorem 15.1 $p=2$, then one can take $q=r=2$.) The proof of the uniqueness of u in $\mathcal{W}^{\frac{3}{2},2}$ goes as follows. Let $v \in \mathcal{W}^{\frac{3}{2}}$ be some weak solution of (15.9). Then, selecting $\varphi = (\mu-\Delta)^{-\frac{1}{2}}\eta$, $\eta \in C_c^\infty$, we have

$$\langle (\mu-\Delta)^{\frac{3}{4}}v, (\mu-\Delta)^{\frac{3}{4}}\eta\rangle + \langle S(\mu-\Delta)^{\frac{3}{4}}u, H^*(\mu-\Delta)^{\frac{3}{4}}\eta\rangle = \langle f, (\mu-\Delta)^{-\frac{1}{2}}\eta\rangle,$$

where $H = (\mu-\Delta)^{-\frac{1}{4}}|b|^{\frac{1}{2}}$, $S = b^{\frac{1}{2}}\cdot\nabla(\mu-\Delta)^{-\frac{3}{4}}$ are bounded on L^2 and $\|S\|_{2\to 2}$, $\|H^*\|_{2\to 2} \leq \sqrt{\delta}$ (for all $\mu \geq \lambda_\delta$ and thus for all $\mu \geq \kappa_d\lambda$), see (14.11) and the discussion after that formula. Thus, the quadratic form

$$\tau[v,\eta] := \langle (\mu-\Delta)^{\frac{3}{4}}v, (\mu-\Delta)^{\frac{3}{4}}\eta\rangle + \langle S(\mu-\Delta)^{\frac{3}{4}}v, H^*(\mu-\Delta)^{\frac{3}{4}}\eta\rangle$$

is bounded and coercive on $\mathcal{W}^{\frac{3}{2},2}$ (endowed with the norm $\|\eta\|_{\mathcal{W}^{\frac{3}{2},2}} := \|(\mu-\Delta)^{\frac{3}{2}}\eta\|_2$):

$$|\tau[v,\eta]| \leq (1+\delta)\|v\|_{\mathcal{W}^{\frac{3}{2},2}}\|\eta\|_{\mathcal{W}^{\frac{3}{2},2}}$$

and

$$|\tau[v,\eta]| \geq (1-\delta)\|v\|_{\mathcal{W}^{\frac{3}{2},2}}\|\eta\|_{\mathcal{W}^{\frac{3}{2},2}},$$

(the assumptions of Theorem 15.1 imply, of course, that $\delta<1$). Extending τ to $\mathcal{W}^{\frac{3}{2},2}$, and returning to the discussion of the weak solution $u \in \mathcal{W}^{\frac{3}{2},2}$ constructed in (15.8), we obtain that $\tau[u-v,\eta]=0$ for all $\eta \in \mathcal{W}^{\frac{3}{2},2}$, but on the other hand, by coercivity, $\tau[u-v,u-v] \geq (1-\delta)\|u-v\|^2_{\mathcal{W}^{\frac{3}{2},2}}$. Thus, $u=v$.

Remark 15.4 The operator-valued function $\mu \mapsto \Theta_p(\mu, b) \in \mathcal{B}(L^p)$ in Theorem 15.1 is the resolvent of the generator of a *quasi bounded* semigroup in L^p. This was proved in [38].

Remark 15.5 There is a non-trivial difference between the resolvent representations Φ_2 and Θ_p. For instance, Θ_p is nonlinear in $|b|$ even if $p = 2$, however Φ_2 is linear $|b|$. This circumstance was used in [39] to extend Theorem 15.1 to measure-valued drifts of the form

$$b = \mathsf{f}dx + \mathsf{h}, \tag{15.10}$$

where f is a vector field in $\mathbf{F}_\delta^{1/2}$ and h is a $\mathbb{R}^d$-valued measure in the Kato class $\bar{\mathbf{K}}_\delta^{d+1}$, i.e.

$$\sup_{x \in \mathbb{R}^d} \int_{\mathbb{R}^d} (\lambda - \Delta)^{-\frac{1}{2}}(x, y) |\mathsf{h}|_1(dy) \leq \delta \tag{15.11}$$

for some $\lambda = \lambda_\delta$. Here $|\mathsf{h}|_1$ denotes the sum of total variations of the components of h. This class contains, of course, the Kato class of vector fields $\mathbf{K}_\delta^{d+1}$ defined in the examples above. We could also define the class of weakly form-bounded measure-valued drifts. Indeed, Definition 14.1 can be stated as

$$\int_{\mathbb{R}^d} |b(x)|_1 \big[(\lambda - \Delta)^{-\frac{1}{4}} f(x)\big]^2 dx \leq \delta \|f\|_2^2, \quad f \in \mathcal{S},$$

where $|b|_1$ is the sum of absolute values of the components of b. Replacing absolute values by total variations, we arrive at a more general condition

$$\int_{\mathbb{R}^d} |b(dx)|_1 \big[(\lambda - \Delta)^{-\frac{1}{4}} f(x)\big]^2 \leq \delta \|f\|_2^2. \tag{15.12}$$

An example of a measure-valued b satisfying (15.12) is (15.10). In [39], the additional constraint that b must be of the form (15.10) comes from the construction of a regularization of b that preserves weak form-bound δ. Although from purely analytic point of view (15.12) is an L^1 condition on $|b|$, from the operator-theoretic point of view (15.12) is still an L^2 condition, i.e. is an (L^2, L^2) operator norm inequality. Thus, (15.10) may be viewed as a way that (15.12) filters out the singular measure component h, because the latter satisfies, in the dual formulation of the Kato class, an (L^1, L^1) operator norm inequality (cf. (14.4)).

Let us add that, at the level of SDEs, Bass-Chen [6] and Kim-Song [36] considered Kato class measure-valued drifts.

2. We now state our result on weak well-posedness of SDEs with weakly form-bounded drifts.

Theorem 15.2 *Let $b \in \mathbf{F}_\delta$ with $\delta < \frac{4(d-2)}{(d-1)^2}$. Let $e^{-t\Lambda_{C_\infty}(b)}$ be the Feller semigroup constructed in Theorem 15.1. Fix $T > 0$. The following is true:*

(*i*) *There exist probability measures* $\{\mathbb{P}_x\}_{x\in\mathbb{R}^d}$ *on the canonical space* $(C[0,T],\mathcal{B}_t)$ *such that*

$$\mathbb{E}_{\mathbb{P}_x}[f(X_t)] = (e^{-t\Lambda_{C_\infty}(b)}f)(x), \quad f\in C_\infty, \quad x\in\mathbb{R}^d.$$

For every $x\in\mathbb{R}^d$ *the measure* $\mathbb{P}_x$ *is a weak solution to SDE*

$$X_t = x - \int_0^t b(X_r)dr + \sqrt{2}W_t, \quad 0\le t\le T. \tag{15.13}$$

(*ii*) *If* $\{\mathbb{Q}_x\}_{x\in\mathbb{R}^d}$ *is another weak solution to* (6.15) *such that*

$$\mathbb{Q}_x = w\text{-}\lim_n \mathbb{P}_x(\tilde{b}_n) \quad \textit{for every } x\in\mathbb{R}^d,$$

for some $\{\tilde{b}_n\}\subset\mathbf{F}_{\delta_1}$ *with* $\delta<\frac{4(d-2)}{(d-1)^2}$ *and* λ_δ *independent of* n, *then* $\{\mathbb{Q}_x\}_{x\in\mathbb{R}^d} = \{\mathbb{P}_x\}_{x\in\mathbb{R}^d}$.

This result was proved in [46]. In fact, in the series of papers on well-posedness of SDEs with form-bounded and form-bounded-type drifts that are discussed in this work [46] appeared first. In turn, [46] was born out of the attempts to obtain a more detailed description of the corresponding Feller semigroup (i.e. the one constructed earlier in [38]).

Remark 15.6 There are two reasons why one might want to assume form-boundedness of $b=b(x)$ and not its weak form-boundedness: the possibility to include discontinuous diffusion coefficients as in Sect. 11 and less restrictive assumptions on δ. (One can compare, using $\mathbf{F}_\delta\subset\mathbf{F}^{1/2}_{\sqrt{\delta}}$, the assumptions on δ in Theorem 6.2 and in Theorem 15.2)

The pointwise estimate (15.4) is also valid for the resolvent of $-\nabla\cdot a\cdot\nabla$ provided that the uniformly elliptic matrix a is Hölder continuous. If we were to extend Theorem 15.2 to non-constant diffusion coefficients in the spirit of Sect. 11, then we could require that a has Hölder continuous entries whose derivatives are weakly form-bounded.

For a form-bounded drift $b\in\mathbf{F}_\delta$, we had two types of gradient bounds on solution u to $(\mu-\Delta+b\cdot\nabla)u=f$, $f\in C_c^\infty$ (let us assume here, for simplicity, that b is bounded and smooth, so we discuss gradient bounds with constants that do not depend on boundedness or smoothness of b but depend only on δ and λ_δ). That is, we had

$$\|(\mu-\Delta)^{\frac{1}{2}+\frac{1}{q}}u\|_p^p \le K_1\|f\|_p^p, \quad q>p, \tag{15.14}$$

(Theorem 6.1(*ii*)) and

$$\|\nabla|\nabla u|^{\frac{p}{2}}\|_2^2 \le K_2\|f\|_p^p, \tag{15.15}$$

proved in [60] using test function $\varphi = -\nabla \cdot (\nabla u |\nabla u|^{p-2})$. In both estimates $p \in [2, \frac{2}{\sqrt{\delta}}[$. These estimates were discussed in Remark 6.1.

Theorem 15.2(*ii*) provides an analogue of (15.14) for weakly form-bounded $b \in \mathbf{F}_\delta^{1/2}$:

$$\|(\mu - \Delta)^{\frac{1}{2}+\frac{1}{2q}} u\|_p^p \leq K_1 \|f\|_p^p, \quad q > p, \tag{15.16}$$

for $p \in I_\delta$. Does there exist an analogue of (15.15) for weakly form-bounded $b \in \mathbf{F}_\delta^{1/2}$? The answer is "yes". Let us note first that, by the solution representation (15.8),

$$(1 - c_p m_d \delta) \| |b|_1^{-\frac{1}{p'}} (\mu - \Delta) u \|_p \leq \| |b|_1^{-\frac{1}{p'}} f \|_p, \quad |b|_1 := |b| + 1 \tag{15.17}$$

(cf. Remark 6.2), where $c_p m_d \delta < 1$ since $p \in I_\delta$. Without loss of generality, p is rational with odd denominator, so that we can raise functions taking negative values to power p. Also, since all our assumptions on δ are strict inequalities, we may assume, without loss of generality, that $\|T_p(\mu, |b|_1)\|_{p \to p} \leq m_d c_p \delta$ for μ sufficiently large. (Here $T_p(\mu, |b|_1) = |b|_1^{1/p} (\mu - \Delta)^{-\frac{1}{2}} |b|_1^{1/p'}$). Now, we multiply equation $(\mu - \Delta + b \cdot \nabla) u = f$ by test function

$$\varphi := [(\mu - \Delta)^{\frac{1}{2}} u]^{p-1}$$

and integrate:

$$\langle (\mu - \Delta) u, [(\mu - \Delta)^{\frac{1}{2}} u]^{p-1} \rangle + \langle b \cdot \nabla, [(\mu - \Delta)^{\frac{1}{2}} u]^{p-1} \rangle = \langle f, [(\mu - \Delta)^{\frac{1}{2}} u]^{p-1} \rangle.$$

We treat each term separately:

1. We have

$$\langle b \cdot \nabla u, [(\mu - \Delta)^{\frac{1}{2}} u]^{p-1} \rangle = \langle b^{\frac{1}{p}} \cdot \nabla (\mu - \Delta)^{-1} |b|_1^{\frac{1}{p'}} |b|_1^{-\frac{1}{p'}} (\mu - \Delta) u,$$
$$\left[|b|^{\frac{1}{p}} (\mu - \Delta)^{\frac{1}{2}} u \right]^{p-1} \rangle$$
$$= \langle b^{\frac{1}{p}} \cdot \nabla (\mu - \Delta)^{-1} |b|_1^{\frac{1}{p'}} |b|_1^{-\frac{1}{p'}} (\mu - \Delta) u,$$
$$\left[|b|^{\frac{1}{p}} (\mu - \Delta)^{-\frac{1}{2}} |b|_1^{\frac{1}{p'}} |b|_1^{-\frac{1}{p'}} (\mu - \Delta) u \right]^{p-1} \rangle$$

(we use (15.4) and apply Hölder's inequality)

$$\leq m_d \| |b|^{\frac{1}{p}} (\kappa_d^{-1} \mu - \Delta)^{-\frac{1}{2}} |b|_1^{\frac{1}{p'}} \|_{p \to p} \| |b|^{\frac{1}{p}} (\mu - \Delta)^{-\frac{1}{2}} |b|_1^{\frac{1}{p'}} \|_{p \to p}^{p-1}$$
$$\| |b|_1^{-\frac{1}{p'}} (\mu - \Delta) u \|_p^p$$

$$\leq m_d \|T_p(\kappa_d^{-1}\mu, |b|_1)\|_{p\to p} \|T_p(\mu, |b|_1)\|_{p\to p}^{p-1}$$

$$\||b|_1^{-\frac{1}{p'}}(\mu - \Delta)u\|_p^p$$

$$\leq m_d c_p^p \delta^p \||b|_1^{-\frac{1}{p'}}(\mu - \Delta)u\|_p^p$$

(we apply (15.17))

$$\leq m_d c_p^p \delta^p (1 - c_p m_d \delta)^{-p} \|(|b| + 1)^{-\frac{1}{p'}} f\|_p^p \leq m_d c_p^p \delta^p (1 - c_p m_d \delta)^{-p} \|f\|_p^p.$$

2. Next,

$$\langle (\mu - \Delta)u, [(\mu - \Delta)^{\frac{1}{2}}u]^{p-1}\rangle = \langle (\mu - \Delta)^{\frac{1}{2}} (\mu - \Delta)^{\frac{1}{2}}u, [(\mu - \Delta)^{\frac{1}{2}}u]^{p-1}\rangle$$

$((\mu - \Delta)^{\frac{1}{2}}$ is a symmetric Markov generator, so we apply (15.2))

$$\geq \frac{4}{pp'}\|(\mu - \Delta)^{\frac{1}{4}}[(\mu - \Delta)^{\frac{1}{2}}u]^{\frac{p}{2}}\|_2^2$$

$$\geq \frac{2}{pp'}\|(\lambda - \Delta)^{\frac{1}{4}}[(\mu - \Delta)^{\frac{1}{2}}u]^{\frac{p}{2}}\|_2^2 + \frac{C}{pp'}\|(\lambda - \Delta)^{\frac{1}{2}}u\|_p^p,$$

where constant C is from the fractional Sobolev embedding theorem.

3. Also,

$$\langle f, [(\lambda - \Delta)^{\frac{1}{2}}u]^{p-1}\rangle \leq \|f\|_p \|(\lambda - \Delta)^{\frac{1}{2}}u\|_p^{p-1}.$$

Combining 1–3, we obtain the following result. Let $b \in \mathbf{F}_\delta^{1/2}$ with $m_d\delta < 1$. Let $p \in I_\delta$. Then

$$\|(\lambda - \Delta)^{\frac{1}{4}}[(\lambda - \Delta)^{\frac{1}{2}}u]^{\frac{p}{2}}\|_2^2 \leq K\|f\|_p^p \tag{15.18}$$

for λ sufficiently large. The estimate (15.18) is the analogue of (15.15) for $b \in \mathbf{F}_\delta^{1/2}$. Note that it gives the same Hölder continuity of u as (15.16), cf. Remark 6.1.

The proof of Lemma 15.1, and hence the proof of Theorem 15.1, use inequalities for *symmetric* Markov generators, see Remark 15.2, and thus depend in an essential manner on the fact that we are working in the elliptic setting. Below we will treat time-inhomogeneous drifts at expense of restricting the class $\mathbf{F}_\delta^{1/2}$, but, from some points of view, not by much. At the same time, we will substantially strengthen all aspects of Theorems 9.1 and 10.1 except for their assumptions on δ.

16 Time-Inhomogeneous Drifts in Morrey Class

In this section we consider drifts in the parabolic Morrey class E_q with integrability parameter $q > 1$ that can be chosen arbitrarily close to 1. Define parabolic cylinder

$$C_r(t,x) := \{(s,y) \in \mathbb{R}^{d+1} \mid t \le s \le t + r^2, |x-y| \le r\}$$

and, given a vector field $b : \mathbb{R}^{d+1} \to \mathbb{R}^d$ with components in $L^q_{\rm loc}(\mathbb{R}^{d+1})$, $q \in [1, d+2]$, set

$$\begin{aligned}\|b\|_{E_q} &:= \sup_{r>0, z\in\mathbb{R}^{d+1}} r\left(\frac{1}{|C_r|}\int_{C_r(z)} |b(t,x)|^q dtdx\right)^{\frac{1}{q}} \\ &= \sup_{r>0, z\in\mathbb{R}^{d+1}} r\left(\frac{1}{|C_r|}\int_{C_r(z)} |b(-t,x)|^q dtdx\right)^{\frac{1}{q}}.\end{aligned}$$

Definition We say that a vector field b belongs to the parabolic Morrey class E_q if $\|b\|_{E_q} < \infty$.

One has

$$\|b\|_{E_q} \le \|b\|_{E_{q_1}} \quad \text{if } q < q_1.$$

So, the smaller is q the larger is Morrey class E_q.

If above $b = b(x)$, then one obtains the usual elliptic Morrey class M_q defined earlier.

Examples

1. The critical Ladyzhenskaya-Prodi-Serrin class

$$|b| \in L^l(\mathbb{R}, L^p(\mathbb{R}^d)), \quad p \ge d, l \ge 2, \quad \frac{d}{p} + \frac{2}{l} \le 1$$

is contained in E_q. To prove the inclusion it suffices to consider only the cases $l = 2$, $p = \infty$ and $l = \infty$, $p = d$ (see the argument in Appendix B(3)). In the former case the inclusion is trivial, in the latter case the inclusion follows using Hölder's inequality.

This example is strengthened in the next two examples.

2. Let $|b| \in L^{2,w}(\mathbb{R}, L^\infty(\mathbb{R}^d))$. Here and below, $L^{p,w}$ denotes weak Lebesgue spaces (Appendix B(4)). Then $b \in E_q$, $1 < q < 2$. Indeed, by a well

known characterization of weak Lebesgue spaces, we have, setting $\tilde{b}(t) := \|b(t,\cdot)\|_{L^\infty(\mathbb{R}^d)}$,

$$r\left(\frac{1}{|C_r|}\int_{C_r}|b|^q dz\right)^{\frac{1}{q}} \leq Cr\left(\frac{1}{r^2}\int_t^{t+r^2}|\tilde{b}|^q ds\right)^{\frac{1}{q}} \leq C\|\tilde{b}\|_{L^{2,w}(\mathbb{R})}.$$

Hence, for example, a vector field b that satisfies

$$\|b(t,\cdot)\|_{L^\infty(\mathbb{R}^d)} \leq \frac{C}{\sqrt{t}}, \quad t>0 \tag{16.1}$$

(and defined to be zero for $t \leq 0$) is in E_q with $1<q<2$.

This example shows that the parabolic Morrey class E_q with $1<q<2$ contains vector fields that can have stronger singularities in the time variable than the vector fields in $L^\infty \mathbf{F}_\delta + L^2_{\text{loc}}(\mathbb{R})$ considered in the previous section. Namely, if, for simplicity, b depends only on time, that it will be in E_q, $1<q<2$ if e.g. $|b(t)| \in L^{2,w}(\mathbb{R})$ (as (16.1) above). However, to be in $L^\infty \mathbf{F}_\delta + L^2_{\text{loc}}(\mathbb{R})$ it would have to satisfy more restrictive condition $|b(t)| \in L^2_{\text{loc}}(\mathbb{R})$.

3. By the well known inclusion of the weak Lebesgue space $L^{d,w}(\mathbb{R}^d)$ in M_q,

$$|b| \in L^\infty(\mathbb{R}, L^{d,w}(\mathbb{R}^d)) \quad \Rightarrow \quad b \in E_q \ \text{ with } 1<q \leq d.$$

4. For every $\varepsilon>0$, one can find $b \in E_q$ such that $|b|$ is not in $L^{q+\varepsilon}_{\text{loc}}(\mathbb{R}^{d+1})$. So, selecting $q>1$ close to 1, one obtains $b \in E_q$ that are not in $L^{1+\epsilon}_{\text{loc}}(\mathbb{R}^{d+1})$, $\epsilon>0$.
5. In view of the inclusion $\mathbf{F}_\delta$ (with $c_\delta=0$) in M_2, we obtain that $\mathbf{F}_\delta \subset E_q$ with $1<q\leq 2$. Furthermore, combining this with example 2, we obtain that

$$L^\infty \mathbf{F}_\delta + L^2_{\text{loc}}(\mathbb{R}) \subset E_q, \quad 1<q\leq 2.$$

We now state our results for drifts b in E_q, $1<q<2$. Set for $0<\alpha\leq 2$

$$(\lambda - \partial_t - \Delta)^{-\frac{\alpha}{2}} h(t,x) := \int_t^\infty \int_{\mathbb{R}^d} e^{-\lambda(s-t)} \frac{1}{(4\pi(s-t))^{\frac{d}{2}}} \frac{1}{(s-t)^{\frac{2-\alpha}{2}}} e^{-\frac{|x-y|^2}{4(s-t)}} h(s,y)\,ds\,dy, \tag{16.2}$$

$$(\lambda + \partial_t - \Delta)^{-\frac{\alpha}{2}} h(t,x) := \int_{-\infty}^t \int_{\mathbb{R}^d} e^{-\lambda(t-s)} \frac{1}{(4\pi(t-s))^{\frac{d}{2}}} \frac{1}{(t-s)^{\frac{2-\alpha}{2}}} e^{-\frac{|x-y|^2}{4(t-s)}} h(s,y)\,ds\,dy, \tag{16.3}$$

where $\lambda \geq 0$. By a standard result, if $\lambda>0$, then these operators are bounded on $L^p(\mathbb{R}^{d+1})$, $1\leq p\leq\infty$, with operator norm $\lambda^{-\frac{\alpha}{2}}$. If $\lambda>0$, then $(\lambda \pm \partial_t - \Delta)^{-1}$

is the resolvent of a Markov generator on $L^p(\mathbb{R}^{d+1})$, $1 \leq p < \infty$, which we will denote by $\lambda \pm \partial_t - \Delta$, respectively. In particular, one has well defined fractional powers $(\lambda \pm \partial_t - \Delta)^{\frac{\alpha}{2}}$. We refer to [4, 29] for the properties of these operators.

Define for $p \in]1, \infty[$

$$G_p := b^{\frac{1}{p}}(\lambda + \partial_t - \Delta)^{-\frac{1}{2p}}, \qquad R_p := b^{\frac{1}{p}} \cdot \nabla(\lambda + \partial_t - \Delta)^{-\frac{1}{2}-\frac{1}{2p}},$$

$$Q_p \upharpoonright \mathcal{E} := (\lambda + \partial_t - \Delta)^{-\frac{1}{2p'}}|b|^{\frac{1}{p'}}$$

and

$$T_p \upharpoonright \mathcal{E} := R_p Q_p,$$

where $\mathcal{E} := \cup_{\varepsilon>0} e^{-\varepsilon|b|}L^p(\mathbb{R}^{d+1})$, a dense subspace of $L^p(\mathbb{R}^{d+1})$.

The following is an analogue of Lemmas 6.1 and 15.1.

Lemma 16.1 *Let $b = b_{\mathfrak{s}} + b_{\mathfrak{b}}$, satisfy* (16.10). *Then, for every $p \in]1, \infty[$, for all $\lambda > 0$, the operators G_p, R_p, Q_p admit extension to L^p by continuity, and thus so does T_p. Moreover,*

$$\|G_p\|_{L^p(\mathbb{R}^{d+1})\to L^p(\mathbb{R}^{d+1})}, \|R_p\|_{L^p(\mathbb{R}^{d+1})\to L^p(\mathbb{R}^{d+1})}$$
$$\leq C_{d,p,q}\|b_{\mathfrak{s}}\|_{E_q}^{\frac{1}{p}} + c\lambda^{-\frac{1}{2p}}\|b_{\mathfrak{b}}\|_{L^\infty(\mathbb{R}^{d+1})}^{\frac{1}{p}} \tag{16.4}$$

$$\|Q_p\|_{L^p(\mathbb{R}^{d+1})\to L^p(\mathbb{R}^{d+1})} \leq C'_{p,q}\|b_{\mathfrak{s}}\|_{E_q}^{\frac{1}{p'}} + c'\lambda^{-\frac{1}{2p'}}\|b_{\mathfrak{b}}\|_{L^\infty(\mathbb{R}^{d+1})}^{\frac{1}{p'}}. \tag{16.5}$$

Remark 16.1 The operators G_p and R_p in Lemma 16.1 correspond operators $G_p(q)$, $R_p(r)$ in Lemmas 6.1 and 15.1 with "$q = r = p$", which is impossible in Lemmas 6.1 and 15.1 if one is dealing with form-bounded and weakly form-bounded drifts (they require $r < p < q$). As a consequence, Lemma 16.1 deals with the operator T_p very easily, since T_p is now a product of two bounded operators in L^p. Such decomposition of T_p is impossible for larger classes of form-bounded and weakly form-bounded drifts, see proofs of Lemmas 6.1 and 15.1.

The first estimate (16.4) follows from the boundedness of parabolic Riesz transforms (see [29]) and the following result: let $|b| \in E_q$ for some $q > 1$ close to 1, then, for every $p \in]1, \infty[$, there exists a constant $c_{p,q}$ such that

$$\||b|^{\frac{1}{p}}(\pm\partial_t - \Delta)^{-\frac{1}{2p}}\|_{L^p(\mathbb{R}^{d+1})\to L^p(\mathbb{R}^{d+1})} \leq c_{p,q}\|b\|_{E_q}^{\frac{1}{p}}. \tag{16.6}$$

In the time homogeneous case $b = b(x)$, the estimate on $\||b|^{\frac{1}{p}}(\lambda - \Delta)^{-\frac{1}{2p}}\|_{L^p(\mathbb{R}^d)\to L^p(\mathbb{R}^d)}$ in terms of the elliptic Morrey norm of $|b|$ is due to [1, Theorem 7.3]. Similar estimates in the parabolic case were obtained in [64].

Lemma 16.1 is proved in [41] by adapting the arguments from [65, proof of Prop. 4.1].

The estimate (16.5) follows from (16.6) by duality.

Define

$$b_n := \mathbf{1}_n b, \tag{16.7}$$

where $\mathbf{1}_n$ is the indicator of the set $\{(t,x) \in \mathbb{R}^{d+1} \mid |(t,x)| \leq n, |b(t,x)| \leq n\}$. We can additionally mollify b_n to obtain a C^∞ smooth approximation of b such that the Morrey norm of the approximating vector field does not exceed $(1+\varepsilon)\|b\|_{E_q}$ for any fixed $\varepsilon > 0$. However, regularization (16.7) of b will suffice. In particular, we will be able to apply Itô's formula to solutions of parabolic equations with drift b_n.

Armed with Lemma 16.1, one obtains the following result [41]. For every $p \in]1,\infty[$, there exist constants $c_{d,p,q}$ and $\lambda_{d,p,q}$ such that if

$$\|b_{\mathfrak{S}}\|_{E_q} < c_{d,p,q},$$

then, for every $\lambda \geq \lambda_{d,p,q}$, solutions $u_n \in L^p(\mathbb{R}^{d+1})$ to the approximating parabolic equations

$$(\lambda + \partial_t - \Delta + b_n \cdot \nabla)u_n = f, \quad f \in L^p(\mathbb{R}^{d+1})$$

converge in[16] $\mathbb{W}^{1+\frac{1}{p},p}(\mathbb{R}^{d+1})$ to

$$u := (\lambda + \partial_t - \Delta)^{-1} f - (\lambda + \partial_t - \Delta)^{-\frac{1}{2}-\frac{1}{2p}} Q_p (1+T_p)^{-1} \times R_p (\lambda + \partial_t - \Delta)^{-\frac{1}{2p'}} f. \tag{16.8}$$

Moreover, this u is a unique weak solution to $(\lambda + \partial_t - \Delta + b\cdot\nabla)u = f$, appropriately defined, see [40]. If above $p > d+1$, then, by (16.8) and by the parabolic Sobolev embedding, the convergence is uniform on $\mathbb{R}^{d+1}$ and $u \in C_\infty(\mathbb{R}^{d+1})$.

Let us now construct a Feller evolution family. Let $\delta_{s=r}$ denote the delta-function in the time variable s. Put

$$(\lambda + \partial_t - \Delta)^{-1}\delta_{s=r} g(t,x) := \mathbf{1}_{t\geq r} e^{-\lambda(t-r)} (4\pi(t-r))^{-\frac{d}{2}} \int_{\mathbb{R}^d} e^{-\frac{|x-y|^2}{4(t-r)}} g(y)dy,$$

$$\nabla(\lambda + \partial_t - \Delta)^{-\frac{1}{2}-\frac{1}{2p'}} \delta_{s=r} g := \mathbf{1}_{t\geq r} e^{-\lambda(t-r)} (t-r)^{-\frac{1}{2}+\frac{1}{2p'}} (4\pi(t-r))^{-\frac{d}{2}} \int_{\mathbb{R}^d} \nabla_x e^{-\frac{|x-y|^2}{4(t-r)}} g(y)dy.$$

[16] Here $\mathbb{W}^{\alpha,p}(\mathbb{R}^{d+1}) := (\lambda + \partial_t - \Delta)^{-\frac{\alpha}{2}} L^p(\mathbb{R}^{d+1})$ endowed with the norm $\|h\|_{\mathbb{W}^{\alpha,p}} := \|(\lambda + \partial_t - \Delta)^{-\frac{\alpha}{2}} h\|_p$.

Fix $T > 0$. For given $n = 1, 2, \dots$ and $0 \le r < T$, let v_n denote the classical solution to Cauchy problem

$$\begin{cases} (\lambda + \partial_t - \Delta + b_n(t,x) \cdot \nabla) v_n = 0 & (t,x) \in]r,T] \times \mathbb{R}^d, \\ v_n(r,\cdot) = g(\cdot) \in C_\infty, \end{cases} \tag{16.9}$$

where b_n's are defined by (16.7). By a standard result, for every n, the operators

$$U_n^{t,r} g := v_n(t), \quad 0 \le r \le t \le T$$

constitute a Feller evolution family on C_∞. Recall $D_T = \{0 \le r \le t \le T\}$.

Theorem 16.1 *Let $b = b_{\mathfrak{s}} + b_{\mathfrak{b}}$, where*

$$|b_{\mathfrak{s}}| \in E_q \text{ for some } q > 1 \text{ close to } 1, \text{ and } |b_{\mathfrak{b}}| \in L^\infty(\mathbb{R}^{d+1}) \tag{16.10}$$

(indices $\mathfrak{s}$ and $\mathfrak{b}$ stand for "singular" and "bounded", respectively). Fix $p > d+1$. There exist constants $c_{d,p,q}$ and $\lambda_{d,p,q}$ such that if $\|b_{\mathfrak{s}}\|_{E_q} < c_{d,p,q}$, then the following are true:

(i) The limit

$$U^{t,r} := s\text{-}C_\infty(\mathbb{R}^d)\text{-}\lim_n U_n^{t,r} \quad \text{uniformly in } (r,t) \in D_T$$

exists and determines a Feller evolution family on $C_\infty(\mathbb{R}^d)$.

(ii) For every initial function $g \in C_\infty(\mathbb{R}^d) \cap W^{1,p}(\mathbb{R}^d)$, $v(t) := U^{t,r} g$, where $(r,t) \in D_T$, has representation

$$v = (\lambda + \partial_t - \Delta)^{-1} \delta_{s=r} g - (\lambda + \partial_t - \Delta)^{-\frac{1}{2} - \frac{1}{2p}} Q_p (1 + T_p)^{-1} G_p S_p g, \tag{16.11}$$

where $S_p g := \nabla (\lambda + \partial_t - \Delta)^{-\frac{1}{2} - \frac{1}{2p'}} \delta_{s=r} g$ satisfies

$$\|S_p g\|_{L^p(\mathbb{R}^{d+1})} \le C_{p,d} \|\nabla g\|_{L^p(\mathbb{R}^d)}.$$

(iii) As a consequence of (16.11) and the parabolic Sobolev embedding, we obtain

$$\sup_{(r,t) \in D_T, x \in \mathbb{R}^d} |v(t,x;r)| \le C \|g\|_{W^{1,p}(\mathbb{R}^d)}.$$

Theorem 16.1 was proved in [41].

Define backward Feller evolution family ($0 \le t \le r \le T$)

$$P^{t,r}(b) = U^{T-t,T-r}(\tilde{b}), \quad \tilde{b}(t,x) = b(T-t,x),$$

where $U^{t,s}$ is the Feller evolution family from Theorem 16.1. As was explained in the previous section, weak well-posedness of SDE

$$X_t = x - \int_0^b b(s, X_s)ds + \sqrt{2}W_t, \quad t \geq 0, \tag{16.12}$$

follows from appropriate regularity results for the corresponding inhomogeneous parabolic equation (10.5). Indeed, the solution representations (16.8) and (16.11) can be combined and, furthermore, localized, which yields an analogue of gradient estimates (10.3) and thus allows to prove (see [41]) the following result:

Theorem 16.2 *Under the assumptions of Theorem 16.1, the following are true:*

(*i*) *The backward Feller evolution family* $\{P^{t,r}\}_{0\leq t\leq r\leq T}$ *is conservative, i.e. the density* $P^{t,r}(x,\cdot)$ *satisfies*

$$\langle P^{t,r}(x,\cdot)\rangle = 1 \quad \textit{for all } x \in \mathbb{R}^d,$$

and determines probability measures $\mathbb{P}_x$, $x \in \mathbb{R}^d$ *on* $(C([0,T],\mathbb{R}^d), \mathcal{B}_t)$, *such that*

$$\mathbb{E}_x[f(\omega_r)] = P^{0,r}f(x), \quad 0 \leq r \leq T, \quad f \in C_\infty(\mathbb{R}^d).$$

(*ii*) *For every* $x \in \mathbb{R}^d$, *the probability measure* $\mathbb{P}_x$ *is a weak solution to* (16.12).

(*iii*) *For every* $x \in \mathbb{R}^d$ *and* f *satisfying* (16.10), *given a* $p > d+1$ *as in Theorem 16.1 (generally speaking, the larger* p *is the smaller* $\|b_{\mathfrak{S}}\|_{E_q}$ *has to be), there exists constant* c *such that for all* $h \in C_c(\mathbb{R}^{d+1})$

$$\mathbb{E}_x \int_0^T |\mathsf{f}(r,\omega_r)h(r,\omega_r)|dr \leq c\|\mathbf{1}_{[0,T]}|\mathsf{f}|^{\frac{1}{p}}h\|_{L^p(\mathbb{R}^{d+1})} \tag{16.13}$$

(in particular, one can take $\mathsf{f} = b$*). On the other hand, any martingale solution* $\mathbb{P}'_x$ *to* (16.12) *that satisfies for some* $p > d+1$ *as in Theorem 16.1 the estimate* (16.13) *for* $\mathsf{h} := b$ *coincides with* $\mathbb{P}_x$.

(*iv*) *For every* $x \in \mathbb{R}^d$, *given a* $\nu > \frac{d+2}{2}$, *there exists a constant* c *such that for all* $h \in C_c(\mathbb{R}^{d+1})$ *the following Krylov-type bound is true:*

$$\mathbb{E}_x \int_0^T |h(r,\omega_r)|dr \leq c\|\mathbf{1}_{[0,T]}h\|_{L^\nu(\mathbb{R}^{d+1})}. \tag{16.14}$$

On the other hand, if additionally $|b| \in L^{\frac{d+2}{2}+\varepsilon}_{\rm loc}(\mathbb{R}^{d+1})$ *for some* $\varepsilon > 0$ *and* $\|b_{\mathfrak{S}}\|_{E_q}$ *is sufficiently small, then any martingale solution* $\mathbb{P}'_x$ *to* (16.12) *that satisfies* (16.14) *for some* $\nu > \frac{d+2}{2}$ *sufficiently close to* $\frac{d+2}{2}$ *(depending on how small* ε *is) coincides with* $\mathbb{P}_x$.

We compared the uniqueness results of type (*iii*), (*iv*) in Remark 10.1.

Perhaps, the closest to our Theorems 16.1–16.2 results are contained in recent papers by Krylov [66–68], which also allow to deal with discontinuous (VMO) diffusion coefficients, see literature review in the introduction.

The estimates (16.13), (16.14) follow from the same kind of solution representations as (16.8), (16.11) above, see [41]. The proofs of the uniqueness results in (*iii*), (*iv*) are similar. Let us prove the uniqueness result in (*iv*). Suppose that we have two martingale solutions $\mathbb{P}_x^1$, $\mathbb{P}_x^2$ of (16.12) that satisfy, for $\nu > \frac{d+2}{2}$ close to $\frac{d+2}{2}$,

$$\mathbb{E}_x^i \int_0^T |h(t,\omega_t)|dt \leq c\|\mathbf{1}_{[0,T]}h\|_\nu, \quad h \in C_c(\mathbb{R}^{d+1}) \tag{16.15}$$

with constant c independent of h ($i = 1, 2$). Here and below, $\mathbb{E}_x^1 := \mathbb{E}_{\mathbb{P}_x^1}$, $\mathbb{E}_x^2 := \mathbb{E}_{\mathbb{P}_x^2}$. Our goal is to show: for every $f \in C_c(\mathbb{R}^{d+1})$,

$$\mathbb{E}_x^1[\int_0^T f(t,\omega_t)dt] = \mathbb{E}_x^2[\int_0^T f(t,\omega_t)dt], \tag{16.16}$$

which implies $\mathbb{P}_x^1 = \mathbb{P}_x^2$.

So, let us prove (16.16). Let $u_n \in C([0,T], C_\infty(\mathbb{R}^d))$ be the solution to

$$(\partial_t + \Delta + b_n \cdot \nabla)u_n = f, \quad u_n(T,\cdot) = 0, \tag{16.17}$$

where, recall, $b_n = \mathbf{1}_n b$, and $\mathbf{1}_n$ is the indicator of $\{|b| \leq n\}$. Set $\tau_R := \inf\{t \geq 0 \mid |\omega_t| \geq R\}$, $R > 0$. By Itô's formula

$$\begin{aligned}\mathbb{E}_x^i u_n(T \wedge \tau_R, \omega_{T\wedge\tau_R}) &= u_n(0,x) + \mathbb{E}_x^i \int_0^{T\wedge\tau_R} f(t,\omega_t)dt \\ &\quad + \mathbb{E}_x^i \int_0^{T\wedge\tau_R} \big[(b-b_n)\cdot\nabla u_n\big](t,\omega_t)dt \end{aligned}\tag{16.18}$$

($i = 1, 2$). We have

$$\left|\mathbb{E}_x^i \int_0^{T\wedge\tau_R} \big[(b-b_n)\cdot\nabla u_n\big](t,\omega_t)dt\right| \leq \mathbb{E}_x^i \int_0^{T\wedge\tau_R} \big[|b|(1-\mathbf{1}_n)|\nabla u_n|\big](t,\omega_t)dt$$

(we are applying (16.15))

$$\leq c\|\mathbf{1}_{[0,T]\times B_R(0)}|b|(1-\mathbf{1}_n)|\nabla u_n|\|_\nu$$

$$\leq c\|\mathbf{1}_{[0,T]\times B_R(0)}|b|(1-\mathbf{1}_n)\|_{s'}\|\nabla u_n\|_s, \qquad \frac{1}{s}+\frac{1}{s'}=\frac{1}{q}.$$

At this point we note that $\tilde{u}_n(t) := e^{\lambda(T-t)}u_n(t)$ satisfies

$$(\lambda + \partial_t + \Delta + b_n \cdot \nabla)u_n = \mathbf{1}_{[0,T]}e^{\lambda(T-t)}f,$$

so a solution representation of type (16.8) (i.e. additionally taking into account the terminal value condition), see [41], and the parabolic Sobolev embedding theorem, yield

$$\|\nabla u_n\|_s \leq C\|f\|_p \quad \text{for } s < \frac{d+2}{d+1}p \ \text{ close to } \frac{d+2}{d+1}p.$$

Assuming that the Morrey norm $\|b\|_{E_q}$ is sufficiently small, we can select p sufficiently large to make s' close to ν and hence close to $\frac{d+2}{2}$. To be more precise, we have by our assumption $|b| \in L^{\frac{d+2}{2}+\varepsilon}$ for some $\varepsilon > 0$, so we need $s' \geq \frac{d+2}{2} + \varepsilon$. Now, since $1 - \mathbf{1}_n \to 0$ a.e. on $\mathbb{R}^{d+1}$ as $n \to \infty$, we have $\|\mathbf{1}_{[0,T]\times B_R(0)}|b|(1-\mathbf{1}_n)\|_{s'} \to 0$ as $n \to \infty$. Therefore,

$$\mathbb{E}^i_x \int_0^{T\wedge\tau_R} \big[(b - b_n)\cdot\nabla u_n\big](t,\omega_t)dt \to 0 \quad (n \to \infty).$$

We are left to note, using the convergence result in [41, Cor. 2] (of the same type as Theorem 16.1(i)), that solutions u_n converge to a function $u \in C([0,T], C_\infty(\mathbb{R}^d))$. Therefore, we can pass to the limit in (16.18), first in n and then in $R \to \infty$, to obtain

$$0 = u(0,x) + \mathbb{E}^i_x \int_0^T f(t,\omega_t)dt \quad i = 1, 2,$$

which gives (16.16). □

17 SDEs Driven by α-Stable Process

In this section we deal with the SDE

$$X_t = x - \int_0^t b(X_s)ds + Z_t - Z_0, \quad t \geq 0, \quad x \in \mathbb{R}^d, \tag{17.1}$$

where Z_t be a rotationally symmetric α-stable process, $1 < \alpha < 2$, i.e. a Lévy process with characteristic function

$$\mathbb{E}[\exp(i\varkappa \cdot (Z_t - Z_0))] = \exp(-t|\varkappa|^\alpha) \quad \text{for every } \varkappa \in \mathbb{R}^d.$$

The drift $b : \mathbb{R}^d \to \mathbb{R}^d$ is in general locally unbounded.

Recall that a weak solution to (17.1) is a process X_t defined on some probability space having a.s. right continuous trajectories with left limits, such that $\int_0^t |b(X_s)|ds < \infty$ a.s. for every $t > 0$, and such that X_t satisfies (17.1) a.s. for a symmetric α-stable process Z_t.

A weak solution to (17.1), when it exists (e.g. if $|b| \in L^\infty$, see [59]), is called α-stable process with drift b. It plays a central role in the study of stochastic processes which, in contrast to the Brownian motion, can have long range interactions.

The operator behind SDE (17.1) is the non-local operator $(-\Delta)^{\frac{\alpha}{2}} + b \cdot \nabla$, i.e. one expects that the transition density of X_t solves the corresponding parabolic equation for $(-\Delta)^{\frac{\alpha}{2}} + b \cdot \nabla$.

We are interested in the same question as in the previous sections: what are the minimal assumptions on the local singularities of the vector field b, not assuming additional structure such as symmetry or existence of the divergence, such that, for an arbitrary initial point, there exists a unique (in appropriate sense) weak solution to (17.1)? This question has been extensively studied in the literature. By the results of Portenko [86] and Podolynny-Portenko [87], if

$$|b| \in L^p + L^\infty, \qquad \text{for some } p > \frac{d}{\alpha - 1}, \tag{17.2}$$

then there exists a unique in law weak solution to (17.1). Although the exponent $\frac{d}{\alpha-1}$ is the best possible on the Lebesgue scale, the class (17.2) is far from being the maximal admissible: this result has been strengthened in [16, 18, 36] where the authors consider b in the Kato class of vector fields $\mathbf{K}_\delta^{d,\alpha-1}$ (with δ arbitrarily small), i.e. $|b| \in L^1_{\rm loc}$ and

$$\left\| \left(\lambda + (-\Delta)^{\frac{\alpha}{2}}\right)^{-\frac{\alpha-1}{\alpha}} |b| \right\|_\infty \leq \delta$$

for some $\lambda = \lambda_\delta \geq 0$. This class contains some vector fields b with $|b| \notin L^{1+\varepsilon}_{\rm loc}$, $\varepsilon > 0$, however, it does not contains the class $|b| \in L^{\frac{d}{\alpha-1}} + L^\infty$.

The Kato class $\mathbf{K}_\delta^{d,\alpha-1}$ with δ sufficiently small provides the standard bounds on the heat kernel of the fractional Kolmogorov operator

$$\Lambda(b) \supset (-\Delta)^{\frac{\alpha}{2}} + b \cdot \nabla,$$

i.e.

$$C^{-1} e^{-t(-\Delta)^{\frac{\alpha}{2}}}(x, y) \leq e^{-t\Lambda(b)}(x, y) \leq C e^{-t(-\Delta)^{\frac{\alpha}{2}}}(x, y), \tag{17.3}$$

for all $x, y \in \mathbb{R}^d$ and $0 < t < T$ for a constant $C = C_T$ [11]. It was established in [18], among many other results, that the probability measures $\{\mathbb{P}_x\}_{x \in \mathbb{R}^d}$ determined by $e^{-t\Lambda(b)}$ solve the martingale problem for $(-\Delta)^{\frac{\alpha}{2}} + b \cdot \nabla$ with test functions in C_c^∞. The uniqueness in law of the weak solution to SDE (17.1) with $b \in \mathbf{K}_\delta^{d,\alpha-1}$ (with δ

arbitrarily small) was established in [16]. Let us also mention that [36] considered SDE (17.1) with Kato class measure-valued drift and established the corresponding heat kernel bounds.

We consider a larger class of *weakly form-bounded vector fields*:

Definition 17.1 A vector field $b : \mathbb{R}^d \to \mathbb{R}^d$ with entries in $L^1_{\rm loc}$ is said to be weakly form-bounded if there exist $\delta > 0$ such that

$$\| |b|^{\frac{1}{2}} (\lambda + (-\Delta)^{\frac{\alpha}{2}})^{-\frac{\alpha-1}{2\alpha}} \|_{2\to 2} \leq \sqrt{\delta}$$

for some $\lambda = \lambda_\delta > 0$.

This will be written as $b \in \mathbf{F}_\delta^{\frac{\alpha-1}{2}}$. This definition extends Definition 14.1 from the previous section corresponding to the case $\alpha = 2$.

Our assumptions concerning δ will involve only strict inequalities, so (using e.g. the Spectral theorem) we can re-state our hypothesis on the drift as

$$\| |b|^{\frac{1}{2}} (\lambda - \Delta)^{-\frac{\alpha-1}{4}} \|_{2\to 2} \leq \sqrt{\delta}$$

for some $\lambda = \lambda_\delta > 0$.

Examples

1. Using the fractional Sobolev inequality, it is not difficult to show that

$$|b| \in L^{\frac{d}{\alpha-1}} + L^\infty \quad \Rightarrow \quad b \in \mathbf{F}_\delta^{\frac{\alpha-1}{2}},$$

where $\delta > 0$ can be chosen arbitrarily small. More generally, vector fields with entries in $L^{\frac{d}{\alpha-1},w}$ (the weak $L^{\frac{d}{\alpha-1}}$ class) are weakly form-bounded:

$$|b| \in L^{\frac{d}{\alpha-1},\infty} + L^\infty \quad \Rightarrow \quad b \in \mathbf{F}_\delta^{\frac{\alpha-1}{2}},$$

with

$$\sqrt{\delta} = \Omega_d^{\frac{\alpha-1}{2d}} \frac{2^{-\frac{\alpha-1}{2}} \Gamma\left(\frac{d-\alpha+1}{4}\right)}{\Gamma\left(\frac{d+\alpha-1}{4}\right)} \|b_1\|^{\frac{1}{2}}_{\frac{d}{\alpha-1},w},$$

where Ω_d is the volume of the unit ball $B(0,1) \subset \mathbb{R}^d$. The proof is obtained easily using [61, Corollary 2.9].

2. In particular, by the fractional Hardy inequality,

$$b(x) = \pm\sqrt{\delta}\kappa_{\alpha,d}|x|^{-\alpha}x, \tag{17.4}$$

where

$$\kappa_{\alpha,d} := 2^{\frac{\alpha-1}{2}} \frac{\Gamma(\frac{d+\alpha-1}{4})}{\Gamma(\frac{d-\alpha+1}{4})},$$

is in $\mathbf{F}_\delta^{\frac{\alpha-1}{2}}$ (with $\lambda = 0$).

The drift (17.4) destroys the standard heat kernel bounds (17.3) (and so it is not in the Kato class). However, for such b sharp heat kernel bounds on $e^{-t\Lambda(b)}(x, y)$ exist but they depend explicitly on δ via an additional factor $\varphi_t(y)$,

$$C^{-1}e^{-t(-\Delta)^{\frac{\alpha}{2}}}(x, y)\varphi_t(y) \le e^{-t\Lambda(b)}(x, y) \le Ce^{-t(-\Delta)^{\frac{\alpha}{2}}}(x, y)\varphi_t(y), \quad x, y \in \mathbb{R}^d, \quad y \neq 0.$$

The factor $\varphi_t(y)$ either explodes at the origin or vanishes, depending on the sign of δ [52, 58].

3. The Kato class vector fields are weakly form-bounded:

$$b \in \mathbf{K}_\delta^{d,\alpha-1} \quad \Rightarrow \quad b \in \mathbf{F}_\delta^{\frac{\alpha-1}{2}}.$$

Indeed, if $b \in \mathbf{K}_\delta^{d,\alpha-1}$, then by duality $\||b|(\lambda + (-\Delta)^{\frac{\alpha}{2}})^{-\frac{\alpha-1}{\alpha}}\|_{1\to 1} \le \delta$, and so by interpolation $\||b|^{\frac{1}{2}}(\lambda + (-\Delta)^{\frac{\alpha}{2}})^{-\frac{\alpha-1}{\alpha}}|b|^{\frac{1}{2}}\|_{2\to 2} \le \delta$, hence $b \in \mathbf{F}_\delta^{\frac{\alpha-1}{2}}$.

4. The elliptic Morrey class:

$$|b|^{\frac{1}{\alpha-1}} \in M_{1+\varepsilon} \quad \Rightarrow \quad b \in \mathbf{F}_\delta^{\frac{\alpha-1}{2}}$$

with δ depending on the Morrey norm of $|b|^{\frac{1}{\alpha-1}}$ (see definition (14.2)). Indeed, by [1, Theorem 7.3], one has $\||b|^{\frac{1}{2(\alpha-1)}}(\lambda - \Delta)^{-\frac{1}{4}}\|_{2\to 2} \le \delta^{\frac{1}{2(\alpha-1)}}$. Then, by the Heinz-Kato inequality (i.e. raising operators $|b|^{\frac{1}{2(\alpha-1)}}$ and $(\lambda - \Delta)^{-\frac{1}{4}}$ to power $\alpha - 1 < 1$), we obtain $\||b|^{\frac{1}{2}}(\lambda - \Delta)^{-\frac{\alpha-1}{4}}\|_{2\to 2} \le \sqrt{\delta}$, i.e. $b \in \mathbf{F}_\delta^{\frac{\alpha-1}{2}}$. This examples contains examples 1 and 2.

Remark 17.1 There is a rich literature on weak and strong well-posedness of SDE (17.1) (and its generalizations) in the case $0 < \alpha \le 1$, in which case $|b|$ is assumed to be (locally) Hölder continuous, say, with exponent β, and satisfy the balance condition $\alpha + \beta > 1$ (sub-critical) or $\alpha + \beta = 1$ (critical). See Zhao [112], Song-Xie [95] who considered the case $\alpha + \beta = 1$. Regarding the corresponding heat kernel bounds, we refer to Xie-Zhang [100] and Menozzi-Zhang [75] who proved the two-sided bound on the heat kernel of $(-\Delta)^{\frac{\alpha}{2}} + b \cdot \nabla$ in the case $\alpha + \beta > 1$. Let us add that in the case $\alpha + \beta = 1$ the behaviour of the heat kernel changes drastically, for instance, it can vanish, see [56] who considered the heat kernel of operator $\Lambda = (-\Delta)^{\frac{\alpha}{2}} - \kappa|x|^{-\alpha}x \cdot \nabla$, $\kappa > 0$ (Hardy-type drift) and proved upper bound of the form

$$0 \leq e^{-t\Lambda}(x, y) \leq Ct^{-\frac{d}{\alpha}}[1 \wedge t^{-\frac{\gamma}{\alpha}}|y|^{\gamma}], \quad t \in]0, 1],$$

where the order of vanishing $\gamma \in]0, \alpha[$ is an explicit function of κ.

For a given vector field $b \in \mathbf{F}_\delta^{\frac{\alpha-1}{2}}$, we fix a C^∞ smooth approximation

$$b_n := c_n \eta_{\varepsilon_n} * (\mathbf{1}_n b), \quad \varepsilon_n \downarrow 0, \quad n = 1, 2, \dots,$$

where $\mathbf{1}_n$ is the indicator of $\{x \in \mathbb{R}^d \mid |x| \leq n, |b(x)| \leq n\}$, η_ε is the Friedrichs mollifier. Selecting $\varepsilon_n \downarrow$ sufficiently rapidly and $c_n \uparrow 1$ sufficiently slow, one obtains that

$$b_n \in \mathbf{F}_\delta^{\frac{\alpha-1}{2}}, \quad n = 1, 2, \dots$$

with λ independent of n.

Fix constant $m_{d,\alpha}$ by the pointwise inequality

$$\left|\nabla_x \left(\mu + (-\Delta)^{\frac{\alpha}{2}}\right)^{-1}(x, y)\right| \leq m_{d,\alpha}\left(\kappa^{-1}\mu + (-\Delta)^{\frac{\alpha}{2}}\right)^{-\frac{\alpha-1}{\alpha}}(x, y) \qquad (17.5)$$

for all $x, y \in \mathbb{R}^d$, $x \neq y$, and all $\mu > 0$, for some $\kappa = \kappa_{d,\alpha} > 0$. The following result was proved in [44] (one can find there an elementary estimate on $m_{d,\alpha}$ from above).

Theorem 17.1 *Let $b \in \mathbf{F}_\delta^{\frac{\alpha-1}{2}}$ with $\delta < m_{d,\alpha}^{-1} 4\left[\frac{d-\alpha}{(d-\alpha+1)^2} \wedge \frac{\alpha(d+\alpha)}{(d+2\alpha)^2}\right]$. The following is true.*

(i) The limit

$$s\text{-}C_\infty\text{-}\lim_n e^{-t\Lambda_{C_\infty}(b_n)} \quad (\textit{loc. uniformly in } t \geq 0),$$

where

$$\Lambda_{C_\infty}(b_n) := (-\Delta)^{\frac{\alpha}{2}} + b_n \cdot \nabla \text{ with domain } D(\Lambda_{C_\infty}(b_n)) = \left(1 + (-\Delta)^{\frac{\alpha}{2}}\right)^{-1} C_\infty,$$

exists and determines a Feller semigroup $T^t =: e^{-t\Lambda_{C_\infty}(b)}$. Its generator Λ_{C_∞} is an operator realization of the formal operator $(-\Delta)^{\frac{\alpha}{2}} + b \cdot \nabla$ in C_∞.

(ii) There exists $\mu_0 \geq 0$ such that for all $\mu \geq \mu_0$, for every $p \in [2, p_+[$, $p_+ = \frac{2}{1-\sqrt{1-m_{d,\alpha}\delta}}$, and all $1 < r < p < q < \infty$,

$$(\mu + \Lambda_{C_\infty}(b))^{-1} \upharpoonright C_\infty \cap L^p \text{ extends by continuity to } \mathcal{B}\left(\mathcal{W}^{-\frac{\alpha-1}{r'}, p}, \mathcal{W}^{1+\frac{\alpha-1}{q}, p}\right).$$

In particular, if $p > d - \alpha + 1$, then

$$\big(\mu + \Lambda_{C_\infty}(b)\big)^{-1}[C_\infty \cap L^p] \subset C^{0,\gamma}, \quad \gamma < 1 - \frac{d-\alpha+1}{p}.$$

Also,

$$\big(\mu + \Lambda_{C_\infty}(b)\big)^{-1} \upharpoonright C_\infty \cap L^2 \text{ extends by continuity to } \mathcal{B}(\mathcal{W}^{-\frac{\alpha-1}{2},2}, \mathcal{W}^{\frac{\alpha+1}{2},2}). \tag{17.6}$$

(*iii*) *$e^{-t\Lambda_{C_\infty}(b)}$ is conservative, i.e. $\int_{\mathbb{R}^d} e^{-t\Lambda_{C_\infty}(b)}(x,y)dy = 1 \ \forall x \in \mathbb{R}^d$.*
Let $\{\mathbb{P}_x\}_{x\in\mathbb{R}^d}$ be the probability measures on the canonical space $(D[0,T], \mathcal{B}'_t)$ determined by $e^{-t\Lambda_{C_\infty}(b)}$, i.e.

$$\mathbb{E}_{\mathbb{P}_x}[f(X_t)] = (e^{-t\Lambda_{C_\infty}(b)}f)(x), \quad f \in C_\infty, \quad x \in \mathbb{R}^d.$$

(*iv*) *For every $x \in \mathbb{R}^d$ and $t > 0$, $\mathbb{E}_{\mathbb{P}_x}\int_0^t |b(X_s)|ds < \infty$ and there exists a process Z_t with trajectories in $D(\mathbb{R}_+, \mathbb{R}^d)$, which is a symmetric α-stable process under each $\mathbb{P}_x$, such that*

$$X_t = x - \int_0^t b(X_s)ds + Z_t - Z_0, \quad t \geq 0.$$

(*v*) *The Feller property and property* (17.6) *determine $\{\mathbb{P}_x\}_{x\in\mathbb{R}^d}$ uniquely. That is, suppose that for every $x \in \mathbb{R}^d$ we are given a weak solution $\mathbb{Q}_x$ to SDE* (17.1). *Define for every $f \in C_c^\infty$*

$$R_\mu^Q f(x) := \mathbb{E}_{\mathbb{Q}_x}\int_0^\infty e^{-\mu s} f(X_s)ds,$$
$$X_s \in D(\mathbb{R}_+, \mathbb{R}^d), \quad x \in \mathbb{R}^d, \quad \mu > \lambda_\delta.$$

If $R_\mu^Q C_c^\infty \subset C_b$ and $R_\mu^Q \upharpoonright C_c^\infty$ admits extension by continuity to $\mathcal{B}(\mathcal{W}^{-\frac{\alpha-1}{2},2}, L^2)$, then

$$\{\mathbb{Q}_x\}_{x\in\mathbb{R}^d} = \{\mathbb{P}_x\}_{x\in\mathbb{R}^d}.$$

We stated assertions (*i*), (*ii*) in a form that is somewhat different from Theorems 6.1 or 15.1, but we could have stated it in the form of these theorems as well. The construction of the Feller semigroup (*i*) goes as in Theorems 6.1 (or rather as in Theorem 15.1 since we use pointwise bound (17.5)). Same for the embedding properties (*ii*), i.e. we can write an explicit operator-valued function representation for the resolvent as in the cited two theorems.

The proof of the probability conservation property in (*iii*) uses weighted estimates, similarly to the proof of Theorem 6.1 dealing with the local case $\alpha = 2$ (see Sect. 7, estimate (7.7) there). Set

$$\eta(x) := (1 + |x|^2)^\nu, \quad 0 < \nu < \frac{\alpha}{2}.$$

Denote $L^p_\eta := L^p(\mathbb{R}^d, \eta^2 dx)$, $\| \cdot \|^p_{p,\eta} := \langle | \cdot |^p \eta^2 \rangle$.

Proposition 17.1 *Let* $d \geq 3$, $b \in \mathbf{F}_\delta^{\frac{\alpha-1}{2}}$ *with* $\delta < m_{d,\alpha}^{-1} 4\big[\frac{d-\alpha}{(d-\alpha+1)^2} \wedge \frac{\alpha(d+\alpha)}{(d+2\alpha)^2}\big]$. *There exist* $0 < \nu < \alpha/2$, $p > (d - \alpha + 1) \vee (\frac{d}{2\nu} + 2)$ *and* $\mu_0 > 0$ *such that for every* $h \in C_c$, $\mu \geq \mu_0$

$$\|\eta^{-1}(\mu + \Lambda_{C_\infty}(b))^{-1} \eta h\|_\infty \leq K_1 \|h\|_{p,\eta}, \tag{E_1}$$

$$\|\eta^{-1}(\mu + \Lambda_{C_\infty}(b))^{-1} \eta |b_m| h\|_\infty \leq K_2 \||b_m|^{\frac{1}{p}} h\|_{p,\eta}, \tag{E_2}$$

$$\|\eta^{-1} |b_m|^{\frac{1}{p}} (\mu + \Lambda_{C_\infty}(b))^{-1} \eta |b_m| h\|_{p,\eta} \leq K_3 \||b_m|^{\frac{1}{p}} h\|_{p,\eta}, \tag{E_3}$$

where $K_i > 0$, $i = 1, 2, 3$, *do not depend on* $m = 1, 2, \dots$ *The constant* K_3 *can be chosen arbitrarily small at expense of increasing* μ_0.

The proof of these weighted estimates in [44] is, however, quite different from the proof of Theorem 6.1 where one can control easily the commutator of the weight and the Laplacian. [44] provides a different, rather interesting approach to the proof of Proposition 17.1, but we will not discuss it here. Note that if b has compact support then we can take $\eta \equiv 1$.

Let us describe the proof of (*iv*) in [44], which uses the approach of [86, 87] but in a weighted L^p space. Set

$$Z_t := X_t - X_0 - \int_0^t b(X_s) ds, \quad t \geq 0.$$

Our goal is to prove that under $\mathbb{P}_x$ the process Z_t is a symmetric α-stable process starting at 0. We use notation introduced in the beginning of the previous section. For brevity, write $e^{t\Lambda(b)} = e^{-t\Lambda_{C_\infty}(b)}$.

1. *Define*

$$w(t, x, \varkappa) = \mathbb{E}_x\Big[e^{i\varkappa \cdot \left(X_t - \int_0^t b(X_s)ds\right)}\Big], \quad t \geq 0, \quad \varkappa \in \mathbb{R}^d.$$

Then w is a bounded solution to integral equation

$$w(t,x,\varkappa) = \int_{\mathbb{R}^d} e^{i\varkappa\cdot y} e^{-t\Lambda(b)}(x,y)dy \\ - i\int_0^t \int_{\mathbb{R}^d} e^{-(t-s)\Lambda(b)}(x,z)(\varkappa\cdot b(z))w(s,z,\varkappa)dzds. \quad (17.7)$$

Indeed, in view of

$$e^{-i\cdot\varkappa\int_0^t b(X_\tau)d\tau} = 1 - i\int_0^t (\varkappa\cdot b(X_s))e^{-i\cdot\varkappa\int_s^t b(X_\tau)d\tau},$$

one has

$$\begin{aligned} w(t,x,\varkappa) &= \mathbb{E}_x\Big[e^{i\varkappa\cdot X_t}\Big] - i\int_0^t \mathbb{E}_x\Big[e^{i\varkappa\cdot X_t}(\varkappa\cdot b(X_s))e^{-i\cdot\varkappa\int_s^t b(X_\tau)d\tau}\Big]ds \\ &= \mathbb{E}_x\Big[e^{i\varkappa\cdot X_t}\Big] - i\int_0^t \mathbb{E}_x\Big[(\varkappa\cdot b(X_s))w(t-s,X_s,\varkappa)\Big]ds \\ &= \int_{\mathbb{R}^d} e^{i\varkappa\cdot y}e^{-t\Lambda(b)}(x,y)dy \\ &\quad - i\int_0^t\int_{\mathbb{R}^d} e^{-s\Lambda(b)}(x,z)(\varkappa\cdot b(z))w(t-s,z,\varkappa)dzds. \end{aligned}$$

2. *Set $\tilde{w}(t,x,\varkappa) := e^{i\varkappa\cdot x - t|\varkappa|^\alpha}$. This is another bounded solution to* (17.7). Indeed, multiplying the Duhamel formula

$$e^{-t\Lambda}(x,y) = e^{-t(-\Delta)^{\frac{\alpha}{2}}}(x,y) + \int_0^t \langle e^{-(t-s)\Lambda}(x,\cdot)b(\cdot)\cdot\nabla.e^{-s(-\Delta)^{\frac{\alpha}{2}}}(\cdot,y)\rangle ds$$

(which is proved in [44, Corollary 1(*iv*)]) by $e^{i\varkappa\cdot y}$ and then integrating in y, we obtain the required.

Next, let us show that a bounded solution to (17.7) is unique. We will need

3. *For every $\varkappa\in\mathbb{R}^d$ there exists $T = T(\varkappa) > 0$ such that the mapping*

$$(Hv)(t,x) := -i\int_0^t\int_{\mathbb{R}^d} e^{-(t-s)\Lambda(b)}(x,z)(\varkappa\cdot b(z))v(s,z)dsdz, \\ (t,x)\in[0,T]\times\mathbb{R}^d,$$

is a contraction on $L^p(\mathbb{R}^d, |b|\eta^{-p+2}dx; L^\infty[0,T])$ (i.e. functions taking values in $L^\infty[0,T]$) for p as in Proposition 17.1.

Indeed, we have

$$
\begin{aligned}
|Hv(t,x)| &\le \left| \int_0^t \langle e^{-(t-s)\Lambda(b)}(x,\cdot)(\varkappa \cdot b(\cdot))v(s,\cdot)\rangle ds \right| \\
&\le |\varkappa| \int_0^t \langle e^{-(t-s)\Lambda(b)}(x,\cdot)|b(\cdot)|^{\frac{1}{p'}}|b(\cdot)|^{\frac{1}{p}}|v(s,\cdot)|\rangle ds \\
&\le |\varkappa| \int_0^t \langle e^{-(t-s)\Lambda(b)}(x,\cdot)|b(\cdot)|^{\frac{1}{p'}}|b(\cdot)|^{\frac{1}{p}} \sup_{\tau\in[0,T]} |v(\tau,\cdot)|\rangle ds \qquad (*)
\end{aligned}
$$

Let us note that, for every $x \in \mathbb{R}^d$,

$$
|b(x)|^{\frac{1}{p}}\eta^{-1}(x) \sup_{t\in[0,T]} \int_0^t \langle e^{-(t-s)\Lambda(b)}(x,\cdot)|b(\cdot)|^{\frac{1}{p'}}\eta(\cdot)\rangle ds
$$

(we are applying the Dominated Convergence Theorem)

$$
|b(x)|^{\frac{1}{p}}\eta^{-1}(x) \sup_{t\in[0,T]} \lim_m \int_0^t \langle e^{-(t-s)\Lambda(b)}(x,\cdot)|b_m(\cdot)|^{\frac{1}{p'}}\eta(\cdot)\rangle ds,
$$

where, in turn, the last term

$$
|b|^{\frac{1}{p}}\eta^{-1} \sup_{t\in[0,T]} \lim_m \int_0^t e^{-(t-s)\Lambda(b)}|b_m|^{\frac{1}{p'}}\eta ds
$$

$$
\le |b|^{\frac{1}{p}}\eta^{-1}e^{\mu T} \lim_m (\mu + \Lambda_{C_\infty}(b))^{-1}|b_m|^{\frac{1}{p'}}\eta \in \mathcal{B}(L^p_\eta) \quad \text{by Proposition 17.1}(E_3).
$$

Also by Proposition 17.1(E_3), selecting μ sufficiently large, and then selecting T sufficiently small, the $L^p_\eta \to L^p_\eta$ norm of the last operator can be made arbitrarily small. Applying this in (∗), we obtain that H is indeed a contraction on $L^p(\mathbb{R}^d, |b|\eta^{-p+2}dx; L^\infty[0,T])$.

We have $L^\infty([0,T]\times\mathbb{R}^d) \subset L^p(\mathbb{R}^d, |b|\eta^{-p+2}dx; L^\infty[0,T])$ since (see [44, Lemma 5.1]) $|b|\eta^{-p+2} \in L^1(\mathbb{R}^d)$. Combining the assertions of Steps 1–3, we obtain that for every $\varkappa \in \mathbb{R}^d$

$$
w(t,x,\varkappa) = \tilde{w}(t,x,\varkappa) \quad \text{in } L^p(\mathbb{R}^d, |b|\eta^{-p+2}dx; L^\infty[0,T]),
$$

and thus

$$
w(t,x,\varkappa) = \tilde{w}(t,x,\varkappa) \quad \text{for a.e. } x \in \mathbb{R}^d
$$

(although $t < T(\varkappa)$, one can get rid of this constraint using the reproduction property of $e^{-t\Lambda(b)}$, so without loss of generality $T \neq T(\varkappa)$). Now, applying the continuity of $\int_0^t e^{-s\Lambda_{C_\infty}(b)} b \cdot w ds$ on $\mathbb{R}_+ \times \mathbb{R}^d$ (this is [44, Corollary 1(*iii*)]) in the

RHS of (17.7), we obtain that for every $\varkappa \in \mathbb{R}^d$ $w(t,x,\varkappa)$ is continuous in t and x, and so $w = \tilde{w}$ everywhere. Thus, for all $t \leq T$, $x \in \mathbb{R}^d$

$$\mathbb{E}_x\left[e^{i\varkappa\cdot\left(X_t - X_0 - \int_0^t b(X_s)ds\right)}\right] = e^{-\varkappa\cdot x} w(t,x,\varkappa) = e^{-t|\varkappa|^\alpha}.$$

By a standard result, Z_t is a symmetric α-stable process. The proof of assertion (*iv*) is completed.

Appendix A: Proof of Lemma 6.1

The following proof was given in [40].

Step 1. Let us show that

$$\|T_p\|_{p\to p} \leq c_{\delta,p}, \quad \mu \geq \mu_0,$$

which will give us assertion (*i*). We will also prove that operators

$$G_p = b^{\frac{2}{p}} \cdot \nabla(\mu - \Delta)^{-1}, \quad Q_p = (\mu - \Delta)^{-1}|b|^{1-\frac{2}{p}}$$

satisfy

$$\|G_p\|_{p\to p} \leq C_1 \mu^{-\frac{1}{2}+\frac{1}{p}}, \quad \|Q_p\|_{p\to p} \leq C_2 \mu^{-\frac{1}{2}-\frac{1}{p}}$$

The latter will be needed to prove assertions (*ii*) and (*iii*).

We will be using the operator-norm formulation of the form-boundedness condition:

$$\|b(\lambda - \Delta)^{-\frac{1}{2}}\|_{2\to 2} \leq \delta$$

for some $\lambda = \lambda_\delta$, see (3.2).

(**a**) Set $v := (\mu - \Delta)^{-1}|b|^{1-\frac{2}{p}} f$, $0 \leq f \in L^p$. Then

$$\begin{aligned}
\|T_p f\|_p^p &= \|b^{\frac{2}{p}} \nabla v\|_p^p = \langle |b|^2 |\nabla v|^p \rangle \\
&= \||b|(\lambda - \Delta)^{-\frac{1}{2}}(\lambda - \Delta)^{\frac{1}{2}}|\nabla v|^{\frac{p}{2}}\|_2^2 \qquad (\lambda = \lambda_\delta) \\
&\leq \||b|(\lambda - \Delta)^{-\frac{1}{2}}\|_{2\to 2}^2 \|(\lambda - \Delta)^{\frac{1}{2}}|\nabla v|^{\frac{p}{2}}\|_2^2 \\
&= \delta\|(\lambda - \Delta)^{\frac{1}{2}}|\nabla v|^{\frac{p}{2}}\|_2^2 = \delta\big(\lambda\|\nabla v\|_p^p + \|\nabla|\nabla v|^{\frac{p}{2}}\|_2^2\big).
\end{aligned}$$

It remains to prove the principal inequality

$$\delta\big(\lambda\|\nabla v\|_p^p + \|\nabla|\nabla v|^{\frac{p}{2}}\|_2^2\big) \leq c_{\delta,p}^p\|f\|_p^p, \tag{$*$}$$

and conclude that $\|T_p\|_{p\to p} \leq c_{\delta,p}$.

First, we prove an a priori variant of ($*$), i.e. for $v := (\mu - \Delta)^{-1}|b|^{1-\frac{2}{p}} f$ with $b = b_n$. Since our assumptions on δ involve only strict inequalities, we may assume, upon selecting appropriate $\varepsilon_n \downarrow 0$, that $b_n \in \mathbf{F}_\delta$ with the same $\lambda = \lambda_\delta$ for all n.

Set

$$w := \nabla v, \quad I_q := \sum_{r=1}^d \langle(\nabla_r w)^2|w|^{p-2}\rangle, \quad J_q := \langle(\nabla|w|)^2|w|^{p-2}\rangle.$$

We multiply $(\mu - \Delta)v = |b|^{1-\frac{2}{p}} f$ by $\phi := -\nabla\cdot(w|w|^{p-2})$ and integrate by parts to obtain

$$\mu\|w\|_p^p + I_p + (p-2)J_p = \langle|b|^{1-\frac{2}{p}} f, -\nabla\cdot(w|w|^{p-2})\rangle, \tag{A.1}$$

where

$$\langle|b|^{1-\frac{2}{p}} f, -\nabla\cdot(w|w|^{p-2})\rangle = \langle|b|^{1-\frac{2}{p}} f, (-\Delta v)|w|^{p-2} - (p-2)|w|^{p-3}w\cdot\nabla|w|\rangle$$

(use the equation $-\Delta v = -\mu v + |b|^{1-\frac{2}{p}} f$)

$$= \langle|b|^{1-\frac{2}{p}} f, \big(-\mu v + |b|^{1-\frac{2}{p}} f\big)|w|^{p-2}\rangle - (p-2)\langle|b|^{1-\frac{2}{p}} f, |w|^{p-3}w\cdot\nabla|w|\rangle.$$

Remark A.1 Here we work with the same test function $\phi = -\nabla\cdot(w|w|^{p-2})$ as in [60].

We have

(1) $\langle|b|^{1-\frac{2}{p}} f, (-\mu v)|w|^{p-2}\rangle \leq 0$,
(2) $|\langle|b|^{1-\frac{2}{p}} f, |w|^{p-3}w\cdot\nabla|w|\rangle| \leq \alpha J_p + \frac{1}{4\alpha}N_p$ $(\alpha > 0)$, where $N_p := \langle|b|^{1-\frac{2}{p}} f, |b|^{1-\frac{2}{p}} f|w|^{p-2}\rangle$,

so, the RHS of (A.1) $\leq (p-2)\alpha J_p + \big(1 + \frac{p-2}{4\alpha}\big)N_p$, where, in turn,

$$N_p \leq \langle|b|^2|w|^p\rangle^{\frac{p-2}{p}}\langle f^p\rangle^{\frac{2}{p}}$$
$$\leq \frac{p-2}{p}\langle|b|^2|w|^p\rangle + \frac{2}{p}\|f\|_p^p$$

$$(\text{use } b \in \mathbf{F}_\delta \Leftrightarrow \|b\varphi\|_2^2 \leq \delta\|\nabla\varphi\|_2^2 + \lambda\delta\|\varphi\|_2^2, \varphi \in W^{1,2})$$

$$\leq \frac{p-2}{p}\Big(\frac{p^2}{4}\delta J_q + \lambda\delta\|w\|_p^p\Big) + \frac{2}{p}\|f\|_p^p.$$

Thus, applying $I_q \geq J_q$ in the LHS of (A.1), we obtain

$$(\mu - c_0)\|w\|_p^p + \left[p - 1 - (p-2)\left(\alpha + \frac{1}{4\alpha}\frac{p(p-2)}{4}\delta\right) - \frac{p(p-2)}{4}\delta\right]$$

$$\frac{4}{p^2}\|\nabla|\nabla v|^{\frac{p}{2}}\|_2^2 \leq \left(1 + \frac{p-2}{4\alpha}\right)\frac{2}{p}\|f\|_p^p,$$

where $c_0 = \frac{p-2}{p}\lambda\delta\big(1 + \frac{p-2}{4\alpha}\big)$. It is now clear that one can find a sufficiently large $\mu_0 = \mu_0(d, p, \delta) > 0$ so that, for all $\mu > \mu_0$, $(*)$ (with $b = b_n$) holds with

$$c_{\delta,p}^p = \delta\frac{p^2}{4}\frac{\left(1 + \frac{p-2}{4\alpha}\right)\frac{2}{p}}{p - 1 - (p-2)\left(\alpha + \frac{1}{4\alpha}\frac{p(p-2)}{4}\delta\right) - \frac{p(p-2)}{4}\delta} \quad \Big(\text{we select } \alpha = \frac{p}{4}\sqrt{\delta}\Big)$$

$$= \frac{\frac{p}{2}\delta + \frac{p-2}{2}\sqrt{\delta}}{p - 1 - (p-1)\frac{p-2}{2}\sqrt{\delta} - \frac{p(p-2)}{4}\delta},$$

as claimed. Finally, we pass to the limit $n \to \infty$ using Fatou's Lemma. The proof of $(*)$ is completed.

Remark A.2 It is seen that $\sqrt{\delta} < \frac{2}{p} \Rightarrow c_{\delta,p} < 1$. We also note that the above choice of α is the best possible.

(**b**) Set $v = (\mu - \Delta)^{-1} f$, $0 \leq f \in L^p$. Then

$$\|G_p f\|_p^p = \|b^{\frac{2}{p}} \cdot \nabla v\|_p^p$$

(we argue as in (**a**))

$$\leq \delta\big(\lambda\|\nabla v\|_p^p + \|\nabla|\nabla v|^{\frac{p}{2}}\|_2^2\big),$$

where, clearly, $\|\nabla v\|_p^p \leq \mu^{-\frac{p}{2}}\|f\|_p^p$. In turn, arguing as in (**a**), we arrive at $\mu\|w\|_p^p + I_p + (p-2)J_p = \langle f, -\nabla \cdot (w|w|^{p-2}\rangle$ $(w = \nabla v)$,

$$\mu\|w\|_p^p + (p-1)J_p \leq \langle f^2, |w|^{p-2}\rangle + (p-2)\langle f, |w|^{p-3}w \cdot \nabla|w|\rangle),$$

$$\mu\|w\|_p^p + (p-1)J_p \leq \langle f^2, |w|^{p-2}\rangle + (p-2)\big(\varepsilon J_p + \frac{1}{4\varepsilon}\langle f^2, |w|^{p-2}\rangle\big), \quad \varepsilon > 0.$$

Selecting ε sufficiently small, we obtain

$$J_p \le C_0 \|w\|_p^{p-2} \|f\|_p^2.$$

Now, applying $\|w\|_p \le \mu^{-\frac{1}{2}} \|f\|_p$, we arrive at $\|\nabla|\nabla v|^{\frac{p}{2}}\|_2^2 \le C\mu^{-\frac{p}{2}+1}\|f\|_p^p$. Hence, $\|G_p f\|_p \le C_1 \mu^{-\frac{1}{2}+\frac{1}{p}} \|f\|_p$ for all $\mu \ge \mu_0$.

(c) Set $v = (\mu - \Delta)^{-1}|b|^{1-\frac{2}{p}} f$ $(= Q_p f)$, $0 \le f \in L^p$. Then, multiplying $(\mu - \Delta)v = |b|^{1-\frac{2}{p}} f$ by v^{p-1}, we obtain

$$\mu\|v\|_p^p + \frac{4(p-1)}{p^2}\|\nabla v^{\frac{p}{2}}\|_2^2 = \langle |b|^{1-\frac{2}{p}} f, v^{p-1}\rangle,$$

where we estimate the RHS using Young's inequality:

$$\langle |b|^{1-\frac{2}{p}} v^{\frac{p}{2}-1}, f v^{\frac{p}{2}}\rangle \le \varepsilon^{\frac{2p}{p-2}} \frac{p-2}{2p}\langle |b|^2 v^p\rangle + \varepsilon^{-\frac{2p}{p+2}} \frac{p+2}{2p}\langle f^{\frac{2p}{p+2}} v^{\frac{p^2}{p+2}}\rangle \quad \varepsilon > 0.$$

Using $b \in \mathbf{F}_\delta$ and selecting $\varepsilon > 0$ sufficiently small, we obtain that for any $\mu_1 > 0$ there exists $C > 0$ such that

$$(\mu - \mu_1)\|v\|_p^p \le C\langle f^{\frac{2p}{p+2}} v^{\frac{p^2}{p+2}}\rangle, \qquad \mu > \mu_1.$$

Therefore, $(\mu - \mu_1)\|v\|_p^p \le C\langle f^p\rangle^{\frac{2}{p+2}}\langle v^p\rangle^{\frac{p}{p+2}}$, so $\|v\|_p \le C_2 \mu^{-\frac{1}{2}-\frac{1}{p}}\|f\|_p$.

Step 2: We now use the results of Step 1 to prove assertions (*ii*) and (*iii*). Below we use the following formula: For every $0 < \alpha < 1$, $\mu > 0$,

$$(\mu - \Delta)^{-\alpha} = \frac{\sin \pi\alpha}{\pi}\int_0^\infty t^{-\alpha}(t + \mu - \Delta)^{-1} dt.$$

We have

$$\begin{aligned}\|Q_p(q) f\|_p \le\ & \|(\mu - \Delta)^{-\frac{1}{2}+\frac{1}{q}} |b|^{1-\frac{2}{p}} |f|\|_p \\ & \le k_q \int_0^\infty t^{-\frac{1}{2}+\frac{1}{q}} \|(t + \mu - \Delta)^{-1} |b|^{1-\frac{2}{p}} |f|\|_p dt \\ & \text{(we use (c))} \\ & \le k_q C_2 \int_0^\infty t^{-\frac{1}{2}+\frac{1}{q}} (t+\mu)^{-\frac{1}{2}-\frac{1}{p}} dt\ \|f\|_p = K_{2,q}\|f\|_p, \quad f \in \mathcal{E},\end{aligned}$$

where, clearly, $K_{2,q} < \infty$ due to $q > p$.

It suffices to consider the case $r > 2$. We have

$$\|G_p(r)f\|_p \leq k_r \int_0^\infty t^{-\frac{1}{2}-\frac{1}{r}} \|b^{\frac{2}{p}} \cdot \nabla(t+\mu-\Delta)^{-1} f\|_p dt$$

(we use (**b**))

$$\leq k_r C_1 \int_0^\infty t^{-\frac{1}{2}-\frac{1}{r}} (t+\mu)^{-\frac{1}{2}+\frac{1}{p}} dt \; \|f\|_p = K_{1,r}\|f\|_p, \quad f \in \mathcal{E},$$

where, clearly, $K_{1,r} < \infty$ due to $r < p$. The proof of Lemma 6.1 is completed. □

Appendix B: Some Examples of Form-Bounded Vector Fields

Below we list some sub-classes of the class of form-bounded vector fields, defined in elementary terms.

1. Let us prove that

$$b \in L^\infty(\mathbb{R}_+, L^d + L^\infty) \quad \Rightarrow \quad b \in L^\infty \mathbf{F}_\delta + L^2_{\rm loc}(\mathbb{R}_+)$$

for appropriate δ and g (see Definition 9.1). Here we have, by definition, $b = b_1 + b_2$, where $b_1 \in L^\infty(\mathbb{R}_+, L^d)$, $b_2 \in L^\infty(\mathbb{R}_+, L^\infty)$. By Hölder's inequality, for a.e. $t \in \mathbb{R}_+$ and all $\varphi \in C_c^\infty$,

$$\|b(t)\varphi\|_2^2 \leq (1+\varepsilon)\|b_1(t)\|_d^2 \|\varphi\|^2_{\frac{2d}{d-2}} + (1+\varepsilon^{-1})\|b_2(t)\|_\infty^2 \|\varphi\|_2^2 \qquad (\varepsilon > 0)$$

(apply the Sobolev embedding theorem)

$$\leq C_S(1+\varepsilon)\|b_1(t)\|_d^2 \|\nabla\varphi\|_2^2 + (1+\varepsilon^{-1})\|b_2(t)\|_\infty^2 \|\varphi\|_2^2.$$

Thus, $b \in L^\infty \mathbf{F}_\delta + L^2_{\rm loc}(\mathbb{R}_+)$ with

$$\delta := C_S(1+\varepsilon)\|b_1\|^2_{L^\infty(\mathbb{R}_+, L^d)}, \quad g(t) := (1+\varepsilon^{-1})\|b_2(t,\cdot)\|_\infty^2$$

(in this paper we mostly care about the value of δ, so ε should be chosen sufficiently small).

2. Next, let us show that

$$b \in C(\mathbb{R}_+, L^d + L^\infty) \quad \Rightarrow \quad b \in \mathbf{F}_\delta$$

with δ that can be chosen arbitrarily small.

Without loss of generality, let us carry out the proof for $b \in C(\mathbb{R}_+, L^d)$.

First, let $b = b(x)$. Since $|b| \in L^d$, one can represent for every $\varepsilon > 0$ $b = b_1 + b_2$, where $\|b_1\|_d < \varepsilon$ and $\|b_2\|_\infty < \infty$. (For instance, $b_2 = b\mathbf{1}_{|b|\leq m}$ and $b_1 = b - b_2$, so by the Dominated convergence theorem $\|b_1\|_d$ can be made arbitrarily small by selecting m sufficiently large.) Now the previous example applies and yields the required.

In the general case $b \in C(\mathbb{R}_+, L^d)$, the continuity of b in time allows us to represent $b(t,\cdot) = b_1(t,\cdot) + b_2(t,\cdot)$, where $\|b_1(t,\cdot)\|_d < \varepsilon$ for all $t \in [0,1]$ and b_2 is bounded on $[0,1]\times\mathbb{R}^d$. Repeating this on every interval $[n, n+1]$ ($n \geq 1$), one obtains $\|b_1\|_{L^\infty(\mathbb{R}_+,L^d)} < \varepsilon$ and $b_2 \in L^\infty_{\rm loc}(\mathbb{R}_+, L^\infty)$. In fact, the continuity in time is not necessary for the smallness of δ, e.g. consider $b(t,x) = c(t)b_0(x)$ where $c \in L^\infty(\mathbb{R}_+)$ is discontinuous and $|b_0| \in L^d$.

3. Any vector field

$$b \in L^p(\mathbb{R}_+, L^q), \quad \frac{d}{q} + \frac{2}{p} \leq 1, \quad p \geq 2, \quad q \geq d$$

is in $L^\infty\mathbf{F}_\delta + L^2_{\rm loc}(\mathbb{R}_+)$ with appropriate δ. Indeed, e.g. in the more difficult case $\frac{d}{q} + \frac{2}{p} = 1$, we have by Young's inequality

$$|b(t,x)| = \frac{|b(t,x)|}{\langle|b(t,\cdot)|^q\rangle^{\frac{1}{q}}}\langle|b(t,\cdot)|^q\rangle^{\frac{1}{q}} \leq \frac{d}{q}\left(\frac{|b(t,x)|^q}{\langle|b(t,\cdot)|^q\rangle}\right)^{\frac{1}{d}} + \frac{2}{p}\big(\langle|b(t,\cdot)|^q\rangle^{\frac{1}{q}}\big)^{\frac{p}{2}},$$

where the first term is in $L^\infty(\mathbb{R}_+, L^d)$ (and so by the first example it is form-bounded) and the second term is in $L^2(\mathbb{R}_+, L^\infty)$ (the second term squared is to be absorbed by the function g). If $p < \infty$, $q > d$, then one can argue as in the previous example to show that δ can be chosen arbitrarily small.

4. The class $\mathbf{F}_\delta$ contains vector fields $b = b(x)$ with $|b|$ in $L^{d,w}$, i.e. the weak L^d class (Sect. 2). Indeed, by [61, Prop. 2.5, 2.6, Cor. 2.9], if $|b| \in L^{d,w}$, then $b \in \mathbf{F}_{\delta_1}$ with

$$\begin{aligned}\sqrt{\delta_1} &= \||b|(\lambda-\Delta)^{-\frac{1}{2}}\|_{2\to 2} \\ &< \|b\|_{d,w}\Omega_d^{-\frac{1}{d}}\||x|^{-1}(\lambda-\Delta)^{-\frac{1}{2}}\|_{2\to 2} \leq \|b\|_{d,w}\Omega_d^{-\frac{1}{d}}\frac{2}{d-2},\end{aligned}$$

where $\Omega_d = \pi^{\frac{d}{2}}\Gamma(\frac{d}{2}+1)$ is the volume of $B_1(0) \subset \mathbb{R}^d$.

5. The Chang-Wilson-Wolff class CW^2 consists of vector fields $b = b(x)$ such that

$$|b| \in L^2_{\rm loc}(\mathbb{R}^d)$$

and

$$\|b\|_{\mathrm{CW}^2} := \left(\sup_Q \frac{1}{|Q|}\int_Q |b(x)|^2 l(Q)^2 \xi\big(|b(x)|^2 l(Q)^2\big)dx < \infty\right)^{\frac{1}{2}} < \infty,$$

where $|Q|$ and $l(Q)$ are the volume and the side length of cube $Q \subset \mathbb{R}^d$, respectively, $\xi : \mathbb{R}_+ \to [1, \infty[$ is an increasing function such that

$$\int_1^\infty \frac{dx}{x\xi(x)} < \infty.$$

One has, for every $\varepsilon > 0$,

$$M_{2+\varepsilon} \quad \subsetneq \quad \mathrm{CW}^2 \quad \subsetneq \quad \mathbf{F}_\delta$$

with $\delta = \delta\big(\|b\|_{\mathrm{CW}^2}\big)$, see [13].

6. More generally, vector fields in $L^\infty(\mathbb{R}_+, M_q)$, $q > 2$, or $L^\infty(\mathbb{R}_+, \mathrm{CW}^2)$ are form-bounded.

Appendix C: Smooth Approximations of Form-Bounded (-Type) Vector Fields

We construct two kinds of smooth approximations (or regularizations) of a vector field b: defined by mollifying a cutoff of b, or by mollifying b directly.

1. If $b = b(x)$ is either in the class $\mathbf{F}_\delta$ (form-bounded) or in the class $\mathbf{F}_\delta^{1/2}$ (weakly form-bounded), it is clear that multiplying b by the indicator $\mathbf{1}_m$ of $\{x \in \mathbb{R}^d \mid |b(x)| \leq m, |x| \leq m\}$ leaves us in the corresponding class with the same δ and λ_δ. This gives us a bounded, compact-support approximation $\{\mathbf{1}_m b\}_{m=1}^\infty$ of b (in L^2_{loc} or in L^1_{loc}, respectively).

We can also go one step further and apply to $\mathbf{1}_m b$ a mollifier, which will still give us a uniformly form-bounded (or uniformly weakly form-bounded) approximation of b by vector fields whose components are in the Schwartz space $\mathcal{S}$. Below we provide the details of this simple construction for a time-inhomogeneous form-bounded vector field $b \in L^\infty\mathbf{F}_\delta + L^2_{\mathrm{loc}}(\mathbb{R}_+)$:

$$\|b(t,\cdot)\varphi\|_2^2 \leq \delta\|\nabla\varphi\|_2^2 + g(t)\|\varphi\|_2^2, \quad \varphi \in W^{1,2} \tag{$\bullet$}$$

for a.e. $t \in \mathbb{R}$, for a $0 \leq g \in L^1_{\mathrm{loc}}(\mathbb{R})$ (Definition 9.1).

Let us extend b to $\{t < 0\}$ by 0 and set

$$b_m := c_m E_\varepsilon(\mathbf{1}_m b), \tag{$\star$}$$

where $E_\varepsilon \equiv E_\varepsilon^{d+1}$ is the De Giorgi or Friedrichs mollifier on $\mathbb{R} \times \mathbb{R}^d$ (Sect. 2), $\mathbf{1}_m$ is the indicator of $\{(t,x) \in \mathbb{R}^{1+d} \mid |b(t,x)| \leq m, |x| \leq m, |t| \leq m\}$, and $\varepsilon_m \downarrow 0$ and $c_m \uparrow 1$ are to be chosen. Clearly,

$$b_m \in L^\infty \cap C^\infty(\mathbb{R}^{1+d}, \mathbb{R}^d), \tag{C.1}$$

also, provided that $\varepsilon_m \downarrow 0$ sufficiently rapidly,

$$b_m \to b \quad \text{in } L^2_{\text{loc}}(\mathbb{R}_+ \times \mathbb{R}^d, \mathbb{R}^d). \tag{C.2}$$

and, provided that $c_m \uparrow 1$ sufficiently slow,

$$\|b_m(t)\varphi\|_2^2 \leq \delta\|\nabla\varphi\|_2^2 + g(t)\|\varphi\|_2^2, \quad t \geq 0, \quad \varphi \in W^{1,2}, \tag{C.3}$$

i.e. $\{b_m\}$ are uniformly form-bounded.

Proof of (C.3) First, define $\tilde{b}_m = E_\varepsilon(\mathbf{1}_m b)$ and write

$$\tilde{b}_m = \mathbf{1}_m b + \big(\tilde{b}_m - \mathbf{1}_m b\big). \tag{$*$}$$

Clearly, the first term satisfies

$$\|\mathbf{1}_m b(t)\varphi\|_2^2 \leq \delta\|\nabla\varphi\|_2^2 + g(t)\|\varphi\|_2^2,$$

In turn, since $\mathbf{1}_m b$ has compact support and is in $L^\infty(\mathbb{R}_+, L^r)$ for any $r > d$, given any $\gamma_m \downarrow 0$ we can select $\varepsilon_m \downarrow 0$ so that $\|\tilde{b}_m - \mathbf{1}_m b\|^2_{L^r(\mathbb{R}_+ \times \mathbb{R}^d)} \leq \gamma_m$, so, in view of example 3 in Appendix B, the second term in $(*)$ satisfies

$$\|\big(\tilde{b}_m(t) - \mathbf{1}_m b(t)\big)\varphi\|_2^2 \leq C\gamma_m\|\nabla\varphi\|_2^2.$$

Therefore,

$$\|\tilde{b}_m(t)\varphi\|_2^2 \leq \delta_m\|\nabla\varphi\|_2^2 + g(t)\|\varphi\|_2^2$$

with $\delta_m = (\sqrt{\delta} + \sqrt{C\gamma_m})^2$. Now, multiplying $\tilde{b}_m$ by $c_m = \frac{\delta}{\delta_m}$ (clearly, $c_m \uparrow 1$) and recalling that $b_m = c_m\tilde{b}_m$, we obtain (C.3).

2. In fact, do not need the cutoff function in $(\star)$ to construct a smooth approximation of a $b = b(x)$, $b \in \mathbf{F}_\delta$:

$$\|b\varphi\|_2^2 \leq \delta\|\nabla\varphi\|_2^2 + c_\delta\|\varphi\|_2^2, \quad \varphi \in W^{1,2}.$$

This observation is important e.g. if one needs to control the divergence of the approximating vector fields. In [49] the authors defined

$$b_\varepsilon := E_\varepsilon b, \tag{$\star\star$}$$

where $E_\varepsilon \equiv E_\varepsilon^d := e^{\varepsilon\Delta}$ is the De Giorgi's mollifier on $\mathbb{R}^d$ and $\varepsilon \downarrow 0$ (at any rate). (We can also use Friedrichs' mollifier, see remark below.) We have

$$b_\varepsilon \in L^\infty \cap C^\infty(\mathbb{R}^d, \mathbb{R}^d), \tag{C.4}$$

$$\|b_\varepsilon\varphi\|_2^2 \leq \delta\|\nabla\varphi\|_2^2 + c_\delta\|\varphi\|_2^2, \qquad \text{i.e. } b_\varepsilon \in \mathbf{F}_\delta \text{ with the same } c_\delta, \tag{C.5}$$

$$b_\varepsilon \to b \quad \text{in } L^2_{\rm loc}(\mathbb{R}^d, \mathbb{R}^d). \tag{C.6}$$

Proof of (C.4)–(C.6) We repeat the argument from [49].

To prove (C.4), we represent $b_\varepsilon = E_{\varepsilon/2}E_{\varepsilon/2}b$, so it suffices to only prove that $|b_\varepsilon| \in L^\infty$. Indeed, we have, using Fatou's lemma,

$$\begin{aligned}|b_\varepsilon(x)| &\leq \liminf_n \langle e^{\varepsilon\Delta}(x,\cdot)\mathbf{1}_{B_n(0)}(\cdot)|b(\cdot)|\rangle \\ &\leq \liminf_n \langle e^{\varepsilon\Delta}(x,\cdot)\mathbf{1}_{B_n(0)}(\cdot)|b(\cdot)|^2\rangle^{\frac{1}{2}} \leq \big(\delta\langle|\nabla\sqrt{e^{\varepsilon\Delta}(x,\cdot)}|^2\rangle + c_\delta\big)^{\frac{1}{2}},\end{aligned}$$

where $\left|\nabla_y\sqrt{e^{\varepsilon\Delta}(x,y)}\right| = (4\pi\varepsilon)^{-\frac{d}{4}}\frac{|x-y|}{4\varepsilon}e^{-\frac{|x-y|^2}{8\varepsilon}} \leq C\varepsilon^{-\frac{d}{4}-\frac{1}{2}}e^{-\frac{c|x-y|^2}{\varepsilon}}$, and so $|b_\varepsilon| \in L^\infty$ for each $\varepsilon > 0$.

Let us prove (C.5). Indeed, $|b_\varepsilon| \leq \sqrt{E_\varepsilon|b|^2}$, and so

$$\begin{aligned}\|b_\varepsilon\varphi\|_2^2 &\leq \langle E_\varepsilon|b|^2, \varphi^2\rangle = \|b\sqrt{E_\varepsilon\varphi^2}\|_2^2 \\ &\leq \delta\|\nabla\sqrt{E_\varepsilon\varphi^2}\|_2^2 + c_\delta\|\varphi\|_2^2, \quad \varphi \in W^{1,2},\end{aligned}$$

where

$$\begin{aligned}\|\nabla\sqrt{E_\varepsilon\varphi^2}\|_2 &= \left\|\frac{E_\varepsilon(|\varphi||\nabla|\varphi|)}{\sqrt{E_\varepsilon\varphi^2}}\right\|_2 \\ &\leq \|\sqrt{E_\varepsilon|\nabla|\varphi||^2}\|_2 = \|E_\varepsilon|\nabla|\varphi||^2\|_1^{\frac{1}{2}} \\ &\leq \|\nabla|\varphi|\|_2 \leq \|\nabla\varphi\|_2,\end{aligned} \tag{**}$$

i.e. $b_\varepsilon \in \mathbf{F}_\delta$. (The fact that $\|b\sqrt{E_\varepsilon\varphi^2}\|_2 < \infty$ follows from $\mathbf{1}_{\{|b|\leq n\}}b \in \mathbf{F}_\delta$ and the inequality $\|\mathbf{1}_{\{|b|\leq n\}}b\sqrt{E_\varepsilon\varphi^2}\|_2^2 \leq \delta\|\nabla\varphi\|_2^2 + c_\delta\|\varphi\|_2^2$, using Fatou's lemma).

Regarding the proof of (C.6), let us only demonstrate that $b_\varepsilon - b \to 0$ in $L^2(B_1)$, $B_1 \equiv B_1(0)$. To that end, we fix some $R > 1$ and represent on B_1:

$$b_\varepsilon - b = I_1 + I_2, \qquad I_1 := E_\varepsilon(1-\mathbf{1}_{B_R})b, \quad I_2 := E_\varepsilon(\mathbf{1}_{B_R}b) - \mathbf{1}_{B_R}b$$

(of course, on B_1, one has $b = \mathbf{1}_{B_R}b$ since $R > 1$). Then $I_2 \to 0$ in $L^2(B_1)$ since $\mathbf{1}_{B_R}b$ has compact support. In turn, $I_1 \to 0$ in $L^2(B_1)$ by the separation property of the Gaussian density, i.e. $e^{\varepsilon\Delta}(x,y) \to 0$ uniformly in $x \in B_1$ if $y \in \mathbb{R}^d - B_R$ (here we have used $R > 1$).

Remark C.1 If we were to use the Friedrichs mollifier in ($\star\star$), then we would get smoothness of b_ε and convergence (C.6) from the usual properties of Friedrichs mollifiers. But we would have to be slightly more careful in (∗∗) in order to avoid division by zero, e.g. replace $|f|^2$ by $|f|^2+e^{-k|x|^2}$, carry out the estimates and then take $k \to \infty$.

3. The regularization (⋆⋆) can also be used to handle time-inhomogeneous form-bounded drifts $b \in L^\infty \mathbf{F}_\delta + L^2_{\rm loc}(\mathbb{R})$. That is, we can put

$$b_\varepsilon := E^1_\varepsilon E^d_\varepsilon b, \quad \varepsilon \downarrow 0,$$

where E^d_ε is the De Giorgi or Friedrichs mollifier on $\mathbb{R}^d$ (in the spatial variables) and E^1_ε is the Friedrichs mollifier on $\mathbb{R}$ (in the time variable; we use the Friedrichs mollifier here since, in the time variable, b is in only locally in $L^1(\mathbb{R})$, as is determined by our assumption on g, cf. (•)).

This regularization $\{b_\varepsilon\}$ satisfies

$$|b_\varepsilon| \in L^\infty_{\rm loc}(\mathbb{R}, L^\infty(\mathbb{R}^d)), \quad b_\varepsilon \text{ are } C^\infty \text{ smooth,} \tag{C.7}$$

$$\|b_\varepsilon(t,\cdot)\varphi\|_2^2 \leq \delta\|\nabla\varphi\|_2^2 + g_\varepsilon(t)\|\varphi\|_2^2 \quad \text{for all } t \in \mathbb{R}, \tag{C.8}$$

$$\sup_{\varepsilon>0}\int_{t_0}^{t_1} g_\varepsilon(s)ds < \infty \text{ for all finite } t_0, t_1, \text{ where } g_\varepsilon := E^1_\varepsilon g, \tag{C.9}$$

$$b_\varepsilon \to b \quad \text{in } L^2_{\rm loc}(\mathbb{R}^{1+d}, \mathbb{R}^d). \tag{C.10}$$

Proof of (C.8) First, we regularize b only in the spatial variable: for every $t \in \mathbb{R}$, put

$$\tilde{b}_\varepsilon(t,\cdot) := E^d_\varepsilon b(t,\cdot).$$

Then, since for a.e. t $b(t,\cdot)$ is form-bounded on $\mathbb{R}^d$, we have by (C.5)

$$\|\tilde{b}_\varepsilon(t,\cdot)\varphi\|_2^2 \leq \delta\|\nabla\varphi\|_2^2 + g(t)\|\varphi\|_2^2 \tag{$*\,*\,*$}$$

for a.e. $t \in \mathbb{R}$. Next, recalling that $b_\varepsilon = E^1_\varepsilon \tilde{b}_\varepsilon$ and noting a pointwise inequality

$$|b_\varepsilon(t)|^2 \leq (E^1_\varepsilon|\tilde{b}_\varepsilon|^2)(t), \quad t \in \mathbb{R},$$

we estimate, for every $\varphi \in W^{1,2}(\mathbb{R}^d)$,

$$\begin{aligned}\langle|b_\varepsilon(t)|^2\varphi^2\rangle &\leq E^1_\varepsilon\langle|\tilde{b}_\varepsilon|^2\varphi^2\rangle(t) \\ &\quad \text{(we are applying } E^1_\varepsilon \text{ to both sides of the inequality } (*\,*\,*)) \\ &\leq \delta\|\nabla\varphi\|_2^2 + g_\varepsilon(t)\|\varphi\|_2^2.\end{aligned}$$

Thus, we arrive at (C.8).

A similar construction was considered in [54].

Appendix D: Trotter's Approximation Theorem

Consider a sequence $\{e^{-tA_k}\}_{k=1}^{\infty}$ of C_0 semigroups on a (complex) Banach space Y.

Theorem D.1 (H.F. Trotter [35, Ch. IX]) *Let* $\sup_k \|(\mu + A_k)^{-m}\|_{Y \to Y} \leq M(\mu - \omega)^{-m}$, $m = 1, 2, \dots$, $\mu > \omega$, *and*

$$s\text{-}\lim_{\mu \to \infty} \mu(\mu + A_k)^{-1} = 1 \quad \textit{uniformly in } k,$$

and let $s\text{-}\lim_k (\zeta + A_k)^{-1}$ *exist for some* ζ *with* $\operatorname{Re} \zeta > \omega$. *Then there is a* C_0 *semigroup* e^{-tA} *such that*

$$(z + A_k)^{-1} \overset{s}{\to} (z + A)^{-1} \quad \textit{for every } \operatorname{Re} z > \omega,$$

and

$$e^{-tA_k} \overset{s}{\to} e^{-tA}$$

uniformly in any finite interval of $t \geq 0$.

The first condition of the theorem is satisfied if e.g.

$$\sup_k \|(\mu + A_k)^{-1}\|_{Y \to Y} \leq (\mu - \omega)^{-1}, \quad \mu > \omega$$

(obviously) of if

$$\sup_k \|(z + A_k)^{-1}\|_{Y \to Y} \leq C|z - \omega|^{-1}, \quad \operatorname{Re} z > \omega,$$

see [35, IX.6.1]. The second condition is what can be verified in practice when one is dealing with quasi bounded semigroups.

References

1. Adams, D.: Weighted nonlinear potential theory. Trans. Am. Math. Soc. **297**, 73–94 (1986)
2. Adams, R., Fournier, J.: Sobolev Spaces, 2nd edn. Elsevier, Amsterdam (2003)
3. Albrighton, D., Dong, H.: Regularity properties of passive scalars with rough divergence-free drifts. Arch. Rat. Mech. Anal. **247** (2023). Article no. 75. arXiv:2107.12511
4. Bagby, R.J.: Lebesgue spaces of parabolic potentials. Illinois J. Math. **15**, 610–634 (1971)

5. Baras, P., Goldstein, J.A.: The heat equation with a singular potential. Trans. Am. Math. Soc. **284**, 121–139 (1984)
6. Bass, R., Chen, Z.-Q.: Brownian motion with singular drift. Ann. Probab. **31**, 791–817 (2003)
7. Beck, L., Flandoli, F., Gubinelli, M., Maurelli, M.: Stochastic ODEs and stochastic linear PDEs with critical drift: regularity, duality and uniqueness. Electr. J. Probab. **24**, 72 pp. (2019). Paper No. 136. arXiv:1401.1530
8. Belyi, A.G., Semenov, Yu.A.: On the L^p-theory of Schrödinger semigroups. II. Sibirsk. Math. J. **31**, 16–26 (1990). English transl. in Siberian Math. J. **31**, 540–549 (1991)
9. Birman, M.S.: On the spectrum of singular boundary-value problems. Mat. Sbornik **55**(97), 125–174 (1961) (in Russian)
10. Blumenthal, R.M., Getoor, R.K.: Markov Processes and Potential Theory. Pure and Applied Mathematics, vol. 29. Academic Press, New York (1968)
11. Bogdan, K., Jakubowski, T.: Estimates of heat kernel of fractional Laplacian perturbed by gradient operators. Commun. Math. Phys. **271**, 179–198 (2007)
12. Bresch, D., Jabin, P.-E., Wang, Z.: Mean field limit and quantitative estimates with singular attractive kernels. Duke Math. J. **172**(13), 2591–2641 (2023). arXiv:2011.08022
13. Chang, S.Y.A., Wilson, J.M., Wolff, T.H.: Some weighted norm inequalities concerning the Schrödinger operator. Comment. Math. Helvetici **60**, 217–246 (1985)
14. Chaudru de Raynal, P.-É., Menozzi, S.: On multi-dimensional stable driven stochastic differential equations with Besov drift. Electron. J. Probab. **27**, 52 pp. (2022). Paper No. 163. arXiv:1907.12263
15. Chaudru de Raynal, P.-É., Jabir, J.-F., Menozzi, S.: Multidimensional stable driven McKean-Vlasov SDEs with distributional interaction kernel: a regularization by noise perspective. arXiv:2205.11866
16. Chen, Z.-Q., Wang, L.: Uniqueness of stable processes with drift. Proc. Am. Math. Soc. **144**, 2661–2675 (2017). arXiv:1309.6414
17. Chen, Z.-Q., Fitzsimmons, P.J., Kuwae, K., Zhang, T.-S.: Perturbation of symmetric Markov processes. Probab. Theory Relat. Fields **140**, 239–275 (2008)
18. Chen, Z.-Q., Kim, P., Song, R.: Dirichlet heat kernel estimates for fractional Laplacian with gradient perturbation. Ann. Prob. **40**, 2483–2538 (2012)
19. Cherny, A.S.: On the uniqueness in law and the pathwise uniqueness for stochastic differential equations. Theory Probab. Appl. **46**(3), 406–419 (2002)
20. Cherny, A.S., Engelbert, H.-J.: Singular Stochastic Differential Equations. LNM, vol. 1858. Springer, Berlin (2005)
21. Chiarenza, F., Frasca, M.: A remark on a paper by C. Fefferman. Proc. Am. Math. Soc. **108**, 407–409 (1990)
22. Fefferman, C.: The uncertainty principle. Bull. Am. Math. Soc. **9**, 129–206 (1983)
23. Fitzsimmons, P.J., Kuwae, K.: Non-symmetric perturbations of symmetric Dirichlet forms. J. Funct. Anal. **208**, 140–162 (2004)
24. Flandoli, F., Gubinelli, M., Priola, E.: Well-posedness of the transport equation by stochastic perturbation. Invent. Math. **180**, 1–53 (2010)
25. Flandoli, F., Issoglio, E., Russo, F.: Multidimensional stochastic differential equations with distributional drift. Trans. Am. Math. Soc. **369**, 1665–1688 (2017). arXiv:1401.6010
26. Fournier, N., Jourdain, B.: Stochastic particle approximation of the Keller-Segel and two-dimensional generalization of Bessel process. Ann. Appl. Probab. **27**, 2807–2861 (2017)
27. Gess, B., Maurelli, M.: Well-posedness by noise for scalar conservation laws. Commun. Partial Differ. Equ. **43**(12), 1702–1736 (2018)
28. Goldstein, J.A., Zhang, Qi.S.: Linear parabolic equation with strongly singular potentials. Trans. Am. Math. Soc. **355**, 197–211 (2003)
29. Gopala Rao, V.R.: A characterization of parabolic function spaces. Am. J. Math. **99**, 985–993 (1977)
30. Grafakos, L., Oh, S.: The Kato-Ponce inequality. Commun. Partial Differ. Equ. **39**, 1128–1157 (2014)

31. Gulisashvili, A., van Casteren, J.A.: Non-autonomous Kato Classes and Feynman-Kac Propagators. World Scientific, Singapore (2006)
32. Hara, T.: A refined subsolution estimate of weak subsolutions to second order linear elliptic equations with a singular vector field. Tokyo J. Math. **38**(1), 75–98 (2015)
33. Heinz, E.: Beiträge zur Störungstheorie der Spektralzerlegung. Math. Ann. **123**, 415–438 (1951)
34. Jin, P.: Brownian motion with singular time-dependent drift. J. Theor. Probab. **30**, 1499–1538 (2017). arXiv:1710.05227
35. Kato, T.: Perturbation Theory for Linear Operators. Springer, Berlin (1995)
36. Kim, P., Song, R.: Stable process with singular drift. Stoc. Proc. Appl. **124**, 2479–2516 (2014)
37. Kinzebulatov, D.: A new approach to the L^p-theory of $-\Delta + b \cdot \nabla$, and its applications to Feller processes with general drifts. Ann. Sc. Norm. Sup. Pisa (5) **17**, 685–711 (2017). arXiv:1502.07286
38. Kinzebulatov, D.: Feller evolution families and parabolic equations with form-bounded vector fields. Osaka J. Math. **54**, 499–516 (2017). arXiv:1407.4861
39. Kinzebulatov, D.: Feller generators with measure-valued drifts. Potential Anal. **48**, 207–222 (2018)
40. Kinzebulatov, D.: Regularity theory of Kolmogorov operator revisited. Can. Bull. Math. **64**, 725–736 (2021). arXiv:1807.07597
41. Kinzebulatov, D.: Parabolic equations and SDEs with time-inhomogeneous Morrey drift. arXiv:2301.13805
42. Kinzebulatov, D.: Laplacian with singular drift in a critical borderline case. arXiv:2309.04436
43. Kinzebulatov, D., Madou, K.R.: Stochastic equations with time-dependent singular drift. J. Differential Equ. **337**, 255–293 (2022). arXiv:2105.07312
44. Kinzebulatov, D., Madou, K.R.: On admissible singular drifts of symmetric α-stable process. Math. Nachr. **295**(10), 2036–2064 (2022). arXiv:2002.07001
45. Kinzebulatov, D., Madou, K.R.: Strong solutions of SDEs with singular (form-bounded) drift via Roeckner-Zhao approach. arXiv:2306.04825
46. Kinzebulatov, D., Semënov, Yu.A.: Brownian motion with general drift. Stoch. Proc. Appl. **130**, 2737–2750 (2020). arXiv:1710.06729
47. Kinzebulatov, D., Semënov, Yu.A.: On the theory of the Kolmogorov operator in the spaces L^p and C_∞. Ann. Sc. Norm. Sup. Pisa (5) **21**, 1573–1647 (2020). arXiv:1709.08598
48. Kinzebulatov, D., Semënov, Yu.A.: Feller generators and stochastic differential equations with singular (form-bounded) drift. Osaka J. Math. **58**, 855–883 (2021). arXiv:1904.01268
49. Kinzebulatov, D., Semënov, Yu.A.: Heat kernel bounds for parabolic equations with singular (form-bounded) vector fields. Math. Ann. **384**, 1883–1929 (2022)
50. Kinzebulatov, D., Semënov, Yu.A.: Kolmogorov operator with the vector field in Nash class. Tohoku Math. J. **74**(4), 569–596 (2022). arXiv:2012.02843
51. Kinzebulatov, D., Semënov, Yu.A.: Sharp solvability for singular SDEs. Electr. J. Probab. **28**, 1–15 (2023). Article no. 69. arXiv:2110.11232
52. Kinzebulatov, D., Semënov, Yu.A.: Fractional Kolmogorov operator and desingularizing weights. Publ. Res. Inst. Math. Sci. Kyoto Univ. **59**, 339–391 (2023). arXiv:2005.11199
53. Kinzebulatov, D., Semënov, Yu.A.: Regularity for parabolic equations with singular non-zero divergence vector fields. J. Differential Equ. arXiv:2205.05169 (to appear)
54. Kinzebulatov, D., Semënov, Yu.A.: Remarks on parabolic Kolmogorov operator. arXiv:2303.03993
55. Kinzebulatov, D., Vafadar, R.: On divergence-free (form-bounded type) drifts. Discrete Contin. Dyn. Syst. Ser. S. arXiv:2209.04537 (to appear). https://doi.org/10.3934/dcdss.2023073
56. Kinzebulatov, D., Madou, K.R., Semënov, Yu.A.: On the supercritical fractional diffusion equation with Hardy-type drift. J. Anal. Math. arXiv:2112.06329 (to appear). https://doi.org/10.1007/s11854-023-0300-5
57. Kinzebulatov, D., Semënov, Yu.A., Song, R.: Stochastic transport equation with singular drift. Ann. Inst. Henri Poincaré (B) Probab. Stat. **60**(1), 731–752 (2024)

58. Kinzebulatov, D., Semënov, Yu.A., Szczypkowski, K.: Heat kernel of fractional Laplacian with Hardy drift via desingularizing weights. J. Lond. Math. Soc. **104**, 1861–1900 (2021). arXiv:1904.07368
59. Komatsu, T.: On the martingale problem for generators of stable processes with perturbations. Osaka J. Math. **21**, 113–132 (1984)
60. Kovalenko, V.F., Semënov, Yu.A.: C_0-semigroups in $L^p(\mathbb{R}^d)$ and $C_\infty(\mathbb{R}^d)$ spaces generated by differential expression $\Delta + b \cdot \nabla$. (Russian) Teor. Veroyatnost. i Primenen. **35**, 449–458 (1990). Translation in Theory Probab. Appl. **35**, 443–453 (1990)
61. Kovalenko, V.F., Perelmuter, M.A., Semënov, Yu.A.: Schrödinger operators with $L_w^{\frac{1}{2}}(R^l)$-potentials. J. Math. Phys. **22**, 1033–1044 (1981)
62. Krylov, N.V.: On diffusion processes with drift in L_d. Probab. Theory Relat. Fields **179**(1–2), 165–199 (2021). arXiv:2001.04950
63. Krylov, N.V.: On strong solutions of Itô's equations with $A \in W^{1,d}$ and $B \in L^d$. Ann. Probab. **49**(6), 3142–3167 (2021). arXiv:2007.06040
64. Krylov, N.V.: On parabolic Adams's, the Chiarenza-Frasca theorems, and some other results related to parabolic Morrey spaces. Math. Eng. **5**(2), 1–20 (2022). arXiv:2110.09555
65. Krylov, N.V.: On strong solutions of Itô's equations with $D\sigma$ and b in Morrey classes containing L^d. Ann. Probab. **51**(5), 1729–1751 (2023). arXiv:2111.13795
66. Krylov, N.V.: On weak solutions of time-inhomogeneous Itô's equations with VMO diffusion and Morrey drift. arXiv:2303.11238
67. Krylov, N.V.: Once again on weak solutions of time-inhomogeneous Itô's equations with VMO diffusion and Morrey drift. arXiv:2304.04634
68. Krylov, N.V.: On parabolic equations in Morrey spaces with VMO a and Morrey b, c. arXiv:2304.03736
69. Krylov, N.V., Röckner, M.: Strong solutions of stochastic equations with singular time dependent drift. Probab. Theory Relat. Fields **131**, 154–196 (2005)
70. Kunita, H.: Stochastic Flows and Stochastic Differential Equations. Cambridge Studies in Advanced Mathematics, vol. 24. Cambridge University Press, Cambridge (1990)
71. Liskevich, V.A., Semënov, Yu.A.: Some problems on Markov semigroups. In: Demuth, M., et al. (eds.) Schrödinger Operators, Markov Semigroups, Wavelet Analysis, Operator Algebras. Mathematical Topics: Advances in Partial Differential Equations, vol. 11, pp. 163–217. Akademie Verlag, Berlin (1996)
72. Liskevich, V., Zhang, Q.S.: Extra regularity for parabolic equations with drift terms. Manuscr. Math. **113**, 191–209 (2004)
73. Majda, A.J., Kramer, P.R.: Simplified models for turbulent diffusion: theory, numerical modelling, and physical phenomena. Phys. Rep. **314**, 237–574 (1999)
74. Mazya, V.G., Verbitsky, I.E.: Form boundedness of the general second-order differential operator. Commun. Pure Appl. Math. **59**, 1286–1329 (2006)
75. Menozzi, S., Zhang, X.: Heat kernel of supercritical non-local operators with unbounded drift. J. Éc. Polytech. Math. **9**, 537–579 (2022). arXiv:2012.14475
76. Metafune, G., Negro, L., Spina, C.: Sharp kernel estimates for elliptic operators with second-order discontinuous coefficients. J. Evol. Equ. **18**, 467–514 (2018)
77. Meyer-Brandis, T., Proske, F.: Construction of strong solutions of SDE's via Malliavin calculus. J. Funct. Anal. **258**(11), 3922–3953 (2010)
78. Mohammed, A., Nilssen, T., Proske, F.: Sobolev differentiable stochastic flows for SDEs with singular coefficients: applications to the transport equation. Ann. Probab. **43**(3), 1535–1576 (2015)
79. Nam, K.: Stochastic differential equations with critical drifts. Stoch. Proc. Appl. **130**, 5366–5393 (2020). arXiv:1802.00074
80. Nazarov, A.I., Uraltseva, N.N.: The Harnack inequality and related properties for solutions to elliptic and parabolic equations with divergence-free lower order coefficients. Algebra i Analiz **23**, 136–168 (2011)

81. Ouhabaz, E.-M.: Analysis of Heat Equations on Domains. Princeton University Press, Princeton (2005)
82. Ouhabaz, E.-M., Stollmann, P., Sturm, K.-Th., Voigt, J.: The Feller property for absorption semigroups. J. Funct. Anal. **138**, 351–378 (1996)
83. Perkowski, N., van Zuiljen, W.: Quantitative heat kernel estimates for diffusions with distributional drift. Potential Anal. (2022). https://doi.org/10.1007/s11118-021-09984-3
84. Phan, T.: Local $W^{1,p}$-regularity estimates for weak solutions of parabolic equations with singular divergence-free drifts. Electr. J. Differ. Equ., 22 pp. (2017). Paper No. 75
85. Portenko, N.I.: Generalized Diffusion Processes. AMS, Providence (1990)
86. Portenko, N.I.: Some perturbations of drift-type for symmetric stable processes. Random Oper. Stoch. Equ. **2**, 211–224 (1994)
87. Podolynny, S.I., Portenko, N.I.: On multidimensional stable processes with locally unbounded drift. Random Oper. Stoch. Equ. **3**, 113–124 (1995)
88. Priola, E.: Pathwise uniqueness for singular SDEs driven by stable processes. Osaka J. Math. **49**, 421–447 (2012)
89. Rezakhanlou, F.: Regular flows for diffusions with rough drifts. arXiv:1405.5856
90. Röckner, M., Zhao, G.: SDEs with critical time dependent drifts: weak solutions. Bernoulli **29**, 757–784 (2023). arXiv:2012.04161
91. Röckner, M., Zhao, G.: SDEs with critical time dependent drifts: strong solutions. arXiv:2103.05803
92. Semënov, Yu.A.: On perturbation theory for linear elliptic and parabolic operators; the method of Nash. In: Proceedings of the Conference on Applied Analysis, April 19–21 (1996), Bâton-Rouge. Contemp. Math., vol. 221, pp. 217–284 (1999)
93. Semënov, Yu.A.: Regularity theorems for parabolic equations. J. Funct. Anal. **231**, 375–417 (2006)
94. Simon, B.: Schrödinger semigroups. Bull. Am. Math. Soc. **7**(3), 447–526 (1982)
95. Song, R., Xie, L.: Weak and strong well-posedness of critical and supercritical SDEs with singular coefficients. J. Differ. Equ. **362**, 266–313 (2023). arXiv:1806.09033
96. Veretennikov, A.Yu.: Strong solutions and explicit formulas for solutions of stochastic integral equations. Matematicheski Sbornik **111**(3), 434–452 (1980) (in Russian). English translation in Math. USSR-Sbornik **39**(3), 387–403 (1981)
97. Wei, J., Lv, G., Wu, J.-L.: On weak solutions of stochastic differential equations with sharp drift coefficients. arXiv:1711.05058
98. Williams, R.J.: Brownian motion with polar drift. Trans. Am. Math. Soc. **292**, 225–246 (1985)
99. Xia, P., Xie, L., Zhang, X., Zhao, G.: $L^q(L^p)$-theory of stochastic differential equations. Stoch. Proc. Appl. **130**, 5188–5211 (2020). arXiv:1908.01255
100. Xie, L., Zhang, X.: Heat kernel estimates for critical fractional diffusion operators. Studia Math. **224**(3), 221–263 (2014)
101. Yang, S., Zhang, T.: Strong existence and uniqueness of solutions of SDEs with time dependent Kato class coefficients. arXiv:2010.11467
102. Zhang, Q.S.: Gaussian bounds for the fundamental solutions of $\nabla(A\nabla u) + B\nabla u - u_t = 0$. Manuscripta Math. **93**, 381–390 (1997)
103. Zhang, Q.S.: A strong regularity result for parabolic equations. Commun. Math. Phys. **244**, 245–260 (2004)
104. Zhang, X.: Strong solutions of SDEs with singular drift and Sobolev diffusion coefficients. Stoch. Proc. Appl. **115**(11), 1805–1818 (2005)
105. Zhang, X.: Stochastic homeomorphism flows of SDEs with singular drifts and Sobolev diffusion coefficients. Electr. J. Prob. **16**, 1096–1116 (2011)
106. Zhang, X.: Stochastic homeomorphism flows of SDEs with singular drifts and Sobolev diffusion coefficients. Electron. J. Probab. **16**(38), 1096–1116 (2011)
107. Zhang, X.: Stochastic differential equations with Sobolev drifts and driven by α-stable processes. Ann. Inst. Henri Poincaré (B) Probab. Stat. **49**(4), 1057–1079 (2013)
108. Zhang, X.: Stochastic differential equations with Sobolev diffusion and singular drift and applications. Ann. Appl. Probab. **26**(5), 2697–2732 (2016)

109. Zhang, X., Zhao, G.: Stochastic Lagrangian path for Leray solutions of $3D$ Naiver-Stokes equations. Commun. Math. Phys. **381**(2), 491–525 (2021)
110. Zhang, X., Zhao, G.: Heat kernel and ergodicity of SDEs with distributional drifts. arXiv:1710:10537
111. Zhao, G.: Stochastic Lagrangian flows for SDEs with rough coefficients. arXiv:1911.05562
112. Zhao, G.: Weak uniqueness for SDEs driven by supercritical stable processes with Hölder drifts. Proc. Am. Math. Soc. **147**, 849–860 (2019)
113. Zvonkin, A.K.: A transformation of the phase space of a diffusion process that removes the drift. Math. USSR Sbornik **22**, 129–149 (1974)

About the Regularity of Degenerate Non-local Kolmogorov Operators Under Diffusive Perturbations

L. Marino, S. Menozzi, and E. Priola

Abstract We study here the effects of a time-dependent second order perturbation to a degenerate Ornstein-Uhlenbeck type operator whose diffusive part can be either local or non-local. More precisely, we establish that some estimates, such as the Schauder and Sobolev ones, already known for the non-perturbed operator still hold, and with the same constants, when we perturb the Ornstein-Uhlenbeck operator with second order diffusions with coefficients only depending on time in a measurable way. The aim of the current work is two-fold: we weaken the assumptions required on the perturbation in the local case which has been considered already in Marino et al. (Stud Math 267(3):321–346, 2022) and we extend the approach presented therein to a wider class of degenerate Kolmogorov operators with non-local diffusive part of symmetric stable type.

1 Introduction

1.1 General Aims and Scopes

In this work we are interested in establishing estimates, namely Schauder and L^p-type, for some perturbation of degenerate Ornstein-Uhlenbeck operators, which satisfy a Kalman type condition (see assumption **[K]** below). For non degenerate diffusion operators it has been proved in [11] that for some suitable second order per-

L. Marino
Institute of Mathematics, Polish Academy of Sciences, Warsaw, Poland
e-mail: lmarino@inpan.pl

S. Menozzi
Laboratoire de Modélisation Mathématique d'Evry (LaMME), UMR CNRS 8071, Université Paris-Saclay, Université d'Evry Val d'Essonne, Evry, France
e-mail: stephane.menozzi@univ-evry.fr

E. Priola (✉)
Dipartimento di Matematica, Università di Pavia, Pavia, Italy
e-mail: enrico.priola@unipv.it

S. Menozzi et al. (eds.), *Kolmogorov Operators and Their Applications*,
Springer INdAM Series 56, https://doi.org/10.1007/978-981-97-0225-1_8

turbations, some well known estimates from the parabolic/elliptic theory remained true with the *very* same constants. This also allowed the authors to prove therein that the Schauder and L^p estimates for the heat equation actually did not depend on the dimension. The strategy therein relies on a probabilistic technique which consists in introducing a random source of Poisson type. Once averaged those discontinuities make appear a finite difference operator in the associated PDE. The final estimates then follow by some compactness arguments. The Poisson process actually allows very naturally to perform those operations and up to now, to the best of our knowledge, this is the only approach which consents to derive such result. The question to obtain a purely analytic proof of the indicated stability results remains open.

In [15] we managed to prove that the strategy developed in [11] was sufficiently robust to extend to degenerate Ornstein-Uhlenbeck operators satisfying a Kalman condition provided the second order spatial perturbation has coefficients which are continuous in time. We manage to get rid in the current work of this assumption and establish the aforementioned stability under the mere time measurability and boundedness for the perturbations. We also extend such approach to the non-local degenerate context. Namely, we consider degenerate stable type Ornstein-Uhlenbeck operators and perturb these with the same kind of second order diffusion considered above.

1.2 The Model

Let us describe the general framework we are going to consider here. Fixed N in $\mathbb{N}$, we denote by $\mathcal{S}(\mathbb{R}^N)$ the family of symmetric matrix in $\mathbb{R}^N \otimes \mathbb{R}^N$. We can then consider its subset $\mathcal{S}_0(\mathbb{R}^N)$, which stands for the closed convex cone of non-negative definite matrices, and its interior $\mathcal{S}_+(\mathbb{R}^N)$ which corresponds to the open subset of positive definite matrices.

Let us introduce two matrices A, B in $\mathbb{R}^N \otimes \mathbb{R}^N$ such that B belongs to $\mathcal{S}_0(\mathbb{R}^N)$ and satisfying the following Kalman rank condition:

[K] There exists a non-negative integer k such that

$$\text{Rank}[B, AB, \dots, A^k B] = N,$$

where $[B, AB, \dots, A^k B]$ is the $\mathbb{R}^N \otimes \mathbb{R}^{N(k+1)}$ matrix whose blocks are $B, AB, \dots, A^k B$.

It has been shown (cf. [12]) that, up to a possible change of variables, the space $\mathbb{R}^N$ can be decomposed as $\mathbb{R}^{d_0} \times \mathbb{R}^{d_1}$ (with $d_0 \geq 1$) such that

$$B = \begin{pmatrix} B_0 & 0_{d_0 \times d_1} \\ 0_{d_1 \times d_0} & 0_{d_1 \times d_1} \end{pmatrix}, \tag{1.1}$$

for some matrix B_0 in $\mathbb{S}_+(\mathbb{R}^{d_0})$. Moreover, if we denote by $\kappa_2 > 0$ the smallest eigenvalue of the matrix B_0, it then follows that

$$B_0\xi \cdot \xi \geq \kappa_2|\xi|^2, \tag{1.2}$$

for any ξ in $\mathbb{R}^{d_0}$. Above, "$\cdot$" denotes the usual inner product in $\mathbb{R}^{d_0}$. From the non-degeneracy of B_0, the above condition **[K]** amounts to say that the vectors

$$\{e_1, \dots, e_{d_0}, Ae_1, \dots, Ae_{d_0}, \dots, A^k e_1, \dots, A^k e_{d_0}\} \quad \text{generate } \mathbb{R}^N,$$

where $\{e_i : i \in \{1, \cdots, d_0\}\}$ are the first d_0 vectors of the canonical basis for $\mathbb{R}^N$. For simplicity, let us also denote by σ in $\mathbb{R}^N \otimes \mathbb{R}^{d_0}$ the following matrix:

$$\sigma = \begin{pmatrix} \sqrt{B_0} \\ 0_{d_1 \times d_0} \end{pmatrix},$$

so that $\sigma\sigma^* = B$.

Fixed α in $(0, 2]$, we can now introduce an operator $\mathcal{L}_\alpha$ on $C_b^\infty(\mathbb{R}^N)$, the space of all the bounded, smooth functions on $\mathbb{R}^N$ with bounded derivatives of any order, given by:

$$\begin{aligned} \mathcal{L}_\alpha\phi(x) &:= \mathbb{1}_{\{\alpha=2\}}\mathrm{Tr}(BD_x^2\phi(x)) + \\ \mathbb{1}_{\{\alpha\neq 2\}} &\int_{\mathbb{R}^{d_0}} \left[\phi(x+\sigma z) - \phi(x) - \langle D_x\phi(x), \sigma z\rangle \mathbb{1}_{B(0,1)}(\sigma z)\right] \nu_\alpha(dz) \end{aligned} \tag{1.3}$$

where ν_α is a symmetric, non-degenerate α-stable measure on $\mathbb{R}^{d_0}$, i.e.

$$\nu_\alpha(C) = \int_0^\infty \int_{\mathbb{S}^{d_0-1}} \mathbb{1}_C(r\theta)\mu(d\theta)\frac{dr}{r^{1+\alpha}}, \quad C \in \mathcal{B}(\mathbb{R}^{d_0}), \tag{1.4}$$

for a symmetric, finite measure μ on $\mathbb{S}^{d_0-1}$ for which the following *non-degeneracy* condition holds:

[ND] there exists $\kappa_\alpha > 0$ such that

$$\int_{\mathbb{S}^{d_0-1}} |\lambda \cdot \theta|^\alpha \,\mu(d\theta) \geq \kappa_\alpha|\lambda|^\alpha, \quad \lambda \in \mathbb{R}^{d_0}.$$

In (1.3), $\langle \cdot, \cdot \rangle$ denotes the usual inner product in $\mathbb{R}^N$ and $D_x^2\phi$ represents the full Hessian matrix in $\mathbb{R}^N \otimes \mathbb{R}^N$ with respect to x. From now on, we will say that assumption [**A**] holds when the above conditions [**K**] and [**ND**] are in force.

We will use, as an underlying *proxy* operator, a degenerate Ornstein-Uhlenbeck operator $\mathcal{L}_\alpha^{\mathrm{ou}}$ of the form

$$\mathcal{L}_\alpha^{\mathrm{ou}}\phi(x) := \mathcal{L}_\alpha\phi(x) + \langle Ax, D_x\phi(x)\rangle, \tag{1.5}$$

for x in $\mathbb{R}^N$. In the Gaussian case (i.e. when $\alpha = 2$), assumption **[K]** (which also often appears in control theory; see e.g. [18]) is equivalent to the Hörmander condition on the commutators (c.f. [7]) ensuring the hypoellipticity of the operator $\partial_t - \mathcal{L}_\alpha^{\mathrm{ou}}$. In particular, it implies the existence and the smoothness of a distributional solution to the following equation:

$$\begin{cases} \partial_t u(t,x) = \mathcal{L}_\alpha^{\mathrm{ou}} u(t,x) + f(t,x) & \text{on } \mathbb{R}^N_T; \\ u(0,x) = 0 & \text{on } \mathbb{R}^N, \end{cases} \tag{1.6}$$

where $\mathbb{R}^N_T := (0,T) \times \mathbb{R}^N$ for some fixed final time $T > 0$ and f is a function in $C_c^\infty(\mathbb{R}^N_T)$, the set of all the smooth functions on $\mathbb{R}^N_T$ with compact support.

From this point further, we assume to have fixed a time dependent matrix $S\colon (0,T) \to \mathbb{S}_0(\mathbb{R}^N)$ which is bounded measurable in $(0,T)$. We are then interested in the following perturbation of $\mathcal{L}_\alpha^{\mathrm{ou}}$:

$$\mathcal{L}_{\alpha,t}^{\mathrm{pert}}\phi(x) := \mathcal{L}_\alpha^{\mathrm{ou}}\phi(x) + \mathrm{Tr}(S(t)D_x^2\phi(x)). \tag{1.7}$$

Our main interest here is to understand how such a perturbation influences some well-known estimates for the Ornstein-Uhlenbeck operator $\mathcal{L}_\alpha^{\mathrm{ou}}$. For example, when considering the diffusive setting $\alpha = 2$, Theorem 3 in [2] (but see also [15, Section 2.3]), showed that for any fixed p in $(1, +\infty)$, there exists $c_p := c_p(T, A, B, d_0, d_1)$ such that the solution u to Cauchy problem (1.6) satisfies

$$\|D_{x_0}^2 u\|_{L^p(\mathbb{R}^N_T)} \le c_p \|f\|_{L^p(\mathbb{R}^N_T)}, \tag{1.8}$$

where for any $(t,x) = (t, x_0, x_1)$ in $\mathbb{R}^N_T = (0,T) \times \mathbb{R}^{d_0} \times \mathbb{R}^{d_1}$, $D_{x_0}^2 u(t,x)$ stands for the Hessian matrix in $\mathbb{R}^{d_0} \otimes \mathbb{R}^{d_0}$ with respect to the variable x_0 and $L^p(\mathbb{R}^N_T)$ is the standard L^p-space with respect to the Lebesgue measure. Under some additional regularity assumptions, namely the continuity in time for $S(t)$, we showed in [15] that the above estimates are indeed stable under a diffusive perturbation $\mathrm{Tr}(S(t)D_x^2)$. The main objective of the present paper is to extend the previous results in [15] to possibly fractional proxy operators and to diffusive perturbations where $S(t)$ is only bounded measurable. We will also show some applications of the previous arguments to elliptic or more general parabolic equations.

The article is organised as follows. We introduce in Sect. 2 the notion of solution considered as well as some useful notations about the *natural* function spaces associated with the specific degenerate operators described above. In that section, we also establish a maximum principle for a large class of operators, see

Theorem 2.2, which we think has independent interest. Section 3 is then devoted to the statement of our main results, see Theorem 3.1, and some related extensions (parabolic operators with potential and elliptic counterpart). Section 4 is eventually dedicated to the proof of the main result. As we mentioned before, the approach therein is based on the Poisson perturbative approach firstly considered in [11].

2 Useful Notations

2.1 *Definition of Solution*

Let us consider in this section a slightly more general framework. We are interested in the following Cauchy problem:

$$\begin{cases} \partial_t v(t,x) = L_t v(t,x) + f(t,x) & \text{on } \mathbb{R}^N_T; \\ v(0,x) \;=\; 0 & \text{on } \mathbb{R}^N. \end{cases} \tag{2.1}$$

Above, the operator L_t on $C^\infty_b(\mathbb{R}^N)$ is given by

$$\begin{aligned} L_t\phi(x) &:= \operatorname{Tr}\Big(Q(t)D^2_x\phi(x)\Big) + \langle b(t,x), D_x\phi(x)\rangle - c(t)\phi(x) + \mathcal{N}_t\phi(x), \\ \mathcal{N}_t\phi(x) &:= \int_{\mathbb{R}^{d_0}} \big[\phi(x+\gamma(t)z) - \phi(x) - \langle D_x\phi(x), \gamma(t)z\rangle \mathbb{1}_{B(0,1)}(z)\big]\, \nu_\alpha(dz), \end{aligned} \tag{2.2}$$

where $Q\colon (0,T) \to \mathcal{S}_0(\mathbb{R}^N)$, $b\colon (0,T)\times\mathbb{R}^N \to \mathbb{R}^N$, $c\colon (0,T) \to [0,+\infty)$ and $\gamma\colon (0,T) \to \mathbb{R}^N \otimes \mathbb{R}^d$ are Borel measurable and ν_α is the α-stable Lévy measure given in (1.4).

We will denote by $B_b\big(0,T; C^\infty_c(\mathbb{R}^N)\big)$ the space of all Borel bounded functions $\phi\colon \mathbb{R}^N_T \to \mathbb{R}$ such that $\phi(t,\cdot)$ is smooth and compactly supported for any t in $(0,T)$, for any n in $\mathbb{N}$ the $C^n(\mathbb{R}^N)$-norms of $\phi(t,\cdot)$ are bounded in time and the supports of the functions $\phi(t,\cdot)$ are contained in the same ball. We remark that the choice of such space, which turns out to be rather natural when considering sources that are possibly discontinuous in time, is strictly related to the proof technique used in [11] and based on the Poisson process (cf. proof of Lemma 4.3).

For a bounded, Borel measurable function $f\colon \mathbb{R}^N_T \to \mathbb{R}$, we are going to interpret Cauchy Problem (2.1) in an *integral in time* form:

$$u(t,x) \;=\; \int_0^t [f(s,x) + L_s u(s,x)]\, ds. \tag{2.3}$$

Definitionn 2.1 (Integral Solution for the Cauchy Problem) We say that a function $u\colon [0,T]\times\mathbb{R}^N \to \mathbb{R}$ is a solution to Eq. (2.1) if $D^\varsigma_x u$ is bounded continuous on $[0,T]\times\mathbb{R}^N$ for any ς in $\mathbb{N}^N$ such that $|\varsigma| \le 2$ and (2.3) holds for any (t,x) in $[0,T]\times\mathbb{R}^N$.

We recall that for a given multi-index ς in $\mathbb{N}^N$, $D^\varsigma_x u$ denotes the iterated spatial derivatives of u of order ς, i.e.

$$D^\varsigma_x u := D^{\varsigma_1}_{x_1} D^{\varsigma_2}_{x_2} \dots D^{\varsigma_n}_{x_n} u.$$

Under some additional assumptions (see Theorem 2.2 below), it is indeed possible to show that, when it exists, an integral in time solution u to Cauchy problem (2.1) is unique and the following maximum principle holds:

$$\sup_{(t,x)\in[0,T]\times\mathbb{R}^N} |u(t,x)| \le T \sup_{(t,x)\in\mathbb{R}^N_T} |f(t,x)|. \tag{2.4}$$

While such estimates are well-known in the diffusive setting (see e.g. [10, Theorem 4.1] or [15]), we could not find a precise proof in the fractional case. For this reason, we now present it for future reference.

Theorem 2.2 (Maximum Principle) *Let* $\alpha \in (0,2)$. *Assume in addition that* $Q\colon (0,T) \to \mathcal{S}_0(\mathbb{R}^N)$ *and* $\gamma\colon (0,T) \to \mathbb{R}^N \otimes \mathbb{R}^d$ *are bounded. Let u be a solution to Cauchy problem* (2.1) *in the sense of Definition 2.1, with* $c = 0$ *and* $b = 0$. *Then, the maximum principle* (2.4) *holds for u.*

Proof By considering u and $-u$, it is clearly enough to show that

$$\sup_{(t,x)\in\mathbb{R}^N_T} u(t,x) \le T \sup_{(t,x)\in\mathbb{R}^N_T} |f(t,x)|.$$

In order to mimic the proof technique in [10] (see Theorem 4.1 therein in the diffusive setting), we firstly notice that the function $\tilde{u}\colon [-T,0]\times\mathbb{R}^N \to \mathbb{R}$ given by

$$\tilde{u}(t,x) := e^t \left(u(-t,x) + t\|f\|_\infty\right),$$

solves the following Cauchy problem:

$$\begin{cases} \partial_t \tilde{u}(t,x) + \tilde{L}_t \tilde{u}(t,x) = \tilde{f}(t,x) & \text{on } (-T,0)\times\mathbb{R}^N; \\ \tilde{u}(0,x) = 0 & \text{on } \mathbb{R}^N. \end{cases} \tag{2.5}$$

where $\tilde{f}(t,x) := e^t(\|f\|_\infty - f(-t,x)) \ge 0$ and $\tilde{L}_t$ is the operator given in (2.2) with coefficients $\tilde{Q}(t) = Q(-t)$, $\tilde{c}(t) = 1$ and $\tilde{\gamma}(t) = \gamma(-t)$. For notational convenience, let us also denote $\tilde{\mathcal{N}}_t := \mathcal{N}_{-t}$. In order to conclude the proof, it is then enough to show that

$$\sup_{(t,x)\in[-T,0]\times\mathbb{R}^N} \tilde{u}(t,x) \le 0.$$

Let us introduce now the following barrier function $w(t,x) = \exp(-Ct)\ell_\beta(x)$ where $\ell_\beta(x) = (1+|x|^2)^{\frac{\beta}{2}}$ for some $\beta < \alpha \wedge 1$ and $C > 0$. One can observe that ℓ_β is smooth and that there exists $\bar{C} \ge 1$ such that for any multi-index ζ in $\mathbb{N}^n$ such that $|\zeta| \le 2$,

$$|D_x^\zeta \ell_\beta(x)| \le \bar{C}(1+|x|^2)^{\frac{\beta}{2}-\frac{|\zeta|}{2}}.$$

In particular, by our choice of β, $D_x\ell_\beta$ is bounded. Moreover, it is possible to choose $C > 0$ large enough so that

$$(\partial_t + \tilde{L}_t)w(t,x) \le 0. \tag{2.6}$$

Indeed, one can notice that $\partial_t w(t,x) = -Ce^{-Ct}\ell_\beta(x)$ and

$$|\mathrm{Tr}\left(\tilde{Q}(t)D_x^2 w(t,x)\right)| \le e^{-Ct}\bar{C}.$$

Furthermore, one can infer from the β-Hölder regularity of ℓ_β that

$$|\tilde{\mathbb{N}}_t w(t,x)| = \left|e^{-Ct}\mathrm{p.v.}\int_{\mathbb{R}^d}[\ell_\beta(x+\gamma(t)z) - \ell_\beta(x)]\,\nu_\alpha(dz)\right| \le \bar{C}e^{-Ct}.$$

Now, the point is to prove that for any fixed $\eta > 0$ and any $(t,x) \in [-T,0]\times\mathbb{R}^N$,

$$v(t,x) := e^{-t/2}(\tilde{u}(t,x) - \eta w(t,x)) \le 0,$$

where $w(t,x)$ is the previous barrier which satisfies (2.6). The statement indeed follows for $\tilde{u}$ letting η go to zero and using that u is bounded in $\mathbb{R}^N_T$.

Since $\tilde{u}$ is bounded and w goes to infinity with $|x|$, we deduce that the supremum of v in $[-T,0]\times\mathbb{R}^N$ has to be attained at some point $(t_0,x_0) \in [-T,0]\times\mathbb{R}^N$. We can assume without loss of generality that t_0 is in $[-T,0)$. Moreover, since v is smooth in space, $D_x v(t_0,x_0) = 0$ and $D_x^2 v(t_0,x_0)$ is non-positive definite. Then, by the space-time continuity of v and its spatial derivatives, we notice that for any $\tau > 0$, there exists $\theta := \theta(\tau)$ such that for any $(t,z) \in [-T,0)\times\mathbb{R}^N$ such that $|t-t_0| \vee |z-x_0| < \theta$,

$$v(t,x_0) \ge 0, \qquad |D_x v(t,z)| \le \tau, \qquad [D_x^2 v(t,z)]_{i,j} \le \tau\delta_{i,j}. \tag{2.7}$$

We can now write for any $0 \leq t - t_0 \leq \theta$,

$$0 \geq v(t, x_0) - v(t_0, x_0) = \int_{t_0}^{t} \partial_s v(s, x_0) ds.$$

We then observe that for any (s, x) in $[-T, 0] \times \mathbb{R}^N$, we have that

$$\begin{aligned} \partial_s v(s, x) &= -\frac{1}{2} v(s, x) + e^{-s/2}(\partial_s \tilde{u}(s, x) - \eta \partial_s w(s, x)) \\ &= -\frac{1}{2} v(s, x) - \tilde{L}_s v(s, x) + e^{-s/2} \tilde{f}(s, x) \\ &\quad - \eta e^{-s/2} \left(\partial_s + \tilde{L}_s\right) w(s, x) \\ &\geq -\frac{1}{2} v(s, x) - \tilde{L}_s v(s, x), \end{aligned} \tag{2.8}$$

exploiting, in the last step, (2.6). Hence, by the definition of the operator $\tilde{L}_s$ in (2.2) (with the choice of coefficients described after (2.5)), it holds that

$$\int_{t_0}^{t} v(s, x_0)\, ds \leq 2 \int_{t_0}^{t} \left[\mathrm{Tr}\left(\tilde{Q}(s) D_x^2 v(s, x_0)\right) + \tilde{\mathcal{N}}_s v(s, x_0) \right] ds. \tag{2.9}$$

Taking $t - t_0 < \theta$ and s in $[t, t_0]$, one can easily conclude from (2.7) that

$$\mathrm{Tr}\left(\tilde{Q}(s) D_x^2 v(s, x_0)\right) \leq c\tau. \tag{2.10}$$

On the other hand, we can split $\tilde{\mathcal{N}}_s v$ into $I_s^1 v + I_s^2 v$, where

$$\begin{aligned} I_s^1 v(s, x_0) &:= \int_{|z| \leq t - t_0} \left[v(s, x_0 + \tilde{\gamma}(s) z) \right. \\ &\qquad \left. - v(s, x_0) - \langle \nabla v(s, x_0), \gamma(s) z \rangle \right] \nu_\alpha(dz), \\ I_s^2 v(s, x_0) &:= \int_{|z| > t - t_0} \left[v(s, x_0 + z) - v(s, x_0) \right] \nu_\alpha(dz). \end{aligned}$$

Using again (2.7), one can then derive that

$$\begin{aligned} I_s^1 v(s, x_0) &= \int_0^1 \int_{|z| \leq t - t_0} \langle D_x^2 v(s, x_0 + \lambda \gamma(s) z) \gamma(s) z, \gamma(s) z \rangle \, \nu_\alpha(dz) d\lambda \\ &\leq c\tau \int_{|z| \leq 1} |z|^2 \, \nu_\alpha(dz) \leq c\tau. \end{aligned} \tag{2.11}$$

On the other hand, since $v(t_0, x_0)$ is a maximum, so that $v(t_0, x_0+z)-v(t_0, x_0) \leq 0$, we also have that

$$
\begin{aligned}
&I_s^2 v(s, x_0) \\
&\quad = \int_{|z|>t-t_0} [v(s, x_0+\gamma(s)z) \pm v(t_0, x_0+\gamma(s)z) \pm v(t_0, x_0)-v(s, x_0)]\ \nu_\alpha(dz) \\
&\quad \leq \int_{|z|>t-t_0} [v(s, x_0+\gamma(s)z)-v(t_0, x_0+\gamma(s)z)+v(t_0, x_0) - v(s, x_0)]\ \nu_\alpha(dz) \\
&\quad = \int_{t_0}^{s} \int_{|z|>t-t_0} [\partial_r v(r, x_0+\gamma(s)z) - \partial_r v(r, x_0)]\ \nu_\alpha(dz)dr.
\end{aligned}
$$

Similarly to (2.8), we note that

$$
\partial_s v(s, x) = e^{-s/2}\left[\eta\tilde{C}w(s, x) + \tilde{f}(s, x) - \tilde{L}_s\tilde{u}(s, x)) - \frac{1}{2}\tilde{u}(s, x)\right],
$$

where the constant $\tilde{C} := C + 1/2$ and C defined in (2.6). Hence,

$$
\begin{aligned}
I_s^2 v(s, x_0) \leq \int_{t_0}^{s} \int_{|z|>t-t_0} \Big| &\left(\eta\tilde{C}w + \tilde{f} - \tilde{L}_r\tilde{u} - \frac{1}{2}\tilde{u}\right)(r, x_0+\gamma(s)z) \\
&- \left(\eta\tilde{C}w + \tilde{f} - \tilde{L}_r\tilde{u} - \frac{1}{2}\tilde{u}\right)(r, x_0)\Big|\ \nu_\alpha(dz)dr
\end{aligned}
$$

When α is in $(0, 1)$, we can use the boundedness of $\tilde{u}$ and $\tilde{f}$ and the β-Hölder continuity of ℓ_β to infer that

$$
\begin{aligned}
I_s^2 v(s, x_0) &\leq c(s-t_0)\int_{|z|\geq t-t_0} \left(1+|z|^\beta\right) \nu_\alpha(dz) \\
&\leq c(s-t_0)\int_{t-t_0}^{+\infty} r^{-(1+\alpha)}(1+r^\beta)\, dr \\
&\leq c(s-t_0)(t-t_0)^{\ \alpha}.
\end{aligned}
\tag{2.12}
$$

If instead $\alpha \in (1, 2)$, we need to apply an additional Taylor expansion:

$$
\begin{aligned}
I_s^2 v(s, x_0) \leq \int_0^1 \int_{t_0}^{s} \int_{|z|\geq t-t_0} \Big| &\left\langle\left(\eta\tilde{C}D_x w + D_x\tilde{f}\right)(r, x_0+\lambda\gamma(s)z), \gamma(s)z\right\rangle \\
&-\left\langle\left(D_x\left(\tilde{L}_r\tilde{u}\right) + \frac{1}{2}D_x\tilde{u}\right)(r, x_0+\lambda\gamma(s)z), \gamma(s)z\right\rangle\Big|\ \nu_\alpha(dz)drd\lambda.
\end{aligned}
$$

Using again the boundedness of the functions involved, i.e. $D_x\tilde{u}$, $D_x w$ and $D_x\tilde{f}$, and noting that $D_x\left(\tilde{L}_r\tilde{u}\right) = \tilde{L}_r\,(D_x\tilde{u})$, we can conclude in the same way as above that

$$I_s^2 v(s,x_0) \,\le\, c(s-t_0)(t-t_0)^{1-\alpha}. \tag{2.13}$$

Applying estimates (2.10), (2.11) and (2.12) or (2.13) inside (2.9), we finally get that

$$\int_{t_0}^{t} v(s,x_0)\,ds \,\le\, c\left[\tau + (t-t_0)^{\varepsilon}\right](t-t_0),$$

for some $\varepsilon > 0$. Using the continuity of v, we can divide both sides by $t - t_0$ and let t go to t_0 so that

$$v(t_0,x_0) \,\le\, c\tau.$$

Since $\tau > 0$ is arbitrary, we have concluded the proof. □

2.2 Definition of the Anisotropic Norms

We firstly recall from [12] that the state space $\mathbb{R}^N$ can be split into $\mathbb{R}^{d_0} \times \mathbb{R}^{d_1}$ so that, under such decomposition, the matrix B assumes the form in (1.1). More precisely, in the aforementioned work, it was established that Assumption **[K]** is equivalent to the fact that there exists $k \in \mathbb{N}$ and positive integers $\{\mathfrak{d}_i : i \in \{1,\cdots,k\}\}$ such that $\sum_{i=1}^{k} \mathfrak{d}_i = d_1$ and for all $i \in \{1,\cdots,k\}$, setting $\mathfrak{d}_0 = d_0$ and $\sum_{m=0}^{-1} = 0$, the matrices

$$\mathscr{A}^i := (A_{j,\ell})_{(j,\ell)\in\{\sum_{m=0}^{i-1}\mathfrak{d}_m+1,\cdots,\sum_{m=0}^{i}\mathfrak{d}_m\}\times\{\sum_{m=1}^{i-1}\mathfrak{d}_m+1,\cdots,\sum_{m=1}^{i}\mathfrak{d}_m\}},$$

have rank $\mathfrak{d}_i$. Moreover, the matrix A writes:

$$A = \begin{pmatrix} * & * & \cdots & \cdots & * \\ \mathscr{A}^1 & * & \ddots & \ddots & \vdots \\ 0_{\mathfrak{d}_2,d_0} & \mathscr{A}^2 & * & \ddots & \vdots \\ \vdots & \ddots & \ddots & \ddots & * \\ 0_{\mathfrak{d}_k,d_0} & \cdots & 0_{\mathfrak{d}_k,\mathfrak{d}_{k-1}} & \mathscr{A}^k & * \end{pmatrix}. \tag{2.14}$$

We can then write $x \in \mathbb{R}^N$ as $x = (x_0, x_1, \cdots, x_k)$ with $x_i \in \mathbb{R}^{\mathfrak{d}_i}$, $i \in \{0,\cdots,k\}$. In order to properly introduce the anisotropic functional space we are going to consider, we follow [8] by introducing the orthogonal projection $p_i : \mathbb{R}^N \to \mathbb{R}^{\mathfrak{d}_i}$ such that $p_i(x) = x_i$ and denoting its adjoint by $E_i : \mathbb{R}^{\mathfrak{d}_i} \to \mathbb{R}^N$, for any i in $[\![0, k]\!]$. For notational simplicity, let us denote

$$\alpha_i := \frac{\alpha}{2}\frac{1}{1+\alpha i}. \tag{2.15}$$

The threshold in (2.15) might seem awkward at first sight. While the first term $\alpha/2$ relates to the maximal regularity associated with the fractional Laplacian Δ^α, the second one actually corresponds to the index needed to get the invariance by dilations for the harmonic functions associated with the principal part of the Ornstein-Uhlenbeck operator $\mathcal{L}^{\mathrm{ou}}_\alpha$. Namely, one considers the operator $\mathcal{L}^{\mathrm{ou}}_{0,\alpha}$ given in (1.5) with respect to the matrix

$$A_0 = \begin{pmatrix} 0_{\mathfrak{d}_0,\mathfrak{d}_0} & 0_{\mathfrak{d}_0,\mathfrak{d}_1} & \dots & \dots & 0_{\mathfrak{d}_0,\mathfrak{d}_k} \\ \mathscr{A}^1 & 0_{\mathfrak{d}_1,\mathfrak{d}_1} & \ddots & \ddots & \vdots \\ 0_{\mathfrak{d}_2,\mathfrak{d}_0} & \mathscr{A}^2 & 0_{\mathfrak{d}_2,\mathfrak{d}_2} & \ddots & \vdots \\ \vdots & \ddots & \ddots & \ddots & \vdots \\ 0_{\mathfrak{d}_k,\mathfrak{d}_0} & \dots & 0_{\mathfrak{d}_k,\mathfrak{d}_{k-1}} & \mathscr{A}^k & 0_{\mathfrak{d}_k,\mathfrak{d}_k} \end{pmatrix}. \tag{2.16}$$

Note that A_0, B again satisfy **[K]**. If $(\partial_t - \mathcal{L}^{\mathrm{ou}}_{0,\alpha})u(t,x) = 0$ then for all $\lambda > 0$ $(\partial_t - \mathcal{L}^{\mathrm{ou}}_{0,\alpha})u\big(\delta_\lambda(t,x)\big) = 0$ where the dilation operator

$$\delta_\lambda(t,x) = (\lambda^{1/\alpha}t, \lambda x_0, \lambda^{1/(1+\alpha)}x_1, \cdots, \lambda^{1/(1+\alpha k)}x_k),$$

precisely exhibits the exponents in (2.15) for the spacial components.

Since the main focus of the present work is on Schauder and Sobolev type estimates for solutions to Cauchy problem (1.6), we now briefly recall the definition of Sobolev and Hölder norms in our anisotropic context. Fixed β in (0, 1), let us denote by $\Delta^\beta_{x_i}$ the β-fractional Laplacian along the i-th direction, i.e.

$$\Delta^\beta_{x_i}\phi(x) := \text{p.v.}\int_{\mathbb{R}^{\mathfrak{d}_i}} [\phi(x + E_i z) - \phi(x)]\frac{dz}{|z|^{\mathfrak{d}_i+2\beta}}, \quad x \in \mathbb{R}^N,$$

for any smooth enough function $\phi\colon \mathbb{R}^N \to \mathbb{R}$. Given p in $[1, +\infty)$, we can now define the *homogeneous* Sobolev space $\dot{W}^{2,p}_d(\mathbb{R}^N_T)$ as the family of all the functions $\phi\colon \mathbb{R}^N_T \to \mathbb{R}$ in $L^p(\mathbb{R}^N_T)$ such that for any i in $[\![0, k]\!]$, $\Delta^{\alpha_i}_{x_i}\phi(t,x)$ is well defined for almost every (t,x) in $\mathbb{R}^N_T$ and

$$\Delta^{\alpha_i}_{x_i}\phi(t,x) := \Delta^{\alpha_i}_{x_i}\phi(t,\cdot)(x) \text{ belongs to } L^p(\mathbb{R}^N_T).$$

It is endowed with the natural *semi*-norm $[\cdot]_{\dot{W}^{\alpha,p}_d(\mathbb{R}^N_T)}$ given by:

$$[\phi]^p_{\dot{W}^{\alpha,p}_d(\mathbb{R}^N_T)} = \sum_{i=0}^{k} \|\Delta^{\alpha_i}_{x_i}\phi\|^p_{L^p(\mathbb{R}^N_T)}. \tag{2.17}$$

Following Krylov [9], for some fixed ℓ in $\mathbb{N}_0 := \mathbb{N} \cup \{0\}$ and β in $(0, 1]$, we introduce for a function $\phi\colon \mathbb{R}^N \to \mathbb{R}$ the Zygmund-Hölder semi-norm as

$$[\phi]_{C^{\ell+\beta}} := \begin{cases} \sup_{|\vartheta|=\ell} \sup_{x\neq y} \frac{|D^\vartheta\phi(x)-D^\vartheta\phi(y)|}{|x-y|^\beta}, & \text{if } \beta \neq 1; \\ \sup_{|\vartheta|=\ell} \sup_{x\neq y} \frac{\left|D^\vartheta\phi(x)+D^\vartheta\phi(y)-2D^\vartheta\phi(\frac{x+y}{2})\right|}{|x-y|}, & \text{if } \beta = 1, \end{cases}$$

(we are using usual multi-indices ϑ for the partial derivatives). Consequently, the Zygmund-Hölder space $C^{\ell+\beta}_b(\mathbb{R}^N)$ is the family of bounded functions $\phi\colon \mathbb{R}^N \to \mathbb{R}$ such that ϕ and its derivatives up to order ℓ are continuous and the norm

$$\|\phi\|_{C^{\ell+\beta}_b} := \sum_{i=0}^{\ell} \sup_{|\vartheta|=i} \|D^\vartheta\phi\|_\infty + [\phi]_{C^{\ell+\beta}} \text{ is finite.}$$

We can now define the anisotropic Zygmund-Hölder spaces associated with the current setting and which again reflect the various scales already introduced in (2.15). Let $\gamma \in (0, 3)$, the space $C^\gamma_{b,d}(\mathbb{R}^N)$ is the family of functions $\phi\colon \mathbb{R}^N \to \mathbb{R}$ such that for any i in $[\![0, k]\!]$ and any x_0 in $\mathbb{R}^N$, the real function

$$y \in \mathbb{R}^{\mathfrak{d}_i} \to \phi(x_0 + E_i(y)) \text{ belongs to } C^{\gamma/(1+\alpha i)}_b\left(\mathbb{R}^{\mathfrak{d}_i}\right),$$

with a norm bounded by a constant independent from x_0.

We will also consider the corresponding natural semi-norm for the related homogeneous space

$$[\phi]_{C^\gamma_d} := \sum_{i=0}^{k} \sup_{x_0\in\mathbb{R}^N} [\phi(x_0 + E_i(\cdot))]_{C^{\gamma/(1+\alpha i)}(\mathbb{R}^{\mathfrak{d}_i})}. \tag{2.18}$$

We finally remark that we have denoted by C^γ_d and $\dot{W}^{2,p}_d$ the anisotropic functional spaces because the regularity exponents reflect again the multi-scale features of the system. In particular, the Hölder norm could equivalently be defined through the corresponding spatial parabolic distance d defined as follows: For any $x, x' \in \mathbb{R}^N$:

$$d(x, x') := \sum_{i=0}^{k} |x_i - x_i'|^{\frac{1}{1+\alpha i}},$$

where the exponents are again those who appeared in (2.15).

3 Main Results

Let f be in $B_b\left(0, T; C_c^\infty(\mathbb{R}^N)\right)$ and u the unique solution to the corresponding Cauchy problem (1.6).

In [8], see also [5] and [16] where time inhomogeneous coefficients are considered as well, it has been proven that if A, B satisfy **[K]** and the diagonal and the strictly upper diagonal elements of A in (2.14) are equal to zero (i.e., $A = A_0$ in (2.16)) then the following Sobolev estimates hold (with $p \in (1, +\infty)$):

$$[u]_{\dot{W}_d^{\alpha,p}(\mathbb{R}_T^N)} \le C_p \|f\|_{L^p(\mathbb{R}_T^N)}, \tag{3.1}$$

for some positive constant $C_p := C(p, \alpha, A, B, d_0, d_1)$. We remark that the specific structure assumed on A is actually due to the fact that for such matrices there is an underlying homogeneous space structure which makes easier to establish maximal regularity estimates (see e.g. [6] in this general setting). If A, B satisfy **[K]** with a general A as in (2.14), having non-zero strictly upper diagonal entries, we believe that the approach in [2] could extend to show that (3.1) still holds in this general setting. In particular, non-zero entries in the diagonal do not create additional difficulties but they introduce a dependence in the final time T in the previous estimates. However, the estimates in this more general framework have not been, up to our best knowledge, proven yet.

Let us now fix β in (0, 1) such that $\beta < \alpha$. Under Kalman condition **[K]**, Lemmas 9 and 10 in [14] (but see also [3, 13]) show that the solution u also verifies the following anisotropic Schauder estimates:

$$\sup_{t \in [0,T]} [u(t, \cdot)]_{C_d^{\alpha+\beta}} \le C_\beta \sup_{t \in (0,T)} [f(t, \cdot)]_{C_d^\beta}, \tag{3.2}$$

for some positive constant $C_\beta := C(\beta, \alpha, A, B, d_0, d_1, T)$. Moreover, it is possible to show (see again [14] in a fractional setting) that, when considering a dilation invariant matrix A, i.e. when $A = A_0$ in (2.16), the constants c_p, C_β appearing in the estimates (1.8)-(3.2) are indeed independent from the final time T (see also [4]).

Our main result shows that indeed the above estimates are invariant under time dependent diffusive perturbations.

Theorem 3.1 *Under [A], let f be in $B_b(0, T; C_c^\infty(\mathbb{R}^N))$. Then, there exists a unique solution u to the perturbed Cauchy problem*

$$\begin{cases} \partial_t u(t,x) = \mathcal{L}^{pert}_{\alpha,t} u(t,x) + f(t,x) & \text{on } \mathbb{R}^N_T; \\ u(0,x) = 0 & \text{on } \mathbb{R}^N, \end{cases} \tag{3.3}$$

where $\mathcal{L}^{pert}_{\alpha,t}$ is defined in (1.7) *for a bounded measurable perturbation $S(t)$. Moreover, for any $p > 1$ and any β in $(0, \alpha \wedge 1)$, it holds that*

$$\begin{aligned} \sup_{t\in[0,T]} [u(t,\cdot)]_{C_d^{\alpha+\beta}} &\le C_\beta \sup_{t\in(0,T)} [f(t,\cdot)]_{C_d^\beta}; \\ \text{if } A = A_0 \text{ in (2.16)}, \qquad [u]_{\dot{W}_d^{\alpha,p}(\mathbb{R}^N_T)} &\le C_p \, \|f\|_{L^p(\mathbb{R}^N_T)}, \end{aligned} \tag{3.4}$$

where C_β, C_p are the same constants as in (3.2) *and* (3.1) *respectively.*

In the diffusive setting (i.e. when $\alpha = 2$) the above result can be exploited to show that the estimates do not depend on the intensity of the matrix B but only on the ellipticity constant κ_2 in (1.2). Indeed, under our hypothesis, it is possible to rewrite $B = \kappa_2 B_I + \tilde{B}$ where

$$B_I := \begin{pmatrix} I_{d_0} & 0_{d_0,d_1} \\ 0_{d_1,d_0} & 0_{d_1,d_1} \end{pmatrix}$$

and $\tilde{B} := B - \kappa_2 B_I$ belongs to $\mathcal{S}_0(\mathbb{R}^N)$. Then, Theorem 3.1 implies that the constants appearing in (3.4) do not depend on $\|B\|$ but only linearly on κ_2^{-1}.

Once we will obtain the above result, it is easy to generalize it to more general operators. Indeed, let us consider the operator

$$\mathscr{L}_{\alpha,t}\phi(x) := \mathcal{L}^{\text{pert}}_{\alpha,t}\phi(x) + \langle a(t), D_x\phi(x)\rangle - c(t)\phi(x), \tag{3.5}$$

where $c\colon (0,T) \to [0,\infty)$, $a\colon (0,T) \to \mathbb{R}^N$ are two integrable functions and, we recall, $\mathcal{L}^{\text{pert}}_{\alpha,t}$ is defined in (1.7) for a bounded measurable perturbation $S(t)$. For any sufficiently regular function $\phi\colon [0,T] \times \mathbb{R}^N \to \mathbb{R}$, we are going to denote

$$\mathcal{T}\phi(t,x) := e^{-\int_0^t c(s)\,ds}\phi\left(t, x + \int_0^t a(s)\,ds\right). \tag{3.6}$$

It is not difficult to check that the "operator" $\mathcal{T}$ transforms solutions of the Cauchy Problem (3.3) to solutions of the Cauchy Problem driven by $\mathscr{L}_{\alpha,t}$, even if for a modified source $\mathcal{T}f$.

Corollary 3.2 *Under [A], let f be in $B_b(0,T;C_c^\infty(\mathbb{R}^N))$. Then, there exists a unique solution v to the Cauchy Problem*

$$\begin{cases} \partial_t v(t,x) = \mathscr{L}_{\alpha,t} v(t,x) + f(t,x) & \text{on } \mathbb{R}^N_T; \\ v(0,x) = 0 & \text{on } \mathbb{R}^N. \end{cases} \tag{3.7}$$

Moreover, for any $p>1$ and any β in $(0,1\wedge\alpha)$, it holds that

$$\begin{aligned} \sup_{t\in[0,T]} [v(t,\cdot)]_{C_d^{\alpha+\beta}(\mathbb{R}^N)} &\le C_\beta \sup_{t\in(0,T)} [f(t,\cdot)]_{C_d^\beta(\mathbb{R}^N)}; \\ \text{if } A = A_0 \text{ in (2.16)}, \qquad [v]_{\dot{W}_d^{\alpha,p}(\mathbb{R}^N_T)} &\le C_p e^{\int_0^T c(s)\,ds} \| f\|_{L^p(\mathbb{R}^N_T)}, \end{aligned} \tag{3.8}$$

where C_β, C_p are the same constants appearing in Theorem 3.1.

Proof We will use the following notation:

$$\tilde{c}(t) := \int_0^t c(s)\,ds, \quad \tilde{a}(t) := \int_0^t a(s)\,ds.$$

As explained before, it is not difficult to check that if v is a solution to Cauchy Problem (3.7), then the function

$$u(t,x) := \mathcal{T}^{-1}v(t,x) = e^{\tilde{c}(t)}v(t,x-\tilde{a}(t))$$

is the unique solution to (3.3) with $\tilde{f}$ instead of f, where

$$\tilde{f}(t,x) := e^{\tilde{c}(t)} f(t,x-\tilde{a}(t)), \quad (t,x)\in\mathbb{R}^N_T.$$

Moreover, we have that $\tilde{f}$ is in $B_b\big(0,T;C_c^\infty(\mathbb{R}^N)\big)$.

Considering $t\le T$, we then notice from Theorem 3.1 (applied to $[0,t]\times\mathbb{R}^N$) that

$$[u(t,\cdot)]_{C_d^{\alpha+\beta}(\mathbb{R}^N)} \le C_\beta \sup_{s\in(0,t)} [\tilde{f}(s,\cdot)]_{C_d^\beta(\mathbb{R}^N)}.$$

Using now the invariance of the Hölder norm under translations, we can show that

$$\begin{aligned} [v(t,\cdot)]_{C_d^{\alpha+\beta}(\mathbb{R}^N)} &\le C_\beta e^{-\tilde{c}(t)} \sup_{s\in(0,t)} [e^{\tilde{c}(s)} f(s,\cdot)]_{C_d^\beta(\mathbb{R}^N)} \\ &\le C_\beta \sup_{s\in(0,t)} [f(s,\cdot)]_{C_d^\beta(\mathbb{R}^N)}, \end{aligned}$$

where in the last step we exploited that $\tilde{c}(t)$ is non-decreasing. Taking the supremum with respect to t on both sides of the above inequality, we obtain the first inequality in (3.8). For the second one, notice from Theorem 3.1, that

$$\sum_{i=0}^{k}\int_0^T e^{p\tilde{c}(t)}\|\Delta_{x_i}^{\alpha_i}v(t,\cdot)\|_{L^p(\mathbb{R}^N)}^p\,dt \le C_p\int_0^T e^{p\tilde{c}(t)}\|f(t,\cdot)\|_{L^p(\mathbb{R}^N)}^p\,dt$$

$$\le C_p\, e^{p\tilde{c}(T)}\int_0^T \|f(t,\cdot)\|_{L^p(\mathbb{R}^N)}^p\,dt.$$

Using the fact that $e^{\tilde{c}(t)} \geq 1$ for all $t \in [0, T]$, we notice that

$$\sum_{i=0}^{k}\int_0^T \|\Delta_{x_i}^{\alpha_i}v(t,\cdot)\|_{L^p(\mathbb{R}^N)}^p\,dt \le C_p\, e^{p\tilde{c}(T)}\int_0^T \|f(t,\cdot)\|_{L^p(\mathbb{R}^N)}^p\,dt,$$

and we have concluded. □

We now show how the results in Theorem 3.1 in the parabolic framework can be adapted to establish a-priori estimates for the elliptic one. More precisely, let us introduce

$$\mathscr{L}_\alpha\phi(x) := \mathcal{L}_\alpha^{\text{ou}}\phi(x) + \langle a, D_x\phi(x)\rangle + \text{Tr}(SD_x^2\phi(x)), \tag{3.9}$$

for a matrix S in $\mathbb{S}_0(\mathbb{R}^N)$ and a in $\mathbb{R}^N$. Let us mention that the Schauder and L^p estimates are known for the *unperturbed* operator in the diffusive case, i.e. the one obtained taking $\alpha = 2$ and $S = 0$ in (3.9). We can e.g. refer to the works [2, 13] respectively. We strongly believe that under the current assumptions, the computations performed in [8] could be extended to derive such estimates when $\alpha \neq 2$ and the matrix A is invariant under dilations. Finally, we remark that when the matrix A is invariant under dilations (i.e. $A = A_0$ as in (2.16)), the constants C_β, C_p in the estimates do not depend on the final time T.

Corollary 3.3 *Under [**A**], let g be in $C_c^\infty(\mathbb{R}^N)$ and $A = A_0$ as in* (2.16)*. Assume that there exists a bounded, continuous (classic) solution $u\colon \mathbb{R}^N \to \mathbb{R}$ to the following elliptic equation:*

$$\mathscr{L}_\alpha u(x) = g(x), \quad \textit{on } \mathbb{R}^N. \tag{3.10}$$

Then, such solution u is unique and for any $p > 1$ and any β in $(0, 1 \wedge \alpha)$, it holds that

$$[u]_{C_d^{2+\beta}(\mathbb{R}^N)} \le C_\beta[g]_{C_d^\beta(\mathbb{R}^N)};$$

$$[u]_{\dot{W}_d^{2,p}(\mathbb{R}^N)} \le C_p\|g\|_{L^p(\mathbb{R}^N)},$$

where C_β and C_p are the same constants appearing in Theorem 3.1.

Proof We start with the issue of uniqueness. Let us consider a solution $u\colon \mathbb{R}^N \to \mathbb{R}$ to the elliptic equation (3.10). Fixed a final time $T > 0$, it is easy to check that the function $v\colon \mathbb{R}^N_T \to \mathbb{R}$ given by $v(t,x) := u(x)t/T$ is then a solution of the following Cauchy Problem:

$$\begin{cases} \partial_t v(t,x) = \mathscr{L}_\alpha v(t,x) + f(t,x), & \text{on } \mathbb{R}^N_T; \\ v(0,x) = 0, & \text{on } \mathbb{R}^N, \end{cases}$$

where $f(t,x) = u(x)/T - g(x)t/T$. Thus, the uniqueness of u follows immediately from the uniqueness of solutions in Theorem 3.1. Moreover, since $A = A_0$, we know that v satisfies Estimates (3.4) for constants C_p, C_β independent from T. Hence, we have that

$$\begin{aligned} C^{\alpha+\beta}_d(\mathbb{R}^N) &= \sup_{t\in[0,T]} [v(t,\cdot)]_{C^{\alpha+\beta}_d(\mathbb{R}^N)} \le C_\beta \sup_{t\in(0,T)} [f(t,\cdot)]_{C^{\beta}_d(\mathbb{R}^N)} \\ &\le C_\beta \left(\frac{[u]_{C^{\beta}_d(\mathbb{R}^N)}}{T} + [g]_{C^{\beta}_d(\mathbb{R}^N)} \right). \end{aligned}$$

We can now let T go to infinity in the equation above. Recalling that C_β is independent from T, we conclude that

$$[u]_{C^{\alpha+\beta}_d(\mathbb{R}^N)} \le C_\beta [g]_{C^{\beta}_d(\mathbb{R}^N)}.$$

In order to prove the Sobolev estimates, we firstly notice that for any $i \in [\![0,k]\!]$,

$$\|\Delta^{\alpha_i}_{x_i} v\|^p_{L^p(\mathbb{R}^N_T)} = \int_0^T \left(\frac{t}{T}\right)^p \int_{\mathbb{R}^N} |\Delta^{\alpha_i}_{x_i} u(z)|^p \, dz dt = \frac{T}{p+1} \|\Delta^{\alpha_i}_{x_i} u\|^p_{L^p(\mathbb{R}^N)}.$$

Using Estimates (3.4) for v, we then show that

$$\begin{aligned} \sum_{i=0}^d \|\Delta^{\alpha_i}_{x_i} u\|^p_{L^p(\mathbb{R}^N_T)} &= \frac{p+1}{T} \sum_{i=0}^d \|\Delta^{\alpha_i}_{x_i} v\|^p_{L^p(\mathbb{R}^N_T)} \le C_p^p \frac{p+1}{T} \|f\|^p_{L^p(\mathbb{R}^N_T)} \\ &\le C_p^p \frac{p+1}{T} \int_0^T \int_{\mathbb{R}^N} \left| \frac{u(z)}{T} - g(z)\frac{t}{T} \right|^p dz dt \\ &= C_p^p (p+1) \int_0^1 \int_{\mathbb{R}^N} \left| \frac{u(z)}{T} - s g(z) \right|^p dz ds, \end{aligned}$$

where, in the last step, we applied the change of variables $s = t/T$. Letting T go to infinity, we finally notice that

$$\sum_{i=0}^{d} \|\Delta_{x_i}^{\alpha_i} u\|_{L^p(\mathbb{R}^N)}^p \le C_p^p \|g\|_{L^p(\mathbb{R}^N)}^p \int_0^1 (p+1)s^p\, ds = C_p^p \|g\|_{L^p(\mathbb{R}^N)}^p.$$

We have thus concluded the proof of Corollary 3.3. □

4 Proof of the Main Result

Let $u\colon [0,T] \times \mathbb{R}^N \to \mathbb{R}$ be the solution to the initial (i.e. non-perturbed) Cauchy problem (1.6). We can then introduce a function $v(t,x) := u(t, e^{-tA}x)$. Since u is Lipschitz continuous in $t \in (0,T)$, it is possible to differentiate the function $u(t,x) = v(t, e^{tA}x)$ with respect to t, for almost every t in $[0,T]$. It then follows that v is the solution to

$$\begin{cases} \partial_t v(t,x) = \tilde{\mathcal{L}}_{\alpha,t} v(t,x) + \tilde{f}(t,x); \\ v(0,x) = 0, \end{cases} \tag{4.1}$$

where we denoted $\tilde{f}(t,z) := f(t, e^{-tA}z)$ and

$$\begin{aligned} \tilde{\mathcal{L}}_{\alpha,t}\phi(x) := {} & \mathbb{1}_{\{\alpha=2\}} \mathrm{Tr}\left(e^{tA}Be^{tA^*}D_x^2 v(t,x)\right) \\ & + \mathbb{1}_{\{\alpha\neq 2\}} \int_{\mathbb{R}^d} \Big[v(t, x+e^{tA}\sigma z) - v(t,x) \\ & \qquad - \langle D_x v(t,x), e^{tA}\sigma z\rangle \mathbb{1}_{B(0,1)}(\sigma z)\Big]\, \nu_\alpha(dz). \end{aligned}$$

Noticing that $\det(e^{A_0 t}) = 1$ so that $\|f\|_{L^p(\mathbb{R}^N_T)} = \|\tilde{f}\|_{L^p(\mathbb{R}^N_T)}$, the known estimates on u in (2.4), (3.1) and (3.2) can be rewritten in terms of the function v as:

$$\begin{aligned} & \sup_{(t,x)\in[0,T]\times\mathbb{R}^N} |v(t,x)| \le T \sup_{(t,x)\in\mathbb{R}^N_T} |\tilde{f}(t,x)|; \\ & \sup_{t\in[0,T]} [v(t,\cdot)]_{C^{\alpha+\beta}_{d,A}(\mathbb{R}^N)} \le C_\beta \sup_{t\in(0,T)} [\tilde{f}(t,\cdot)]_{C^{\beta}_{d,A}(\mathbb{R}^N)}; \\ & \text{if } A = A_0 \text{ in (2.16)},\ [v]_{\dot{W}^{\alpha,p}_{d,A}(\mathbb{R}^N_T)} \le C_p \|\tilde{f}\|_{L^p(\mathbb{R}^N_T)}, \end{aligned} \tag{4.2}$$

where we have denoted

$$[v(t,\cdot)]_{C^{\alpha+\beta}_{d,A}(\mathbb{R}^N)} := [v(t,e^{At}\cdot)]_{C^{\alpha+\beta}_d(\mathbb{R}^N)}, \qquad [v]^p_{\dot{W}^{\alpha,p}_{d,A}(\mathbb{R}^N_T)}$$
$$:= \sum_{i=0}^{k} \|\Delta^{\alpha_i,A}_{x_i} v\|^p_{L^p(\mathbb{R}^N_T)}$$

and

$$\Delta^{\alpha_i,A}_{x_i} v(t,x) := \text{p.v.}\int_{\mathbb{R}^{d_i}} [v(t,x+e^{tA}E_i z)) - v(t,x)]\frac{dz}{|z|^{d_i+2\alpha_i}}.$$

Clearly, it is possible to rewrite (4.1) in the form of Cauchy problem (2.1) with $Q(t) = \mathbb{1}_{\{\alpha=2\}}e^{tA}Be^{tA^*}$ and $\gamma(t) = \mathbb{1}_{\{\alpha\neq 2\}}e^{tA}\sigma$, exploiting the symmetry of the measure ν_α. By the arguments before and the maximum principle (Theorem 2.2, we already know that for any $\tilde{f}$ in $B_b(0,T;C^\infty_c(\mathbb{R}^N))$, there exists a unique solution v to Cauchy problem (4.1) and it satisfies the estimates in (4.2).

Let us consider again the bounded measurable perturbation $S\colon (0,T) \to \mathcal{S}_0(\mathbb{R}^N)$. We would like to show that there exists a unique solution $w\colon [0,T]\times\mathbb{R}^N \to \mathbb{R}$ to

$$\begin{cases} \partial_t w(t,x) = \tilde{\mathcal{L}}_{\alpha,t} w(t,x) + \text{Tr}\left(e^{tA}S(t)e^{tA^*}D^2_x w(t,x)\right) + \tilde{f}(t,x); \\ w(0,x) = 0, \end{cases} \tag{4.3}$$

and that the estimates in (4.2) hold for w as well, with the same constants C_β, C_p appearing before. Indeed, the backward change of variables argument explained at the beginning of the section will then allow us to conclude the proof of Theorem 3.1. More precisely, we will prove at the end of the current section the following, more general, result:

Proposition 4.1 *Let $\tilde{f}$ be in $B_b(0,T;C^\infty_c(\mathbb{R}^N))$ and $\tilde{S}\colon (0,T) \to \mathcal{S}_0(\mathbb{R}^N)$ bounded measurable. Then, there exists a unique solution w to:*

$$\begin{cases} \partial_t w(t,x) = \tilde{\mathcal{L}}_{\alpha,t} w(t,x) + \text{Tr}\left(\tilde{S}(t)D^2_x w(t,x)\right) + \tilde{f}(t,x), \\ w(0,x) = 0. \end{cases} \tag{4.4}$$

Moreover, the estimates in (4.2) *hold again with w instead of v and with the same constants appearing before.*

It is clear that Theorem 3.1 will then follow from the previous result by choosing $\tilde{S}(t) := e^{tA}S(t)e^{tA^*}$. The crucial observation in the proof of the above result is that, at least formally, we can approximate the diffusive perturbation $\tilde{S}(t)$ as follows:

$$
\begin{aligned}
\frac{1}{2}\mathrm{Tr}\left(\tilde{S}(t)D_x^2\phi(x)\right) &\approx \frac{1}{2}\sum_{i=1}^{N}\frac{1}{\varepsilon^2}\Big[\phi(x+\varepsilon\sqrt{\tilde{S}(t)}e_i) \\
&\qquad -2\phi(x)+\phi(x-\varepsilon\sqrt{\tilde{S}(t)}e_i)\Big] \\
&= \sum_{i=1}^{N}\lambda_i\left(\delta_{l_i(t)}-\delta_{-l_i(t)}\right)\phi(x) \;=:\; J^{\tilde{S}}_{t,\varepsilon}\phi(x),
\end{aligned}
\tag{4.5}
$$

where $\lambda_i = \frac{1}{2}\varepsilon^{-2}$, $l_i(t) = \varepsilon\sqrt{\tilde{S}(t)}e_i$ and $\delta_y\phi(\cdot) := \phi(\cdot + y) - \phi(\cdot)$ is the first order finite difference operator.

In order to apply the above heuristic argument in our case, when $\tilde{S}(t)$ is only Borel measurable, we firstly need to show the existence of a Borel measurable principal square root for $\tilde{S}(t)$. While this fact is well-known if $\tilde{S}(t)$ is positive-definite, a proof of the following result could be found in [17, Result 5.4]. [1]

Lemma 4.2 *Let $Q\colon (0,T) \to \mathcal{S}_0(\mathbb{R}^N)$ be a measurable function. Then, there exists a measurable function $\sqrt{Q}\colon (0,T) \to \mathcal{S}_0(\mathbb{R}^N)$ such that $\sqrt{Q}(t)$ is a square root of $Q(t)$ for any fixed t in $(0,T)$.*

As a first step in our method of proof, we will show that it is possible to add a finite difference operator, like the one appearing in (4.5), to the Cauchy problem without modifying the estimates. This is indeed the key point in the proof method of [11] and it is performed through a probabilistic argument which relies on the use of a suitable corresponding Poisson process. More precisely, we have the following result:

Lemma 4.3 *Let λ be a real number, $l\colon (0,T) \to \mathbb{R}^N$ a bounded measurable function and $\tilde{f}$ in $B_b(0,T;C_c^\infty(\mathbb{R}^N))$. Then, there exists a unique solution w to*

$$
\begin{cases}
\partial_t w(t,x) = \tilde{\mathcal{L}}_{\alpha,t}w(t,x) + \lambda\delta_{l(t)}w(t,x) + \tilde{f}(t,x); \\
w(0,x) = 0,
\end{cases}
\tag{4.6}
$$

[1] For the sake of completeness of the work, we also highlight that the measurability of the principal square root of the matrix $\tilde{S}(t)$ could be obtained from the following integral representation

$$
\sqrt{\tilde{S}(t)} = c\int_0^{+\infty}\theta^{-3/2}(I - \exp(-\theta\tilde{S}(t)))\,d\theta,
$$

where $c > 0$ is a normalising constant, noticing that the exponential of a measurable matrix is still measurable.

Moreover, the estimates in (4.2) *hold for w as well, with the* same *constants appearing therein.*

Proof Let us assume for the moment that $\lambda > 0$. We can define

$$X_t = \int_0^t l(r)\, d\pi_r$$

where $\{\pi_t\}_{t\geq 0}$ is a scalar Poisson process of intensity λ defined on a probability space $(\Omega, \mathcal{F}, \mathbb{P})$. We notice that the random shifted source $\bar{f}(t,x) := \tilde{f}(t, x - X_t)$ is again in $B_b\big(0, T; C_0^\infty(\mathbb{R}^N)\big)$. We have omitted to write the dependence on the random parameter ω for notational simplicity. We again emphasise that, from the very nature of the Poisson process involved, the modified (random) source $\bar{f}$ has jumps in time. Then, for (almost every) fixed ω in Ω, we already know that there exists a unique solution v to (4.1), replacing therein $\tilde{f}$ with the random source $\bar{f}$, depending on ω as parameter. Moreover, thanks to the invariance for translations of the involved norms, it follows from (4.2) that

$$\begin{aligned}
\sup_{(t,x)\in[0,T]\times\mathbb{R}^N} |v(t,x)| &\leq T \sup_{(t,x)\in\mathbb{R}^N_T} |\tilde{f}(t,x)|;\\
\sup_{t\in[0,T]} [v(t,\cdot)]_{C^{\alpha+\beta}_{d,A}(\mathbb{R}^N)} &\leq C_\beta \sup_{t\in(0,T)} [\tilde{f}(t,\cdot)]_{C^{\beta}_{d,A}(\mathbb{R}^N)}; \qquad (4.7)\\
\text{if } A = A_0, \qquad [v]_{\dot{W}^{\alpha,p}_{d,A}(\mathbb{R}^N_T)} &\leq C_p \|\tilde{f}\|_{L^p(\mathbb{R}^N_T)}.
\end{aligned}$$

It is not difficult to check (cf. [1]) that the solution v is given by

$$v(t,x) = \int_0^t \int_{\mathbb{R}^N} \tilde{f}(s, x - X_s + y)\, \mu_{s,t}(dy) ds,$$

where $\mu_{s,t}$ is the law of the stochastic integral

$$I_{s,t} := \int_s^t e^{rA}\sigma\, dZ_r,$$

with $\{Z_t\}_{t\geq 0}$ a Brownian motion if $\alpha = 2$ or an α-stable process with Lévy measure ν_α, otherwise.

For each $x \in \mathbb{R}^N$, the stochastic process $(v(t,x))_{t\in[0,T]}$ has continuous paths ($\mathbb{P}$-a.s.) and it is $\mathcal{F}_t$-adapted where $\mathcal{F}_t$ is the completed σ-algebra generated by the random variables π_s, $0 \leq s \leq t$. For fixed $x \in \mathbb{R}^N$, let us introduce the process $(v(t, x + X_t))_{t\in[0,T]}$ which is given by

$$v(t, x + X_t) = \int_0^t \int_{\mathbb{R}^N} [\tilde{f}(s, x + X_t - X_s + y) \mu_{s,t}(dy) ds.$$

It is not difficult to check that it is $\mathcal{F}_t$-adapted and it has càdlàg paths.

Denoting by $\sigma_n,\ n \in \{0, \dots, \pi_t\}$ the jump times of the process $(X_s)_{s\in[0,t]}$ and setting as well $\sigma_0 = 0$, we have that X_s is constant for $s \in [\sigma_n, \sigma_{n+1} \wedge t)$. One then derives that:

$$\begin{aligned}
v(t, x + X_t) &= v(t, x + X_t) - v(0, x)\\
&= \sum_{n=0}^{\pi_t - 1} \Big[v(\sigma_{n+1}^-, x + X_{\sigma_{n+1}^-}) - v(\sigma_n, x + X_{\sigma_n}) + v(t, x + X_t)\\
&\quad - v(\sigma_{\pi_t}, x + X_{\sigma_{\pi_t}}) + v(\sigma_{n+1}, x + X_{\sigma_{n+1}})\\
&\quad - v(\sigma_{n+1}^-, x + X_{\sigma_{n+1}^-}) \Big]\\
&= \int_0^t \Big[\tilde{\mathcal{L}}_{\alpha,t} v(s, x + X_s)) + \tilde{f}(s, x) \Big] ds + \int_0^t g(s, x)\, d\pi_s, \qquad (4.8)
\end{aligned}$$

where $g(s, x) = v(s, x + l(s) + X_{s-}) - v(s, x + X_{s-})$ is precisely the contribution associated with the jump times. Here, $X_{s-} = \lim_{t\uparrow s} X_t,\ s > 0$. It is clear that $g(s, x) \neq 0$ if and only if π_s has a jump at time s. We then have:

$$\mathbb{E}\left[\int_0^t g(s, x)\, d\pi_s \right] = \lambda \int_0^t [w(s, x + l(s)) - w(s, x)]\, ds, \qquad (4.9)$$

where $w(s, x) = \mathbb{E}[v(s, x + X_s)]$. The above equality can be easily proven when g is piece-wise constant and then extended to our more general context by standard approximation arguments (cf. Lemma 2.1 in [11]). Taking the expectation on both sides of Eq. (4.8), we find out that w is an integral solution to (4.6). Moreover by (4.7) we obtain (using also the Jensen inequality and the Fubini theorem)

$$\begin{aligned}
[w]^p_{\dot{W}^{\alpha,p}_{d,A}(\mathbb{R}^N_T)} &= \sum_{i=0}^{k} \int_{\mathbb{R}^N_T} |\Delta^{\alpha_i, A}_{x_i} w(t, x)|^p\, dxdt\\
&= \sum_{i=0}^{k} \int_{\mathbb{R}^N_T} \left| \mathbb{E}\left[\Delta^{\alpha_i, A}_{x_i} v(t, x + X_t) \right] \right|^p dxdt\\
&\le \sum_{i=0}^{k} \int_{\mathbb{R}^N_T} \mathbb{E}\left[\left| \Delta^{\alpha_i, A}_{x_i} v(t, x + X_t) \right|^p \right] dxdt\\
&= \sum_{i=0}^{k} \mathbb{E}\left[\int_{\mathbb{R}^N_T} \left| \Delta^{\alpha_i, A}_{x_i} v(t, x + X_t) \right|^p dxdt \right]\\
&= \sum_{i=0}^{k} \mathbb{E}\left[\int_{\mathbb{R}^N_T} |\Delta^{\alpha_i, A}_{x_i} v(t, x')|^p\, dx'dt \right] \le (C_p)^p \|\tilde{f}\|^p_{L^p(\mathbb{R}^N_T)}.
\end{aligned}$$

A similar argument works as well for the Schauder estimates and the maximum principle. Uniqueness of solutions to (4.6) follows by Theorem 2.2. Indeed, if we assume for the moment that $\lambda T \leq 1/4$, a solution w to (4.6) is also a solution to Cauchy problem (4.1) with source $\tilde{f}(t,x) + \lambda[w(t,x+l(t)) - w(t,x)]$. Since $\tilde{f}$ is regular enough, we can then apply Theorem 2.2 and, by a circular argument, conclude that

$$\begin{aligned}\sup_{(t,x)\in[0,T]\times\mathbb{R}^N} |w(t,x)| &\leq T\Big(\sup_{(t,x)\in\mathbb{R}^N_T} |\tilde{f}(t,x)| + 2\lambda \sup_{(t,x)\in[0,T]\times\mathbb{R}^N} |w(t,x)|\Big)\\ &\leq 2T \sup_{(t,x)\in\mathbb{R}^N_T} |\tilde{f}(t,x)|,\end{aligned}$$

when precisely $\lambda T \leq 1/4$. The above estimates clearly imply the uniqueness of solutions. By iteration of the above procedure by steps of size $1/(4\lambda)$, we can claim the uniqueness of solutions over the full interval $[0, T]$.

By considering the opposite random shift $\bar{f}(t,x) := \tilde{f}(t, x + X_t)$, the same arguments above allows to obtain the result when $\lambda < 0$. □

Now, by an easy iterative argument using the above Lemmas 4.3 and 4.2, it is not difficult to show that

Proposition 4.4 *Fixed $\varepsilon > 0$, let $\tilde{f}$ be in $B_b(0,T; C^\infty_c(\mathbb{R}^N))$. Then, there exists a unique solution w_ε to*

$$\begin{cases}\partial_t w_\varepsilon(t,x) = \tilde{\mathcal{L}}_{\alpha,t} w_\varepsilon(t,x) + J^{\tilde{S}}_{t,\varepsilon} w_\varepsilon(t,x) + \tilde{f}(t,x);\\ w_\varepsilon(0,x) = 0,\end{cases} \tag{4.10}$$

where, we recall, the operator $J^{\tilde{S}}_{t,\varepsilon}$ has been defined in (4.5). *Moreover, the estimates in* (4.2) *hold again with w_ε instead of v and with the same constants appearing before.*

In order to conclude and show Proposition 4.1, we are interested now in understanding what happens when we let ε goes to zero in (4.10). In particular, we want the corresponding second order term in (4.5) to appear at the limit.

Proof of Proposition 4.1 Let us recall from Proposition 4.4 that there exists a family $\{w_\varepsilon\}_{\varepsilon>0}$ such that $w_\varepsilon\colon [0,T]\times\mathbb{R}^N \to \mathbb{R}$ is the unique solution to (4.10). Moreover, it holds that

$$\sup_{(t,x)\in[0,T]\times\mathbb{R}^N} |w_\varepsilon(t,x)| \leq T \sup_{(t,x)\in\mathbb{R}^N_T} |f(t,x)|. \tag{4.11}$$

For a multi-index γ in $\mathbb{N}^N$ and $h > 0$, let us denote by Δ_h^γ the iteration of the finite difference operators

$$\Delta_h^i \phi(x) = \frac{\phi(x + he_i) - \phi(x)}{h}, \quad h > 0, \, i \in [\![1, N]\!].$$

We then notice that $\Delta_h^\gamma w_\varepsilon(t, x)$ solves again (4.10) with $\Delta_h^\gamma f$. Using that f is in $B_b(0, T; C_c^\infty(\mathbb{R}^N))$ and the maximum principle in (4.11), it then follows that w_ε is smooth in space and $D_x^\gamma w_\varepsilon$ is bounded in t, x, ε. Equation (4.10), to be understood in its integral form, together with (4.11) then gives that those derivatives are themselves Lipschitz continuous in time, uniformly in x, ε. This precisely gives that the family $\{w_\varepsilon : \varepsilon > 0\}$ is equi-Lipschitz continuous on any compact subset K of $[0, T] \times \mathbb{R}^N$. Similar arguments hold as well for the spatial derivatives of w_ε.

We can now apply the Arzelà-Ascoli theorem to w_ε showing the existence of a sub-sequence $\{w_{\varepsilon_n}\}_{n\in\mathbb{N}}$ which converges uniformly on any compact set to a function $w\colon [0, T] \times \mathbb{R}^N \to \mathbb{R}$. Moreover, w is smooth in space and all its space derivatives are Lipschitz continuous on any compact set. Similarly, any derivative in space of w_{ε_n} tends to the respective derivative of w, uniformly on the compact sets. Passing to the limit as $n \to \infty$ along the sub-sequence $\{\varepsilon_n\}_{n\in\mathbb{N}}$ in the estimates in (4.2), we immediately find out that they hold again with w instead of v and with the same constants appearing before. Letting ε_n goes to zero in Cauchy problem (4.10) (written in the integral form), we can also conclude that w solves

$$\partial_t w(t, x) = \tilde{\mathcal{L}}_{\alpha,t} w(t, x) + \mathrm{Tr}\left(\tilde{S}(t) D_x^2 w(t, x)\right) + \tilde{f}(t, x).$$

Indeed, the dominated convergence theorem immediately implies that

$$\lim_{n\to+\infty} \int_0^t J_{t,\varepsilon_n}^{\tilde{S}} w_{\varepsilon_n}(s, x)\, ds = \int_0^t \left[\mathrm{Tr}\left(\tilde{S}(s) D_x^2 w(s, x)\right)\right] ds.$$

Similarly, it also holds that

$$\lim_{n\to+\infty} \int_0^t \left[\mathrm{Tr}\left(B D_x^2 w_{\varepsilon_n}(s, x)\right)\right] ds = \int_0^t \left[\mathrm{Tr}\left(B D_x^2 w(s, x)\right)\right] ds,$$

when $\alpha = 2$. On the other hand (i.e. $\alpha \neq 2$), we have that

$$\int_0^t \tilde{\mathcal{L}}_{\alpha,t} w_{\varepsilon_n}(s, x)\, ds = \int_0^t \int_{|\sigma z|\geq 1} \left[w_\varepsilon(s, x + e^{sA}\sigma z) - w_\varepsilon(s, x)\right] \nu_\alpha(dz)$$
$$+ \int_0^t \int_{|\sigma z|<1} \left[w_\varepsilon(s, x + e^{sA}\sigma z) - w_\varepsilon(s, x) - \langle D_x w_\varepsilon(s, x), e^{sA}\sigma z\rangle\right] \nu_\alpha(dz).$$

While the first integral is clearly convergent by the bounded convergence theorem, we notice that

$$|w_\varepsilon(s, x + e^{sA}\sigma z) - w_\varepsilon(s, x) - \langle D_x w_\varepsilon(s, x), e^{tA}\sigma z\rangle| \leq |e^{sA}\sigma z|^2 \|D_x^2 w_\varepsilon\|_\infty$$

and the dominated convergence theorem can thus be applied to the second integral, thanks to a uniform estimate for $D_x^2 w_\varepsilon$ (cf. (4.11)). Finally, the uniqueness of such solution w follows immediately from the maximum principle (Theorem 2.2). □

References

1. Applebaum, D.: Lévy Processes and Stochastic Calculus. Cambridge Studies in Advanced Mathematics, vol. 116, 2nd edn. Cambridge University Press, Cambridge (2009)
2. Bramanti, M., Cupini, G., Lanconelli, E., Priola, E.: Global L^p estimates for degenerate Ornstein-Uhlenbeck operators. Math. Z. **266**(4), 789–816 (2010)
3. Chaudru de Raynal, P.E., Honoré, I., Menozzi, S.: Sharp Schauder estimates for some degenerate Kolmogorov equations. Ann. Sc. N. Sup. Pisa Cl. Sci. (5) **22**(3), 989–1089 (2021)
4. Chaudru de Raynal, P.E., Menozzi, S., Priola, E.: Schauder estimates for drifted fractional operators in the supercritical case. J. Funct. Anal. **278**(8), 108425 (2020)
5. Chen, Z.Q., Zhang, X.: Propagation of regularity in L^p-spaces for Kolmogorov-type hypoelliptic operators. J. Evol. Equ. **19**(4), 1041–1069 (2019)
6. Coifman, R., Weiss, G.: Analyse Harmonique Non-commutative sur Certains Espaces Homogènes. Lecture Notes in Mathematics, vol. 242. Springer, New York (1971). Étude de certaines intégrales singulières
7. Hörmander, L.: Hypoelliptic second order differential equations. Acta Math. **119**, 147–171 (1967)
8. Huang, L., Menozzi, S., Priola, E.: L^p estimates for degenerate non-local Kolmogorov operators. J. Math. Pures Appl. **121**, 162–215 (2019)
9. Krylov, N.V.: Lectures on Elliptic and Parabolic Equations in Hölder Spaces. Graduate Studies in Mathematics, vol. 12. AMS, Providence (1996)
10. Krylov, N.V., Priola, E.: Elliptic and parabolic second-order PDEs with growing coefficients. Commun. Partial Differ. Equ. **35**(1), 1–22 (2010)
11. Krylov, N.V., Priola, E.: Poisson stochastic process and basic Schauder and Sobolev estimates in the theory of parabolic equations. Arch. Ration. Mech. Anal. **225**(3), 1089–1126 (2017)
12. Lanconelli, E., Polidoro, S.: On a class of hypoelliptic evolution operators. Rend. Sem. Mat. Univ. Politec. Torino **52**(1), 29–63 (1994)
13. Lunardi, A.: Schauder estimates for a class of degenerate elliptic and parabolic operators with unbounded coefficients in $\mathbf{R}^n$. Ann. Scuola Norm. Sup. Pisa Cl. Sci. **24**(1), 133–164 (1997)
14. Marino, L.: Schauder estimates for degenerate stable Kolmogorov equations. Bull. Sci. Math. **162**, 102885 (2020)
15. Marino, L., Menozzi, S., Priola, E.: Poisson process and sharp constants in L^p and Schauder estimates for a class of degenerate Kolmogorov operators. Stud. Math. **267**(3), 321–346 (2022)
16. Menozzi, S.: Martingale problems for some degenerate Kolmogorov equations. Stoch. Process. Appl. **128**(3), 756–802 (2018)
17. Reid, W.T.: Some elementary properties of proper values and proper vectors of matrix functions. SIAM J. Appl. Math. **18**, 259–266 (1970)
18. Zabczyk, J.: Mathematical Control Theory: An Introduction. Systems & Control: Foundations & Applications. Birkhäuser, Boston (1992)

Integration by Parts Formula for Exit Times of One Dimensional Diffusions

Noufel Frikha, Arturo Kohatsu-Higa, and Libo Li

Abstract In line with the methodology introduced in Frikha et al. (Electron J Probab 24(95):1–44, 2019) for formulating probabilistic representations of integration by parts involving killed diffusion, we establish an integration by parts formula for the first exit time of one-dimensional diffusion processes. However, our approach diverges from the conventional differential calculus applied to the associated space Markov chain; instead, we employ calculus techniques that focus on the underlying time variables.

1 Introduction

In this article, we consider the process denoted as X, which is defined as the unique strong solution to the following one dimensional stochastic differential equation (SDE for short) with the following dynamics:

$$X_t = x + \int_0^t b(X_s)\,\mathrm{d}s + \int_0^t \sigma(X_s)\,\mathrm{d}W_s, \tag{1}$$

where the coefficients $b, \sigma : \mathbb{R} \to \mathbb{R}$ are smooth and bounded functions and $(W_t)_{t\geq 0}$ denotes a standard one-dimensional Brownian motion within a given filtered probability space $(\Omega, \mathcal{F}, (\mathcal{F}_t)_{t\geq 0}, \mathbb{P})$.

N. Frikha (✉)
Université Paris 1 Panthéon-Sorbonne, Centre d'Economie de la Sorbonne, Paris, France
e-mail: noufel.frikha@univ-paris1.fr

A. Kohatsu-Higa
Department of Mathematical Sciences, Ritsumeikan University, Kusatsu, Shiga, Japan
e-mail: khts00@fc.ritsumei.ac.jp

L. Li
University of New South Wales, School of Mathematics and Statistics, Sydney, NSW, Australia
e-mail: libo.li@unsw.edu.au

S. Menozzi et al. (eds.), *Kolmogorov Operators and Their Applications*,
Springer INdAM Series 56, https://doi.org/10.1007/978-981-97-0225-1_9

The aim of the present paper is to derive an integration by parts (IBP) formula for the first hitting time of a given level L by the process X. To elaborate further, for a fixed starting point $x \geq L$, we define τ as the first time X crosses the level L: $\tau = \inf\{t \geq 0 : X_t = L\}$. For any given finite time horizon $T > 0$, our focus here is on establishing a explicit probabilistic representation for the IBP formula pertaining to the quantity:

$$\mathbb{E}[f'(\tau \wedge T)]$$

where f is a real-valued smooth function defined on $\mathbb{R}_+$.

Numerous theoretical properties concerning these exit times have been documented. While not exhaustive, one can refer, for instance, to [18] and [19]. Nonetheless, as mentioned in [20]: "The distribution of the exit time can be expressed using the sum of an infinite sequence of independent exponential variables. While such expansions are of interest in a number of contexts, the corresponding representations of the density or cumulative distribution function of the exit time can be difficult to work with".

The application of Malliavin Calculus to random variables associated with exit times remains somewhat fragmented. In [6] and [2], the authors establish certain regularity properties of the stopped diffusion process at the boundary of a general smooth domain. In contrast, in [1], the authors prove that, in general, exit times can not be Malliavin differentiable even once. Therefore, to the best of our knowledge, the pursuit of an integration by parts formula for exit times using Malliavin Calculus appears to be a daunting task.

In this article, we first establish a probabilistic representation formula in the following format

$$\mathbb{E}[f(\tau \wedge T)] = \mathbb{E}[f(\bar{\tau} \wedge T) \prod_{i=1}^{N_T+1} \theta^i].$$

Here $\bar{\tau}$ is an approximation of the exit time τ and the sequence of random variables $(\theta^i)_{1 \leq i \leq N_T+1}$ is derived from a Markov chain that approximates the trajectory of the underlying diffusion process. Additionally, $(N_t)_{0 \leq t \leq T}$ represents a Poisson process, and f is a bounded measurable function. The proof employs similar arguments as those utilized in [14].

Our starting point in establishing our IBP formula is based on the probabilistic representation formula mentioned earlier. The representation proposed in this context is associated with a general perturbation theory of Markov processes, commonly employed for investigating the smoothness properties of probability density functions. This theory, which can be seen as an intermediate step between approximations and their limits, has been extensively developed in various directions. without being exhaustive, references to this theory can be found in works such as [5, 11] and [7–9, 12–14].

To derive IBP formulas in our context, we initially observe that the problem bears a certain 'dual' relationship with the one pertaining to one-dimensional killed diffusion processes. In a loose sense, space and time variables undergo an interchange. With this perspective in mind, our approach involves considering the conditional distribution of the underlying time variables with respect to the space variables. Subsequently, we apply the IBP formulas with respect to the time variables. Given that these formulas involve differentiation of the time random variables, we also anticipate that the degree of degeneracy will be more pronounced compared to the case of killed diffusion.

When we carry out the conditioning process with respect to the space variables as described above, the distribution of the jump times undergoes a transformation from exponential to generalized inverse Gaussian. This change occurs because, in addition to the exponential density, the spatial Gaussian transition densities must also be taken into account as components of the conditional distribution of the jump times. Consequently, based on the preceding discussion, it becomes necessary to introduce dual operators that are associated with the jump times of Poisson processes conditioned on space variables.

The literature regarding integration by parts formulas concerning the jump times of a Poisson process is extensive. In all cases known to the authors, the general approach involves employing IBP formulas with respect to exponential distributions, which may inherently introduce boundary conditions. Dealing with these terms can pose difficulties, necessitating the use of localization techniques centered around the boundary values of these time random variables to address this technical challenge. For further insights into this topic, please consult references such as [4, 10], and [3].

In the current configuration, we do not seek to circumvent these boundary terms; rather, we include them in our calculations. Since the calculation is carried out conditionally, based on the values of space variables, within a random partition of the time interval-assumed, without loss of generality, to be all distinct-we ascertain that all boundary terms will ultimately disappear.

Lastly, following a similar approach to our companion paper [13], where we employed a time stabilization argument for stopped processes, in this article, we employ a spatial stabilization technique. This approach leads to a Malliavin variance quantity that relies on the squares of distances between points within the underlying Markov chain structure. The central result of this paper is presented in Theorem 14, where we establish the IBP formula for the first hitting time of X. We achieve this using the probabilistic representation derived in Theorem 6

Notations We use $A \overset{\mathbb{E}}{=} B$ to denote that $\mathbb{E}[A] = \mathbb{E}[B]$. We will be working on a Wiener space with all necessary additional independent r.v.'s added to it. Products $\prod_{j=i}^{k} \theta^j$ are defined to be 1 when $i > k$. We also use $\bar{\mathbb{N}} = \mathbb{N} \cup \{0\}$ and $\bar{\mathbb{N}}_{n+1} = \{0, \ldots, n+1\}$.

2 Preliminaries

2.1 Assumptions

Throughout the article, we work on the filtered probability space $(\Omega, \mathcal{F}, (\mathcal{F}_t)_{t\geq 0}, \mathbb{P})$ which is assumed to be rich enough to support all random variables that will be considered in what follows. In addition, we will suppose that the following assumptions on the coefficients are in force:

Assumption (H)

(i) The coefficients of SDE (1) are smooth, bounded with bounded derivatives. That is, $b \in \mathscr{C}_b^\infty(\mathbb{R})$ and $a := \sigma^2 \in \mathscr{C}_b^\infty(\mathbb{R})$.

(ii) The function σ is uniformly elliptic and bounded. That is, there exist $\underline{a}, \overline{a} > 0$ such that for any $x \in \mathbb{R}$, $\underline{a} \leq a(x) \leq \overline{a}$.

2.2 A Reflection Principle

As in [13], our probabilistic representation involves the following approximation process

$$\bar{X}_t^{s,x} \equiv \bar{X}_t^{s,x}(\rho) = \rho x + (1-\rho)(2L - x) + \sigma(x)(W_t - W_s), \tag{2}$$

where ρ is a Bernoulli random with parameter $1/2$, independent of W, namely $\mathbb{P}(\rho = 1) = \mathbb{P}(\rho = 0) = 1/2$. This random variable is related to a reflection principle. We may use the simplified form $\bar{X}_t$ for $\bar{X}_t^{0,x}$. The above approximation process $\bar{X}$ is the main building block in the forthcoming probabilistic representation formula of the couple $(\tau \wedge T, X_{\tau\wedge T})$. The next lemma follows from the reflection principle, see e.g. Karatzas and Shreve [17], or from the explicit density of the Brownian motion killed at level L given for $x, y > L$ by

$$\frac{1}{\sqrt{2\pi\sigma^2 T}}\left(\exp\left(-\frac{(y-x)^2}{2\sigma^2 T}\right) - \exp\left(-\frac{(y-(2L-x))^2}{2\sigma^2 T}\right)\right).$$

Lemma 1 *Consider the process*

$$\bar{Y}_t = x + \sigma(x) W_t, \quad x \geq L,$$

together with its hitting time $\bar{\tau} := \inf\left\{t \geq 0 : \bar{Y}_t = L\right\}$. *Then, for any bounded and measurable function* f, *it holds that*

$$\begin{aligned}\mathbb{E}\left[f(\bar{Y}_T)\mathbf{1}_{\{\bar{\tau}>T\}}\right] =& \mathbb{E}\left[f(\bar{Y}_T)\mathbf{1}_{\{\bar{Y}_T\geq L\}}\right] - \mathbb{E}\left[f(2L-\bar{Y}_T)\mathbf{1}_{\{\bar{Y}_T<L\}}\right]\\ =& 2\mathbb{E}\left[(2\rho-1)f(\bar{X}_T)\mathbf{1}_{\{\bar{X}_T\geq L\}}\right].\end{aligned} \tag{3}$$

2.3 The Underlying Markov Chain and Associated Simplified Malliavin Calculus

This section serves as an introduction to the foundational Markov chain that will underpin our probabilistic representation and facilitate the development of our IBP formulas. Furthermore, we shall present the essential material required for our differential calculus computations. It is noteworthy that a substantial portion of this section has been drawn directly from [13] for reference.

We consider a Poisson process with parameter $\lambda > 0$, independent of the one-dimensional Brownian motion W with jump times T_i and we set $\zeta_i := T_i \wedge T$, for $i \in \mathbb{N}$, with the convention that $\zeta_0 = T_0 = 0$. We let π be the partition of $[0, T]$ given by $\pi := \{0 =: \zeta_0 < \cdots < \zeta_{N_T} \leq T\}$. Associated with this set, we define the simplex $A_n := \{t \in (0, T]^n; 0 < t_1 < \cdots < t_n \leq T\}$. For instance, on the set $\{N_T = n\}$, $n \in \mathbb{N}$, we have $(\zeta_1, \ldots, \zeta_n) \in A_n$ and $\zeta_{n+1} = T$. In particular, for the set $\{N_T = 0\}$ (i.e. $n = 0$), we let $\pi := \{0, T\}$ and $A_0 = \emptyset$. In this sense, we will use through the rest of the paper the index $n \in \bar{\mathbb{N}}$ without any further mention of its range of values.

Let $\bar{X} := (\bar{X}_i)_{i\in\bar{\mathbb{N}}}$ be the discrete time Markov chain starting at time 0 from $\bar{X}_0 = x$ and evolving according to

$$\bar{X}_{i+1} := \rho_{i+1}\bar{X}_i + (1 - \rho_{i+1})(2L - \bar{X}_i) + \sigma_i Z_{i+1}, \qquad i \in \bar{\mathbb{N}}_{N_T}, \tag{4}$$

where for simplicity we set $\sigma_i := \sigma(\bar{X}_i)$, $Z_{i+1} := W_{\zeta_{i+1}} - W_{\zeta_i} = \sigma_i^{-1}\left(\bar{X}_{i+1} - \rho_{i+1}\bar{X}_i - (1 - \rho_{i+1})(2L - \bar{X}_i)\right)$ and $\{\rho_i; i \in \mathbb{N}\}$ is an i.i.d. sequence of Bernoulli(1/2) random variables such that W, N and $\{\rho_i; i \in \mathbb{N}\}$ are mutually independent. In what follows, we use the notation $h_i \equiv h(\bar{X}_i)$, $i \in \bar{\mathbb{N}}_{n+1}$ for any function $h : \mathbb{R} \to \mathbb{R}$. In particular, the reader may have noticed that we already used this notation in the above formula for $h = \sigma$. We also associate to the Markov chain $\bar{X}$ the following sets

$$D_{i,n} := \{\bar{X}_i \geq L, N_T = n\}, \qquad \text{for } i \in \bar{\mathbb{N}}_{n+1}. \tag{5}$$

Instead of using an infinite dimensional calculus as it is usually done in the literature, the approach developed below is based on a finite dimensional calculus for which the dimension is given by the number of jumps of the underlying Poisson process involved in the Markov chain $\bar{X}$.

We now introduce the following space of smooth random variables.

Definition 2 For $i \in \bar{\mathbb{N}}_n$, we let $\mathbb{S}_{i+1,n}(\bar{X})$ be the subset of random variables $H \in \mathbb{L}^0$ such that there exists a measurable function $h : \mathbb{R}^2 \times \{0, 1\} \times A_2 \to \mathbb{R}$ satisfying

1. $H = h(\bar{X}_i, \bar{X}_{i+1}, \rho_{i+1}, \zeta_i, \zeta_{i+1})$ on the set $\{N_T = n\}$.
2. For any $(s, t) \in A_2$ and $r \in \{0, 1\}$, $h(\cdot, \cdot, r, s, t) \in \mathscr{C}_p^\infty(\mathbb{R}^2)$.

For a r.v. $H \in \mathbb{S}_{i+1,n}(\bar{X})$, $i \in \bar{\mathbb{N}}_n$, we may sometimes simply write

$$H \equiv H(\bar{X}_i, \bar{X}_{i+1}, \rho_{i+1}, \zeta_i, \zeta_{i+1}), \tag{6}$$

that is the same symbol H may denote the r.v. or the function in the set $\mathbb{S}_{i+1,n}(\bar{X})$. Also note that $\mathbb{S}_{n+1,n}(\bar{X})$ is well defined.

One can easily define the flow derivatives for $H \in \mathbb{S}_{i+1,n}(\bar{X})$ as follows:

$$\begin{aligned}
\partial_{\bar{X}_{i+1}} H :=& \partial_2 h(\bar{X}_i, \bar{X}_{i+1}, \rho_{i+1}, \zeta_i, \zeta_{i+1}),\\
\partial_{\bar{X}_i} H :=& \partial_1 h(\bar{X}_i, \bar{X}_{i+1}, \rho_{i+1}, \zeta_i, \zeta_{i+1}) + \partial_2 h(\bar{X}_i, \bar{X}_{i+1}, \rho_{i+1}, \zeta_i, \zeta_{i+1}) \partial_{\bar{X}_i} \bar{X}_{i+1},
\end{aligned} \tag{7}$$

$$\partial_{\bar{X}_i} \bar{X}_{i+1} := (2\rho_{i+1} - 1) + \sigma_i' Z_{i+1}.$$

We now define the derivative and integral operators for $H \in \mathbb{S}_{i+1,n}(\bar{X})$, $i \in \bar{\mathbb{N}}_n$, as

$$\mathcal{I}_{i+1}(H) := H \frac{Z_{i+1}}{\sigma_i(\zeta_{i+1} - \zeta_i)} - \mathcal{D}_{i+1} H, \qquad \mathcal{D}_{i+1} H := \partial_{\bar{X}_{i+1}} H. \tag{8}$$

Note that due to the above definitions and Assumption (**H**), we also have that $\mathcal{I}_{i+1}(H), \mathcal{D}_{i+1} H \in \mathbb{S}_{i+1,n}(\bar{X})$ so that we can define iterations of the above operators, namely $\mathcal{I}^{\ell+1}_{i+1}(H) = \mathcal{I}_{i+1}(\mathcal{I}^{\ell}_{i+1}(H))$ and similarly $\mathcal{D}^{\ell+1}_{i+1}(H) = \mathcal{D}_{i+1}(\mathcal{D}^{\ell}_{i+1} H)$, $\ell \in \bar{\mathbb{N}}$, with the convention $\mathcal{I}^0_{i+1}(H) = \mathcal{D}^0_{i+1}(H) = H$.

We define the filtration $\mathcal{G} := (\mathcal{G}_i)_{i \in \bar{\mathbb{N}}}$ where $\mathcal{G}_i := \sigma(Z^i, \zeta^i, \rho^i)$ with the notation $a^i := (a_1, \ldots, a_i)$ for $a = Z, \ \zeta, \ \rho, \ i \in \mathbb{N}$. We let $\mathcal{G}_0$ be defined as the trivial σ-field. We assume that the σ-fields $\mathcal{G}_i$ are complete. Throughout this article, we will use the following notation for a certain type of conditional expectation that will appear frequently. For any $X \in \mathbb{L}^1$ and any $i \in \bar{\mathbb{N}}_n$,

$$\mathbb{E}_{i,n}[X] := \mathbb{E}[X \mid \mathcal{G}_i, T^{n+1}, \rho^{n+1}, N_T = n].$$

Here again we have used the vector notation $S^{n+1} = (S_1, \ldots, S_{n+1})$ for $S = T, \rho$. Having the above definitions at hand, the following duality formula[1] is satisfied for any $f \in \mathscr{C}^1_p(\mathbb{R})$ and any $(i, \ell) \in \bar{\mathbb{N}}_n \times \mathbb{N}$:

$$\mathbb{E}_{i,n}\left[\mathcal{D}^{\ell}_{i+1} f(\bar{X}_{i+1}) H\right] = \mathbb{E}_{i,n}\left[f(\bar{X}_{i+1}) \mathcal{I}^{\ell}_{i+1}(H)\right]. \tag{9}$$

[1] This duality is obtained using the Gaussian density of $\bar{X}_{i+1}$ while in classical Malliavin calculus it is based on the density of the Wiener process. Therefore the derivative and integral, $\mathcal{D}_{i+1}$ and $\mathcal{I}_{i+1}$, $i \in \bar{\mathbb{N}}_n$ defined in the formula (9) are renormalizations of the usual duality principle in Malliavin calculus. In our case this notation simplifies greatly many equations.

In order to obtain explicit norm estimates for random variables in $\mathbb{S}_{i+1,n}(\bar{X})$, it is useful to define for $H \in \mathbb{S}_{i+1,n}(\bar{X})$, $i \in \bar{\mathbb{N}}_n$ and $p \geq 1$

$$\|H\|_{p,i,n}^{p} := \mathbb{E}_{i,n}\left[|H|^{p}\right].$$

We will also use at several places the following *extraction formula* for $H_1, H_2 \in \mathbb{S}_{i+1,n}(\bar{X})$:

$$\mathcal{I}_{i+1}^{\ell}(H_1 H_2) = \sum_{j=0}^{\ell}(-1)^{j}\binom{\ell}{j}\mathcal{I}_{i+1}^{\ell-j}(H_1)\mathcal{D}_{i+1}^{j} H_2, \tag{10}$$

whose proof follows by induction. By iteration, one also obtains that $\mathcal{I}_i^{\ell}(1) \in \mathbb{S}_{i,n}(\bar{X})$ and it satisfies $\mathcal{I}_i^{\ell+1}(1) = \mathcal{I}_i^{\ell}(1)\mathcal{I}_i(1) - \ell\mathcal{I}_i^{\ell}(1)$ which, in particular, implies:

$$\mathcal{I}_i^{\ell}(1) = (-1)^{\ell}\mathcal{H}_{\ell}(a_{i-1}(\zeta_i - \zeta_{i-1}), \sigma_{i-1}Z_i), \tag{11}$$

where $\mathcal{H}_{\ell}$ stands for the Hermite polynomial of order ℓ, $\ell \in \mathbb{N}$, defined by $\mathcal{H}_{\ell}(t,x) = (g(t,x))^{-1}\partial_x^{\ell} g(t,x)$. Here, $g(t,x)$ denotes the density of the Gaussian law $\mathcal{N}(0,t)$. Using (10) for $H_1 = 1$ and $H_2 = H$ as well as (8), the following norm bound for stochastic integrals is clearly satisfied for $i \in \bar{\mathbb{N}}$ and any measurable set $A \in \mathcal{F}$

$$\left\|1_A \mathcal{I}_{i+1}^{\ell}(H)\right\|_{p,i,n} \leq C_{\ell,p}\sum_{j=0}^{\ell}(\zeta_{i+1} - \zeta_i)^{\frac{j-\ell}{2}}\|1_A \mathcal{D}_{i+1}^{j} H\|_{p,i,n}. \tag{12}$$

We will frequently use the following Hölder inequality for smooth random variables $H_1, H_2 \in \mathbb{S}_{i+1,n}(\bar{X})$: for any $i \in \bar{\mathbb{N}}$, any $p, p_1, p_2 \geq 1$ satisfying $p^{-1} = p_1^{-1} + p_2^{-1}$ and any $A \in \mathcal{F}$:

$$\|1_A H_1 H_2\|_{p,i,n} \leq C\|1_A H_1\|_{p_1,i,n}\|1_A H_2\|_{p_2,i,n}. \tag{13}$$

Quantities such as $\mathbb{E}_{i,n}[\delta_L(\bar{X}_{i+1})H]$ for $H \in \mathbb{S}_{i+1,n}(\bar{X})$ have a clear meaning due to the IBP formula (9) (see also the theory introduced in Chapter V9 [16] for a much more general framework) or in the sense of conditional laws

$$\begin{aligned}\mathbb{E}_{i,n}[\delta_L(\bar{X}_{i+1})H] &= \mathbb{E}_{i,n}[1_{D_{i+1,n}}\mathcal{I}_{i+1}(H)] \\ &= \mathbb{E}_{i,n}[H|\bar{X}_{i+1} = L]g(a_i(\zeta_{i+1} - \zeta_i), L - \bar{X}_i),\end{aligned}$$

where the set $D_{i,n}$ is defined by (5). We finally introduce the following space of random variables which satisfies certain time (ir)regularity estimates.

Definition 3 For $\ell \in \mathbb{Z}$, $i \in \bar{\mathbb{N}}_n$, the space $\mathbb{M}_{i+1,n}(\bar{X}, \ell/2)$ is the set of random variables $H \in \mathbb{L}^0$ satisfying the property *(1)* in Definition 2 and such that

$$\mathbb{1}_{D_{i,n}} \|\mathbb{1}_{D_{i+1,n}} H\|_{p,i,n} \leq C(\zeta_{i+1} - \zeta_i)^{\ell/2},$$

for some deterministic constant C independent of (p, i, n).

Again we remark that the definition of the space $\mathbb{M}_{i+1,n}(\bar{X}, \ell/2)$ uses the conditional norm $\mathbb{E}_{i,n}[.]$ and property *(1)* in Definition 2, so that this property is always stated on $\{N_T = n\}$, $n \in \mathbb{N}$.

A straightforward consequence of Eqs. (10) and (12) is the following property.

Lemma 4 *For $j \in \{0, 1\}$, $i \in \bar{\mathbb{N}}$, and $k \in \mathbb{N}$, if $H_1 \in \mathbb{M}_{i+1,n}(\bar{X}, j/2) \cap \mathbb{S}_{i+1,n}(\bar{X})$ with $\|\mathcal{D}^k_{i+1} H_1\|_{p,i,n} \leq C$ then $\mathcal{I}^k_{i+1}(H_1) \in \mathbb{M}_{i+1,n}(\bar{X}, (j-k)/2)$. Furthermore, if $H_2 \in \mathbb{M}_{i+1,n}(\bar{X}, k/2)$ then the product $H_1 H_2 \in \mathbb{M}_{i+1,n}(\bar{X}, (j+k)/2)$.*

Lemma 5 *Let $H \equiv H(\bar{X}_i, \bar{X}_{i+1}, \rho_{i+1}, \zeta_i, \zeta_{i+1}) \in \mathbb{S}_{i+1,n}(\bar{X})$ with $i \in \bar{\mathbb{N}}$, then the following chain rule type formula holds*

$$\partial_{\bar{X}_i} \mathcal{I}_{i+1}(H) = \mathcal{I}_{i+1}(\partial_{\bar{X}_i} H) - \frac{\sigma_i'}{\sigma_i} \mathcal{I}_{i+1}(H).$$

Proof From the extraction formula (10), the usual chain rule and the fact that $\partial_{\bar{X}_i} \mathcal{I}_{i+1}(1) = -\frac{\sigma_i'}{\sigma_i} \mathcal{I}_{i+1}(1)$, we have that

$$\partial_{\bar{X}_i} \mathcal{I}_{i+1}(H) = -\frac{\sigma_i'}{\sigma_i} \mathcal{I}_{i+1}(1) H + \mathcal{I}_{i+1}(1) \partial_{\bar{X}_i} H - \partial_{\bar{X}_i} \mathcal{D}_{i+1} H$$

$$\mathcal{I}_{i+1}(\partial_{\bar{X}_i} H) = \mathcal{I}_{i+1}(1) \partial_{\bar{X}_i} H - \mathcal{D}_{i+1} \partial_{\bar{X}_i} H.$$

Note that $\partial_{\bar{X}_i}$ and $\mathcal{D}_{i+1}$ do not commute. Indeed, by the usual chain rule, one has

$$\begin{aligned} \partial_{\bar{X}_i} \mathcal{D}_{i+1} h - \mathcal{D}_{i+1} \partial_{\bar{X}_i} h &= \partial_1 \partial_2 h + \partial_2^2 h \frac{\partial \bar{X}_{i+1}}{\partial \bar{X}_i} - \mathcal{D}_{i+1}(\partial_1 h + \partial_2 h \frac{\partial \bar{X}_{i+1}}{\partial \bar{X}_i}) \\ &= -\partial_2 h \frac{\sigma_i'}{\sigma_i} \end{aligned}$$

where in the last equality we used the fact that $\mathcal{D}_{i+1} \frac{\partial \bar{X}_{i+1}}{\partial \bar{X}_i} = \frac{\sigma_i'}{\sigma_i}$. By combining the above computations we obtain

$$\partial_{\bar{X}_i} \mathcal{I}_{i+1}(H) - \mathcal{I}_{i+1}(\partial_{\bar{X}_i} H) = -\frac{\sigma_i'}{\sigma_i} (\mathcal{I}_{i+1}(1) - \mathcal{D}_{i+1} H) = -\frac{\sigma_i'}{\sigma_i} \mathcal{I}_{i+1}(H).$$

□

3 The Probabilistic Representation for the First Hitting Time

3.1 Preliminaries

The probabilistic representation formula for the first hitting time τ of X is obtained in a similar way as in the case of the killed process treated in [14]. We briefly explain the main arguments and deliberately omit some technical details.

In order to introduce the associated semigroup, we let $\tau^{s,x} := \inf\{t \geq s : X_t^{s,x} = L\}$ where $X^{s,x}$ stands for the unique solution to (1) starting from x at time s and for any real-valued bounded and measurable function f defined on $\mathbb{R}_+ \times \mathbb{R}$, we let $P_t f(u,x) = \mathbb{E}\big[f(u + \tau^{u,x} \wedge t, X^{u,x}_{u+\tau^{u,x}\wedge t})\big]$. We will sometimes omit the dependence of (τ, X) with respect to (u,x) when there is no confusion and simply write $P_t f(u,x) = \mathbb{E}[f(u + \tau \wedge t, X_{\tau\wedge t})]$.

Under Assumption (**H**), for f a smooth function, we will assume for the sake of the argument that $Pf \in \mathscr{C}^{1,2}(\mathbb{R}_+ \times (\mathbb{R}_+ \times [L,\infty)))$ with $\sup_{0\leq t\leq T} |\partial_x^\ell P_t f|_\infty \leq C$, for $\ell = 1, 2$, for some positive constant $C := C(T, f, a, b)$ and that it satisfies $\partial_t P_t f = \mathcal{L} P_t f$ on $[L,\infty)$, $t > 0$, where the differential operator $\mathcal{L}$ is given by

$$\mathcal{L} f(u,x) = \partial_u f(u,x) + b(x)\,\partial_x f(u,x) + \frac{1}{2} a(x)\,\partial_x^2 f(u,x).$$

Removing the above hypothesis requires some technical arguments which can be found in Proposition 3.1 in [13].

The argument used to obtain the probabilistic representation formula starts by applying Itô's formula to $(P_{T-t} f(u + t \wedge \bar{\tau}, \bar{Y}_{t\wedge\bar{\tau}}))_{t\in[0,T]}$, recalling that $\bar{Y}_t = x + \sigma(x) W_t$ and $\bar{\tau}$ is its associated exit time from $[L,\infty)$. Hence, from Itô's rule and Lemma 1, it holds

$$\begin{aligned}
& f(T \wedge \bar{\tau}, \bar{Y}_{T\wedge\bar{\tau}}) \\
& \overset{\mathbb{E}}{=} P_T f(u,x) + \int_0^T \Big(-\partial_r P_r f(u+s, \bar{Y}_s)\Big|_{r=T-s} + \partial_u P_{T-s} f(u+s, \bar{Y}_s) \\
& \quad + \frac{1}{2} a(x) \partial_x^2 P_{T-s} f(u+s, \bar{Y}_s) \Big) \mathsf{1}_{\{\bar{\tau} > s\}}\, \mathrm{d}s \\
& \overset{\mathbb{E}}{=} P_T f(u,x) + \int_0^T \Big(\frac{1}{2} \big(a(x) - a(\bar{Y}_s)\big)\, \partial_x^2 P_{T-s} f(u+s, \bar{Y}_s) \\
& \quad - b(\bar{Y}_s) \partial_x P_{T-s} f(u+s, \bar{Y}_s)\Big) \mathsf{1}_{\{\bar{\tau} > s\}}\, \mathrm{d}s \\
& \overset{\mathbb{E}}{=} P_T f(u,x) + 2(2\rho - 1) \int_0^T \Big(\frac{1}{2} \big(a(x) - a(\bar{X}_s)\big)\, \partial_x^2 P_{T-s} f(u+s, \bar{X}_s) \\
& \quad - b(\bar{X}_s) \partial_x P_{T-s} f(u+s, \bar{X}_s)\Big) \mathsf{1}_{\{\bar{X}_s \geq L\}}\, \mathrm{d}s.
\end{aligned}$$

We now rewrite the previous representation using the Markov chain $(\bar{X}_i)_{0\leq i\leq N_T+1}$ defined by (4) together with the Poisson process N. From the previous identity, we get

$$\begin{aligned}
P_T f(u,x) &\overset{\mathbb{E}}{=} f(u+T\wedge\bar{\tau}, \bar{Y}_{T\wedge\bar{\tau}}) \\
&\quad + 2(2\rho-1)\int_0^T \left(\frac{1}{2}\left(a(\bar{X}_s)-a(x)\right)\partial_x^2 P_{T-s} f(u+s,\bar{X}_s)\right. \\
&\quad \left. + b(\bar{X}_s)\partial_x P_{T-s} f(u+s,\bar{X}_s)\right) \mathbf{1}_{\{\bar{X}_s\geq L\}}\,\mathrm{d}s \qquad (14)\\
&\overset{\mathbb{E}}{=} f(u+T\wedge\bar{\tau}, \bar{Y}_{T\wedge\bar{\tau}})e^{\lambda T}\mathbf{1}_{\{N_T=0\}} + e^{\lambda T}\lambda^{-1}2(2\rho_{N_T}-1) \\
&\quad \times\left\{\frac{1}{2}(a(\bar{X}_1)-a(x))\partial_x^2 P_{T-\zeta_1} f(u+\zeta_1,\bar{X}_1)\right. \\
&\quad \left. + b(\bar{X}_1)\partial_x P_{T-\zeta_1} f(u+\zeta_1,\bar{X}_1)\right\} \mathbf{1}_{\{\bar{X}_1\geq L\}}\mathbf{1}_{\{N_T=1\}}.
\end{aligned}$$

We now apply the IBP formula (9) with respect to the random variables $\bar{X}_1$ for the last term appearing in the right-hand side of the above expression. The IBP formula is applied once to the terms associated with the drift coefficient b and two times with respect to the terms related to the diffusion coefficient a. In order to do that one first has to take the conditional expectation $\mathbb{E}_{0,1}[.]$. However, one must be cautious insofar as these IBPs involve the indicator function $\mathbf{1}_{\{\bar{X}_1\geq L\}}$. Hence,

$$\begin{aligned}
P_T f(u,x) &\overset{\mathbb{E}}{=} f(u+T\wedge\bar{\tau}, \bar{Y}_{T\wedge\bar{\tau}})e^{\lambda T}\mathbf{1}_{\{N_T=0\}} + e^{\lambda T}\lambda^{-1}2(2\rho_{N_T}-1) \\
&\quad \times\left(\frac{1}{2}\mathcal{I}_1\left(\left(a(\bar{X}_1)-a(x)\right)\mathbf{1}_{\{\bar{X}_1\geq L\}}\right)\partial_x P_{T-\zeta_1} f(\bar{X}_1)\right. \\
&\quad \left. + \mathcal{I}_1\left(b(\bar{X}_1)\mathbf{1}_{\{\bar{X}_1\geq L\}}\right) P_{T-\zeta_1} f(\bar{X}_1)\right)\mathbf{1}_{\{N_T=1\}} \\
&\overset{\mathbb{E}}{=} f(u+T\wedge\bar{\tau}, \bar{Y}_{T\wedge\bar{\tau}})e^{\lambda T}\mathbf{1}_{\{N_T=0\}} + e^{\lambda T}2\lambda^{-1}(2\rho_{N_T}-1) \\
&\quad \times\left(\frac{1}{2}\mathcal{I}_1\left(\left(a(\bar{X}_1)-a(x)\right)\right)\partial_x P_{T-\zeta_1} f(u+\zeta_1,\bar{X}_1)\right. \\
&\quad \left. + \mathcal{I}_1\left(b(\bar{X}_1)\right) P_{T-\zeta_1} f(u+\zeta_1,\bar{X}_1)\right)\mathbf{1}_{\{\bar{X}_1\geq L\}}\mathbf{1}_{\{N_T=1\}},
\end{aligned}$$

where we used the extraction formula (10) applied to the r.v. $\mathbf{1}_{\{\bar{X}_s\geq L\}}$ and Lemma 10.5 [14] (taking first the conditional expectation w.r.t ζ_1) for $\ell = 1,\ k = 0$ for the last equality.

We apply again the IBP formula (9) in order to remove the first order derivative on $P_{T-\zeta_1} f$. We then use again the extraction formula (10). We obtain

$$
\begin{aligned}
P_T f(u,x) &\stackrel{\mathbb{E}}{=} f(u+T\wedge\bar{\tau}, \bar{Y}_{T\wedge\bar{\tau}})e^{\lambda T}\mathbb{1}_{\{N_T=0\}} + e^{\lambda T}2\lambda^{-1}(2\rho_{N_T}-1) \\
&\times\left(\frac{1}{2}\mathcal{I}_1^2\left(a(\bar{X}_1)-a(x)\right)P_{T-\zeta_1}f(u+\zeta_1,\bar{X}_1)+\mathcal{I}_1\left(b(\bar{X}_1)\right)P_{T-\zeta_1}f(u+\zeta_1,\bar{X}_1)\right) \\
&\quad \mathbb{1}_{\{\bar{X}_1\geq L\}}\mathbb{1}_{\{N_T=1\}} \qquad (15) \\
&- e^{\lambda T}2\lambda^{-1}(2\rho_{N_T}-1)\frac{1}{2}\mathcal{I}_1^1\left(a(\bar{X}_1)-a(x)\right)P_{T-\zeta_1}f(u+\zeta_1,\bar{X}_1)\delta_L(\bar{X}_1)\mathbb{1}_{\{N_T=1\}}.
\end{aligned}
$$

Using the extraction formula and then Lemma 10.5 in [14], we get

$$
\begin{aligned}
& e^{\lambda T}2\lambda^{-1}\mathbb{E}[(2\rho_{N_T}-1)\mathcal{I}_1^1\left(a(\bar{X}_1)-a(x)\right)P_{T-\zeta_1}f(u+\zeta_1,\bar{X}_1)\delta_L(\bar{X}_1)\mathbb{1}_{\{N_T=1\}}] \\
&\quad = e^{\lambda T}\lambda^{-1}\mathbb{E}[(2\rho_{N_T}-1)(a(L)-a(x))\mathcal{I}_1^1(1) \\
&\qquad P_{T-\zeta_1}f(u+\zeta_1,\bar{X}_1)\delta_L(\bar{X}_1)\mathbb{1}_{\{N_T=1\}}] \\
&\quad = \int_0^T \mathbb{E}[(2\rho-1)(a(L)-a(x))\frac{(L-x)}{a(x)s}\delta_L(\bar{Y}_s)P_{T-s}f(u+s,L)]\,\mathrm{d}s \\
&\quad = -\frac{a(L)-a(x)}{a(x)}\mathbb{E}[f(u+\bar{\tau},L)\mathbb{1}_{\{\bar{\tau}\leq T\}}].
\end{aligned}
$$

Plugging the above identity into (15), we get

$$
\begin{aligned}
P_T f(u,x) \stackrel{\mathbb{E}}{=} f(u+T\wedge\bar{\tau}, \bar{Y}_{T\wedge\bar{\tau}})e^{\lambda T}\mathbb{1}_{\{N_T=0\}} &+ e^{\lambda T}\bar{\theta}^1\mathbb{1}_{\{\bar{X}_1\geq L\}} \\
& P_{T-\zeta_1}f(u+\zeta_1,\bar{X}_1)\mathbb{1}_{\{N_T=1\}} \qquad (16) \\
& + \frac{a(L)-a(x)}{a(x)}f(u+\bar{\tau},L)\mathbb{1}_{\{\bar{\tau}\leq T\}},
\end{aligned}
$$

where for $i=1,\cdots,n$

$$
\bar{\theta}^i := 2\lambda^{-1}(2\rho_i-1)\left(\mathcal{I}_i^2(c_2^i)+\mathcal{I}_i(c_1^i)\right), \qquad (17)
$$

with

$$
c_1^i := b_i, \quad c_2^i := \frac{1}{2}(a_i-a_{i-1}), \qquad (18)
$$

recalling that $a_i = a(\bar{X}_i)$ and $b_i = b(\bar{X}_i)$. Let us emphasize that, under assumption **(H)**, the following time estimates hold: for all $p\geq 1$, there exists

$C := C(a, b, T, p)$ such that

$$\|\mathbb{1}_{\{\bar{X}_1 \geq L\}}\mathcal{I}_1(a(\bar{X}_1) - a(x))\|_{p,0,1} \leq C, \tag{19}$$

$$\|\mathbb{1}_{\{\bar{X}_1 \geq L\}}\mathcal{I}_1^2(a(\bar{X}_1) - a(x))\|_{p,0,1} + \|\mathbb{1}_{\{\bar{X}_1 \geq L\}}\mathcal{I}_1(b(\bar{X}_1))\|_{p,0,1} \leq C\zeta_1^{-1/2},$$

which in turn, by using the fact that on $\mathbb{P}(N_T = 1, \zeta_1 \in dt) = \lambda e^{-\lambda T} dt$ on $[0, T]$, lead to the integrability of the second term appearing in the right-hand side of (16).

The main idea to obtain the following result consists in iterating the identity (16) by following similar lines of reasoning as those employed in [9, 13, 14]. Omitting the remaining technical details, we obtain the following result.

Theorem 6 *Let $T > 0$ and suppose that Assumption **(H)** is satisfied. For any $n \in \mathbb{N}$, define on the set $\{N_T = n\}$ the following random variables*

$$\Gamma_n = \begin{cases} \prod_{i=1}^n \theta^i & \text{if } n \geq 1, \\ 1 & \text{if } n = 0, \end{cases}$$

and

$$\bar{\Gamma}_n = \frac{a(L) - a_n}{a_n}\Gamma_n,$$

where,

$$\begin{aligned}\theta^i :=& \mathbb{1}_{\{\bar{X}_i > L\}}\bar{\theta}^i = 2(2\rho_i - 1)\lambda^{-1}\left(\mathcal{I}_i(c_1^i) + \mathcal{I}_i^2(c_2^i)\right)\mathbb{1}_{\{\bar{X}_i > L\}} \\ =& \mathbb{1}_{\{\bar{X}_i > L\}}2(2\rho_i - 1)\lambda^{-1}\left\{c_2^i\mathcal{I}_i^2(1) + (c_1^i - 2\mathcal{D}_i c_2^i)\mathcal{I}_i(1) + \mathcal{D}_i^2 c_2^i + \mathcal{D}_i c_1^i\right\}.\end{aligned} \tag{20}$$

Then, for any bounded and measurable function f defined on $\mathbb{R}_+ \times \mathbb{R}$, the following probabilistic representation holds

$$\begin{aligned}\mathbb{E}[f(T \wedge \tau, X_{T\wedge\tau})] =& e^{\lambda T}\mathbb{E}\left[f((\zeta_{N_T} + \bar{\tau}^{N_T}) \wedge T, \bar{Y}_{(\zeta_{N_T}+\bar{\tau}^{N_T})\wedge T})\Gamma_{N_T}\right] \\ &+ e^{\lambda T}\mathbb{E}\left[f(\zeta_{N_T} + \bar{\tau}^{N_T}, L)\mathbb{1}_{\{\zeta_{N_T}+\bar{\tau}^{N_T} \leq T\}}\bar{\Gamma}_{N_T}\right].\end{aligned} \tag{21}$$

Importantly note that we have used the extraction formula in order to obtain (20). For this, we refer the reader to (10) and to the proof of Lemma 2.5 [14] to see how it is applied. In order to avoid long expressions in the above identity, we have used the shortened notation $\bar{\tau}^{N_T}$ for $\bar{\tau}^{\zeta_{N_T}, \bar{X}_{N_T}}$ where we recall that $\bar{\tau}^{s,x}$ stands for the first hitting time of the level L associated to the approximation process $\bar{Y}$ starting from x at time s. That is,

$$\bar{\tau}^{s,x} := \inf\{u \geq s : \bar{Y}_u^{s,x} = L\}.$$

In particular, $\bar{\tau}^{N_T}$ is sampled independently of N and W. Similarly, we used the shorten notation $\bar{Y}_{(\zeta_{N_T}+\bar{\tau}^{N_T})\wedge T}$ for $\bar{Y}^{\zeta_{N_T},\bar{X}_{N_T}}_{(\zeta_{N_T}+\bar{\tau}^{N_T})\wedge T}$ where $\bar{Y}^{s,x}_t = x+\sigma(x)(W_t-W_s)$. We also recall that N is a λ-Poisson process with jump times T_n, $n \in \mathbb{N}$, so that $\Delta T_i = T_i - T_{i-1}$ is a sequence of i.i.d. exponential random variables with parameter λ and that on the set $\{N_T = n\}$, we have $\pi = \{0 = \zeta_0 < \ldots < \zeta_n < \zeta_{n+1} = T\}$ recalling that $\zeta_i = T_i \wedge T$.

The second term on the right-hand side of (21) containing $\bar{\Gamma}$ stems from the boundary terms appearing when one performs the integration by parts formulas in the expansion. In particular, this term does not appear in the case of the killed diffusion investigated in [14].

Note that both terms appearing on the right-hand side of (21) have the same form if we assume that $\zeta_{N_T} + \bar{\tau}^{N_T} < T$ in which case $\bar{Y}_{(\zeta_{N_T}+\bar{\tau}^{N_T})\wedge T} = L$. Also note that $\mathbb{P}\left(\zeta_{N_T} + \bar{\tau}^{N_T} = T\right) = 0$.

4 IBP Formulas with Respect to the Exit Time

4.1 IBP with Respect to Generalized Inverse Gaussian Times

In this section, we shall introduce the general transfer formula employing IBP with respect to the interarrival times of the Poisson process. A transfer formula, as explained in reference [13], is a formula that transfers the derivative from outside the expectation to inside, making it subsequently amenable to iterative procedures.

The principal difference lies in our current consideration of the conditional distribution of the random variables with respect to the spatial variables. Furthermore, given our perpetual focus on compact time intervals, it follows that test functions are bounded.

As our objective is to elucidate the technique, as opposed to pursuing utmost generality and for the sake of streamlining expressions, we restrict our consideration to $f(t,x) \equiv f(t)$, subject to the simplifying condition that $f(T) = 0$, where $T > 0$ is an arbitrary time horizon. In other words, the test function f solely relies on the time variable and vanishes at the time interval's boundary. As is customary, in more general circumstances, one may consider $\bar{f}(t) = f(t) - f(T)$ instead of f.

The overarching structure of the IBP formula, when dealing with the stopped process $X_{T\wedge\tau}$, was derived through an initial conditioning on the Poisson process. Then, one performs IBP operations with respect to the spatial variables and eventually integrates with respect to the time variables. In this scenario, it becomes imperative to consider conditional expectations with respect to the spatial variables. Consequently, the entities that replace the time differences in this context are the spatial-related quantities:

$$
\begin{aligned}
(\Delta_i \bar{X})^2 :=& (\bar{X}_i - \bar{X}_{i-1}\rho_i - (1-\rho_i)(2L - \bar{X}_{i-1}))^2, \; i \le N_T, \\
(\Delta_{N_T+1} \bar{X})^2 :=& (L - \bar{X}_{N_T})^2.
\end{aligned}
$$

The rationale behind introducing the aforementioned quantity is rooted in the fact that when the derivative operator is applied to each term in (21), it will diminish the time order by a factor of 1. Consequently, to maintain stability and balance in the analysis, we introduce a stabilization mechanism that relies on spatial variables possessing an equivalent order given by $(\Delta_i \bar{X})^2$. To achieve this equilibrium, we employ the square of the distance between points.

The subsequent result, which serves as the starting point of our analysis, stems directly from Theorem 6.

Corollary 7 *Under the same conditions as in Theorem 6 and assuming that f does not depend on the space component, i.e. $f(s, y) \equiv f(s)$, $y \in \mathbb{R}$ with $f(T) = 0$, we have*

$$
\mathbb{E}[f(\tau \wedge T)] = e^{\lambda T} \mathbb{E}\left[f(\zeta_{N_T} + \bar{\tau}^{\zeta_{N_T}}) \mathbb{1}_{\{\bar{\tau}^{N_T} \le T - \zeta_{N_T}\}} \widehat{\theta}^{N_T} \prod_{i=1}^{N_T} \theta^i \right], \tag{22}
$$

$$
\textit{with} \quad \widehat{\theta}^{N_T} := 1 + \frac{a(L) - a_{N_T}}{a_{N_T}}.
$$

4.2 Conditioning on the Space Variables

In this section, our objective is to state the conditional distribution of the jump times of the Poisson process, with respect to the spatial variables. In pursuit of this objective, we introduce the notations $\bar{X} := \{\bar{X}_i;\ i \le N_T\}$ and $\rho := \{\rho_i;\ i \le N_T\}$. It is worth noting that, with probability one, $\Delta_i \bar{X} \neq 0$ for all $i \le N_T + 1$. We will rely on this fact without reiterating it in the subsequent discussions. Additionally, it is pertinent to observe that the "weight" random variable $\bar{\Gamma}_{N_T} \in \sigma(\bar{X}_i, T_i; i \le N_T)$.

In order to give proper definitions of how our differential operators will be defined, we first need to reconstruct the sequences of random variables $\bar{X}_i$ and T_i, $i \in \mathbb{N}$, as follows.

Lemma 8 *Let $\tilde{Z}_i$ be a sequence of identically and independently distributed symmetric exponential random variables with parameter*[2] $1/\sqrt{2\lambda}$. *Inductively,*

[2] That is $\tilde{Z}_i = (-1)^{\eta_i} F_i$, where $F_i \sim \text{Exp}(1/\sqrt{2\lambda})$ and $\eta_i \sim \text{Ber}(1/2)$.

define the following sequence of random variables:

$$Y_i = \rho_i Y_{i-1} + (1-\rho_i)(2L - Y_{i-1}) + \sigma(Y_{i-1})\tilde{Z}_i, \quad Y_0 = \bar{X}_0$$

Here $\{\rho_i; i \in \mathbb{N}\}$ *is assumed to be independent of* $\{\tilde{Z}_i; i \in \mathbb{N}\}$. *Given* $\{Y_i; i \in \mathbb{N}\}$, *define* τ_i, $i \in \mathbb{N}$, *to be an independently distributed sequence of random variables such that* τ_i *follows a Generalized Inverse Gaussian distribution*[3] *with parameters* $a = 2\lambda$, $b = 2\lambda\mu_i^2$, $p = \frac{1}{2}$, *denoted by* $GIG(2\lambda, 2\lambda\mu_i^2, \frac{1}{2})$, *where with some abuse of notation, we let* $\mu_i = \frac{|\Delta Y_i|}{\sigma_{i-1}\sqrt{2\lambda}}$ *where* $\sigma_{i-1} = \sigma(Y_{i-1})$. *Then, the following equality in law is satisfied* $\{(\rho_i, \bar{X}_i, \Delta T_i) : i \in \mathbb{N}\} \overset{\mathcal{L}}{=} \{(\rho_i, Y_i, \tau_i) : i \in \mathbb{N}\}$ *where* $\Delta T_i = T_i - T_{i-1}$.

Proof Let ρ be a Bernoulli random variable, independent of the exponential random variable T_1 and the Wiener process W which is also independent of T_1. Note that the joint density of $(\rho x + (1-\rho)(2L - x) + \sigma_0 W_{T_1}, T_1)$ conditioned to ρ is given by

$$\frac{\lambda}{\sqrt{2\pi\sigma_0^2 s}} \exp\left(-\frac{(y - \rho x - (1-\rho)(2L - x))^2}{2\sigma_0^2 s} - \lambda s\right) 1_{\{s>0\}}. \tag{23}$$

First, we compute the marginal density of $\rho x + (1-\rho)(2L - x) + \sigma_0 W_{T_1}$. This is done using characteristic functions as follows. Using that T_1 is exponentially distributed and independent from W we have

$$\mathbb{E}[e^{\theta W_{T_1}}] = \mathbb{E}[\mathbb{E}[e^{\theta W_s}|T_1]\big|_{s=T_1}] = \mathbb{E}[e^{\frac{1}{2}\theta^2 T_1}] = \int_0^\infty \lambda e^{(\frac{1}{2}\theta^2 - \lambda)s} ds = \frac{\lambda}{\lambda - \frac{1}{2}\theta^2}.$$

This corresponds to the Laplace transform of symmetric exponential distribution with parameter $1/\sqrt{2\lambda}$. From direct computation, the density of $\tilde{Z}_1$ is given by

$$f_{W_{T_1}}(y) = f_{\tilde{Z}_1}(y) = \frac{1}{2}\sqrt{2\lambda}e^{-\sqrt{2\lambda}|y|}, \quad y \in \mathbb{R}.$$

Then, the conditional density function of Y_1 given ρ is given by

$$f_{Y_1}(y) = \frac{1}{2\sigma_0}\sqrt{2\lambda}e^{-\frac{|y-\rho x-(1-\rho)(2L-x)|}{\sigma_0}\sqrt{2\lambda}} = \frac{1}{2\sigma_0}\sqrt{2\lambda}e^{-2\lambda\mu_1}, \quad y \in \mathbb{R},$$

[3] We denote by $GIG(a, b, p)$ the probabilistic distribution with density $f(x) = \frac{(a/b)^{p/2}}{2K_p(\sqrt{ab})} x^{(p-1)} e^{-(ax+b/x)/2}$, $x > 0$, where $a, b > 0$ and $p \in \mathbb{R}$. Here K denotes the modified Bessel functions of the second kind . For more information, see the Appendix.

with $\mu_1 = \frac{|y-\rho x-(1-\rho)(2L-x)|}{\sigma_0\sqrt{2\lambda}}$. To obtain the conditional density of τ_1 given Y_1 and ρ, we see that

$$f_{\tau_1|\mathrm{Y}_1}(s|y) = \frac{\lambda}{\sqrt{2\pi\sigma_0^2 s}} \exp\left(-\frac{(y-\rho x-(1-\rho)(2L-x))^2}{2\sigma_0^2 s} - \lambda s\right)$$

$$1_{\{s>0\}} 2\sigma_0 \frac{1}{\sqrt{2\lambda}} e^{2\lambda\mu_1}$$

$$= \frac{\sqrt{\lambda}}{\sqrt{\pi s}} \exp\left(-\lambda\mu_1^2 s^{-1} - \lambda s\right) 1_{\{s>0\}} \exp\left(2\lambda\mu_1\right).$$

By using the fact that $K_{\frac{1}{2}}(z) = \sqrt{\frac{\pi}{2}} \frac{e^{-z}}{\sqrt{z}}$, where $K_\nu(z)$ is a Bessel function of the second kind, we can deduce that the above is a Generalized Inverse Gaussian density with parameters (a, b, p) where $a = 2\lambda$, $b = 2\lambda\mu_1^2$, $p = \frac{1}{2}$. In order to obtain the result, it is enough to do it by partitioning the expectation using $1_{\{T_1<\ldots<T_n<T<T_{n+1}\}}$ which leads to the announced result. Actually, given the above result, it is not difficult to obtain

$$\mathbb{E}\left[f(\bar{X}_i, \Delta T_i)1_{\{N_T\geq i\}} \,\middle|\, \bar{X}_{i-1}, \rho_{i-1}, \Delta T_j,\ j = 1\ldots i-1\right]$$

$$= \mathbb{E}\left[f\big(\mathrm{Y}_i, \tau_i\big)1_{\{N_T^\tau\geq i\}} \,\middle|\, \mathrm{Y}_{i-1} = \bar{X}_{i-1}, \rho_{i-1}, \tau_j = \Delta T_j,\ j = 1\ldots i-1\right].$$

Here N^τ denotes the renewal process generated from the times between jumps given by $\{\tau_i; i \in \mathbb{N}\}$. □

On the set $\{N_T^\tau = n\}$ and for the last interval $[\zeta_n, \bar{\tau}^n]$, we remark that conditioning on the space variables does not imply any change in the law of $\bar{\tau}^n$ except that its parameter becomes $(\Delta_{n+1}\mathrm{Y})^2 := (L - \mathrm{Y}_n)^2$. That is, the law of $\bar{\tau}^n$ is a Lévy distribution with parameter $|\Delta_{n+1}\mathrm{Y}|/\sigma_n > 0$. In other words, the probability density of $\bar{\tau}^n$ is given by $p_{\bar{\tau}^n}(s) = \frac{|\Delta_{n+1}\mathrm{Y}|}{\sqrt{a_n 2\pi s^3}} \exp(-\frac{(\Delta_{n+1}\mathrm{Y})^2}{2a_n s})$, $s > 0$ with $p_{\bar{\tau}^n}(0) = 0$ by continuity. Here, we abuse the notation slightly by letting $\zeta_j = (\sum_{i=1}^j \tau_i) \wedge T$. A similar comment applies to σ_i, $\mathcal{I}_i$ and $\mathbb{E}_{M,k,n}[\cdot]$ which we define for a positive random variable M being measurable with respect to $\sigma(\mathrm{Y}, \zeta, \rho)$ as

$$\mathbb{E}_{M,k,n}[X] := \mathbb{E}[XM^{-1}1_{\{N_T^\tau=n\}} \mid \mathcal{F}_{\zeta_k}, \mathrm{Y}, \rho].$$

Here, $\mathcal{F}_{\zeta_k} := \sigma(\zeta_i; i \leq k)$. In the first part, we will use $M = 1$. Later, M will denote the Malliavin variance whose inverse moment properties are discussed in Sect. 4.4.

This completes our set-up for the consideration of differential calculus on this space. Using this new set-up, one clearly has that the following basic IBP formula is satisfied for $n \in \mathbb{N}_0$ and $f \in \mathscr{C}_b^1(\mathbb{R})$ with $f(T) = 0$:

$$\mathbb{E}_{1,n,n}[f'(\bar{\tau}^n)1_{\{\zeta_n+\bar{\tau}^n<T\}}] = \mathbb{E}_{1,n,n}[f(\bar{\tau}^n)\widehat{\mathcal{I}}_{n+1}(1)1_{\{\zeta_n+\bar{\tau}^n<T\}}], \tag{24}$$
$$\widehat{\mathcal{I}}_{n+1}(G) := G\left\{\frac{3}{2\bar{\tau}^n} - \frac{(\Delta_{n+1}\mathtt{Y})^2}{2a_n(\bar{\tau}^n)^2}\right\} - D_{\bar{\tau}^n}G.$$

In the above formula, conditioned on $\mathtt{Y}$ and ρ, we assume that $G = G(\bar{\tau}^n)$ for $G \in \mathscr{C}_p^\infty(\mathbb{R})$ so that $D_{\bar{\tau}^n}G = G'(\bar{\tau}^n)$. We also remark that the result is applicable if f or G depend on $\zeta_1, \cdots, \zeta_n$ or $\bar{X}$.

Due to Lemma 17 in the Appendix, we have that for a positive non-random constant $C > 0$

$$\mathbb{E}_{1,n,n}\left[\left|\widehat{\mathcal{I}}_{n+1}(G)\right|1_{\{\zeta_n+\bar{\tau}^n<T\}}\right] \le C\frac{\sup_{t\in[0,T]}(|G(t)| + |G'(t)|)}{(\Delta_{n+1}\mathtt{Y})^2}. \tag{25}$$

It is worth noting that the inequality described above may be referred to as a spatial degeneration inequality, mirroring the circumstances encountered when examining IBP formulas in the context of stopped processes, where analogous estimates were denoted as time degeneration inequalities.

In order to give the IBP formula, we need to introduce the following operators. We define[4] for a function $G \in \mathscr{C}_p^1(\mathbb{R})$, the r.v. $G \equiv G(\tau_i)$. We define the operators to $(\mathtt{I}_i, \mathtt{D}_i)$ for $i \le n$ on the set $\{N_T^\tau = n\}$. That is, for $G \equiv G(\tau_i)$,

$$\mathtt{I}_i(G) := G\left(\lambda + (2\tau_i)^{-1} + \frac{(\Delta_i\mathtt{Y})^2}{2a_{i-1}\tau_i^2}\right) - \mathtt{D}_iG$$
$$\mathtt{D}_iG := G'(\tau_i).$$

Remark 9 In the current framework, it is crucial to observe that, based on the construction outlined at the outset of this section, the random variables $\Delta_i\mathtt{Y}$ exhibit no dependence on τ_i. Consequently, it follows that $\mathtt{D}_i\Delta_j\mathtt{Y} = 0$ for any $j \in \mathbb{N}$.

We also have from Lemma 17, the space degeneration estimates for a non-random constant $C > 0$ and $k \in \{0, 1, 2\}$:

$$\mathbb{E}_{1,i-1,n}\left[|\mathtt{I}_i(\tau_i^{-k})|\right] \le C\left(1 + |\Delta_i\mathtt{Y}|^2 + |\Delta_i\mathtt{Y}|^{-(2k+1)}\right) \tag{26}$$

It's important to note that in many situations, we may employ a function $G(\tau_1, \cdots, \tau_n)$ defined on the event $N_T^\tau = n$. In such cases, the definitions provided

[4] One may define the equivalent spaces $\mathbb{S}$ as in the case of stopped diffusions but we refrain to do it in order to simplify the presentation.

above are to be understood conditionally to all times between jumps except τ_i, besides the conditioning involving spatial variables and all ρ_i for $i \in \mathbb{N}$.

Furthermore, we will continue to use $\delta_t(\tau_i)$ to denote the Dirac delta distribution function at the point t.

Similar to the approach taken in [14], we anticipate that employing the symbol $\mathtt{I}_i$ for the dual operator of the derivative operator $\mathtt{D}_i$ will not result in confusion, as the context will consistently clarify the object under discussion. Additionally, the formulas for $\mathcal{I}_i$ and $\mathcal{D}_i$ which are used from now on are restated using $\mathtt{Y}$ instead of $\bar{X}$. That is, we adopt the notation convention of using the same symbol, despite the possibility of being expressed with $\mathtt{Y}$.

Lemma 10 *With the above definitions, we have the following IBP formula for* $0 \leq t_1 < t_2$ *and* $f, G \in \mathscr{C}_p^1(\mathbb{R})$*:*

$$\begin{aligned}\mathbb{E}_{1,i-1,n}\left[f'(\tau_i)G(\tau_i)\mathbf{1}_{\{t_1<\tau_i<t_2\}}\right] =&\mathbb{E}_{1,i-1,n}\left[f(\tau_i)\mathcal{I}_i(G)\mathbf{1}_{\{t_1<\tau_i<t_2\}}\right]\\ &+\mathbb{E}_{1,i-1,n}\left[f(\tau_i)G(\tau_i)(\delta_{t_2}(\tau_i)-\delta_{t_1}(\tau_i))\right].\end{aligned} \tag{27}$$

In the particular case that $t_1 = 0$*, using that* $\mathtt{Y}_{i-1} \neq \mathtt{Y}_i$ *a.s., the above formula reduces to*

$$\begin{aligned}\mathbb{E}_{1,i-1,n}\left[f'(\tau_i)G(\tau_i)\mathbf{1}_{\{0<\tau_i<t_2\}}\right] =& \mathbb{E}_{1,i-1,n}\left[f(\tau_i)\mathcal{I}_i(G)\mathbf{1}_{\{0<\tau_i<t_2\}}\right]\\ &+\mathbb{E}_{1,i-1,n}\left[f(\tau_i)G(\tau_i)\delta_{t_2}(\tau_i)\right].\end{aligned} \tag{28}$$

Proof The proof follows using the usual IBP formula for Lebesgue integrals. □

The next result states the independence of the weight θ^j with respect to the noise in τ^i once they are rewritten using the set-up stated at the beginning of this section. This result simplifies calculations greatly.

Lemma 11 $\mathtt{D}_i\theta^j = 0$ *for* $i \neq j$.

Proof Note that as $\tilde{Z}_j = \sigma_{j-1}^{-1}\Delta_j\mathtt{Y}$, one obtains that θ^j is only a function of $(\mathtt{Y}, \rho, \tau_j)$ by observing (20). Using that $j \neq i$, we get the result. □

Building upon the information provided in the preceding proof and in conjunction with (26) and Lemma 17, we can derive the following space degeneration estimates:

$$\begin{aligned}\mathbb{E}_{1,i-1,n}\left[|\theta^i|\right] \leq& C\left(1+|\Delta_i\mathtt{Y}|^3\right),\\ \mathbb{E}_{1,i-1,n}\left[|\mathtt{I}_i(\theta^i)|\right] \leq& C\left(1+|\Delta_i\mathtt{Y}|^{-2}+|\Delta_i\mathtt{Y}|^5\right).\end{aligned} \tag{29}$$

In fact, using (20) and (8) twice, we see that by linearity it is enough to estimate and upper bound $\mathbb{E}_{1,i-1,n}\left[|\mathtt{I}_i(\mathcal{I}_i^j(c_j^i))|\right]$ for $j = 1, 2$. Now we apply (10), (11) and the linearity of $\mathtt{I}_i(\cdot)$ to obtain

$$\mathbb{E}_{1,i-1,n}\left[|\mathtt{I}_i(\mathcal{I}_i^j(c_j^i))|\right] \leq \sum_{k=0}^{j} \mathbb{E}_{1,i-1,n}\left[|\mathtt{I}_i\left(\mathcal{H}_{j-k}(a_{i-1}\tau_i, \sigma(\mathrm{Y}_{i-1})\tilde{Z}_i)\right)\mathcal{D}_i^k c_j^i|\right].$$

Using explicit expressions for Hermite polynomials, the definition of $\mathtt{I}_i$ and (26), we obtain the second estimate in (29).

4.3 *The Transfer Formula*

The subsequent equalities are all stated with respect to conditional expectations involving the spatial variables. In particular, the symbol $\overset{\mathbb{E}_{M,k,n}}{=}$ denotes equality within the respective conditional expectation context.

With this notation established, we can express the formula for transferring derivatives in the final interval as follows:

Lemma 12 *Let $f \in \mathscr{C}_p^1(\mathbb{R})$ with $f(T) = 0$ then, with $\overleftarrow{\theta}\,^e_{n+1} = \hat{\theta}^n$, it holds*

$$\mathbb{E}_{M,n,n}[f'(\zeta_n + \bar{\tau}^n)\mathbb{1}_{\{\zeta_n+\bar{\tau}^n<T\}}\hat{\theta}^n] = \partial_{\zeta_n}\mathbb{E}_{M,n,n}[f(\zeta_n + \bar{\tau}^n)\mathbb{1}_{\{\zeta_n+\bar{\tau}^n<T\}}\overleftarrow{\theta}\,^e_{n+1}]. \tag{30}$$

Also, we have that, on the set $\{\zeta_n = T\}$, $\mathbb{E}_{M,n,n}[f(\zeta_n + \bar{\tau}^n)\mathbb{1}_{\{\zeta_n+\bar{\tau}^n<T\}}\hat{\theta}^n] = 0$.

Proof Given the conditioning we may assume without loss of generality that $\hat{\theta}^n = 1$. Recall that the IBP formula for the last interval is obtained using a Lévy distribution as stated in (24). Hence, by employing Leibniz's rule for interchanging derivatives and integrals, it can be deduced that the right-hand side of (30) is differentiable with respect to ζ_n, and:

$$f'(\zeta_n + \bar{\tau}^n)\mathbb{1}_{\{\bar{\tau}^n<T-\zeta_n\}} \overset{\mathbb{E}_{M,n,n}}{=} f(\zeta_n + \bar{\tau}^n)\mathbb{1}_{\{\bar{\tau}^n<T-\zeta_n\}}\widehat{\mathcal{I}}_{n+1}(1).$$

For the second result, it is enough to note that $\mathbb{E}_{M,n,n}[f(\zeta_n + \bar{\tau}^n)\mathbb{1}_{\{\zeta_n+\bar{\tau}^n<T\}}]$ is a continuous function of ζ_n. □

We would like to mention that the notation $\overleftarrow{\theta}\,^e_{n+1}$ is adopted from Section 5.1 of [14]. We have retained this notation to emphasize its significance within the tree structure, as explained in [14].

Now, let's present the general transfer formula for the remaining intervals.

Lemma 13 *Let $f \in \mathscr{C}_b^1(\mathbb{R}_0)$ with $f(T) = 0$. It holds*

$$\mathbb{E}_{M,i,n}[f'(\zeta_{i+1})\mathbb{1}_{\{\zeta_{i+1}<T\}}\theta^{i+1}] = \partial_{\tau_i}\mathbb{E}_{M,i,n}[f(\zeta_{i+1})\mathbb{1}_{\{\zeta_{i+1}<T\}}\overleftarrow{\theta}\,^e_{i+1}]. \tag{31}$$

Here, $\overleftarrow{\theta}^{e}_{i+1} := \theta^{i+1}$. *Moreover, the following boundary condition is satisfied*

$$\mathbb{E}_{M,i,n}[f(\zeta_{i+1})\mathbb{1}_{\{\zeta_{i+1}<T\}}\overleftarrow{\theta}^{e}_{i+1}]\big|_{\zeta_i=T} = 0,$$

as well as the following time degeneracy inequality:[5]

$$\|\overleftarrow{\theta}^{e}_{i+1}\|_{0,p} \leq C(\zeta_{i+1}-\zeta_i)^{-1/2}.$$

Proof The proof can be established through straightforward differentiation of the right-hand side of Eq. (31). In particular, note that θ^{i+1} does not depend on τ_i. Concerning the boundary conditions, it's worth noting that we have $f(T) = 0$. Additionally, the time degeneracy estimate has been previously established in [14]. □

4.4 The IBP Formula

The concepts applied here bear a resemblance to those used in the case of stopped processes, albeit with a shift in the roles of time and space.

First, let's focus on the last interval. Without delving into technical details, we can derive the IBP formula on the set $\{N_T = n\}$ from Lemma 10:

$$\begin{aligned}&(\Delta_{n+1}\Upsilon)^2 f'(\zeta_n+\bar{\tau}^n)\mathbb{1}_{\{\bar{\tau}^n\leq T-\zeta_n\}}\hat{\theta}^n\\ &\overset{\mathbb{E}_{M,n,n}}{=} (\Delta_{n+1}\Upsilon)^2 f(\zeta_n+\bar{\tau}^n)\mathbb{1}_{\{\bar{\tau}^n\leq T-\zeta_n\}}\hat{\theta}^n\widehat{\mathcal{I}}_{n+1}(1).\end{aligned}$$

Furthermore, the above expression satisfies the following space degeneration estimate due to (25):

$$\left|(\Delta_{n+1}\Upsilon)^2 f(\zeta_n+\bar{\tau}^n)\hat{\theta}^n\widehat{\mathcal{I}}_{n+1}(1)\right|\mathbb{1}_{\{\zeta_n+\bar{\tau}^n<T\}} \overset{\mathbb{E}_{M,n,n}}{\leq} C\sup_{t\in[0,T]}|f(t)|.$$

Now, to make use of the transfer formula provided in Lemma 13 for all time intervals leading up to the interval where the IBP formula will be applied, we define, for $k \geq i$

$$G_k(\zeta_k) \equiv G_k(n,\zeta_k,\Upsilon,\rho) := \mathbb{E}_{M,k,n}\left[f(\zeta_n+\bar{\tau}^n)\mathbb{1}_{\{\zeta_n+\bar{\tau}^n<T\}}\hat{\theta}^n\prod_{j=k+1}^{n}\theta^j\right].$$

[5] Recall that the norm $\|\cdot\|_{0,p}$ is a norm on the Wiener space conditional to ρ and N^τ.

With the definition provided above and using (29), we can establish that $G_k(T) = 0$, $|G_k(\zeta_k)| \le C \prod_{j=k+1}^{n+1}(1 + |\Delta_j \mathrm{Y}|^3)$. Consequently, the application of the transfer formulas (30) and (31), along with Lemma 10, yields

$$(\Delta_i \mathrm{Y})^2 f'(\zeta_n + \bar{\tau}^n) 1_{\{\zeta_n + \bar{\tau}^n < T\}} \prod_{j=i}^{n} \theta^j \overset{\mathbb{E}_{M,i-1,n}}{=} (\Delta_i \mathrm{Y})^2 G_i'(\zeta_i) 1_{\{\zeta_i \le T\}} \theta^i$$

$$\overset{\mathbb{E}_{M,i-1,n}}{=} (\Delta_i \mathrm{Y})^2 G_i(\zeta_i) 1_{\{\zeta_i \le T\}} \mathrm{I}_i(\theta^i).$$

Again from (29), the space degeneration estimate in this case is

$$\left|(\Delta_i \mathrm{Y})^2 f(\zeta_i)\right| 1_{\{\zeta_i \le T\}} \hat{\theta}^n \prod_{j=i+1}^{n} |\theta^j||\mathrm{I}_i(\theta^i)|$$

$$\overset{\mathbb{E}_{M,i-1,n}}{\le} C \sup_{t \in [0,T]} |f(t)| \prod_{j=i+1}^{n} (1 + |\Delta_j \mathrm{Y}|^3)(1 + |\Delta_i \mathrm{Y}|^7).$$

We can now leverage the fact that, as indicated by Lemma 8, $|\Delta_i \mathrm{Y}| = \sigma(\mathrm{Y}_{i-1})$ $|\tilde{Z}_i|$, $i = 1, \ldots, N_T^\tau$, where $\tilde{Z}_i$ represents an i.i.d. sequence of symmetric exponential random variables characterized by a parameter of $1/\sqrt{2\lambda}$. Hence, for any integer n,

$$\mathbb{E}\Bigg[\left|f(\zeta_n + \bar{\tau}^n)\right| 1_{\{\zeta_n + \bar{\tau}^n \le T\}} |\hat{\theta}^n| \Bigg(\sum_{i=1}^{n} (\Delta_i \mathrm{Y})^2 \prod_{j=i+1}^{n} |\theta^j||\mathrm{I}_i(\theta^i)| \prod_{k=1}^{i-1} |\theta^k|$$

$$+ (\Delta_{n+1} \mathrm{Y})^2 |\widehat{\mathcal{I}}_{n+1}(1)| \prod_{k=1}^{n} |\theta^k| \Bigg) M^{-1} 1_{\{N_T^\tau = n\}} \Bigg]$$

$$\le (n+1) C^n \sup_{t \in [0,T]} |f(t)| \mathbb{P}(N_T = n).$$

Here, $M := \sum_{i=1}^{N_T^\tau + 1} (\Delta_i \mathrm{Y})^2$. We remark here that once all the integration by parts have been performed, expressions for the weights can be written using the original formulation in (2) due to Lemma 8.

As it was done in [14], we may give an algebraic structure to the IBP formula. We therefore refer the reader to that reference for the notation and terminology that follows.

In this case, the structure is easier and it follows the same graph as in the stopped case except that all terms with arrows corresponding to $(\overleftarrow{\theta}^c, \overleftarrow{\theta}^\partial)$ do not appear in this case.[6] In conclusion, from the above calculations one understands that the algebraic structure in this case is formed only of components $I_i^{n+1} = (0, \dots, 0, I, e, \dots, e)$, $i \le N_T + 1$. These components have exchange r.v.'s in all intervals $[\zeta_k, \zeta_{k+1}]$ for $k \in \{i, \dots, N_T\}$, denoted by e, an integration by parts on the interval $[\zeta_{i-1}, \zeta_i]$, denoted by I and no change in weight r.v.'s in all previous intervals $[\zeta_k, \zeta_{k+1}]$ for $k \in \{0, \dots, i-2\}$ which is denoted by 0 in the component I_i^{n+1}.

Putting all the above reasoning together, one obtains the IBP formula for the exit time. For this, we need to define for $i \le n+1$ on the set $\{N_T = n\}$

$$\theta^{I_i} := \hat{\theta}^n \prod_{j=i+1}^{n} \overleftarrow{\theta}_j^e \times \mathtt{I}_i(\theta^i) \times \prod_{l=1}^{i-1} \theta^l = \hat{\theta}^n \prod_{j=i+1}^{n} \theta^j \times \mathtt{I}_i(\theta^i) \times \prod_{l=1}^{i-1} \theta^l$$

$$\theta^{I_{n+1}} := \hat{\theta}^n \hat{\mathcal{I}}_{n+1}(1) \times \prod_{l=1}^{n} \theta^l.$$

Theorem 14 *Let $x > L$ and $f \in \mathscr{C}_b^1(\mathbb{R})$ such that $f(T) = 0$. Then, it holds*

$$\mathbb{E}\left[f'(\tau \wedge T)\right] = e^{\lambda T} \mathbb{E}\left[M^{-1} \sum_{i=1}^{N_T+1} (\Delta_i \bar{X})^2 f(\zeta_{N_T} + \bar{\tau}^{N_T}) \mathbb{1}_{\{\bar{\tau}^{N_T} \le T - \zeta_{N_T}\}} \theta^{I_i}\right].$$

The random variable which appear within the expectation on the right-hand side above is integrable.

The proof of the above result is similar to the case of stopped processes given in Theorem 5.1 of [14]. Algebraically, the proof essentially follows from applying the recursive relation given in (5.2), the transfer formula and the iteration step given in (5.4) of Theorem 5.1 [14]. The structure given here is much simpler compared to [14] since all boundary terms are zero. Furthermore, the necessary inequalities and estimates are provided at the beginning of this section. It is important to note that integrability can be inferred from the space degeneration estimates outlined in Lemma 17. Lastly, the assumption $f(T) = 0$ can be readily eliminated by introducing the function $g(t) = f(t) - f(T)$.

Acknowledgments The research of A. Kohatsu-Higa was supported by KAKENHI grant 20K03666. He would like to express his thanks to the organizers of the meeting "Kolmogorov Operators and Their Applications" for their kind invitation to contribute to this volume. The research of N. Frikha has benefited from the support of the Institut Europlace de Finance.

[6] We refer the reader for details on this "path algebraic structure" to Section 5.1 in [14].

Appendix

The Malliavin Variance

Define the Malliavin variance as

$$M := \sum_{i=1}^{N_T+1} \left(\Delta_i \bar{X}\right)^2.$$

We remark here that an equivalent definition using (ρ, Υ, τ), instead of $(\rho, \bar{X}, T)$ is possible and the study of the inverse moments is equivalent due to Lemma 8 which states the equality in law. We prefer this version as we will be using geometrical properties of the path of $\bar{X}$. The derivation of the IBP formula necessitates the existence of inverse moments of the Malliavin variance on the set $\bar{\tau}^{N_T} \leq T - \zeta_{N_T}$. The reason for the inverse integrability is that each space increment $\Delta_i \bar{X}$ has a positive probability of being close to $\frac{L-x}{N_T}$. This provides a lower bound for the variable M, ensuring that its inverse is integrable in any $L^p(\mathbb{P})$ norm. This result is established in the following section.

Lemma 15 *For any $q > 0$, $k \in \mathbb{N}$ and $x > L$, we have that*

$$\mathbb{E}\left[\left(\sum_{i=1}^{N_T+1} \left(\Delta_i \bar{X}\right)^{2k}\right)^{-q} 1_{\{\zeta_{N_T}+\bar{\tau}^{N_T}<T\}} \prod_{i=1}^{N_T+1} 1_{\{\bar{X}_i>L\}}\right] < \infty.$$

Proof One considers the set

$$A := \left\{\exists i \in \{1, \ldots, N_T+1\};\ \left|\Delta_i \bar{X}\right| \geq \frac{L-x}{N_T+1}\right\}.$$

Note that $\mathbb{P}(A^c, \bar{X}_i > L, i = 1, \cdots, N_T) = 0$ because the following inequality is satisfied

$$\sum_{i=1}^{N_T+1} \left|\Delta_i \bar{X}\right| \geq L - x,$$

on the set $\left\{\bar{X}_i > L, i = 1, \cdots, N_T\right\}$. The proof of the above inequality can be explained easily in geometrical terms. In the case that $\rho_i = 1$ then $|\Delta_i \bar{X}|$ stands for the distance between the two points $\bar{X}_i$ and $\bar{X}_{i-1}$. In the case that $\rho_i = 0$ then the value $\bar{X}_{i-1}$ is reflected with respect to the line $y = L$ but it also has to finish at a point $\bar{X}_i > L$ and therefore $|\bar{X}_i - (2L - \bar{X}_{i-1})|$ is larger than $|\bar{X}_i - \bar{X}_{i-1}|$. As the initial point and final point of the process $\bar{X}$ are fixed at x and L respectively, the above inequality follows.

Then, using Chebyshev's inequality, one obtains:

$$\mathbb{E}\left[\left(\sum_{i=1}^{N_T+1}(\Delta_i\bar{X})^{2k}\right)^{-q} 1_{\{\zeta_{N_T}+\bar{\tau}^{N_T}<T\}}\prod_{i=1}^{N_T+1}1_{\{\bar{X}_i>L\}}\right]$$

$$\leq \mathbb{E}\left[\frac{(N_T+1)^{2qk+1}}{(L-x)^{2qk}}1_A\right]<\infty.$$

□

Auxiliary Lemmas for Lévy and Generalized Inverse Gaussian Distributions

Lemma 16 *Let $f:\mathbb{R}_+\to\mathbb{R}$ be a measurable and bounded function. Then, it holds*

$$\mathbb{E}\left[(2\rho_1-1)\mathcal{I}_1(1)\delta_L(\bar{X}_1)f(\zeta_1)\,|\,N_T=1\right]=-T^{-1}\mathbb{E}[f(\bar{\tau}^x)1_{\{\bar{\tau}^x\leq T\}}].$$

Proof The proof consists of explicit calculations. In fact,

$$\mathbb{E}\left[(2\rho_1-1)\mathcal{I}_1(1)\delta_L(\bar{X}_1)f(\zeta_1)\,|\,N_T=1\right]=T^{-1}\int_0^T\frac{L-x}{\sigma(x)s}g_s(x,L)f(s)\,\mathrm{d}s.$$

The above equals $-T^{-1}\mathbb{E}[f(\bar{\tau}^x)1_{\{\bar{\tau}^x\leq T\}}]$ because the density function of $\bar{\tau}^x$ is explicitly known and given by $\frac{x-L}{\sigma(x)s}g_s(x,L)$. □

Lemma 17 *For $k\in\mathbb{N}$ and $1\leq i\leq n\in\mathbb{N}$,*

$$\mathbb{E}_{1,n,n}[(\bar{\tau}^n)^{-k}1_{\{\zeta_n+\bar{\tau}^n\leq T\}}]=\mathbb{E}[(\bar{\tau}^n)^{-k}1_{\{\zeta_n+\bar{\tau}^n\leq T\}}1_{\{N_T=n\}}\,|\,\mathcal{F}_{\zeta_n},Y,\rho]$$
$$\leq C(\Delta_{n+1}Y)^{-2k},$$

$$\mathbb{E}_{1.i-1,n}\left[\tau_i^{-k}\right]=\mathbb{E}\left[\tau_i^{-k}1_{\{N_T=n\}}\,|\,\mathcal{F}_{\zeta_{i-1}},Y,\rho\right]\leq C\left(1+|\Delta_i Y|^{1-2k}\right).$$

The proof of the above statement is done using explicit integral expressions of inverse moments related to the Lévy distribution and generalized inverse Gaussian distributions on $\mathbb{R}_+$. In fact, the first inequality follows using the explicit density of $\bar{\tau}^n$ together with the inequality $x^{2k}e^{-cx^2}\leq\left(\frac{k}{2c}\right)^k e^{-k}$. Therefore, applying this inequality to

$$\begin{aligned}&\mathbb{E}[(\bar{\tau}^n)^{-k}\mathbb{1}_{\{\zeta_n+\bar{\tau}^n\le T\}}\mid N_T=n,\mathcal{F}_{\zeta_n},\Upsilon,\rho]\\&=\int_0^{T-\zeta_n}s^{-k}\frac{|\Delta_{n+1}\Upsilon|}{\sqrt{2\pi}(a_ns)^{3/2}}\exp\left(-\frac{(\Delta_{n+1}\Upsilon)^2}{2a_ns}\right)\mathrm{d}s,\end{aligned}$$

one obtains the result in the first case using the inequality for $c=(4a_ns)^{-1}$.

The second inequality follows using the explicit form of the GIG distribution. The expectation of τ_i^{-k} can be reinterpreted using the GIG distribution with parameters $p=-k+\frac{1}{2}$, $a=2\lambda$, $b=2\lambda\mu_i^2$ or, in other words, the GIG$(2\lambda,2\lambda\mu_i^2,-k+\frac{1}{2})$ distribution. This introduces a change of normalization constant which gives the inequality. When doing these calculations, one uses the following facts about modified Bessel functions of the second kind $K_{3/2}(x)=(1+x^{-1})K_{1/2}(x)=(1+x^{-1})\sqrt{\frac{\pi}{2x}}e^{-x}$. In general, one may write for $\nu-\frac{1}{2}\in\mathbb{Z}$, (see, for example, page 925 in [15])

$$K_\nu(x)=K_{|\nu|}(x)=K_{1/2}(x)\sum_{j=0}^{\lfloor|\nu|-\frac{1}{2}\rfloor}\frac{(j+|\nu|-\frac{1}{2})!}{j!(-j+|\nu|-\frac{1}{2})!}(2x)^{-j}.$$

In fact, the GIG density multiplied by x^{-k} is given by

$$\begin{aligned}&\frac{1}{x^k}\frac{(a/b)^{p/2}}{K_p(\sqrt{ab})}x^{p-1}e^{-(ax+b/x)/2}\\&=\frac{(a/b)^{k/2}K_{p-k}(\sqrt{ab})}{K_p(\sqrt{ab})}\left[\frac{(a/b)^{(p-k)/2}}{K_{p-k}(\sqrt{ab})}x^{p-k-1}e^{-(ax+b/x)/2}\right].\end{aligned}$$

Here $\sqrt{ab}=2\lambda\mu_i$ and $a/b=1/\mu_i^2$. Hence, recalling that $p=\frac{1}{2}$, the previous quantity is upper-bounded by

$$\mu_i^{-k}\sum_{j=0}^{\lfloor|\nu|-\frac{1}{2}\rfloor}\frac{(j+|\nu|-\frac{1}{2})!}{j!(-j+|\nu|-\frac{1}{2})!}(4\lambda\mu_i)^{\ j}\left[\frac{(a/b)^{(p-k)/2}}{K_{p-k}(\sqrt{ab})}x^{p-k-1}e^{-(ax+b/x)/2}\right].$$

Note that as $|\nu|-\frac{1}{2}=k-1$, the second inequality in the statement of the lemma follows using the fact that $|\Delta_i\Upsilon|^{-j-k}\le C(1+|\Delta_i\Upsilon|^{-2k+1})$ for all $j=0,\ldots,\lfloor|\nu|-\frac{1}{2}\rfloor$.

References

1. Airault, H., Malliavin, P., Ren, J.: Smoothness of stopping times of diffusion processes. J. Math. Pures Appl. **78**(10), 1069–1091 (1999)
2. Arous, G.B., Kusuoka, S., Stroock, D.W.: The Poisson kernel for certain degenerate elliptic operators. J. Funct. Anal. **56**(2), 171–209 (1984)
3. Bally, V., Clément, E.: Integration by parts formula and applications to equations with jumps. Probab. Theory Relat. Fields **151**(3–4), 613–657 (2011)
4. Bally, V., Bavouzet, M.P., Messaoud, M.: Integration by parts formula for locally smooth laws and applications to sensitivity computations. Ann. Appl. Probab. **17**(1), 33–66 (2007)
5. Bally, V., Caramellino, L., Cont, R., Utzet, F., Vives, J.: Stochastic Integration by Parts and Functional Itô Calculus. Advanced Courses in Mathematics - CRM Barcelona. Springer, Berlin (2016)
6. Bismut, J.M.: Last exit decompositions and regularity at the boundary of transition probabilities. Z. Wahrscheinlichkeitst.Verwandte Gebiete **69**(1), 65–98 (1985)
7. Chaudru de Raynal, P.E., Frikha, N.: From the backward kolmogorov PDE on the wasserstein space to propagation of chaos for Mckean-Vlasov SDEs. J. Math. Pures Appl. **156**, 1–124 (2021)
8. Chaudru de Raynal, P.E., Frikha, N.: Well-posedness for some non-linear SDEs and related PDE on the wasserstein space. J. Math. Pures Appl. **159**, 1–167 (2022)
9. Chen, J., Frikha, N., Li, H.: Probabilistic representation of integration by parts formulae for some stochastic volatility models with unbounded drift. ESAIM Probabil. Stat. **26**, 304–351 (2022)
10. Denis, L., Nguyen, T.M.: Malliavin calculus for Markov chains using perturbations of time. Stochastics **88**(6), 813–840 (2016)
11. Fournier, N., Printems, J.: Absolute continuity for some one-dimensional processes. Bernoulli **16**(2), 343–360 (2010)
12. Frikha, N., Li, L.: Weak uniqueness and density estimates for SDEs with coefficients depending on some path-functionals. Ann. l'Inst. Henri Poincaré, Probabil. Stat. **56**(2), 1002–1040 (2020)
13. Frikha, N., Kohatsu-Higa, A., Li, L.: On the first hitting times of one dimensional elliptic diffusions (2016, preprint). arXiv:1609.09327
14. Frikha, N., Kohatsu-Higa, A., Li, L.: Integration by parts formula for killed processes: a point of view from approximation theory. Electron. J. Probab. **24**(95), 1–44 (2019)
15. Gradshteyn, I.S., Ryzhik, I.M.: Table of Integrals, Series, and Products, 7th edn. Elsevier/Academic Press, Amsterdam (2007). Translated from the Russian, Translation edited and with a preface by Alan Jeffrey and Daniel Zwillinger, With one CD-ROM (Windows, Macintosh and UNIX)
16. Ikeda, N., Watanabe, S.: Stochastic Differential Equations and Diffusion Processes. North-Holland Mathematical Library, vol. 24, 2nd edn. North-Holland, Amsterdam/Kodansha, Ltd., Tokyo (1989)
17. Karatzas, I., Shreve, S.E.: Brownian Motion and Stochastic Calculus, 2nd edn. Springer, Berlin (1991)
18. Kent, J.T.: Eigenvalue expansions for diffusion hitting times. Z. Wahrscheinlichkeit. Verwandte Gebiete **52**(3), 309–319 (1980)
19. Kent, J.T.: The spectral decomposition of a diffusion hitting time. Ann. Probabil. **10**, 207–219 (1982)
20. Pitman, J., Yor, M.: Hitting, occupation and inverse local times of one-dimensional diffusions: martingale and excursion approaches. Bernoulli **9**(1), 1–24 (2003)

On Averaged Control and Iteration Improvement for a Class of Multidimensional Ergodic Diffusions

A. Yu. Veretennikov

Abstract The paper is devoted to the derivation of ergodic Bellman's equation for a special case of a uniformly ergodic multidimensional controlled diffusion with variable diffusion and drift coefficients both depending on control; the diffusion coefficient must be a scalar function. The convergence of Howard's iterative reward improvement algorithm to the unique solution of Bellman's equation is also established.

1 Introduction

Bellman's or HJB equations for stochastic differential equations (SDEs) with control are nonlinear analogues of Fokker–Planck–Kolmogorov differential equations for SDEs without control. This paper is a continuation of [1] and [2] where the dimension was equal to one. Consider a d-dimensional stochastic differential equation (SDE) on the probability space $(\Omega, \mathcal{F}, (\mathcal{F}_t), P)$ with a d-dimensional $(\mathcal{F}_t)$ Wiener process $B = (B_t)_{t\ge 0}$ with coefficients b and σ, and with a feedback control function α (called strategy in the sequel)

$$\begin{aligned} dX_t^\alpha &= b(\alpha(X_t^\alpha), X_t^\alpha)\,dt + \sigma(\alpha(X_t^\alpha), X_t^\alpha)\,dW_t, \quad t \ge 0, \\ X_0^\alpha &= x, \end{aligned} \tag{1}$$

This work is funded by the Foundation for the Advancement of Theoretical Physics and Mathematics "BASIS".

A. Yu. Veretennikov (✉)
Kharkevich Institute for Information Transmission Problems, Moscow, Russia
e-mail: ayv@iitp.ru

S. Menozzi et al. (eds.), *Kolmogorov Operators and Their Applications*,
Springer INdAM Series 56, https://doi.org/10.1007/978-981-97-0225-1_10

with the function σ is a scalar one, $\sigma : \mathbb{R}^1_+ \times \mathbb{R}^d \mapsto \mathbb{R}^1$. This is, of course, not the most general case, but still it is a multidimensional one which was not considered in earlier papers on the subject with sigma depending on control.

Note that according to [15] the Eq. (1) has a weak solution which is a homogeneous Markov process. In the general situation weak uniqueness of this solution is not claimed. However, this weak uniqueness will follow under the assumptions of the main result which is Theorem 1 of Sect. 3.

Let a compact set $U \subset \mathbb{R}$ be a set where any strategy takes its values. The functions b and σ on $U \times \mathbb{R}$ are assumed Borel measurable; later on some further conditions will be imposed, but we note straight away that σ will be assumed non-degenerate. Denote the class of all Borel functions α with values in U by $\mathcal{A}$. For $u \in U$ and $\alpha(\cdot) \in \mathcal{A}$ denote

$$L^u(x) = b(u, x)\nabla_x + \frac{1}{2}\sigma^2(u, x) \sum_i \frac{\partial^2}{\partial x_i^2},$$

and

$$L^\alpha(x) = b(\alpha(x), x)\nabla_x + \frac{1}{2}\sigma^2(\alpha(x), x) \sum_i \frac{\partial^2}{\partial x_i^2}, \qquad x \in \mathbb{R}^d.$$

Although this is not the most general form of the (non-divergent) second order differential operator, however, this restriction on L with the scalar multiplier σ^2 makes sense because in most media the heat coefficient is, indeed, scalar. Of course, there exist anisotropic media where the scalar description of the diffusion term does not suffice.

An additional restriction will be imposed on the drift b to guarantee the uniform recurrence. This is, again, not the most general ergodic case in the ergodic control setting, as shown in the series of works [3–10], et al. Currently, by using the method of this paper we cannot tackle the near-monotone case, which is available when σ does not depend on control.

The following a bit more general case may be considered without essential changes,

$$L^u(x) = b(u, x)\nabla_x + \frac{1}{2}\sigma^2(u, x) \sum_{ij} a_{ij}(x) \frac{\partial^2}{\partial x_i \partial x_j},$$

where $a(x) = (a_{ij}(x))$ is a non-degenerate bounded Hölder-continuous matrix which does not depend on control u, while σ^2 is still a scalar function of its arguments.

Denote by $\mathcal{K}$ the class of functions $f : U \times \mathbb{R}^d \mapsto \mathbb{R}^d$ growing no faster than some polynomial. This will be the class of *running cost*, or *instantaneous cost* functions. The functional

$$\rho^\alpha(x) := \limsup_{T\to\infty} \frac{1}{T} \int_0^T \mathbb{E}_x f(\alpha(X_t^\alpha), X_t^\alpha)\, dt \tag{2}$$

defines the *averaged cost* function corresponding to the strategy $\alpha \in \mathcal{A}$.

For a strategy $\alpha \in \mathcal{A}$ the function $f^\alpha : \mathbb{R}^d \to \mathbb{R},\ f^\alpha(x) = f(\alpha(x), x),\ x \in \mathbb{R}^d$, is defined. Then (2) has an equivalent form

$$\rho^\alpha(x) = \limsup_{T\to\infty} \frac{1}{T} \int_0^T \mathbb{E}_x f^\alpha(X_t^\alpha)\, dt. \tag{3}$$

The functional

$$\rho(x) := \inf_{\alpha\in\mathcal{A}} \limsup_{T\to\infty} \frac{1}{T} \int_0^T \mathbb{E}_x f^\alpha(X_t^\alpha)\, dt \tag{4}$$

is called a *cost function* in this model.

The solution of the Eq. (1) X^α is assumed Markov and uniformly ergodic for every $\alpha \in \mathcal{A}$ in the sense that there exists a limit μ^α in total variation of the marginal measures of X_t^α as $t \to \infty$, and this distribution μ^α does not depend on the initial condition $X_0 = x \in \mathbb{R}^d$; moreover, it is unique and invariant for the generator L^α. The cost function ρ^α—which does not depend on x in this case either—may be then rewritten as

$$\rho^\alpha := \int f^\alpha(y)\, \mu^\alpha(dy) =: \langle f^\alpha, \mu^\alpha \rangle. \tag{5}$$

Then the goal is to compute the minimal value

$$\rho := \inf_{\alpha\in\mathcal{A}} \int f^\alpha(y)\, \mu^\alpha(dy) = \inf_{\alpha\in\mathcal{A}} \langle f^\alpha, \mu^\alpha \rangle. \tag{6}$$

For any $\alpha \in \mathcal{A}$ let

$$v^\alpha(x) := \int_0^\infty \mathsf{E}_x(f^\alpha(X_t^\alpha) - \rho^\alpha)\, dt.$$

Under our assumptions in what follows, this integral will converge.

Computing the value ρ is, of course, a challenging task. One approach to it is to establish the *ergodic Bellman's (or HJB) equation* on the couple (V, ρ) where ρ is a constant, which looks as follows:

$$\inf_{u\in U} [L^u V(x) + f^u(x) - \rho] = 0, \quad x \in \mathbb{R}^d. \tag{7}$$

Here $V : \mathbb{R}^d \mapsto \mathbb{R}$. Equivalently, (7) may be written as

$$\inf_{u\in U}\left[\frac{1}{\sigma^2(u,x)}L^u V(x) + \frac{f^u(x)}{\sigma^2(u,x)} - \frac{\rho}{\sigma^2(u,x)}\right] = 0, \quad x \in \mathbb{R}^d. \tag{8}$$

As it was said, the solution of (7) and (8) is the couple $(V(\cdot), \rho)$. Under certain conditions it may be shown that the second component ρ in this solution is unique and coincides with the cost (6). The function V plays an auxiliary role, yet rather important. The uniqueness of V may also be shown up to an additive constant.

The first component of the solution (V, ρ) is to be found in the class of Borel measurable functions with two Sobolev derivatives locally integrable in any power and with a moderate (no faster than some polynomial) growth at infinity. The Eq. (7) will be understood almost everywhere and in the Sobolev sense. Due to the embedding theorems, the vector of first derivatives $\nabla V(x)$ may be considered continuous. The second derivatives will be assumed Borel measurable, as Borel representatives of Lebesgue's measurable functions.

The Reward Improvement Algorithm (RIA) is an approach which theoretically may serve as a tool to compute the cost ρ. In practice, it is, of course, a new challenging task, but, at least, each step of it combined with a discretisation is doable. This algorithm offers a *monotonic sequence* of values, say, ρ_n which converges to the exact cost value ρ. The challenge is because simultaneously the researcher needs to approximate V either by computing the sequence of integrals using Monte Carlo or some similar method, or by solving linear elliptic equations of the second order. See [16, ch.1, §4] about an approximate RIA.

Notice that, of course, uniqueness of solution of (7) for the first component may only be up to a constant since for any solution (V, ρ) the couple $(V + C, \rho)$ is also a solution for any constant C. It is natural to choose C so that V satisfies a "centering condition", see what follows.

The reader may find some brief account of the history of the problem in [1] as well as in the references in [9]. More recent publications on the close topics include [5, 7, 14, 20]; see also the references therein.

The paper consists of three sections: 1—Introduction, 2—Assumptions and some auxiliaries, 3—Main result and its proof. The convention will be used that arbitrary constants C in the calculus may change from line to line.

2 Assumptions and Auxiliaries

To ensure ergodicity of X^α under any feedback control strategy $\alpha \in \mathcal{A}$, we make the following assumptions on the drift and diffusion coefficients.

(A1) The functions b and σ are Borel measurable and bounded; $|b(u,x)| \le C_b$, $|\sigma(u,x)| \le C_\sigma$, σ is uniformly non-degenerate, $|\sigma(u,x)|^{-1} \le C_\sigma$; the functions $\sigma(u,x)$ and $f^u(x)$ are continuous in u for every x.

(A2)

$$\lim_{|x|\to\infty} \sup_u \langle x, b(u,x)\rangle = -\infty. \tag{9}$$

(A3) The function f belongs to the class $\mathcal{K}$ of functions which are Borel measurable in x for each u and admit a uniform in u polynomial bound: there exist constants $C_1, m_1 > 0$ such that for any x,

$$\sup_{u\in U} |f^u(x)| \le C_1(1+|x|^{m_1}).$$

(A4) The set U is compact.

Lemma 1 *Let the assumptions (A1)–(A3) hold true. Then*

- *For any $C_1, m_1 > 0$ there exist $C, m > 0$ such that for any strategy $\alpha \in \mathcal{A}$ and for any function g growing no faster than $C_1(1+|x|^{m_1})$,*

$$\sup_{t\ge 0} |\mathsf{E}_x g(X_t^\alpha)| \le C(1+|x|^m). \tag{10}$$

- *For any $\alpha \in \mathcal{A}$, the (unique) invariant measure μ^α integrates any polynomial and*

$$\sup_{\alpha\in\mathcal{A}} \int |x|^k \,\mu^\alpha(dx) < \infty, \quad \forall\, k > 0. \tag{11}$$

- *For any strategy $\alpha \in \mathcal{A}$ the function ρ^α is a constant, and*

$$\sup_{\alpha\in\mathcal{A}} |\rho^\alpha| \le C < \infty; \tag{12}$$

moreover, for any $k > 0$ and $f \in \mathcal{K}$ there exist $C, k_0 > 0$ such that

$$\sup_{\alpha\in\mathcal{A}} \left|\mathbb{E}_x f^\alpha(X_t^\alpha) - \rho^\alpha\right| \le C\frac{1+|x|^{k_0}}{1+t^k}, \tag{13}$$

and

$$\sup_{\alpha\in\mathcal{A}} \left|\frac{1}{T}\int_0^T \mathbb{E}_x f^\alpha(X_t^\alpha)\,dt - \rho^\alpha\right| \to 0, \quad T\to\infty. \tag{14}$$

Proof Follows from the calculus quite similar to that in [23] and [21]. □

Corollary 1 *Under the same assumptions,*

$$\sup_{t\ge 0} \mathsf{E}_x 1(|X_t^\alpha| > N) \le \sup_{t\ge 0} \mathsf{E}_x \frac{|X_t^\alpha|^m}{N^m} \le \frac{C(1+|x|^m)}{N^m}. \tag{15}$$

Proof is straightforward by Markov's inequality. □

Lemma 2 *Let the assumptions (A1)–(A3) be satisfied. Then for any strategy $\alpha \in \mathcal{A}$ the cost function v^α has the following properties:*

1. *The function v^α is continuous as well as (∇v^α), and there exist $C, m > 0$ both depending only on the constants in (A1)–(A3) such that*

$$\sup_\alpha (|v^\alpha(x)| + |\nabla v^\alpha(x)|) + |\Delta v^\alpha(x)|) \le C(1+|x|^m). \tag{16}$$

2. *$v^\alpha \in W^2_{p,loc}$ for any $p \ge 1$.*
3. *$v^\alpha \in C^{1,Lip}$ (i.e., (∇v^α) is locally Lipschitz).*
4. *v^α satisfies a Poisson equation in the whole space,*

$$L^\alpha v^\alpha + f^\alpha - \langle f^\alpha, \mu^\alpha\rangle = 0, \tag{17}$$

in the Sobolev sense; in particular, for almost every $x \in \mathbb{R}$

$$L^\alpha(x) v^\alpha(x) + f^\alpha(x) - \langle f^\alpha, \mu^\alpha\rangle = 0, \tag{18}$$

or, equivalently,

$$\frac{1}{\sigma^2(\alpha)} \left(L^\alpha v^\alpha + f^\alpha - \langle f^\alpha, \mu^\alpha\rangle\right) = 0. \tag{19}$$

5. *Solution of the Eq. (17) is unique up to an additive constant in the class of Sobolev solutions $W^2_{p,loc}$ with any $p > 1$ with a no more than some (any) polynomial growth of the solution v^α.*
6. *$\langle v^\alpha, \mu^\alpha\rangle = 0$.*

Proof Denote

$$\bar{L}^u = \frac{1}{\sigma^2(u,x,z)} L^u, \quad \bar{L}^\alpha = \frac{1}{\sigma^2(\alpha(x,z),x,z)} L^\alpha,$$

Let $\gamma(t)$ denote the inverse function for the mapping

$$t \mapsto \int_0^t \sigma^2(\alpha(X_s^\alpha), X_s^\alpha)\, ds =: \phi^\alpha(t),$$

(in what follows we will drop the index α) that is,

$$\gamma(t) = (\phi^\alpha)^{-1}(t),$$

and let

$$\bar{X}_t := X^\alpha_{\gamma(t)}.$$

This random time change $t \mapsto \gamma(t)$—see, for example, [11, Theorem 15.5],—tells us that the process $\bar{X}^\alpha_t$ satisfies the following SDE

$$d\bar{X}_t = \bar{b}(\alpha(\bar{X}_t), \bar{X}_t)\, dt + d\bar{W}_t, \quad \bar{X}^\alpha_0 = x, \tag{20}$$

with a new Wiener process $\bar{W}_t = \int_0^{\gamma(t)} \sigma(\alpha(X^\alpha_s), X^\alpha_s)\, dW_s$ and with the drift $\bar{b}^\alpha(x) = b^\alpha(x)\sigma^{-2}(\alpha(x), x)$. The process $(\bar{X}^\alpha)$ is unique in distribution (more than that, it is, actually, also pathwise unique), ergodic with a unique invariant measure $\bar{\mu}^\alpha(dx)$ and there is a convergence better than any polynomial to it in total variation. In fact, the Lemma 1 is applicable straighforwardly to the process $\bar{X}$, so that the statements similar to (10)–(14) hold true for the process $\bar{X}$ as well.

Notice that

$$\bar{X}_t = X_{\phi_t^{-1}}, \quad \phi(t) = \int_0^t \sigma^2(\alpha(X^\alpha_s), X^\alpha_s)\, ds.$$

Define

$$\psi_t := \int_0^t \sigma^{-2}(\bar{X}_r)dr, \quad \tilde{X}_t := \bar{X}_{\psi_t^{-1}}.$$

It may be seen that $\tilde{X}_t = X_t$; as an immediate corollary, we obtain that the law of X is unique. The key is to show that for any t

$$\psi_t^{-1} = \phi_t \quad \sim \quad t = \psi \circ \phi_t = \phi^{-1} \circ \psi_t^{-1},$$

then it would follow that

$$\tilde{X}_t = \bar{X}_{\psi_t^{-1}} = X_{\phi^{-1} \circ \psi_t^{-1}} = X_t.$$

Indeed, we have,

$$\psi \circ \phi_t = \int_0^{\phi_t} \sigma^{-2}(\bar{X}_r)dr \overset{r=\phi_s^{-1}}{=} \int_{s=0}^{\phi^{-1}\circ\phi_t} \sigma^{-2}(\bar{X}_{\phi_s^{-1}})d\phi_s^{-1}$$

$$= \int_{s=0}^{t} \sigma^{-2}(\bar{X}_{\phi_s^{-1}})\sigma^2(\bar{X}_{\phi_s^{-1}})ds = t,$$

as required. The Lemma 2 follows.

This time change allows to rewrite the definition of v^α as follows,

$$v^\alpha(x) = \int_0^\infty \mathsf{E}_x(f^\alpha(X_t^\alpha) - \rho^\alpha)\, dt = \int_0^\infty \mathsf{E}_x \bar{f}^\alpha(\bar{X}_t^\alpha)\, dt, \tag{21}$$

with

$$\bar{f}^\alpha(x) = \frac{f^\alpha(x) - \rho^\alpha}{\sigma^2(\alpha(x), x)} \equiv \hat{f}^\alpha(x) - \frac{\rho^\alpha}{\sigma^2(\alpha(x), x)}.$$

Indeed, using the random change of variables (see, e.g., [11]) we have

$$v^\alpha(x) = \int_0^\infty \mathsf{E}_x(f^\alpha(X_t^\alpha) - \rho^\alpha)\, dt = \int_0^\infty \mathsf{E}_x \bar{f}^\alpha(X_t^\alpha)\sigma^2(\alpha(X_t^\alpha), X_t^\alpha)\, dt$$

$$= \int_0^\infty \mathsf{E}_x \bar{f}^\alpha(\bar{X}_{\phi^\alpha(t)})\sigma^2(\alpha(\bar{X}_{\phi^\alpha(t)}), \bar{X}_{\phi^\alpha(t)})\, d\phi^\alpha(t) = \int_0^\infty \mathsf{E}_x \bar{f}^\alpha(\bar{X}_t^\alpha)\, dt.$$

Hence, the only option for the integral in (21) to converge is that $\bar{f}^\alpha$ satisfies a centering condition of this function with respect to the measure $\bar{\mu}^\alpha$:

$$\langle \bar{f}^\alpha, \bar{\mu}^\alpha \rangle = 0.$$

Now it follows from (21) in a standard way (see, for example, [21]) that the function v^α is a solution of the Poisson equation

$$\bar{L}^\alpha v^\alpha + \bar{f}^\alpha = 0; \tag{22}$$

recall that here

$$\bar{L}^u g(x) := \bar{b}(u, x)\nabla g(x) + \frac{1}{2}\Delta g(x), \quad x \in \mathbb{R}^d,$$

and

$$\bar{L}^\alpha g(x) := \bar{b}(\alpha(x), x)\nabla g(x) + \frac{1}{2}\Delta g(x).$$

Let us show existence of derivatives v^α_x and v^α_{xx} in a Sobolev sense in the spaces $W^2_{p,loc}$ with any $p > 1$ (we use the standard notations from [19]). For this aim let us denote

$$v^{(T)}(s,x) := \int_0^{T-s} \mathsf{E}_x \bar f^\alpha(\bar X^\alpha_t)\,dt, \qquad 0 \le s \le T,$$

for any $T > 0$. In fact, it turns out to be a bit easier to prove a formally stronger claim than just existence of v^α_x and v^α_{xx}, namely, that

$$(v^{(T)}_x, v^{(T)}_{xx})|_{t=0} \underset{T\to\infty}{\rightarrow} (v^\alpha_x, v^\alpha_{xx}) \quad \text{in the } L_{p,loc} \text{ sense,}$$

or, a bit more precisely, that the left hand side here converges in $L_{p,loc}$, and the limit turns out to be $(v^\alpha_x, v^\alpha_{xx})$. The functions $v^{(T)}$ have been introduced for this aim. In the sequel the following short notations will be used:

$$\bar P_t \bar f^\alpha(x) := \mathsf{E}_x(\bar f^\alpha(\bar X^\alpha_t)).$$

We have,

$$\begin{aligned}
v^{(T)}_x(0,x) &= \partial_x \int_0^T \bar P_t \bar f^\alpha(x)dt = \partial_x \int_0^1 \bar P_t \bar f^\alpha(x)\,dt + \partial_x \int_1^T \bar P_t \bar f^\alpha(x)dt \\
&= \partial_x \int_0^1 \bar P_t \bar f^\alpha(x)dt + \partial_x \int_1^T \bar P_1 \bar P_{t-1} \bar f^\alpha(x)\,dt \\
&= \partial_x \int_0^1 \bar P_t \bar f^\alpha(x)dt + \int_1^T \partial_x \bar P_1 (\bar P_{t-1} \bar f^\alpha)(x)dt.
\end{aligned} \tag{23}$$

The first term here does not change with time (as far as $T \ge 1$) and is well-defined in $L_{p,loc}$ Sobolev sense along with $\partial^2_x \int_0^1 \bar P_t \bar f^\alpha(x)dt$, due to [22, Theorem 5.5, 5.7].

The integrand in the second term admits the bound (see [22, Theorems 5.7 & 5.5] with $T_1 = \epsilon > 0$),

$$\|\bar P_\bullet(\bar P_{t-1}\bar f^\alpha)\|_{W^2_p((\epsilon,1)\times B_R)} \le C_\epsilon(\|\bar P_\bullet(\bar P_{t-1}\bar f^\alpha)\|_{L_p((\epsilon,1)\times B_{R+1})})$$

It follows from ergodic bounds of the SDE solutions similar to those in [23] that the right hand side here decreases to zero faster than any polynomial in time, and, hence, the second integral in the representation (23) converges as $T \to \infty$; this convergence is locally uniform with respect to the initial value x. The same is also true for any partial derivative of the second order $\partial^2_x \int_1^\infty \bar P_t \bar f^\alpha(x))dt$. Thus, we obtain,

$$\nabla_x \int_0^T \bar{P}_t \bar{f}^\alpha(x))dt \to \bar{v}_1(x), \quad \partial_x^2 \int_0^T \bar{P}_t \bar{f}^\alpha(x))dt \to \bar{v}_2(x)$$

as $T \to \infty$, both locally uniformly in x and, hence, also in the $L_{p,loc}$ sense. In a standard manner by integration over x it follows that $\bar{v}_1(x)$ and $\bar{v}_2(x)$ serve as ∇v^α and $(v^\alpha_{x_i x_j})$, respectively. Namely, for any x_1, x_1' we have by the first theorem of the calculus (also known as Newton–Leibnitz formula),

$$v^T(0, x_1, x_2, \ldots) - v^T(0, x_1', x_2, \ldots) = \int_{x_1}^{x_1'} v^T_{x_1}(0, z, x_2, \ldots)\, dz$$

and in the limit as $T \to \infty$ we obtain

$$v^\alpha(x_1, x_2, \ldots) - v^\alpha(x_1', x_2, \ldots) = \int_{x_1}^{x_1'} \bar{v}_1^1(z, x_2, \ldots)\, dz,$$

which means exactly that $\bar{v}_1^1 = v^\alpha_{x_1}$. Similarly for any other variable x_i, and similarly for the second order derivatives: for example,

$$v^T_{x_1}(0, x_1, x_2, \ldots) - v^T_x(0, x_1', x_2, \ldots) = \int_{x_1}^{x_1'} v^T_{x_1 x_1}(0, z, x_2, \ldots)\, dz,$$

and in the limit as $T \to \infty$ we obtain

$$v^\alpha_{x_1}(x_1, x_2, \ldots) - v^\alpha_{x_1}(x_1', x_2, \ldots) = \int_{x_1}^{x_1'} \bar{v}_2^{11}(z, x_2, \ldots)\, dz,$$

which means that $\bar{v}_2^{11} = v^\alpha_{x_1 x_1}$ (a.e.) due to the Lebesgue differentiation theorem, as required. So, indeed, $v^\alpha \in W^2_{p,loc}$.

Now let us show that the function u from the previous step satisfies in the generalised sense the Eq. (22) and, hence, also the equivalent one (17). Indeed, for any smooth test function $g(x)$ with a compact support we have,

$$\langle \bar{L}^\alpha v^\alpha, g\rangle = \langle v^\alpha, \bar{L}^* g\rangle = \lim_{T\to\infty} \langle v^{(T)}|_{t=0}, (\bar{L}^\alpha)^* g\rangle = \lim_{T\to\infty} \langle \bar{L}^\alpha v^{(T)}|_{t=0}, g\rangle$$

$$= \lim_{T\to\infty} \langle \int_0^T \bar{L}^\alpha \mathsf{E}_\bullet \bar{f}^\alpha(\bar{X}_t^\alpha)dt, g\rangle = \lim_{T\to\infty} \langle \int_0^T \partial_t \mathsf{E}_\bullet \bar{f}^\alpha(\bar{X}_t^\alpha)dt, g\rangle$$

$$= \lim_{T\to\infty} \langle \mathsf{E}_\bullet \bar{f}^\alpha(\bar{X}_T^\alpha) - \bar{f}^\alpha(\cdot), g\rangle = -\langle \bar{f}^\alpha, g\rangle,$$

which means that u is a generalised solution of (22); the last equality in this calculus is because of the ergodic properties of the process $(\bar{X}_t^{\alpha})$ from the Lemma 1 and due to the centering property of $\bar{f}^{\alpha}$. Now, since v^{α}, actually, possesses two Sobolev derivatives in x, this function is not just a generalised, but a true Sobolev solution to the Eq. (22), or, equivalently, (17).

The bound

$$\sup_{\alpha} |v^{\alpha}(x)| \le C(1+|x|^m) \tag{24}$$

follows from the inequalities (14) and (11).

The estimate

$$\sup_{\alpha}(|v^{\alpha}(x)| + |\nabla v^{\alpha}(x)|) \le C(1+|x|^m) \tag{25}$$

follows from (24) along with the a priori bound (see, e.g., [13, Theorem 9.11])

$$\|v\|_{W_p^2(B)} \le C(\|v\|_{L_p(B')} + \|\bar{f}^{\alpha}\|_{L_p(B')}), \quad \forall\, B \subset\subset B' \subset \mathbb{R}^d,$$

and from the embedding theorems (see, e.g., [19, lemma 2.3.3]).

The bound (16)

$$\sup_{\alpha}(|v^{\alpha}(x)| + |\nabla v^{\alpha}(x)|) + |\Delta v^{\alpha}(x)|) \le C(1+|x|^m)$$

now follows from (25) and from the Eq. (22).

Now let us show uniqueness of solution of the linear Poisson equation (system)

$$L^{\alpha}v(x) + f^{\alpha}(x) - \rho^{\alpha} = 0$$

in the Sobolev sense up to a constant in the class of functions growing no faster than some polynomial in x. (Clearly, if v is a solution, then $v + C$ is also a solution for any constant C.) Suppose there is a solution v and let us add a constant to it so that to make it centered, i.e.,

$$\langle v, \mu^{\alpha} \rangle = 0.$$

Let us use Itô–Krylov's formula:

$$\mathsf{E}_x v(X_T^{\alpha}) - v(x) = \mathsf{E}_x \int_0^T L^{\alpha} v(X_t^{\alpha}) dt.$$

Since $\mathsf{E}_x v(X_T^\alpha) \to \langle v, \mu^\alpha \rangle = 0$ as $T \to \infty$, then in the limit we get (see (21))

$$v(x) = \int_0^\infty \mathsf{E}_x(f^\alpha(X_t^\alpha) - \rho^\alpha)\,dt = \int_0^\infty \mathsf{E}_x \bar{f}^\alpha(\bar{X}_t^\alpha)\,dt. \tag{26}$$

This shows uniqueness of solution of the Poisson system (17) in the described class of functions.

Now when it has been established that the solution v of the Eq. (17) equals v^α, and since this right hand side satisfies polynomial growth restrictions, the property (16) follows from a similar lemma for an equation in the 1D case (see [2, Lemma 1]).

We have already seen that the centered version of the solution to (17) is equal to v^α; so, v^α itself is μ^α-centered. Equivalently the equality $\langle v^\alpha, \mu^\alpha \rangle = 0$ follows from integration of the right hand side in the definition of v^α with respect to μ^α, due to the centering property of the function $\bar{f}^\alpha$. The Lemma 2 follows. □

In what follows Λ stands for the Lebesgue measure in $\mathbb{R}^d$.

Lemma 3 *Let the assumptions (A1)–(A2) hold true. Then*

$$\Lambda \ll \bar{\mu}^\alpha. \tag{27}$$

In other words, the measure Λ is absolutely continuous with respect to μ^α.

Proof We have, by Chapman–Kolmogorov's equation, where $Q^\alpha(x, dx')$ is the transition kernel of the process $\bar{X}$,

$$\langle g, \bar{\mu}^\alpha \rangle = \int g(x)\mu^\alpha(dx) = \iint g(x')\bar{\mu}^\alpha(dx)Q^\alpha(x, dx')$$
$$= \int \mathsf{E}_x g(\bar{X}_1)\bar{\mu}^\alpha(dx).$$

Let

$$\rho := \exp\left(-\int_0^1 b(\alpha(\bar{X}_s), \bar{X}_s)dW_s - \frac{1}{2}\int_0^1 |b(\alpha(\bar{X}_s), \bar{X}_s)|^2 ds\right).$$

Since b is bounded, by Girsanov's theorem ρ is a probability density [12]. Consider a new probability measure equivalent to P,

$$\tilde{\mathsf{P}}(A) := \mathsf{E}\rho 1(A).$$

We have,

$$\mathsf{E}_x g(\bar{X}_1) = \tilde{\mathsf{E}}_x g(\bar{X}_1)\rho^{-1} = \tilde{\mathsf{E}}_x g(\bar{X}_1) \exp\left(\int_0^1 b(\alpha(\bar{X}_s), \bar{X}_s) dW_s + \frac{1}{2}\int_0^1 |b(\alpha(\bar{X}_s), \bar{X}_s)|^2 ds\right)$$
$$= \tilde{\mathsf{E}}_x g(\bar{X}_1) \exp\left(\int_0^1 b(\alpha(\bar{X}_s), \bar{X}_s) d\tilde{W}_s - \frac{1}{2}\int_0^1 |b(\alpha(\bar{X}_s), \bar{X}_s)|^2 ds\right),$$

where

$$\tilde{W}_t := W_t + \int_0^t b(\alpha(\bar{X}_s), \bar{X}_s) ds$$

is a new Wiener process on the [0, 1] time interval with respect to the measure $\tilde{\mathsf{P}}$, and respectively,

$$W_t = \tilde{W}_t - \int_0^t b(\alpha(\bar{X}_s), \bar{X}_s) ds.$$

Further, by virtue of the Cauchy–Bouniakowskii–Schwarz inequality $\mathsf{E}\xi\eta \geq (\mathsf{E}\sqrt{\xi})^2(\mathsf{E}\eta^{-1})^{-1}$ (assuming $\xi, \eta \geq 0$ and $\mathsf{E}\eta^{-1} \neq 0$), we estimate for any $R > 0$,

$$\mathsf{E}_x g(\bar{X}_1) = \tilde{\mathsf{E}}_x g(\bar{X}_1) \exp(\int_0^1 b(\alpha(\bar{X}_s), \bar{X}_s) d\tilde{W}_s - \frac{1}{2}\int_0^1 |b(\alpha(\bar{X}_s), \bar{X}_s)|^2 ds)$$
$$\geq \left(\tilde{\mathsf{E}}_x g^{1/2}(\bar{X}_1)\right)^2 \left(\tilde{\mathsf{E}}_x \exp(-\int_0^1 b(\alpha(\bar{X}_s), \bar{X}_s) d\tilde{W}_s + \frac{1}{2}\int_0^1 |b(\alpha(\bar{X}_s), \bar{X}_s)|^2 ds)\right)^{-1}$$
$$\geq \left(\tilde{\mathsf{E}}_x g^{1/2}(\bar{X}_1) 1(|\bar{X}_1| \leq R)\right)^2 \left(\tilde{\mathsf{E}}_x \exp(-\int_0^1 b(\alpha(\bar{X}_s), \bar{X}_s) d\tilde{W}_s + \frac{1}{2}\int_0^1 |b(\alpha(\bar{X}_s), \bar{X}_s)|^2 ds)\right)^{-1}.$$

(Recall that only nonnegative g are considered.) Here by virtue of Girsanov's theorem we have,

$$\tilde{\mathsf{E}}_x \exp(-\int_0^1 b(\alpha(\bar{X}_s), \bar{X}_s)d\tilde{W}_s + \frac{1}{2}\int_0^1 |b(\alpha(\bar{X}_s), \bar{X}_s)|^2 ds)$$

$$\leq \exp(\|b\|^2)\,\tilde{\mathsf{E}}_x \exp(-\int_0^1 b(\alpha(\bar{X}_s), \bar{X}_s)d\tilde{W}_s - \frac{1}{2}\int_0^1 |b(\alpha(\bar{X}_s), \bar{X}_s)|^2 ds)$$

$$= \exp(\|b\|^2).$$

So,

$$\mathsf{E}_x g(\bar{X}_1) = \tilde{\mathsf{E}}_x g(\bar{X}_1)\exp(\int_0^1 b(\alpha(\bar{X}_s), \bar{X}_s)d\tilde{W}_s - \frac{1}{2}\int_0^1 |b(\alpha(\bar{X}_s), \bar{X}_s)|^2 ds)$$

$$\geq \left(\tilde{\mathsf{E}}_x g^{1/2}(\bar{X}_1) 1(|\bar{X}_1| \leq R)\right)^2 \exp(-\|b\|^2),$$

or, equivalently,

$$\left(\tilde{\mathsf{E}}_x g^{1/2}(\bar{X}_1) 1(|\bar{X}_1| \leq R)\right)^2 \leq \mathsf{E}_x g(\bar{X}_1)\exp(\|b\|^2).$$

Therefore,

$$\langle g, \bar{\mu}^\alpha \rangle = \int \mathsf{E}_x g(\bar{X}_1)\bar{\mu}^\alpha(dx) \geq \exp(-\|b\|^2)$$

$$\times \int \left(\tilde{\mathsf{E}}_x g^{1/2}(\bar{X}_1) 1(|\bar{X}_1| \leq R)\right)^2 \mu^\alpha(dx)$$

$$\geq \exp(-\|b\|^2) \int \left(\tilde{\mathsf{E}}_x g^{1/2}(\bar{X}_1) 1(|\bar{X}_1| \leq R)\right)^2 1(|x| \leq R)\mu^\alpha(dx)$$

$$\geq \exp(-\|b\|^2) \int \left(\int_{|x'|\leq R} g^{1/2}(x') \frac{1}{(2\pi)^{d/2}} \exp(-(x'+x)^2/2)\right)^2$$

$$\times 1(|x| \leq R)\mu^\alpha(dx)$$

$$\geq \exp(-\|b\|^2) \int \left(\int_{|x'|\leq R} g^{1/2}(x') \frac{1}{(2\pi)^{d/2}} \exp(-(2R)^2/2)dx'\right)^2$$

$$\times 1(|x| \leq R)\mu^\alpha(dx)$$

$$= \exp(-\|b\|^2) \frac{1}{(2\pi)^d} \exp(-(2R)^2) \left(\int_{|x'|\leq R} g^{1/2}(x')dx'\right)^2 \int 1(|x| \leq R)\mu^\alpha(dx).$$

In other words,

$$\left(\int_{|x'|\le R} g^{1/2}(x')dx'\right)^2 \le \langle g, \bar\mu^\alpha\rangle \exp(\|b\|\sqrt{3})\frac{1}{(2\pi)^d}\exp((2R)^2)\mu^\alpha(|x|\le R)^{-1}$$

Choose $R > 0$ so large that such that $\mu^\alpha(|x| \le R) \ge 1/2$ for any α. Then

$$\left(\int_{|x'|\le R} g^{1/2}(x')dx'\right)^2 \le 2\langle g, \bar\mu^\alpha\rangle \exp(\|b\|^2)\frac{1}{(2\pi)^d}\exp((2R)^2). \tag{28}$$

So, in particular, (27) holds. Hence, the Lemma 3 is proved. □

Remark 1 Note that a similar upper bound can be established for the expression $\langle g, \bar\mu^\alpha\rangle$, which implies that, in fact, $\Lambda \sim \bar\mu^\alpha$. However, this will not be used in what follows.

Corollary 2 *Under the assumptions of the Lemma 3, let the sequence* $(g_n \ge 0, n \ge 1)$ *satisfy the bound* $\sup_n g_n(x) \le C(1+|x|^\ell)$ *with the same* C *and* ℓ *for any* x, *and*

$$\langle g_n, \bar\mu^\alpha\rangle \to 0, \quad n \to \infty.$$

Then for any $R > 0$ *and for any* $m \ge 1$,

$$\int_{|x'|\le R} g_n^m(x')dx' \to 0, \quad n \to \infty. \tag{29}$$

Proof We have due to (28), uniformly in n,

$$\int_{|x'|\le R} g_n^m(x')dx' \le \left(2\langle g_n^{2m}, \bar\mu^\alpha\rangle \exp(\|b\|^2)\frac{1}{(2\pi)^{d/2}}\exp((2R)^2)\right)^{1/2}. \tag{30}$$

Note that under the assumptions of the Corollary,

$$\langle g_n^{2m}, \bar\mu^\alpha\rangle \to 0, \quad n \to \infty. \tag{31}$$

Indeed, by virtue of the Lemma 1, inequaity (11), for any $\delta > 0$ the value $R_0 > 0$ can be chosen so large that for any $R \ge R_0$,

$$\int 1(|x|) > R)C^{2m}(1+|x|)^{2m\ell}\bar\mu^\alpha(dx) < \delta, \tag{32}$$

where C is the constant from the assumption $|g_n(x)| \le C(1+|x|)^\ell$. Hence, it implies

$$\sup_n \int 1(|x|) > R)g_n^{2m}(x)\bar\mu^\alpha(dx) < \delta. \tag{33}$$

For the complementary indicator under the integral, for a fixed value of R we have,

$$\int 1(|x| < R) g_n^{2m}(x)\bar{\mu}^\alpha(dx) \le C(1+R)^{(2m-1)\ell}\langle g_n, \bar{\mu}^\alpha\rangle \to 0, \quad n \to \infty. \tag{34}$$

From (33) and (34) the claim (29) follows. The Corollary 2 is proved. □

The sequence ψ_n (or even ψ_{n_j}) for n large enough (or n_j large enough) will be used in the proof of the Theorem as a family of such functions (g) : firstly, we will show that $\langle \psi_n, \mu^\alpha\rangle$ is small enough for n large (see (47) below), and then we deduce from the Corollary 2 that the value $\sup_{n\ge n(\epsilon)} \int_{|x|\le R} \psi_n^d(x)\,dx$ is also small (see (50) in what follows). The latter will be useful for the application of one of Krylov's estimates.

3 Main Results

Repeat that (weak) solution of the SDE (1) with any Markov strategy does exist and is unique in distribution, strong Markov and Markov ergodic. All these claims follow from [17] and from the assumptions (A1) and (A2) (see [24] about ergodicity).

For any couple $(v, \rho) : \ v \in \bigcap_{p>1} W^2_{p,loc}, \ \rho \in \mathbb{R}$, let

$$F[v,\rho](x) := \inf_{u\in U}\left[L^u v(x) + f^u(x) - \rho\right], \quad G[v](x) := \inf_{u\in U}\left[L^u v(x) + f^u(x)\right],$$

and

$$F_1[v', \rho](x) := \inf_{u\in U}[\hat{b}^u v' + \hat{f}^u - \hat{\rho}](x),$$

where

$$a^u(x) = \frac{1}{2}(\sigma^u(x))^2, \quad \hat{b}^u(x) = b^u(x)/a^u(x),$$

$$\hat{f}^u(x) = f^u(x)/a^u(x), \quad \hat{\rho}^u(x) = \rho/a^u(x).$$

The functions v and ∇v may be assumed continuous and absolutely continuous due to the embedding theorems [19, lemma 2.3.3]. The function $F[v,\rho](\cdot)$ is defined by the formula above as a function of the class $L_{p,loc}$ for any $p > 1$; it is Lebesgue measurable and is defined almost everywhere (a.e.) in x. We may and will use a (any) Borel measurable version of it. It will turn out in what follows that the function $F_1[v', \rho](x)$ is continuous in x and locally Lipschitz in the two other variables.

Here is a reminder of what is called the reward improvement algorithm (RIA). Let us choose any initial strategy $\alpha_0 \in \mathcal{A}$. Denote the corresponding cost, the invariant measure, and the auxiliary function by $\rho_0 = \rho^{\alpha_0} = \langle f^{\alpha_0}, \mu^{\alpha_0}\rangle$, and $v_0 = v^{\alpha_0}$, respectively. If for some $n = 0, 1, \dots$ the triple (α_n, ρ_n, v_n) is determined, then the strategy α_{n+1} is defined as follows: for a.e. x the function α_{n+1} is chosen so that for each x

$$\alpha_{n+1}(x) \in \operatorname{Argmin}_{u\in U}\left[L^u v_n(x) + f^u(x)\right].$$

In other words,

$$L^{\alpha_{n+1}} v_n(x) + f^{\alpha_{n+1}}(x) = G[v_n](x), \tag{35}$$

We *assume* that a Borel measurable version of such strategy may be chosen; see the reference in [2, Appendix]. To this strategy α_{n+1} there correspond the invariant measure $\mu^{\alpha_{n+1}}$, the value $\rho_{n+1} := \langle f^{\alpha_{n+1}}, \mu^{\alpha_{n+1}}\rangle$, and the function $v_{n+1} = v^{\alpha_{n+1}}$.

Theorem 1 *Let the assumptions (A1)–(A4) be satisfied. Then:*

1. *For any n, $\rho_{n+1} \le \rho_n$, and there is a limit $\rho_n \downarrow \tilde\rho$.*
2. *The sequence (v_n) is compact in $C^1[-N, N]$ for each $N > 0$, and there exists a bounded sequence of constants β_n such that there is a limit* $\lim_n (v_n(x) + \beta_n) =: \tilde v(x)$.
3. *The couple $(\tilde v, \tilde\rho)$ solves the Eq. (7).*
4. *This solution $(\tilde v, \tilde\rho)$ is unique – up to an additive constant for $\tilde v$—in the class of functions growing no faster than some (any) polynomial and belonging to the class $W^2_{p,loc}$ for any $p > 0$ for the first component and for $\tilde\rho \in \mathbb{R}$.*
5. *The component $\tilde\rho$ in the couple $(\tilde v, \tilde\rho)$ coincides with ρ.*

NB. Such a special organisation of the set of claims was chosen because in this way it corresponds to the natural order in the proof of this theorem. In the first instance it will be shown that there exists a limit, say, $\tilde\rho$, and only in the end of the proof it will follow that this limit equals our cost value ρ. Of course, logically the first claim in a combination with the last one could have been re-phrased as $\rho_n \downarrow \rho$. On the other hand, the numbering of the steps in the proof does not correspond exactly to the numbering of the claims of the theorem, and even these two numbers are different.

Proof

1. Due to (35) and (17), for almost every (a.e.) $x \in \mathbb{R}^d$,

$$\rho_n = L^{\alpha_n} v_n(x) + f^{\alpha_n}(x) \ge G[v_n](x) = L^{\alpha_{n+1}} v_n(x) + f^{\alpha_{n+1}}(x),$$

and also for a.e. $x \in \mathbb{R}^d$,

$$\rho_{n+1} = L^{\alpha_{n+1}} v_{n+1}(x) + f^{\alpha_{n+1}}(x).$$

The first equality in the former formula line and the latter equation here are nothing else but Poisson equation in the whole space for centered $f^{\alpha_n} - \rho_n$ and $f^{\alpha_{n+1}} - \rho_{n+1}$, respectively, see (17) in the Lemma 2. The second inequality in the former line is due to the definition of G. So, by subtracting, we get

$$\rho_n - \rho_{n+1} \overset{a.e.}{\geq} (L^{\alpha_{n+1}} v_n + f^{\alpha_{n+1}})(x) - (L^{\alpha_{n+1}} v_{n+1} + f^{\alpha_{n+1}})(x) \qquad (36)$$
$$= (L^{\alpha_{n+1}} v_n - L^{\alpha_{n+1}} v_{n+1})(x).$$

Let us apply Ito–Krylov's formula (see [16, Chapter 2]) with expectations, also known as Dynkin's formula, to $(v_n - v_{n+1})(X_t^{\alpha_{n+1}})$: we have for any $x \in \mathbb{R}^d$,

$$\begin{aligned} &\mathsf{E}_x \left(v_n(X_t^{\alpha_{n+1}}) - v_{n+1}(X_t^{\alpha_{n+1}})\right) - (v_n - v_{n+1})(x) \\ &= \mathsf{E}_x \int_0^t (L^{\alpha_{n+1}} v_n - L^{\alpha_{n+1}} v_{n+1})(X_s^{\alpha_{n+1}})\, ds \\ &\leq \mathsf{E}_x \int_0^t (\rho_n - \rho_{n+1})\, ds = (\rho_n - \rho_{n+1})\, t. \end{aligned} \qquad (37)$$

Why does it hold for any x: since the functions $v_n \in C$ due to the embedding theorems [19] as Sobolev solutions of Poisson equations, and because $\mathsf{E}_x v_n(X_t^{\alpha_{n+1}})$ and $\mathsf{E}_x v_{n+1}(X_t^{\alpha_{n+1}})$ as functions of x for each $t > 0$ are both Hölder continuous, being solutions of non-degenerate parabolic equations [18]. We also used the fact that the distribution of $X_s^{\alpha_{n+1}}$ for almost all $s > 0$ is absolutely continuous with respect to the Lebesgue measure by virtue of Krylov's estimates [16, Chapter 2]. Further, by virtue of the Lemma 2 the left hand side in (37) is bounded for any fixed x. So, after division by t in the last inequality, we obtain that

$$0 \leq \rho_n - \rho_{n+1},$$

as $t \to \infty$, as required. Therefore, $\rho_n \geq \rho_{n+1}$. Since the sequence ρ_n is bounded from below (as well as from above) for $f \in \mathcal{K}$, see (12) in the Lemma 1, we can see that $\rho_n \downarrow \tilde{\rho}$ with some limiting value $\tilde{\rho}$. This shows the first claim of the theorem.

Apparently,

$$\tilde{\rho} \geq \rho, \qquad (38)$$

since ρ is the infimum over all Markov strategies, while $\tilde{\rho}$ is the inf over a subset of them. In fact, they do coincide, which will be shown later.

2. Now we are to justify the second claim of the theorem. Firstly, we will do it for some subsequence n_j.

Let us show local compactness of the family of functions (v_n) in C^1. The Eq. (7) may be rewritten as

$$\frac{1}{2}\sum_i \frac{d^2}{dx_i^2}v(x) + \inf_{u\in U}\left[\frac{b(u,x)}{a(u,x)}\nabla_x v(x) + \frac{f(u,x)}{a(u,x)} - \frac{\rho}{a(u,x)}\right] = 0. \qquad (39)$$

Similarly, the equation

$$L^{\alpha_{n+1}} v_{n+1}(x) + f^{\alpha_{n+1}}(x) - \rho_{n+1} \overset{a.e.}{=} 0, \qquad (40)$$

may be rewritten as

$$\frac{1}{2}\sum_i \frac{d^2}{dx_i^2}v_{n+1}(x) + \frac{b(\alpha_{n+1}(x),x)}{a(\alpha_{n+1}(x),x)}\nabla_x v_{n+1}(x) + \frac{f(\alpha_{n+1}(x),(x))}{a(\alpha_{n+1}(x),x)} - \frac{\rho_{n+1}}{a(\alpha_{n+1}(x),x)} = 0.$$

By the Lemma 2, the functions ∇v_{n+1} are uniformly locally bounded. Since the sequence ρ_{n+1} is bounded, and due to the assumptions about the uniform local boundedness of the functions $f(\alpha_{n+1}(x), x)$ and uniform nondegeneracy of a, the latter equation implies that (Δv_n) are locally uniformly bounded and satisfy the growth bounds similar to (16) uniformly with respect to n. This shows the local compactness of (v_n) in C^1, as required in the second claim of the theorem.

3. Due to the (local) compactness property showed in the previous step, we may choose a converging in C^1_{loc} subsequence of functions $v_{n_j}(x)$. We want to show that the limit is unique, *up to a constant*. First of all, let us establish that if some $v_{n_j}(x)$ has a limit, say, $\tilde{v}(x)$ (locally in C), then there exists some bounded sequence of real values β_{n_j} such that $v_{n_j+1}(x) + \beta_{n_j}$ has the same limit. Recall that we have,

$$L^{\alpha_n} v_n(x) + f^{\alpha_n}(x) - \rho_n \overset{a.e.}{=} 0,$$

and

$$L^{\alpha_{n+1}} v_{n+1}(x) + f^{\alpha_{n+1}}(x) - \rho_{n+1} \overset{a.e.}{=} 0,$$

and

$$L^{\alpha_{n+1}} v_n(x) + f^{\alpha_{n+1}}(x) - \rho_n =: -\psi_{n+1}(x) \overset{a.e.}{\le} 0. \qquad (41)$$

The latter may be re-phrased as

$$L^{\alpha_{n+1}} v_n(x) \stackrel{a.e.}{=} -(f^{\alpha_{n+1}}(x) + \psi_{n+1}(x) - \rho_n),$$

with some *non-negative* function ψ_{n+1}. Hence, the function v_n is a solution of a Poisson equation with the second order operator $L^{\alpha_{n+1}}$ and with the right hand side $-(f^{\alpha_{n+1}}(x) + \psi_{n+1}(x) - \rho_n)$. In the ergodic case this is only possible if this right hand side is centered with respect to the invariant measure μ^{n+1} (cf., e.g., [21]). Therefore,

$$\langle f^{\alpha_{n+1}} + \psi_{n+1} - \rho_n, \mu^{n+1} \rangle = 0.$$

Thus,

$$\langle \psi_{n+1}, \mu^{n+1} \rangle = \rho_n - \rho_{n+1}. \tag{42}$$

By virtue of the Lemma 3 and from the Definition (41) it follows that

$$\psi_n(x) \le C(1 + |x|^m). \tag{43}$$

4. Let

$$w_n(x) := v_n(x) - v_{n+1}(x).$$

Then

$$L^{\alpha_{n+1}} w_n(x) + \psi_{n+1}(x) - (\rho_n - \rho_{n+1}) \stackrel{a.e.}{=} 0.$$

Hence,

$$w_n(x) - \beta_n = \int_0^\infty \mathsf{E}_x(\psi_{n+1}(X_t^{n+1}) - (\rho_n - \rho_{n+1}))\, dt, \tag{44}$$

with a constant $\beta_n = \langle w_n, \mu^{n+1} \rangle$. Using the time change transfromation, we have,

$$\int_0^\infty \mathsf{E}_x(\psi_{n+1}(X_t^{n+1}) - (\rho_n - \rho_{n+1}))\, dt$$

$$= \int_0^\infty \mathsf{E}_x(\psi_{n+1}(\bar{X}_s^{n+1}) - (\rho_n - \rho_{n+1}))\, \sigma^{-2}(\alpha_{n+1}(\bar{X}_s^{n+1}), \bar{X}_s^{n+1}) ds$$

$$=: \int_0^\infty \mathsf{E}_x \phi_{n+1}(\bar{X}_s^{n+1}) ds,$$

where

$$\phi_{n+1}(z) := (\psi_{n+1}(z) - (\rho_n - \rho_{n+1}))\,\sigma^{-2}(\alpha_{n+1}(z), z).$$

By the same reasons as earlier for the process X, we now obtain

$$\langle \psi_{n+1}(\cdot), \bar{\mu}_{n+1}\rangle - (\rho_n - \rho_{n+1})\langle \sigma^{-2}(\alpha_{n+1}(\cdot), \cdot), \bar{\mu}_{n+1}\rangle = \langle \phi_{n+1}, \bar{\mu}_{n+1}\rangle = 0.$$

5. Let us show that for any $N > 0$,

$$\int_{|x|\le N} \psi_n^{2d}(x)\,dx \to 0, \quad n \to \infty. \tag{45}$$

First of all, the functions ψ_n are uniformly locally bounded and possess no more than some polynomial growth:

$$(0 \le)\ \psi_n(x) \le C(1 + |x|^m), \tag{46}$$

with some constants C, m which are the same for all values of n, see the Lemma 3 and Corollary 2. So are the functions ψ_n, of course, with new values of C and m. Recall that $\psi_n \ge 0$ and that

$$\langle \psi_{n+1}, \mu^{n+1}\rangle = \rho_n - \rho_{n+1} \to 0, \quad n \to \infty. \tag{47}$$

For each n the Eq. (44) may be rewritten by using a *stationary version* of our diffusion process X^{n+1}, say, $\tilde{X}_t^{n+1}$:

$$w_n(x) - \beta_n = \int_0^\infty \mathsf{E}_x(\psi_{n+1}(X_t^{n+1}) - E_{\mu^{n+1}}(\psi_{n+1}(\tilde{X}_t^{n+1}))\,dt. \tag{48}$$

6. Using the coupling method (cf., e.g., [23]), let us consider coupled processes X_t^{n+1} and $\tilde{X}_t^{n+1}$ on the same probability space; here "coupled" means that they intersect with a positive probability. We change the notations neither for the probability space $(\Omega, \mathcal{F}, \mathsf{P})$, nor for the processes X_t^{n+1} and $\tilde{X}_t^{n+1}$ in order not to overload this presentation; in fact, the probability space must be a new one, simply because independent non-degenerate diffusions in $\mathbb{R}^d$ do not intersect in the case $d > 1$, while they still may be coupled since their transition kernels are not singular. The construction of the "markovian coupling" may be found, for example, in [23], or in [25]. In the latter paper such a construction is presented for Markov chains; this is not an obstacle because the sequence (X_n) with $n = 0, 1, \ldots$ instead of $t \in [0, \infty)$ is a Markov chain. The (locally uniform) Markov–Dobrushin condition of non-singularity of transition kernels from different initial states—which is a sufficient condition for the markovian coupling construction—follows from the Harnack inequality in [18].

So, let us denote the time of the first meeting of X_t^{n+1} and $\tilde X_t^{n+1}$ by

$$\tau := \inf(t \ge 0 : \; X_t^{n+1} = \tilde X_t^{n+1}).$$

This is a stopping time. Due to the results in [23], under our recurrence assumptions for any $k > 0$ there exist constants C_k, m such that uniformly with respect to n,

$$\mathsf{E}_{x,\mu^{n+1}} \tau^k \le C_k(1 + |x|^m).$$

Let

$$\hat X_t^{n+1} := 1(t < \tau) X_t^{n+1} + 1(t \ge \tau) \tilde X_t^{n+1}.$$

Since the couple $(X_t^{n+1}, \tilde X_t^{n+1})$ is strong Markov (see [15]), the process $(\hat X_t^{n+1})$ is also strong Markov equivalent to (X_t^{n+1}) in the sense of distributions in the space of trajectories. Hence, (48) may be rewritten as follows,

$$w_n(x) - \beta_n = \int_0^\infty \mathsf{E}_{x,\mu} (\psi_{n+1}(\hat X_t^{n+1}) - \psi_{n+1}(\tilde X_t^{n+1}))\, dt.$$

Using the equality

$$\mathsf{P}(\hat X_t^{n+1} = \tilde X_t^{n+1}, \, t \ge \tau) = 1,$$

we obtain,

$$w_n(x) - \beta_n = \int_0^\infty \mathsf{E}_{x,\mu} 1(t < \tau)(\psi_{n+1}(\hat X_t^{n+1}) - \psi_{n+1}(\tilde X_t^{n+1}))\, dt$$

$$= \int_0^\infty \mathsf{E}_{x,\mu} \sum_{i=0}^\infty 1(i \le \tau < i+1) 1(t < \tau)(\psi_{n+1}(\hat X_t^{n+1}) - \psi_{n+1}(\tilde X_t^{n+1}))\, dt$$

$$= \sum_{i=0}^\infty \mathsf{E}_{x,\mu} \int_0^\infty 1(i \le \tau < i+1) 1(t < \tau)(\psi_{n+1}(\hat X_t^{n+1}) - \psi_{n+1}(\tilde X_t^{n+1}))\, dt.$$

Hence, due to Cauchy-Buniakowskii-Schwarz inequality and Fubini Theorem, we get,

$$|w_n(x) - \beta_n| \le \sum_{i=0}^{\infty} \mathsf{E}_{x,\mu} \int_0^{i+1} 1(\tau > i)|\psi_{n+1}(\hat{X}_t^{n+1}) - \psi_{n+1}(\tilde{X}_t^{n+1})|\, dt$$

$$\le \sum_{i=0}^{\infty} \int_0^{i+1} \mathsf{E}_{x,\mu} 1(\tau > i)(|\psi_{n+1}(\hat{X}_t^{n+1})| + |\psi_{n+1}(\tilde{X}_t^{n+1})|)\, dt$$

$$\le \sum_{i=0}^{\infty} \int_0^{i+1} (\mathsf{E}_{x,\mu} 1(\tau > i))^{1/2} (\mathsf{E}_{x,\mu}(|\psi_{n+1}(\hat{X}_t^{n+1})| + |\psi_{n+1}(\tilde{X}_t^{n+1})|)^2)^{1/2}\, dt$$

$$\le 2\sum_{i=0}^{\infty} (\mathsf{E}_{x,\mu} 1(\tau > i))^{1/2} \int_0^{i+1} (\mathsf{E}_{x,\mu}|\psi_{n+1}(\hat{X}_t^{n+1})|^2$$

$$+ \mathsf{E}_{x,\mu}|\psi_{n+1}(\tilde{X}_t^{n+1})|^2)^{1/2}\, dt$$

$$\le 2\sum_{i=0}^{\infty} (\mathsf{E}_{x,\mu} 1(\tau > i))^{1/2} \int_0^{i+1} [(\mathsf{E}_{x,\mu}(\psi_{n+1}(\hat{X}_t^{n+1}))^2)^{1/2}$$

$$+ (\mathsf{E}_{x,\mu}(\psi_{n+1}(\tilde{X}_t^{n+1}))^2)^{1/2}]\, dt.$$

We estimate, with some $\epsilon > 0$ and using the elementary inequality $\sqrt{a} \le \frac{\epsilon}{2} + \frac{a}{2\epsilon}$,

$$\int_0^{i+1} [(\mathsf{E}_{x,\mu}(\psi_{n+1}(\hat{X}_t^{n+1}))^2)^{1/2} + (\mathsf{E}_{x,\mu}(\psi_{n+1}(\tilde{X}_t^{n+1}))^2)^{1/2}]\, dt$$

$$\le \epsilon(i+1) + \frac{1}{2\epsilon} \int_0^{i+1} [\mathsf{E}_{x,\mu}\psi_{n+1}^2(\hat{X}_t^{n+1}) + \mathsf{E}_{x,\mu}\psi_{n+1}^2(\tilde{X}_t^{n+1})]\, dt.$$

For the term with the stationary process we have,

$$
\begin{aligned}
&\frac{1}{2\epsilon}\int_0^{i+1} \mathsf{E}_{x,\mu}\psi_{n+1}^2(\tilde X_t^{n+1})\,dt \\
&= \frac{1}{2\epsilon}\int_0^{i+1} \mathsf{E}_{x,\mu}\psi_{n+1}^2 1_{|x|\le N}(\tilde X_t^{n+1})\,dt \\
&+ \frac{1}{2\epsilon}\int_0^{i+1} \mathsf{E}_{x,\mu}\psi_{n+1}^2 1_{\mathbb{R}^d\setminus B_N}(\tilde X_t^{n+1})\,dt \\
&+ \frac{1}{2\epsilon}\int_0^{i+1} \mathsf{E}_{x,\mu}\psi_{n+1}^2 1_{B_N}(\hat X_t^{n+1})\,dt + \frac{1}{2\epsilon}\int_0^{i+1} \mathsf{E}_{x,\mu}\psi_{n+1}^2 1_{\mathbb{R}^d\setminus B_N}(\hat X_t^{n+1})\,dt.
\end{aligned}
$$

Since any stationary measure integrates (uniformly with respect to α) any power function, and having in mind the bound (46) the Lemmata 1 and 3, we may find such N that the following bound holds uniformly with respect to n:

$$
\langle C(1+|x|^{2m})1_{\mathbb{R}^d\setminus B_N}, \mu^{n+1}\rangle < \epsilon^2/2. \tag{49}
$$

In addition, we may take $N > \epsilon^{-2}$. Given this N, let us choose $n(\epsilon)$ so that

$$
\sup_{n\ge n(\epsilon)} \int_{|x|\le N} \psi_n^{2d}(x)\,dx < \epsilon^2/2. \tag{50}
$$

Now we evaluate with $n \ge n(\epsilon)$ due to Krylov's estimate [16, Chapter 2], or [17],

$$
\begin{aligned}
&\frac{1}{2\epsilon}\int_0^{i+1} \mathsf{E}_{x,\mu}\psi_{n+1}^2 1_{B_N}(\tilde X_t^{n+1})\,dt \\
&= \frac{1}{2\epsilon}\sum_{k=0}^{i} \mathsf{E}_x \int_k^{k+1} \psi_{n+1}^2 1_{B_N}(\tilde X_t^{n+1})\,dt \\
&\le \frac{i+1}{2\epsilon} K \|\psi_{n+1}^2 1_{|x|\le N}\|_{L^d} \le \frac{(i+1)K\epsilon}{2}.
\end{aligned}
$$

For the term with the non-stationary process, also due to Krylov's estimate, we have similarly,

$$\frac{1}{2\epsilon}\int_0^{i+1} \mathsf{E}_{x,\mu}\psi_{n+1}^2 1_{B_N}(\hat{X}_t^{n+1})\,dt$$

$$= \frac{1}{2\epsilon}\sum_{k=0}^{i} \mathsf{E}\int_k^{k+1} \psi_{n+1}^2 1_{B_N}(\hat{X}_t^{n+1})\,dt$$

$$\le \frac{i+1}{2\epsilon} K\|\psi_{n+1}^2 1_{B_N}\|)_{L^d} \le \frac{(i+1)K\epsilon}{2}.$$

Further,

$$\frac{1}{2\epsilon}\int_0^{i+1} \mathsf{E}_{x,\mu}\psi_{n+1}^2 1_{\mathbb{R}^d\setminus B_N}(\tilde{X}_t^{n+1})\,dt \le \frac{i+1}{2\epsilon}\times\frac{\epsilon^2}{2} = \frac{(i+1)\epsilon}{4}.$$

Finally, using (15), we obtain with some m,

$$\frac{1}{2\epsilon}\int_0^{i+1} \mathsf{E}_{x,\mu}\psi_{n+1}^2 1_{\mathbb{R}^d\setminus B_N}(\hat{X}_t^{n+1})\,dt = \frac{1}{2\epsilon}\int_0^{i+1} \mathsf{E}_{x,\mu}\psi_{n+1}^2 1_{\mathbb{R}^d\setminus B_N}(X_t^{n+1})\,dt$$

$$\le C\frac{i+1}{2\epsilon}\frac{(1+|x|^m)}{N} \le C(i+1)(1+|x|^m)\epsilon.$$

Hence, with some uniformly bounded β_n we get the inequality

$$|w_n(x)-\beta_n| \le C(1+|x|^{2m})\epsilon\sum_{i=0}^{\infty}(i+1)(\mathsf{E}_{x,\mu}1(\tau>i))^{1/2}, \quad n\ge n(\epsilon). \tag{51}$$

From the results in [23] it follows that for any $k>0$ there exist constants $C, m>0$ such that

$$\mathsf{P}_{x,\mu}1(\tau>i) \le C\frac{1+|x|^m}{1+i^k}.$$

So, taking just $k>2$, we get a converging series (51), which leads to the estimate

$$|w_n(x)-\beta_n| \le C(1+|x|^{3m})\epsilon, \quad n\ge n(\epsilon). \tag{52}$$

So, the difference $w_n(x)-\beta_n = v_n - v_{n+1} - \beta_n$ tends to zero locally uniformly as $n\to\infty$. Hence, for any subsequence n_j such that v_{n_j} converges locally uniformly in C^1, the functions $\partial_{x_i} v_{n_j}$ and $\partial_{x_i} v_{n_j+1}$ for any $1 \le i \le d$, must converge to the same limit, that is,

$$\partial_{x_i} v_{n_j} - \partial_{x_i} v_{n_j+1} \to 0, \quad j \to \infty,$$

locally uniformly in x.

7. Now consider the equations

$$L^{\alpha_{n_j+1}} v_{n_j+1}(x) + f^{\alpha_{n_j+1}}(x) - \rho_{n_j+1} \stackrel{a.e.}{=} 0, \quad \& \quad G[v_{n_j}](x) - \rho_{n_j} \le 0,$$

Take any sequence $(n_j, j \to \infty)$ such that v_{n_j} converges in C^1 locally uniformly. It follows from

$$G[v_{n_j}](x) - \rho_{n_j} = L^{\alpha_{n_j+1}} v_{n_j}(x) + f^{\alpha_{n_j+1}}(x) - \rho_{n_j}$$
$$(= \inf_{u\in U} [L^u v_{n_j}(x) + f^u(x) - \rho_{n_j}] \stackrel{a.e.}{\le} 0),$$

and (40) that we get almost everywhere

$$G[v_{n_j}](x) - \rho_{n_j} = L^{\alpha_{n_j+1}} (v_{n_j}(x) - v_{n_j+1}(x)) - (\rho_{n_j} - \rho_{n_j+1}). \tag{53}$$

Let us now show that (53) implies the equation

$$\frac{1}{2}\Delta \tilde{v}(x) + F_1[x, \nabla \tilde{v}(x), \tilde{\rho}] = 0, \tag{54}$$

where

$$F_1[x, v_1, r] := \inf_{u\in U} \left[\frac{b(u,x)}{a(u,x)} v_1 + \frac{f(u,x)}{a(u,x)} - \frac{r}{a(u,x)}\right], \quad \forall x,$$

where $v_1 \in \mathbb{R}^d$ and bv_1 stands for the scalar product.

Note that $G[v_{n_j}](x) - \rho_{n_j} \le 0$ (a.e.). Let us divide both sides of (53) by $a_{n_j+1} = a^{\alpha_{n_j+1}}$ and denote for brevity $\delta := \inf_{u,x} a^u(x) > 0$. It follows with some $K > 0$ almost everywhere,

$$0 \ge \frac{(G[v_{n_j}](x) - \rho_n)}{a_{n_j+1}}$$

$$= (\frac{1}{2}\Delta v_{n_j}(x) - \frac{1}{2}\Delta v_{n_j+1}(x)) + (\hat{b}^{\alpha_{n_j+1}}(\nabla v_{n_j} - \nabla v_{n_j+1})) - \frac{(\rho_{n_j} - \rho_{n_j+1})}{a_{n_j+1}}$$

$$\geq (\frac{1}{2}\sum_{j}(D^2_{x_i} v_{n_j}(x) - D^2_{x_i} v_{n_j+1})(x)) - \frac{K}{\delta}|\nabla v_{n_j}(x)$$

$$- \nabla v_{n_j+1}(x)| - \frac{1}{\delta}(\rho_{n_j} - \rho_{n_j+1}). \tag{55}$$

It means that we have almost everywhere,

$$0 \geq (\frac{1}{2}\Delta v_{n_j}(x) - \frac{1}{2}\Delta v_{n_j+1}(x))$$
$$-\frac{K}{\delta}|\nabla v_{n_j}(x) - \nabla v_{n_j+1}(x)| - \frac{\rho_{n_j} - \rho_{n_j+1}}{\delta}. \tag{56}$$

Further, note that again due to (55), and since $\rho_{n_j} \geq \rho_{n_j+1}$, and because $\delta \leq a \leq C$,

$$0 \overset{a.e.}{\geq} G[v_{n_j}](x) - \rho_{n_j} \geq a_{n_j+1}(\frac{1}{2}\Delta v_{n_j} - \frac{1}{2}\Delta v_{n_j+1})(x)$$
$$- C'|v'_{n_j} - v'_{n_j+1}|(x) - (\rho_{n_j} - \rho_{n_j+1}).$$

As a consequence, we obtain with some $C, c > 0$,

$$0 \overset{a.e.}{\geq} \frac{1}{2}\Delta v_{n_j} + F_1[v'_{n_j}, \rho_{n_j}] \geq ((\frac{1}{2}\Delta v_{n_j} - \frac{1}{2}\Delta v_{n_j+1})$$
$$- C|v'_{n_j} - v'_{n_j+1}|) - c(\rho_{n_j} - \rho_{n_j+1}). \tag{57}$$

In other words,

$$0 \overset{a.e.}{\geq} \frac{1}{2}\Delta v_{n_j} + \inf_{u\in U}\left[\frac{b(u,x)}{a(u,x)}\nabla_x v_{n_j}(x) + \frac{f(u,x)}{a(u,x)} - \frac{\rho_{n_j}}{a(u,x)}\right] \tag{58}$$

$$\geq (\frac{1}{2}\Delta v_{n_j} - \frac{1}{2}\Delta v_{n_j+1})(x) - C|\nabla v_{n_j} - \nabla v_{n_j+1}|(x) - c(\rho_{n_j} - \rho_{n_j+1}).$$

Since $C|\nabla v_{n_j} - \nabla v_{n_j+1}|) - c(\rho_{n_j} - \rho_{n_j+1}) \to 0$ locally uniformly, and because $\nabla v_{n_j} \to \nabla\tilde{v}$ and $\nabla v_{n_j+1} \to \nabla\tilde{v}$ also locally uniformly, (58) and (56) prompt that there exists a (locally uniform) limit

$$\lim_{n_j\to\infty}\frac{1}{2}\Delta v_{n_j+1}(x) = -F_1[x, \nabla\tilde{v}(x), \tilde{\rho}(x)],$$

or, for brevity,

$$\lim_{n_j\to\infty} \frac{1}{2}\Delta v_{n_j+1} = -F_1[\nabla \tilde v, \tilde\rho].$$

It is plausible that in the limit we obtain the equation for the function $\tilde v$:

$$\frac{1}{2}\Delta \tilde v = -F_1[\nabla \tilde v, \tilde\rho]. \tag{59}$$

This will be justified in the next step of the proof.

8. For this justification let us rewrite the double inequality (58) as follows:

$$\frac{1}{2}\Delta v_{n_j+1}(x) + \inf_{u\in U}\left[\frac{b(u,x)}{a(u,x)}\nabla_x v_{n_j}(x) + \frac{f(u,x)}{a(u,x)} - \frac{\rho_{n_j}}{a(u,x)}\right] = \zeta_{n_j}(x), \tag{60}$$

and

$$\frac{1}{2}\Delta v_{n_j}(x) + \inf_{u\in U}\left[\frac{b(u,x)}{a(u,x)}\nabla_x v_{n_j}(x) + \frac{f(u,x)}{a(u,x)} - \frac{\rho_{n_j}}{a(u,x)}\right] = \eta_{n_j}(x), \tag{61}$$

with some functions

$$\zeta_{n_j}(x) \ge -C|\nabla v_{n_j} - \nabla v_{n_j+1}|(x) - c(\rho_{n_j} - \rho_{n_j+1})$$

and

$$\eta_{n_j}(x) \le 0.$$

It maybe rewritten as follows,

$$\frac{1}{2}\Delta v_{n_j+1}(x) + F_1\left[\nabla_x v_{n_j}, \rho_{n_j}\right] = \zeta_{n_j}(x), \tag{62}$$

and

$$\frac{1}{2}\Delta v_{n_j}(x) + F_1\left[\nabla_x v_{n_j}, \rho_{n_j}\right] = \eta_{n_j}(x). \tag{63}$$

Let B_r be any open ball of radius r, and let

$$\tau_r := \inf(t \ge 0:\ x + W_t \not\in B_r).$$

It is known that the stopping time τ_r has some finite exponential moment and, of course, all polynomial moments. From (62) and (63) it follows that for any $x \in B_r$ we get,

$$v_{n_j}(x) = \mathsf{E}\left(\int_0^{\tau_r} \left(F_1\left[\nabla_x v_{n_j}\right](x+W_s) - \eta_{n_j}(x+W_s)\right) ds + v_{n_j}(x+W_{\tau_r})\right) \tag{64}$$

$$\geq \mathsf{E}\left(\int_0^{\tau_r} \left(F_1\left[\nabla_x v_{n_j}\right](x+W_s)\right) ds + v_{n_j}(x+W_{\tau_r})\right),$$

and

$$\begin{aligned} & v_{n_j+1}(x) \\ & = \mathsf{E}\left(\int_0^{\tau_r} \left(F_1\left[\nabla_x v_{n_j}\right](x+W_s) - \zeta_{n_j}(x+W_s)\right) ds + v_{n_j}(x+W_{\tau_r})\right) \\ & \leq \mathsf{E}\left(\int_0^{\tau_r} \left(F_1\left[\nabla_x v_{n_j}\right](x+W_s) + (C|\nabla v_{n_j} - \nabla v_{n_j+1}|(x+W_s)\right.\right. \\ & \left.\left. +c(\rho_{n_j} - \rho_{n_j+1}))\right) ds + v_{n_j}(x+W_{\tau_r})\right). \end{aligned} \tag{65}$$

Indeed, let $T_N := \inf(t \geq 0 : x + W_t \notin B_{r-1/N})$ for $N > 1/r$. Then we have,

$$T_N \uparrow \tau_r, \quad N \to \infty,$$

and

$$\begin{aligned} & v_{n_j}(x) \\ & = \mathsf{E}\left(\int_0^{T_N} \left(F_1\left[\nabla_x v_{n_j}\right](x+W_s) - \eta_{n_j}(x+W_s)\right) ds + v_{n_j}(x+W_{T_N})\right) \\ & \geq \mathsf{E}\left(\int_0^{T_N} \left(F_1\left[\nabla_x v_{n_j}\right](x+W_s)\right) ds + v_{n_j}(x+W_{T_N})\right), \end{aligned} \tag{66}$$

and

$$\begin{aligned} v_{n_j+1}(x) = \mathsf{E}\left(\int_0^{T_N} \left(F_1\left[\nabla_x v_{n_j}\right](x+W_s)\right.\right. \\ \left.\left. -\zeta_{n_j}(x+W_s)\right) ds + v_{n_j}(x+W_{T_N})\right) \end{aligned}$$

$$\leq \mathsf{E}\left(\int_0^{T_N} \left(F_1\left[\nabla_x v_{n_j}\right](x+W_s) + (C|\nabla v_{n_j} - \nabla v_{n_j+1}|(x+W_s)\right.\right.$$
$$\left.\left.+c(\rho_{n_j} - \rho_{n_j+1}))\right) ds + v_{n_j}(x+W_{T_N})\right). \tag{67}$$

As $N \to \infty$, we get the desired inequalities (64) and (65). In turn, (64) and (65) imply in the limit as $n_j \to \infty$ that

$$\tilde{v}(x) \geq \mathsf{E}\left(\int_0^{\tau_r} \left(F_1\left[\nabla_x \tilde{v}\right](x+W_s)\right) ds + \tilde{v}(x+W_{\tau_r})\right),$$

and

$$\tilde{v}(x) \leq \mathsf{E}\left(\int_0^{\tau_r} \left(F_1\left[\nabla_x \tilde{v}\right](x+W_s)\right) ds + \tilde{v}(x+W_{\tau_r})\right).$$

Hence, actually, for each $x \in B_r$

$$\tilde{v}(x) = \mathsf{E}\left(\int_0^{\tau_r} \left(F_1\left[\nabla_x \tilde{v}\right](x+W_s)\right) ds + \tilde{v}(x+W_{\tau_r})\right). \tag{68}$$

On the other hand, if we denote by $\bar{v}(x)$ the right hand side of (68), then due to the standard probabilistic arguments the function $\bar{v}$ is a unique solution of the elliptic PDE (see [13, Corollary 9.18])

$$\frac{1}{2}\Delta\bar{v}(x) + F_1\left[\nabla_x \tilde{v}\right](x) = 0, \quad \bar{v}|_{\partial B_r} = \tilde{v}|_{\partial B_r}, \tag{69}$$

in the Sobolev classes $W^2_{p,loc}(B_r) \cap C(\bar{B}_r)$ with any $p > 1$.

Indeed, let us take the unique solution $\bar{v}$ of the Eq. (69) from this class and apply Itô-Krylov's formula to $\bar{v}(x+W_t)$ for $t < \tau_r$:

$$d\bar{v}(x+W_t) = \frac{1}{2}\Delta\bar{v}(x+W_t)dt + \nabla\bar{v}(x+W_t)dW_t.$$

Using the same stopping time sequence $(T_N, N \geq 1)$ as above, we have in the integral form,

$$\bar{v}(x+W_{T_N}) - \bar{v}(x) = \int_0^{T_N} \frac{1}{2}\Delta\bar{v}(x+W_t)dt + \int_0^{T_N} \nabla\bar{v}(x+W_t)dW_t.$$

Taking expectations of both sides, we obtain

$$\mathsf{E}\bar{v}(x+W_{T_N}) - \bar{v}(x) = \mathsf{E}\int_0^{T_N} \frac{1}{2}\Delta\bar{v}(x+W_t)dt = \mathsf{E}\int_0^{T_N} F_1\left[\nabla_x \tilde{v}\right](x+W_t)dt.$$

Here due to the properties of $\bar{v}$, $F_1 \in C$, we get in the limit as $N \to \infty$,

$$\mathsf{E}\bar{v}(x + W_{\tau_r}) - \bar{v}(x) = \mathsf{E}\int_0^{\tau_r} F_1\left[\nabla_x \tilde{v}\right](x + W_t)dt. \tag{70}$$

Comparing it to (68), we conclude that the function $\tilde{v}$ itself satisfies the Eq. (69) in B_r and belongs to the same class, $\tilde{v} \in W^2_{p,loc}(B_r) \cap C(\bar{B}_r)$ with any $p > 1$ (and for any ball B_r), and the Eq. (59) holds true, as it was promised. It is equivalent to (54).

In the sequel it will follow from the uniqueness of solution of Bellman's equation that actually the whole sequence v_n converges up to additive constants locally uniformly in C^1 to a single limit. However, it is not needed for our proof.

9. *Uniqueness for ρ in (7).* Suppose there are two solutions of the (HJB) equation, (v^1, ρ^1) and (v^2, ρ^2) with $v^i \in \mathcal{K}$, $i = 1, 2$:

$$\inf_{u\in U}(L^u v^1(x) + f^u(x) - \rho^1) = \inf_{u\in U}(L^u v^2(x) + f^u(x) - \rho^2) = 0.$$

According to the steps above, both v^1 and v^2 are classical solutions with locally Lipschitz second derivatives. Denote $w(x) := v^1(x) - v^2(x)$. Let $\alpha_1, \alpha_2 \in \mathcal{A}$ be two strategies such that $\alpha_1(x) \in \mathrm{Argmax}_{u\in U}(L^u w(x))$ and $\alpha_2(x) \in \mathrm{Argmin}_{u\in U}(L^u w(x))$; let X^1_t, X^2_t be solutions of the SDEs corresponding to each strategy α_i, $i = 1, 2$. Note that due to the measurable choice arguments such Borel measurable strategies exist, see the appendix in [2]; then corresponding weak solutions also do exist. Denote

$$h_1(x) := \sup_{u\in U}(L^u w(x) - \rho^1 + \rho^2), \quad h_2(x) := \inf_{u\in U}(L^u w(x) - \rho^1 + \rho^2).$$

We have,

$$h_2(x) = \inf\nolimits_{u\in U}(L^u v^1(x) + f^u(x) - \rho^1 - (L^u v^2(x) + f^u(x) - \rho^2))$$

$$\leq \inf_{u\in U}(L^u v^1(x) + f^u(x) - \rho^1) - \inf_{u\in U}(L^u v^2(x) + f^u(x) - \rho^2) = 0.$$

Also,

$$h_1(x) = -\inf\nolimits_u(L^u(-v^2)(x) - \rho^2 + \rho^1)$$

$$= -\inf\nolimits_u(L^u v^2(x) + f^u(x) + \rho^2 - (L^u v^1(x) + f^u(x) + \rho^1))$$

$$\geq -\left[\inf\nolimits_u(L^u v^2(x) + f^u(x) - \rho^2) - \inf\nolimits_u(L^u v^1(x) + f^u(x) - \rho^1)\right] = 0.$$

Then we have the equations

$$L^{\alpha_2} w(x) = h_2(x) - \rho^2 + \rho^1,$$

and

$$L^{\alpha_1} w(x) = h_1(x) - \rho^2 + \rho^1.$$

Applying Dynkin's formula we get,

$$\mathsf{E}_x w(X_t^1) - w(x) = \mathsf{E}_x \int_0^t L^{\alpha_1} w(X_s^1)\, ds$$

$$= \mathsf{E}_x \int_0^t h_1(X_s^1)\, ds + (\rho^1 - \rho^2)\, t \stackrel{(h_1 \ge 0)}{\ge} (\rho^1 - \rho^2)\, t. \tag{71}$$

Hence, dividing by t and letting $t \to \infty$, by virtue of the Lemma 1 and since the left hand side here is bounded, we obtain,

$$\rho^1 - \rho^2 \le 0.$$

Similarly, for the strategy α_2 we get,

$$\mathsf{E}_x w(X_t^2) - w(x) = \mathsf{E}_x \int_0^t L^{\alpha_2} w(X_s^2)\, ds$$

$$= \mathsf{E}_x \int_0^t h_2(X_s^2)\, ds + (\rho^1 - \rho^2)\, t. \tag{72}$$

Therefore, since the left hand side here is bounded (see the Lemma 1), we obtain,

$$\rho^2 - \rho^1 = \liminf_{t \to 0} (t^{-1} \mathsf{E}_x \int_0^t h_2(X_s^2)\, ds) \stackrel{(h_2 \le 0)}{\le} 0.$$

So, $\rho^1 - \rho^2 \ge 0$ and, hence, it follows that, in fact,

$$\rho^1 = \rho^2.$$

10. *Proof of* $\rho = \tilde{\rho}$. As it was already established, the sequence ρ_n converges to the same value $\tilde{\rho}$, whatever is the initial strategy $\alpha_0 \in \mathcal{A}$. So, let us choose α_0 such that

$$\rho_0 = \rho^{\alpha_0} < \rho + \epsilon.$$

Since the sequence (ρ_n) monotonically decreases, it follows that also

$$\tilde{\rho} = \lim_{n\to\infty} \rho_n < \rho + \epsilon.$$

Since $\epsilon > 0$ is arbitrarily small, we have,

$$\tilde{\rho} \le \rho.$$

The inequality $\tilde{\rho} \ge \rho$ was justified earlier (38). Hence, in fact,

$$\tilde{\rho} = \rho,$$

as required.

11. *Uniqueness for V*. It was already established that the second component in the solution of the Eq. (7), or, equivalently, (8) is unique. So, returning to the formula (71) with the non-negative h_1 and replacing in it $\rho_1 - \rho_2$ by zero, we obtain,

$$\mathsf{E}_x w(X^1_t) - w(x) = \mathsf{E}_x \int_0^t h_1(X^1_s)\, ds. \tag{73}$$

Dividing by t we get as $t \to 0$,

$$0 = \langle h_1, \tilde{\mu}_1 \rangle.$$

So, we get,

$$h_1 = 0, \quad \tilde{\mu}_1 - \text{a.s.}$$

Since $\Lambda \ll \tilde{\mu}_1$ (see the Lemma 3), this implies

$$h_1 = 0, \quad \Lambda - \text{a.s.}$$

By Krylov's estimate we have,

$$0 \le \mathsf{E}_x \int_0^t h_1(X^1_s)\, ds \le N \|h_1\|_{L_d} = 0.$$

Thus, in fact,

$$\mathsf{E}_x w(X^1_t) - w(x) = 0. \tag{74}$$

This holds true for any $t \geq 0$. By virtue of the last statement of the Lemma 1 it follows that

$$w(x) = \lim_{t\to\infty} \mathsf{E}_x w(X_t^1) = \langle w, \mu_1 \rangle.$$

Hence, $w(x)$ is a *constant.* Therefore, the uniqueness of the first component in the solution of the Eq. (7) *up to a constant* is proved,

$$v^1(x) - v^2(x) = \text{const},$$

as required.

12. Let us return to the second statement of the Theorem. By virtue of uniqueness of solution of the HJB equation, we can now see that convergence of the whole sequence (v_n) up to additive constants depending only on n is to the same limiting function v, as required. The theorem is proved.

□

Acknowledgments The author thanks the anonymous referee for useful remarks which helped minimise the number of mistakes and inaccuracies.

References

1. Anulova, S.V., Mai, H., Veretennikov, A.Yu.: On averaged expected cost control as reliability for 1D ergodic diffusions. Reliab.: Theory Appl. **12**(4) (47), 31–38 (2017)
2. Anulova, S.V., Mai, H., Veretennikov, A.Yu.: On iteration improvement for averaged expected cost control for one-dimensional ergodic diffusions. SIAM J. Control Optim. **58**(4), 2312–2331 (2020). https://doi.org/10.1137/19M1271944. Preprint: https://arxiv.org/abs/1812.10665
3. Arapostathis, A.: On the policy iteration algorithm for nondegenerate controlled diffusions under the ergodic criterion. In: Optimization, Control, and Applications of Stochastic Systems. Systems and Control Foundations and Applications, pp. 1–12. Birkhäuser/Springer, New York (2012)
4. Arapostathis, A., Biswas, A., Carroll, J.: On solutions of mean field games with Ergodic cost. J. Math. Pures Appl. **107**(2), 205–251 (2017). https://doi.org/10.1016/j.matpur.2016.06.004
5. Arapostathis, A., Biswas, A., Pradhan, S.: On the policy improvement algorithm for ergodic risk-sensitive control. Proc. R. Soc. Edinb. Section A Math. **151**(4), 1305–1330 (2021). https://doi.org/10.1017/prm.2020.61
6. Arapostatis, A., Borkar, V.S., Fernãndes-Gaucherand, E., Ghosh, M.K., Markus, S.I.: Discrete-time controlled Markov processes with average cost criterion: a survey. SIAM J. Control Optim. **31**(2), 282–344 (1993)
7. Arapostathis, A., Borkar, V.S.: Average cost optimal control under weak ergodicity hypotheses: Relative value iterations (2019/2023). https://doi.org/10.48550/arXiv.1902.01048. arxiv.org/abs/1902.01048
8. Arapostathis, A., Borkar, V.S.: A relative value iteration algorithm for non-degenerate controlled diffusions. SIAM J. Control Optim. **50**(4), 1886–1902 (2012)

9. Arapostathis, A., Borkar, V.S., Ghosh, M.K.: Ergodic Control of Diffusion Processes. Encyclopedia of Mathematics and its Applications, vol. 143. Cambridge Uinversity Press, Cambridge (2012)
10. Borkar, V.S.: Optimal Control of Diffusion Processes. Longman Scientific & Technical/John Wiley & Sons, Harlow/New York (1989)
11. Gikhman, I.I., Skorokhod, A.V.: Stochastic Differential Equations. Springer, Berlin (1972)
12. Girsanov, I.V.: On transforming a class of stochastic processes by absolutely continuous substitution of measures. Theory Probab. Appl. **3**(5), 285–301 (1960)
13. Gilbarg, D., Trudinger, N.S.: Elliptic Partial Differential Equations of Second Order, 2nd edn. Springer, Berlin (1983)
14. Jacquet, Q., van Ackooij, W., Alasseur, C., Gaubert, S.: Ergodic control of a heterogeneous population and application to electricity pricing (2022). https://doi.org/10.48550/arXiv.2204.01410. arXiv:2204.01410
15. Krylov, N.V.: On the selection of a Markov process from a system of processes and the construction of quasi-diffusion processes. Math. USSR Izv. **7**, 691–709, (1973)
16. Krylov, N.V.: Controlled Diffusion Processes, 2nd edn. Berlin, Springer (2009)
17. Krylov, N.V.: On Ito's stochastic integral equations. Theory Probab. Appl. **14**, 330–336 (1969)
18. Krylov, N.V., Safonov, M.V.: A certain property of solutions of parabolic equations with measurable coefficients. Math. USSR Izv. **16**(1), 151–164 (1981)
19. Ladyzhenskaja, O.A., Solonnikov, V.A., Ural'ceva, N.N.: Linear and Quasi-linear Equations of Parabolic Type. AMS, Providence (1968)
20. Mei, H., Wei, Q., Yong, J.: Optimal Ergodic Control of Linear Stochastic Differential Equations with Quadratic Cost Functionals Having Indefinite Weights (2020). https://doi.org/10.48550/arXiv.2004.11088. arXiv:2004.11088
21. Pardoux, É., Veretennikov, A.Yu.: On Poisson equation and diffusion approximation. I. Ann. Probab. **29**(3). 1061–1085, (2001)
22. Solonnikov, V.A.: On boundary value problems for linear parabolic systems of differential equations of general form. Proc. Steklov Inst. Math. **83**, 1–184 (1965)
23. Veretennikov, A.Yu.: On polynomial mixing and convergence rate for stochastic difference and differential equations. Theory Probab. Appl. **44**(2), 361–374 (2000)
24. Veretennikov, A.Yu.: On polynomial mixing for SDEs with a gradient-type drift. Theory Probab. Appl. **45**(1), 160–164 (2001)
25. Veretennikov, A.Yu., Veretennikova, M.A.: On improved bounds and conditions for the convergence of Markov chains. Izv. Math. **86**(1), 92–125 (2022). https://doi.org/10.1070/IM9076

www.ingramcontent.com/pod-product-compliance
Ingram Content Group UK Ltd.
Pitfield, Milton Keynes, MK11 3LW, UK
UKHW022215051026
15624UKWH00001B/166
* 9 7 8 9 8 1 9 7 0 2 2 4 4 *